D1561155

Selected Titles in This Series

(*Continued in the back of this publication*)

Rings and Things and a Fine Array of Twentieth Century Associative Algebra

Mathematical Surveys and Monographs

Volume 65

Rings and Things and a Fine Array of Twentieth Century Associative Algebra

Carl Faith

American Mathematical Society

1991 *Mathematics Subject Classification.* Primary 00-XX, 01-XX, 12-XX, 13-XX, 16-XX; Secondary 03-XX, 04-XX, 06-XX, 08-XX, 14-XX, 15-XX, 18-XX.

ABSTRACT. A survey of aspects of the development of associative rings and modules in the twentieth century including: (1) updates on topics treated in the author's two Springer-Verlag Grundlehren (Foundations) volumes written a quarter of a century ago, (2) a considerable expansion of topics to include exciting new ideas that drive and dominate contemporary research. The title of this book is derived from *The Taming of the Shrew.*

Library of Congress Cataloging-in-Publication Data

Faith, Carl Clifton, 1927–
Rings and things and a fine array of twentieth century associative algebra / Carl Faith.
p. cm. — (Mathematical surveys and monographs, ISSN 0076-5376 ; v. 65)
Includes bibliographical references (p. –) and index.
ISBN 0-8218-0993-8 (acid-free paper)
1. Associative algebras. 2. Associative rings. 3. Modules (Algebra) I. Title. II. Series: Mathematical surveys and monographis ; no. 65.
QA251.5.F355 1998
512′.24—dc21
98-38824
CIP

∞ The paper used in this book is acid-free and falls within the guidelines
established to ensure permanence and durability.
Visit the AMS home page at URL: http://www.ams.org/

10 9 8 7 6 5 4 3 2 1   04 03 02 01 00 99

Dedications

To my wife: Molly Kathleen Sullivan

You are my sun
You are my moon
You are my day
You are my night
My lodestar
My terra incognita
My guiding light
My terra firma
My earth
My sky
My heaven
Mi luna caliente
Mi manzana carnal
Y el pequeño infinito
Tuyo es mi vida!

To the memory and love of Mama: Vila Belle Foster

"So mayest thou, like a ripe fruit, drop in thy mother's lap." (anon.)

For my daughter, Heidi Lee, Numero Uno

Your heroism in saving two Princeton University students from drowning in Lake Carnegie where they fell through the ice when you were just fifteen, won you a Red Cross Medal and taught me what greatness truly is: Nobody I know has ever done anything as great.

To my "little" brother: Frederick Thomas Faith

You taught me the meaning, and the sweetness, of the word–brother: "May all your parachute leaps land you on feather beds."

To my son: Japheth

For showing that minimal algebras of types two and four are not computable in your Berkeley Ph.D. Thesis, May 23, 1997. Congratulations and thank you. You may be the only one in the family who can read this book! But will you?

To my son (El niño): Ezra

For winning honors at your Rutgers graduation, May 22, 1997: Chemistry, Phi Beta Kappa, Hypercube, the Howard Hughes Research Award, and College Honors. And for receiving Fellowship offers for graduate school in environmental chemistry from Berkeley, UCLA, the University of Texas at Austin, University of Washington, Seattle, and the University of Colorado at Boulder. Congratulations! (I'm in awe.)

To a friend: Barbara Lou Miller

You are the sine qua non of this book. Your skill and art in compositing at the computer, and the spunk it takes to do it, are inspirational. You have in jurisprudence terms aided and abetted me on every page (not that writing a book per se is a criminal offense, but maybe the way I write is?)

Contents

Symbols

$\forall$ (= universal quantifier)	$\forall a \in A$
$\exists$ (= existential quantifier)	$\exists a \in A$
$\in$ (= membership)	$a \in A$
$\notin$ (= nonmembership)	$a \notin A$
$\subset$ (= proper containment)	$A \subset B$
$\subseteq$ (= containment)	$A \subseteq B$
$\hookrightarrow$ (= embedding)	$A \hookrightarrow B$
$\Rightarrow$ (= implication)	$A \Rightarrow B$
$=$ (= equals)	$A = B$
$\neq$ (=unequals)	$A \neq B$
$\backslash$ (= backslash)	$A \backslash B$ (complement of B in A)
$\emptyset$ (= empty set)	$A = \emptyset$
$\mathbb{N}$ (= natural numbers)	$1, 2, \ldots$
$\mathbb{Z}$ (= integers)	$0, \pm 1, \pm 2, \ldots$
$\mathbb{Q}$ (= rational numbers)	$a/b, a, b \in \mathbb{Z}, b \neq 0$
$\mathbb{R}$ (= real numbers)	$\sqrt{2}$, pi
$\mathbb{C}$ (= complex numbers)	$a + bi, i^2 = -1, a, b \in \mathbb{R}$
$(,)$ (= ordered pair)	(a, b)
$\bigcup$ (= union)	$A \bigcup B$
$\bigcap$ (= intersection)	$A \bigcap B$
$+$ (= plus)	$a + b$
$-$ (= minus)	$a - b$
$\times$ (= Cartesian product)	$\alpha \times \beta$
$\rightarrow$ (= mapping)	$A \rightarrow B$
$\mapsto$ (= corresponds to)	$a \mapsto b$ (b corresponds to a)
X (= cross)	$X_i A_i$
$\prod$ (= product)	$\prod_i A_i$
$\coprod$ (= coproduct)	$\coprod_i A_i$
$\approx$ (= isomorphism)	$A \approx B$
$\wedge$ (= wedge)	$A \wedge B$
$\vee$ (= vee)	$A \vee B$
$>$ (= greater than)	$a > b$
$<$ (= less than)	$a < b$
$\ltimes$ (= split-null extension)	$A \ltimes B$
$\perp$ (= perpendicular("perp"))	$A^{\perp}$ and ${}^{\perp}A$
$\sum$ (= summation)	$\sum_{i \in I} A_i$
$\oplus$ (= direct sum)	$A \oplus B$
$\otimes$ (= tensor product)	$A \otimes_R B$

mod-R (= category of right R-modules)	$M \in$ mod-R
R-mod (= category of left R-modules)	$M \in R$-mod
$\sim$ (= similarity or Morita Equivalence of rings)	$A \sim B$
$\mid\ \mid$ (= cardinal (also length))	$\lvert A \rvert$
A^I (= exponentiation of cardinals)	(= product $\prod_{i\in I} A_I$ where $A_I + A \quad \forall i$)
$A^{(I)}$ (= all $f : I \to A$ with finite support. Cf. p.1)	
$T :$ mod-$A \rightsquigarrow$ mod-B	functor from mod-A to mod-B
$R\langle x\rangle, R[[x]]$	power series ring over R
$R[x]$	polynomial ring over R

Preface

"There is no royal road to mathematics"
(From Proclus, *Commentary on Euclid*, Prologue)

My two Springer-Verlag volumes, *Algebra I* and *II*, written a quarter of a century ago (see References) are devoted to the development of modern associative algebra and ring and module theory, so here I am faced with the challenging questions of where to begin, what to leave out and how much to add. Nevertheless, I hope the reader will discover that the various topics have an *uncanny* affinity for each other. Or maybe that I had a *canny* affinity for them: *the apples fall near the apple tree* (Russian Proverb).

Maschke's Theorem

We begin with a theorem published a century ago (in 1898) by H. Maschke about the representation theory of a group algebra kG over a field k. For a field k of characteristic not dividing the order of G, it states that every representation for G, that is, any kG-module M, is a direct sum $\oplus V_i$ of "irreducible" representations V_i, where "irreducible" means that V_i has no smaller representations, that is, V_i is a module with no proper submodules. In the terminology introduced below, we say that V_i is a simple kG-module, and that M is a semisimple kG-module. (In §11, we shall come back to Maschke's theorem and group algebras.)

Other Nineteenth Century Theorems

Those of D. Hilbert—the Basis theorem (1888) and the Nullstellensatz (1893)—are taken up in Chapter 2, see 2.20 and 2.30, and their modern forms are scattered throughout the text, e.g. the Generalized or Weak Nullstellensatz is Theorem 3.36B.

In 1893 an Estonian mathematician T. Molien obtained the decomposition of semisimple algebras over the field $\mathbb{C}$ of complex numbers into matrix algebras, fifteen years before Wedderburn's Theorem over arbitrary fields.

Going back even further, a theorem proved in 1878 by G. Frobenius and L. Stickelberger is that every finite commutative (= Abelian) group is a direct sum of primary cyclic groups (Cf. 1.9B and 1.14). The Fundamental Theorem of Abelian Groups (= FTAG) and the Wedderburn-Artinian Theorems (= WAT) are offered as paradigms for algebraic structure theorems, and *inter alia*, both state that finitely generated ($= f \cdot g$) modules are direct sum of cyclic modules! And WAT further states that **all** modules over semisimple rings are direct sums of cyclic modules, and actually every indecomposable cyclic module is simple. Further, WAT not only implies that (1) every module is a direct summand of every over-module, but that (2) every module is a direct summand of a free R-module. Thus, by (1) every module over a semisimple ring is injective (N.B.) and by (2) every module is projective.

Those Twins: Injective and Projective Modules

An R-module E is *injective* if every embedding $E \hookrightarrow F$ into an overmodule splits, i.e., is a direct summand, while a *projective* module P has the dual property: every onto homomorphism $M \to P$ splits in the sense that the kernel is a direct summand.

And so it goes. You **have** to have injectives and you **have** to have projectives in any discussion of direct summands. But if $\{E_a\}_{a\in A}$ is a set of injectives indexed by a set A, it is natural to ask when is their direct sum $\oplus_{a\in A} E_a$ injective? When this is so, then the direct sum splits off in the direct product. This is trivially true when A is finite but it is true for all direct sums of injectives iff R is Noetherian, i.e., R satisfies the ascending chain condition (= acc) on all *right* ideals (assuming the E_a are *right* R-modules (see 3.4B)).

If every direct sum of copies of an injective module E is injective, then E is said to be $\sum$-injective. This happens iff R satisfies the acc on right annihilators in R from subsets of E (see 3.7A).

Another Twin: Acc and Dcc

So you **have** to have ascending chain conditions on certain (right) ideals, and *maybe* the descending chain condition (= dcc) on certain ideals. The latter happens whenever you have direct sum splitting in the direct product of an infinite set $\{M_a\}_{a\in A}$ of modules that are not even injective (1.23,1.24 and 1.25). Furthermore, the dual condition regarding a direct product of projectives also produces chain conditions (see 1.17A and 3.31; Cf. 6.6).

These theorems show the power of the condition that direct sums split off, but other direct sum conditions are also powerful: if every injective R-module is a direct sum of indecomposable modules, then ring R is again right Noetherian (3.4C). Moreover, if we assume every injective R-module is isomorphic to a direct sum of modules from a *given* set of modules, then R is Noetherian (3.5A); and if every module is isomorphic to a direct sum of modules in a given set, then R is Artinian (3.5A), that is, satisfies the dcc on all right ideals.

FGC Rings

Much of the survey is an elaboration of these themes. For example, §5 is devoted to describing the classification of all commutative rings, called *FGC* rings, over which every $f \cdot g$ module is a direct sum of cyclics, and even more generally, in §6, when all finitely presented modules are direct sum of cyclics (6.3). The first question involves the notions of (almost) maximal rings, equivalently (almost) linearly compact rings in the discrete topology, and Bezout domains (**sup.** 5.4B), h-local domains (**sup.** 5.4A), and fractionally self-injective (=FSI) rings (**sup.** 5.9). The FGC Classification Theorem 5.11 states *inter alia* that R is FGC iff FSI and Bezout.

A Companion to the Fundamental Theorem

The companion theorem to FTAG for finitely presented (= $f \cdot p$) modules (the aforementioned Theorem 6.3) involves elementary divisor rings (=EDR's), i.e., rings over which every matrix is equivalent to a diagonal matrix. Thus: every $f \cdot p$ R-module is a direct sum of cyclics iff R is an EDR (Cf. also 6.5B).

FP-Injective Modules and Rings

One might call these latter rings FPC rings. A concept that pops up in this regard is that of FP-injectivity (Cf. 6.2ff.). And the concept of fractionally self-FP-injective (=FSFPI) also appears, and to an extent parallels FSI in the description of FGC rings (6.4).

Every ring R can be embedded in an FP-injective ring (6.21). This is a consequence of the fact that every ring can be embedded in an existentially closed (= EC) ring (6.20). In this connection the conception of EC fields is of interest: every sfield (= skew field) can be embedded in an EC sfield (6.24).

Mal'cev Domains

On the subject of embeddings, Mal'cev domains are not embeddable in sfields (6.27), and moreover, there exists integral domains not embeddable in left Noetherian nor in right Artinian rings (6.34).

IF and QF Rings

On the subject of FP-injective rings, there pops up IF rings, or rings over which every injective R-module is flat. This happens iff R is a coherent FP-injective ring (Cf. 6.9).

The IF rings parallel the QF (= quasi-Frobenius rings) in that QF rings are those over which every injective is projective (3.5B) and similarly over which every projective is injective (3.5C).

Another parallel: R is right IF iff every $f \cdot p$ right R-module embeds in a free module (6.8), whereas R is QF iff every right R-module embeds in a free module. Furthermore, R is QF iff every cyclic right and every cyclic left R-module embeds in a free R-module (3.5D).

Duality via Annihilators

Yet another parallel: a duality by annihilation between one-sided $f \cdot g$ ideals characterizes IF rings (6.99), and QF rings too since every one-sided ideal is $f \cdot g$ (**sup.** 3.5B). Cf. Dual rings in §13.

Pure-injective (algebraically compact) modules, i.e., modules M that are direct summands of any module containing M as a pure submodule are defined in §6, **sup.** 6B (Cf. 1.26).

Krull-Schmidt Theorems and Failure

Any Noetherian (resp. Artinian) module M is decomposable into a finite direct sum of indecomposable modules, but this decomposition need not be unique. The Krull-Schmidt theorem gives uniqueness assuming that M is *both* Noetherian and Artinian. Krull-Schmidt also holds over a complete local Noetherian ring R, i.e., for just Noetherian modules over R. The failure of the Krull-Schmidt theorem for just Artinian modules was proved in 1995, and I have included an account of this and related questions including the decomposition of modules into an arbitrary set of indecomposables in §8. This introduces the concepts of exchange rings and modules, **sup.** 8.4.

Acc on Annihilators

In §9, we find that the acc on annihilators (= acc ⊥) of a ring R is not inherited by the polynomial ring (9.2), but that it is if R contains an uncountable field as a subring (9.3) or if R is locally Noetherian (9.6), or if R has finite Goldie dimension and the quotient ring Q of R has nil Jacobson radical (9.4), e.g. if Q is an algebraic algebra over a field k of cardinality larger than the dimension of Q over k (9.5).

Non-uniqueness of the Coefficient Ring

In §10, we find for a polynomial ring $R[X]$ that the coefficient ring R need not be uniquely determined up to isomorphism, even if R is a Noetherian domain (10.1) but it is if R is a zero dimensional ring (10.2), e.g. a von Neumann regular ring (10.3), or a finite product of local rings (10.4), or a domain of transcendence 1 over a field (10.5), etc. We also list some matrix cancellable rings from Lam's survey [95].

Group Rings

§11 is devoted to various properties of group rings AG; in particular, when AG is QF, self-injective, QF, perfect, VNR, semisimple, etc. Also considered is the question of when the group ring determines the group.

Maximal Quotient Rings, Duality, Krull and Global Dimension, and Polynomial Identities

§12 is on the subject of maximal quotient rings, localizing functors, and torsion theories. §13 is on Morita and other duality and applications. §14 is on classical Krull dimension *dim R* of commutative Noetherian rings, the global dimension, *gl.dim R*, of any ring R, and regular rings (= Noetherian R of finite global dimension, in which case = *dim R*). Also in §14, noncommutative Krull dimension of rings and modules is sketched and various applications given. §15 is on PI-rings, that is, rings with polynomial identities.

Aspects of Commutative Algebra and the Rest of the Story

Chapter 16 is on the subjects in commutative algebra: unions of prime ideals, prime avoidance, associated prime ideals, and the acc on annihilator ideals and irreducible ideals.

Chapter 17 is on the subject of the author's Ph.D. thesis (Purdue 1955): Galois theory and independence of automorphisms. But whereas his thesis was devoted to fields, this chapter is on the subject of papers dating to 1982, on the linear independence of automorphisms of commutative rings, or, as the title suggests: "Dedekind's Theorem Revisited."

The above sketches cover perhaps only twenty-five percent of the text. Since the titles only sketchily indicate the chapter contents, we have included the paragraph headings in Contents to tell "the rest of the story."

Mathematical Commentaries on the Works of Wedderburn, Artin, Noether, and Jacobson

Extensive commentaries on the work of Emmy Noether appear in Brewer-Smith [81], notably Swan on "Galois Theory" (Chap. 6), Gilmer on "Commutative Ring Theory" (Chap. 8), Lam on "Representation Theory" (Chap. 9),[1] and Fröhlich

[1] Lam also discusses (*ibid.*, pp.149–150) the work of T. Molien mentioned on p. v, its influence on Noether, and its applications to representation theory.

on "Algebraic Number Theory" (Chap. 10). Also included is Noether's address to the ICM in 1932 on "Hypercomplex Systems and Their Relations to Commutative Algebra and Number Theory." Also included are personal reminiscences of Emmy Noether by Clark Kimberling, Saunders MacLane, B. L. Van der Waerden, and P. S. Alexandroff. (Also see Jacobson's introduction to Noether's Collected Papers [83].)

Additional commentaries appear in Srinivasan-Sally [82], including Jacobson's "Brauer Factor Sets, Noether Factor Sets, and Crossed Products", Swan's "Noether's Problem in Galois Theory", Sally's "Noether's Normalization", LaDuke's "The Study of Linear Associative Algebras in the United States, 1870–1927" (Cf. Parshall [85]), and personal reminiscences by a number of her students and colleagues. Also see Lang-Tate [65] for a succinct discussion of Artin's life and work. Van der Waerden acknowledged the lectures of E. Artin and E. Noether as a basis for his books [31–48], and for many years these books were the standards for abstract algebra. (The 4th edition in 1959 incorporated the Perlis-Jacobson radical (p. 204ff.).) In his *Collected Mathematical Papers* [89], Jacobson included memoirs of his world travels and his meetings with hundreds of mathematicians.

Krull, Struik, Boyer and Van der Waerden

Krull's book [48], originally published in 1935 as volume 3 of the Ergebnisse der Mathematik by Julius Springer, gives a brief foreword on the work of Dedekind, Hilbert, Kronecker, Lasker, Macaulay, Noether, Van der Waerden, Artin and Prüfer.

Other books outlining the development of modern algebra are those of Struik [87], Boyer-Merzbach [91] and Van der Waerden [85]. For those who thought that Dedekind originated the ring concept-definition, Kleiner [96] has a surprise for you: it was Fraenkel (in 1914) who was better known as a logician. Kronecker is generally credited for the concepts of a module and tensor products.

Kaplansky's Afterthoughts

In the interim, I have read Irving Kaplansky's retiring presidential address "Rings and Things" in January 1996 to the American Mathematical Society (unpublished) where he cites Bourbaki, who *earlier* vouched for Fraenkel. I also have read with the greatest pleasure Kaplansky's "Selected Papers and Other Writings" [95] including his insightful "Afterthoughts," and the introduction by Hyman Bass. In these few pages an entire era of mathematics is highlighted by Kaplansky's mathematical vigor and vision.

Snapshots

In writing "Snapshots" I have tried to share how friendship shapes lives and mathematics. My hope is that people, especially young people, will take note, and forget about accumulating information at the expense of friendship and personal contact. Because of the widespread use of the World Wide Web and e-mail, there is a real peril here in the loss of the art of friendship.

Georgia O'Keefe's epigram, accompanying a sheet of U.S. postage stamps depicting her "Red Poppy, 1927," says it beautifully:

> *No one sees a flower really. It's small and takes time to see, like a friend takes time.*

I wrote the above in a Christmas letter to Jim Huckaba in which I thanked him for taking the time to read "Snapshots," and for his enthusiastic response.

When I asked Claudia Menini what in the book would she like to see changed, she said, "Nothing!" And Jim Huckaba said the same thing (in other words): "I like the way you are writing it."

In addition to Huckaba, Chantal and Greg Cherlin, Sarah Donnelly, Barbara Miller and many people encouraged me, indeed, "aided and abetted," me in writing "Snapshots" (see Acknowledgements). John D. O'Neill has had a benign influence on the entire book. My friendship with John grew out of a letter I wrote to him in Fall 1995 telling him how much I admired his great theorem on direct summands of copies of the integers that had just been published in *Communications in Algebra.* (See Theorem 1.27C.) A lot of the group theory in Chapter 1 was suggested by him, mostly other people's work.

A similar instance occasioned my friendship with the late Pere Menal (see "Snapshots," p.308). There are dozens and dozens of such instances, in fact, everybody mentioned in "Snapshots" is a friend, mathematical or other. Like the rose, some friends are prickly, and may not take kindly to the often too brief mention given them, and other friendships are like violets in Wordsworth's "Lucy" poem:

"Lucy"

......
......
A violet by a mossy stone
Half-hidden from the eye
As fair as a star, when only one
Is shining in the sky.
She lived unknown, and few could know
When Lucy ceased to be;
But she is in her grave, and, oh,
The difference to me!

(from "She Dwelt among the Untrodden Ways" (1800))

Carl Faith
Princeton and New Brunswick
Tibi dabo, 28 April 1998

Acknowledgements

This survey was written during the year 1996–1997 starting in May, and I am hoping to finish it in time for my seventieth birthday (late April).[1] I wish to thank Rutgers University and the Mathematics Department, particularly Deans Joseph A. Potenza, Robert L. Wilson, Chairman Antoni Kosinski and Acting Chairman Richard Falk, for not only their help in arranging my Faculty Academic Leave (despite a late application!) but also for their kind expression of concern during an interim illness that I experienced.

I am also deeply indebted to Barbara Miller of the Rutgers Mathematics Department for her skill and editorial ability, without which this survey might never have seen the light of day and certainly not been nearly as readable! In addition I am grateful to Barbara for her constant, often daily encouragement in the form of her avid interest in "Snapshots", which kept me thinking about the people and places appearing there, many of which she knows from her own wide experiences in life and travel.

As I told Sarah Donnelly of the Acquisitions Department of the American Mathematical Society, this book is the work of *two* septuagenarians—Barbara Miller and me.

I wish to take this opportunity to thank Pat "Patty" Barr, who copied countless drafts and regaled us with her hilariously funny jokes: What a morale booster! She is deserving of thanks for her five years (1990–1995) of service as a Rutgers Central Telephone operator; Pat handled countless telephone calls for me. In "Snapshots" (see Part II), I told the story about Arthur Guy, whom I knew at the Dearborn Navy Radio Materiel School only as a voice. The parallel here was exactly the same, except that Pat knew me as a "name". Was I ever surprised when she came to us in 1995 (after the switchboard was fully automated) as the mathematics department's Xerox secretary and told me the story I just told you...Pat recognized my name from her telephone days.

Furthermore, I am indebted to John D. O'Neill for reading the manuscript in various editions, for making constructive remarks, for additional references and for picking out as many solecisms and barbarisms as I would permit. (John's background as a classics major surely made this an ordeal for him!) Thanks, also, to Toma Albu, Pere Ara, Pete Belluce, Victor Camillo, Ferran Cedó, Gregory Cherlin, Gertrude Ehrlich, Alberto Facchini, José-Luis Gómez Pardo, Ken Goodearl, Ram Gupta, Carolyn Huff, Dinh Van Huynh, Tsit Yuen Lam, Jim Lambek, Richard Lyons, Ahmad Shamsuddin, Stefan Schmidt, Wolmer Vasconcelos, and Weimin Xue for a number of references and/or corrections. I also wish to thank the Ohio Ring

[1] Note: I was off by a whole year!

Theory "ring", consisting of S. K. Jain and Sergio López-Permouth (at Athens) and Tariq Rizvi (at Lima) for numerous constructive suggestions. I also take pleasure in thanking Donald Babbitt, Pat Barr, Greg and Chantal Cherlin, Sarah Donnelly, Sergei Gelfand, Jim Huckaba, Claudia Menini, Barbara Miller, Jaume Moncasi, Keith Nicholson, and John O'Neill for their encouraging words of support for this survey while it was a work-in-progress. In addition, Dr. Rita Csákány (our newest Ph.D.!) has my gratitude for constructing the Register of Names and checking out the Index and Contents.

Mere mention of the people who helped me write this book does not sufficiently express my deep gratitude to those few who went way beyond the call of collegiality and truly became "mathematical friends" by giving the manuscript a thoroughly rigorous reading. They have relieved the prospective reader of the burden of hundreds of typos, dozens of "howlers," and so many *mea culpas*!

I am deeply honored by the beautiful song by Linda York (Undergraduate Secretary Extraordinaire of the Department) composed on the occasion of my retirement in April 1997, and for her kind permission to reprint it in "Snapshots."

And how can I ever repay Billy Reeves for his hilarious poem in the summer of '58 at Penn State: "Carl, You Will Always Have Dumb Students?" See "Snapshots" just preceding "Envoi."

I am also indebted to Antoni Kosinski, the Chair, Judy Lige, Business Manager, and the Mathematics Department for supporting my writing this book after my retirement.

I am grateful to Professor I. Kaplansky for the title of my book, which I filched from his retiring presidential address referred to in "Kap's *Rings and Things*" in " Snapshots". When I wrote for his permission to use it, he replied, "But of course. Anyway, Shakespeare has first claim." (Letter of April 6, 1998)

What can I say about my wife's indulgence that left me time to create this book? As my daughter, Heidi, has said, "Dad, there are *always* tradeoffs!" Well, Molly teaches Latin to six classes of high school students in nearby Hamilton, which keeps her from being a "book widow." *Sic semper magistrae! Et carpe librum!*

And that's not all—Molly's careful reading of "Snapshots" resulted in the addition of so many commas that I nicknamed her the *Kommakazi Kid*!

Carl Faith

Part I

An Array of Twentieth Century Associative Algebra

What's past is prologue!
William Shakespeare
The Tempest, II, i, 261

CHAPTER 1

Direct Product and Sums of Rings and Modules and the Structure of Fields

1.1. General Concepts[1].

The *direct product* $M = \prod_{i \in I} M_i$ of a family $\{M_i\}_{i \in I}$ of right R-modules consists of all sequences $x = \{x_i\}_{i \in I}$ with $x_i \in M_i \quad \forall i \in I$. With equality, addition, and scalar multiplication all defined pointwise, the direct product becomes an R-module with *projections* $p_i : M \to M_i$ and *injections* $u_i : M_i \to M$ defined canonically. The submodule $\oplus_{i \in I} M_i$ consisting of all $x \in \prod_{i \in I} M_i$ with *finite support* (i.e., with $x_i = p_i(x) = 0$ for all but finitely many $i \in I$) is called the *direct sum* of the family $\{M_i\}_{i \in I}$. In either case, $\forall i \in I, M_i$ is said to be a *(direct) summand* of $\oplus_{i \in I} M_i$ (resp. $\prod_{i \in I} M_i$). The direct sum $R^{(I)} = \oplus_{i \in I} R_i$, where $R_i \approx R$ **qua** right R-module for all $i \in I$ is said to be the *free right R-module of rank $|I|$*, where $|I|$ is the cardinality of the set I.

If S is a subset of a right R-module M, then: (1) a **linear combination of elements** of S is a finite sum of elements sr with $s \in S$ and $r \in R$; (2) the set $[S]$ of all such linear combinations is a submodule of M, called the **submodule of M generated** by S. In this case, S is said to be a **basis** or, a **set of generators of** M.

For a nonempty set I let e_i denote the element of $R^{(I)}$ with 1 in the i-position and 0's elsewhere, that is, $e_i : R^{(I)} \to R$ sends $i \in I$ onto 1 and j onto 0 if $j \neq i$. Then $\{e_i\}_{i \in I}$ is a basis of $R^{(I)}$, called a **free basis** of $R^{(I)}$. (N.B.) Cf. Free basis and invariant basis number in the author's book [76], pp.126,430.

If $S = \{s_i\}_{i \in I}$ is a basis of a right R-module M, then there is a canonical epimorphism (= onto homomorphism) $f : R^{(I)} \to M$ such that $f(e_i) = s_i \quad \forall i \in I$. If I is countable (or finite), then M is said to be **countably** (or **finitely**), **generated**, and S is said to a **countable** (or **finite**) **basis** of M. We sometimes write $f \cdot g$ instead of finitely generated. Also, we say that M is **$\aleph$-generated** if M has a basis indexed by a set I of cardinality $\aleph$. A basis S of M is **free** provided for any finite subset $\{s_i\}_{i=1}^n$ of S and elements $\{r_i\}_{i=1}^n$ of R, that

$$\sum_{i=1}^{n} s_i r_i = 0 \Rightarrow r_i = 0 \qquad \forall i.$$

In this case, $M \approx R^{(I)}$ under a mapping

$$\sum_{\text{finite sum}} s_i r_i \mapsto \{r_i\} \in R^{(I)}$$

that is, M is a free module iff M has a free basis.

[1]For expository reasons, paragraph headings are numbered in this (and only this) chapter.

Rings Without 1

REMARKS. (1) We are assuming that a ring R has a multiplicative identity element 1, hence R is a free (left and right) module with free basis 1; (2) A right (left) ideal I of a ring R in some cases may be thought of as a **ring without 1**, notation; **ring-1**. Thus, a ring-1 R satisfies all the requirements of a ring: Abelian additive group $(R,+)$, multiplicative semigroup $(R,\cdot)$, and satisfying the right and left distributive laws; (3) **Example.** If A is any Abelian group, and $ab = 0 \;\; \forall a,b \in A$, then A is a ring-1; (4) **Nilpotent Subrings.** A subring-1 S of a ring R is said to be **nilpotent of index** k if all products of k elements of S are zero, e.g. in (3) if $A \neq 0$, then A is nilpotent of index 2.

1.2. Internal Direct Sums.

A direct sum $M = \oplus_{i \in I} M_i$ is characterized by the properties that there is an R-module embedding $u_i : M_i \to M$ for each $i \in I$ such that every element $m \in M$ has a unique representation as a finite sum of elements in $M_i' = u_i(M_i), i \in I$. If I is a finite set, say $I = \{1, \dots, n\}$, then the direct sum and direct product coincide, and $M = M_1 \oplus \cdots \oplus M_n = M_1 \times \cdots \times M_n$ denotes both $\bigoplus_{i=1}^n M_i$ and $\prod_{i=1}^n M_i$. M is then said to be the *internal* direct sum of the family $\{M_i'\}_{i \in I}$ of submodules: Notation:

$$M = M_1' \oplus \cdots \oplus M_n'$$

and the modules M_i' are *summands*, $\forall i$.

Endomorphism Ring

If M and N are right R-modules, $\mathrm{Hom}_R(M,N)$ denotes the set of all **mappings** (= homomorphisms) $f : M \to N$. It is an Abelian group under pointwise additions of mappings. Further,

$$\mathrm{End} M_R = \mathrm{Hom}_R(M,M)$$

is a ring called the **endomorphism ring** of M, with respect to addition and composition of mappings.

Direct Summands and Idempotents

M is *decomposable* if $M = L \oplus N$ for submodules L and N such that $L \neq 0$ and $L \neq M$. (Then $N \neq 0$ and $N \neq M$.) L and N are then called **direct summands** of M over an associative ring R with unit. We let $A(M) = \mathrm{End}_R M$ denote the endomorphism ring. If $M = M_1 \times M_2$ is a direct product representation of M as a direct product of two submodules then there exist orthogonal idempotents e_1 and e_2 in $A = A(M)$ such that $M_i = e_i(M)$, $i = 1, 2$, and $1 = e_1 + e_2$ is the identity endomorphism. Then $A = e_1 A \times e_2 A = e_1 A \oplus e_2 A$ is a representation of A as a direct sum of two right ideals and $A = Ae_1 \oplus Ae_2$ is the left-right symmetry.

More generally, $M = M_1 \times \cdots \times M_n = M_1 \oplus \cdots \oplus M_n$ is a finite internal direct product (or sum) of R-submodules $M_1, \dots, M_n$ iff there exist n orthogonal idempotents $e_1, \dots, e_n \in A$ such that $e_i(M) = M_i, i = 1, \dots, n$, and such that $1 = e_i + \cdots + e_n$. (*Orthogonal* means $e_i e_j = 0 \quad \forall i \neq j$.) Then $A = e_1 A \oplus \cdots \oplus e_n A = Ae_1 \oplus \cdots \oplus Ae_n$ are direct decompositions of A as a direct sum of right and left ideals. Conversely, any direct sum decomposition of A as a direct sum

$A = A_1 \oplus \cdots \oplus A_n$ of right ideals gives rise to orthogonal idempotents $e_1, \ldots, e_n$ in A with $1 = e_1 + \cdots + e_n$, $A_i = e_iA$, and M is a direct sum

$$M = e_1(M) \oplus \cdots \oplus e_n(M) \tag{1}$$

and, furthermore,

$$A = Ae_1 \oplus \cdots \oplus Ae_n. \tag{2}$$

The sets e_iAe_i are rings with identities e_i, and are canonically isomorphic to $\mathrm{End}(e_iM)$, $i = 1, \ldots, n$.

1.3. Products of Rings and Central Idempotents.

If $A = A_1 \times \cdots \times A_n$ is a direct product of finitely many rings $A_1, \ldots, A_n$ then there exist central idempotents $e_i \in A$ with sum equal to 1 such that $A_i = e_iA = Ae_i$, and the rings A_i are ideals of A, $i = 1, \ldots, n$.

This situation occurs when $A = \mathrm{End}M_R$, and the direct summands M_i of M in (1) are *fully invariant submodules* in the sense that $aM_i \subseteq M_i$ $\forall a \in A, i = 1, \ldots, n$.

1.4. Direct Summands and Independent Submodules.

A submodule S of M is a *summand* if it is a direct summand, that is, there is a submodule T of M so that $M = S + T$ and $S \cap T = 0$. We then say that S *splits(off)* in M.

DEFINITION.

A collection $\{M_i\}_{i \in I}$ of submodules of M is said to be *independent* provided the equivalent conditions hold:

(1) $M_i \cap \sum_{j \in I \setminus \{i\}} M_j = 0$ (= M_i has 0 intersection with the sum of the other M_j).

(2) The sum $\sum_{i \in I} M_i$ is canonically isomorphic to the direct sum $\oplus_{i \in I} M_i$.

To paraphrase: a set of submodules is independent iff their sum is direct (and conversely).

REMARK. Any set $\{V_i\}_{i \in I}$ of non-isomorphic simple submodules of M is independent.

1.5. Dual Modules and Torsionless Modules.

If M is a right R-module, then the dual module $M^\star = \mathrm{Hom}_R(M, R)$ is a left R-module where

$$(rf)(m) = rf(m) \quad \forall r \in R, m \in M, f \in M^\star.$$

A right R-module M is *torsionless* if M satisfies any one of the equivalent conditions:

(1) M embeds in a product R^α of copies of R,

(2) For every $0 \neq m \in M$, there exists $f \in M^\star = \mathrm{Hom}_R(M, R)$ so that $f(m) \neq 0$,

(3) The canonical map $\varphi : M \to M^{\star\star}$ is injective, where

$$f\varphi(m) = f(m) \quad \forall f \in M^\star, m \in M.$$

REMARK. If I is a right ideal of R, then R/I is torsionless (resp. embeds in a free module) iff I is a right annihilator of a subset X (resp. finite subset X of R). Cf. Rosenberg-Zelinsky [59], or the author's **Algebra II**, p.203, Prop. 24.1 and Cor. 24.2.

Obviously, any submodule of the *direct sum* $R^{(\alpha)}$ of copies of R is torsionless inasmuch as $R^{(\alpha)} \hookrightarrow R^\alpha$, i.e., $R^{(\alpha)}$ consists of all $f \in R^\alpha$ with finite support.

1.6. Torsion Abelian Groups.

The *order of a group* A is the number $|A|$ of elements in A. The *order of an element* $a \in A$ is the order of the cyclic group (a) that a generates. If every element of an Abelian group A has finite order, then A is a *torsion* group (also called a *periodic* group). If at the other extreme, every element of A except 0 has infinite order, then A is said to be *torsionfree*. Any subgroup of a direct product $\mathbb{Z}^\alpha$ of copies of $\mathbb{Z}$ (i.e., any torsionless $\mathbb{Z}$-module) is torsionfree. Furthermore, any $f \cdot g$ torsionfree group is torsionless (Exercise). A group is *mixed* if neither torsion nor torsionfree.

Any finite group is torsion, and so is any direct sum of groups of finite orders, but the direct product $\prod_{n=1}^\infty \mathbb{Z}/(p^n)$ of cyclic groups of orders p^n is a mixed group (Exercise). As stated in the Preface, any finite Abelian group is a direct sum of cyclic groups of prime power order (cf. Fundamental Theorem of Abelian Groups $(= FTAG)$).

1.7. Primary Groups.

A group A is *primary for a prime* p if A is a torsion group and every element has order equal to a power of p. Then A is also said to be *p-primary.* The sum of all p-primary subgroups of an Abelian group A is a p-primary group called the *p-primary* part of A.

1.7A THEOREM. *Any torsion Abelian group A is a direct sum of p-primary groups A_p, where*

$$A_p = \{a \in A | p^i a = 0 \quad \textit{for some } i > 0\}.$$

1.8. Bounded Order.

An Abelian group A has *bounded order* n provided that there is an integer $n > 0$ so that $na = 0$ for $\forall a \in A$. Notation: $nA = 0$.

1.9. Theorems of Zippin and Frobenius-Stickelberger.

1.9A ZIPPIN'S THEOREM. *Any Abelian group of bounded order n is a direct sum of cyclic groups of prime power orders.*

1.9B COROLLARY (FROBENIUS AND STICKELBERGER [1878]). *Any finite Abelian group is a direct sum of cyclic groups of prime power orders.*

This theorem generalizes the fundamental theorem $(FTAG)$ in one respect: A need not be finitely generated to be of bounded order. Theorem 1.9A was attributed by Kaplansky ([69], p.74) to Zippin [35] who stated it for a countable group A but Kaplansky remarks that "his proofs did not really make use of countability."

1.10. Divisible Groups.

An Abelian group A is *divisible* provided that $nA = A$ for any integer $n \neq 0$. The field $\mathbb{Q}$ of rational numbers under addition is an example, and so is any vector space over $\mathbb{Q}$.

Another example is $\mathbb{Q}/\mathbb{Z}$ the factor group of $\mathbb{Q}$ modulo $\mathbb{Z}$. Furthermore, the union

$$\mathbb{Z}_{p^\infty} = \bigcup_{n=1}^{\infty} \mathbb{Z}_{p^n}$$

of cyclic groups of order p^n for a prime p, is divisible, and in fact

$$\mathbb{Q}/\mathbb{Z} \approx \bigoplus_{p \text{ prime}} \mathbb{Z}_{p^\infty}$$

(Cf. Theorem 4 of Kaplansky [69], p.10).

1.11 SPLITTING THEOREM FOR DIVISIBLE GROUPS. *If B is a divisible subgroup of an Abelian group A, then B is a direct summand.*

A is **reduced** if 0 is the only divisible subgroup.

1.12 SECOND SPLITTING THEOREM. *For divisible groups every Abelian group A contains a unique maximal divisible subgroup $D = D(A)$. Moreover $\overline{A} = A/D(A)$ is reduced, and $A \approx D \oplus \overline{A}$ (then $\overline{A}$ is called the reduced part of A).*

According to Kaplansky [69], p.74 the first "general statement" of the splitting Theorem appears in a paper of Baer [36], p.766, where it is described as "well-known".

1.13 DECOMPOSITION THEOREM FOR DIVISIBLE GROUPS. *Every divisible Abelian group A is isomorphic to a direct sum $\oplus_{i \in I} A_i$ where either $A_i \approx \mathbb{Q}$, or $A_i \approx \mathbb{Z}_{p^\infty}$ for some prime number p.*

The *torsion subgroup* $t(A)$ of an Abelian group A is the set of all elements of A of finite order.

1.14 TORSION GROUP SPLITS OFF THEOREM. *If A is an Abelian group such that the reduced part of $t(A)$ is a group of bounded order, then $t(A)$ splits off.*

REMARK. Theorem 1.14 implies a theorem of Frobenius and Stickelberger [1878] stating that $t(A)$ splits off of any group A that is a finite direct sum of cyclics, e.g. by FTAG, A can be any $f \cdot g$ Abelian group (Cf. Theorem 4.1D).

In his book, Kaplansky [69] states that Theorems 1–14 of his book hold for modules over a principal ideal domain (= PID), and he actually states "Theorem 16" in this context. This is the next theorem.

1.15. Fundamental Theorem of Abelian Groups and Kulikoff's Subgroup Theorem.

1.15A FUNDAMENTAL THEOREM OF ABELIAN GROUPS. *Over a PID, any $f \cdot g$ module is a direct sum of cyclics.*

The next theorem is "Theorem 17", *ibid*, p.45.

1.15B KULIKOV'S SUBGROUP THEOREM [52]. *If R is a PID and if M is a direct sum of cyclic R-modules, then so is any submodule.*

This implies, e.g., that any subgroup of a free Abelian groups is free. See Kaplansky [69] for other references to Kulikov's papers.

1.16. Corner's Theorem and the Dugas-Göbel Theorem.

1.16 CORNER'S THEOREM [63]. *Every countable ring A whose additive group is reduced and torsionfree is isomorphic to the endomorphism ring of an Abelian group with the same properties.*

An Abelian group A has *no nonzero cotorsion subgroups* if A is reduced, torsionfree, and has no subgroup $\approx \mathbb{Z}_{(p)}$, the group of p-adic integer, for any prime p. In this case, A is said to be **cotorsion-free**. A ring R is cotorsion-free if its additive group is cotorsion-free.

1.16A FIRST DUGAS-GÖBEL THEOREM [82]. *Any cotorsion-free ring $R \approx End_{\mathbb{Z}}G$ for some Abelian group G.*

This theorem is included in the:

1.16B SECOND DUGAS-GÖBEL THEOREM [82]. *Let R be a Dedekind domain not a field, and κ any cardinal number. the following conditions are equivalent:*

(1) *R is not a complete discrete valuation ring.*
(2) *There are indecomposable R-modules of rank ≥ 2.*
(3) *There exist indecomposable R-modules of rank $\geq \kappa$.*
(4) *There are R-modules of rank κ which do not satisfy the test problems of Kaplansky [69].*
(5) *Any cotorsion-free R-algebra A is an endomorphism ring: $A \approx End_R M$ for some R-module M.*

1.17. Direct Products as Summands of Direct Sums.

A submodule S of a right R-module M is $\cap$-pure or, *RD-pure*, provided that $S \cap Ma = Sa \ \forall a \in R$. Below, $|J|$ denotes the cardinality of a set J.

1.17A THEOREM (CHASE [60]). *If there is an infinite set J such that the product R^J of $|J|$ copies of a ring R is an RD-pure submodule of a direct sum of left R-modules $\oplus_{a\in A}M_a$ such that $|M_a| \leq |J|$, then R satisfies the dcc on principal right ideals.*

Cf. Perfect rings, 3.31. Also see 1.22ff.

Note: Since any domain satisfying the dcc on principal right ideals is a field, one has:

1.17B COROLLARY. *The product R^ω of a countable set of copies of an integral domain R not a field is not an RD-pure submodule of a free R-module. In particular R^ω is not free (in fact not projective. see §3).*

1.18 THEOREM (BAER [37]). *$\mathbb{Z}^\omega$ is not free, in fact, $Hom_{\mathbb{Z}}(\mathbb{Z}^\omega, \mathbb{Z})$ is countable.*

1.19. Specker-Nöbeling-Balcerzyk Theorems.

For any cardinal α, the **Specker** group of $\mathbb{Z}^\alpha$ is the set of all bounded sequences.

1.19A THEOREM (SPECKER [50]-NÖBELING [68]). *The Specker group of* $\mathbb{Z}^\alpha$ *is free for any cardinal* α.

See Fuchs [73], Theorem 97.3 and Corollary 97.4.

1.19B THEOREM (SPECKER [50]-BALCERZYK [62]). *For any cardinal* α, $Hom(\mathbb{Z}^\alpha, \mathbb{Z})$ *is free.*

Specker [50] proved 1.19A and 1.19B under additional assumptions, e.g. $\mathbb{Z}^\omega$ is $\aleph_1$-free, i.e., any subgroup A with $|A| < \aleph_1$ is free.

1.20. Dubois' Theorem.

Let $P(R) = R^\omega$ denote the set of all countable sequences of elements of R and $B(R)$ those with sequences such that the entries form a finite set. $B(\mathbb{Z})$ is the set of all bounded sequences in $P(\mathbb{Z})$.

For cardinals m and m_1, let $m \doteq m_1$ mean either both are finite or both are equal. Dubois employed a change of rings argument to generalize 1.19A to rings but restricted to $\alpha = \omega$.

1.20 THEOREM (DUBOIS [66]). *For any ring R, every submodule S of B(R) generated by* $m \leq \aleph_1$ *elements is contained in a free (R, R) submodule with a free basis consisting of* $m_1 \doteq m$ *elements.*

1.21. Balcerzyk, Bialynicki, Birula and Łoś Theorem, Nunke's Theorem, and O'Neill's Theorems.

1.21A THEOREM (BALCERZYK, BIALYNICKI, BIRULA AND ŁOŚ [61] AND NUNKE [62]). *Let* μ *be the least measurable cardinal (assuming existence in ZFC), and let* $\mathbb{Z}^\kappa = A \oplus B$ *for infinite* $\kappa < \mu$. *Then* $A \approx \mathbb{Z}^\alpha$ *and* $B \approx \mathbb{Z}^\beta$ *for* $\alpha, \beta \leq \kappa$.

1.21B THEOREM (O'NEILL [94]). *In Theorem 1.21A, the conclusion still holds if* $\kappa = \mu$.

REMARKS. (1) Nunke proved that any subgroup S of $\mathbb{Z}^\omega$ closed in the product topology is a product (*ibid.*); (2) O'Neill [96a] gives a simpler proof of 1.21A.

O'Neill dramatically generalized Baer's theorem.

1.21C THEOREM (O'NEILL [95]). *If* $\mathbb{Z}^\alpha \approx A \oplus B$ *for subgroups A and B and any infinite cardinal* α, *then either* $A \approx \mathbb{Z}^\alpha$ *or* $B \approx \mathbb{Z}^\alpha$.

This was generalized to any Dedekind ring not a field or a complete discrete valuation ring, in O'Neill [97].

On when a ring R is an **F-ring** (in the sense of Lenzing: if any product of free modules is free), see e.g. O'Neill [93b]. For any positive cardinal number α, there exists a domain R such that, as a left R-module, $R \oplus R \cong R^\alpha$ (O'Neill [91]). If R is a commutative ring, α is infinite and if a direct product of α nonzero R-modules is free of finite rank, then R is a ring-direct product of α nonzero rings (O'Neill [93a]).

1.21D REMARKS. (1) R is a left F-ring iff R^ω is a free left R-module of infinite rank (O'Neill [93b]) and then R is right coherent (§6) and the conditions of Corollary 1.24B are satisfied;

(2) There exists a PID R of arbitrarily large cardinality with free additive group (in fact, R can be a discrete valuation domain (O'Neill [84])). Let $A = \mathbb{Z}[X]$ be the polynomial ring in an arbitrary set X of commuting variables. Then the set R of the quotient field Q of A consisting of $p(x)/q(x)$, where $p(x)$ and $q(x)$ are polynomals, and $q(x)$ is primitive is a ring with the stated property.

(3) (O'Neill [90]) A vector group V is a direct product of subgroups of $\mathbb{Q}$. If $|V| < \mu$ of 1.21A, then any direct summand of V is a vector group also. Cf. Corollary 8 of O'Neill [87].

1.22. Direct Sums as Summands of Their Direct Products.

In general, the direct sum is not a summand of the direct product:

1.23 THEOREM (CAMILLO [85]). *If a direct sum $M = \oplus_{a \in A} M_a$ of right R-modules splits off in their direct product $\prod_{a \in A} M_a$, then there is a cofinite subset B of A, such that*

$$R/\mathit{ann}_R(\oplus_{b \in B} M_b)$$

satisfies the acc on annihilator right ideals.

Cf. $\sum$-injective modules, 3.7.

Sarath and Varadarajan [74] characterized direct sum splitting very differently.

1.24A THEOREM (LENZING [76]). *If the direct sum of infinitely many copies of a faithful R-module M splits off in their direct product then R satisfies the acc on annihilator right ideals ($=$ acc $\perp$).*

1.24B COROLLARY (*loc. cit.*). *If $R^{(\mathbb{N})}$ splits off in $R^{\mathbb{N}}$, then R is a semiprimary ring (see §3) with acc$\perp$.*

Cf. 1.17A, also 6.55.

A right R-module M is **algebraically compact or pure-injective** provided that every finitely solvable system of linear equations over R in M has a simultaneous solution.

1.25 ZIMMERMAN'S THEOREM [77] ON PURE INJECTIVE MODULES. *Every direct sum of copies of a right R-module M splits in the direct product iff every direct sum of copies of M is pure-injective.*

We shall come back to this concept in §6. Cf. 6B-6D and 6.46ff, esp. Σ-pure-injective modules.

1.26. Theorems of Szele, Fuchs, Ayoub and Huynh: When the Additive Torsion Ideal Splits Off.

The torsion subgroup $t(R)$ of the additive group of a ring R is an ideal of R. If $t(R)$ is a ring direct factor of R, then R is said to be **fissile**.

1.26A THEOREM (T. SZELE-L. FUCHS [56], C. AYOUB [77], AND D. V. HUYNH [77]). *If a ring R satisfies the dcc on principal right ideals, then R is fissile; in fact, there is a unique ideal I such that $R = T \times I$, where $T = t(R)$.*

REMARK. A ring R satisfying the dcc on principal right ideals is called **left perfect**. See 3.31. The Szele-Fuchs theorem is for Artinian R, while those of Ayoub and Huynh are for perfect rings. Szele and Fuchs furthermore completely determine the structure of any Artinian ring-1 with no additive subgroups $\approx \mathbb{Z}_{p^\infty}$.

1.27. Kertész-Huynh-Tominaga Torsion Splitting Theorems.

The next theorem not only generalizes Theorem 1.26A but Tominaga gives a short, simple self-contained proof of a half-page length.

1.27A THEOREM (KERTÉSZ [64], HUYNH [76], AND TOMINAGA [79]). *Let A be a subring of a ring R such that*

(1) $(R,+) = (A,+) + B$ *and* $AB = BA = 0$ *for an additive subgroup B of R and such that*
(2) *For each $x \in R$, there exists $f \in R$ with $x - fx \in A$.*

Then, $R = A \times B^2$.

1.27B REMARK. See O'Neill [76,77] for rings whose additive subgroups are subrings or ideals.

1.28. Three Theorems of Steinitz on the Structure of Fields.

We refer to Volume 1 of Van der Waerden's classic *Modern Algebra* [49] for the pertinent concepts and proofs. We shall denote this reference by [VDW].

Let F be a field and K a subfield. The intersection of all subfields of a field F is called the **prime subfield** of F, and is either $\mathbb{Q}$, or $\mathbb{Z}_p$ for a prime p.

F is *algebraic* over K if every $a \in F$ satisfies a nonzero polynomial $f(x)$ with coefficients in K, equivalently the subfield $K(a)$ generated by a and F is a finite dimensional vector space over K. F is **absolutely algebraic** if F is algebraic over the prime subfield. Steinitz's first theorem below is for a non-algebraic, that is, *transcendental* field extension F over K. If $x \in F$ is not algebraic over K, then we say that x is *transcendental* over K, equivalently, the subfield $K(x)$ generated by K and x is isomorphic to the field of all rational functions $f(x)/g(x)$ in the variable x, where $f(x)$ and $g(x)$ are polynomials, and $g(x) \neq 0$. By transfinite induction, or by Zorn's lemma (see 2.17A), there is a maximal set $X = \{x_i\}_{i \in I}$ of algebraically independent elements of L in the sense that for every finite subset $x_1, \dots, x_s$ of X, all finite products $x_1^{n_1} x_2^{n_2} \dots x_s^{n_s}$ are linearly independent over K, equivalently the subfield $K(x_1, \dots, x_n)$ is isomorphic to the field $K(X_1, \dots, X_n)$ of rational functions n variables $X_1, \dots, X_n$ over K. Two different maximal sets of algebraically independent elements have the same cardinality $\alpha = tr.d.(K/F)$ called the *transcendence degree* of F over K. Moreover X is called a *transcendence basis* of K/F. By agreement $X = \emptyset$ if K/F is algebraic, in which case tr.d.$(K/F) = 0$.

1.28A FIRST STEINITZ THEOREM [10,50]. *Let $L \supseteq K$ be fields, and let X be a transcendence basis of L/K. Then $L/K(X)$ is algebraic.*

PROOF. This follows from the fact that X is a maximal set of algebraicaly independent elements of L over K, that is, for any $y \in L$, y is algebraic over $K(X)$. □

Cf. [VDW], p.202, for the case tr.d.$K/F = n < \infty$.

If k is a field, then K is *algebraically closed*, if there are no fields $L \supseteq K$ that are algebraic over K, except $L = K$, equivalently, every polynomial $f(x)$ in the ring of polynomials $K[X]$ over k factors into linear factors over K.

1.28B SECOND STEINITZ THEOREM (*op.cit.*). *(1) Every field K can be embedded in an algebraicaly closed field $\overline{K}$ that is algebraic over K; (2) $\overline{K}$ is called the algebraic closure of K, and any two algebraic closures of K are isomorphic by a mapping that induces the identity on K; (3) If L is an algebraically closed field $\supseteq K$, then L contains an algebraic closure $\overline{K}$ of K.*

PROOF. (1) By Zorn's Lemma there exists a maximal algebraic extension field $\overline{K} \supseteq K$. Moreover, by maximality, $\overline{K}$ is algebraically closed. (2) and (3) are also proved by Zorn's Lemma. See, e.g. vol. 2 of Cohn's Algebra ([77]), p.214.

1.28C REMARKS. (1) If K is countable, then $\overline{K}$ is countable. Via elementary set theory, a countable union of countable sets is countable. However, this fact *too* appears to require the axiom of choice! See Hrbacek and Jech [84], p.81, 3.6ff, for a discussion of this anomaly; (2) The classic theorem of Gauss states that the field of complex numbers $\mathbb{C}$ is algebraically closed. Hence, by 1.29, $\mathbb{C}$ contains an algebraic closure A of $\mathbb{Q}$, namely *the field of all algebraic numbers.*

1.28D THIRD STEINITZ THEOREM [10,50]. *If K and F are uncountable algebraically closed fields, then $K \approx F$ iff char$K$ =char$F$ and $|K| = |F|$.*

PROOF. See *op.cit.* [50].

1.28E REMARKS. (1) If $K = A(X)$ is the field of rational functions over the field A of algebraic numbers, then $\overline{K}$ is countable by Remark 1.28 (1), however $\overline{K} \not\approx A$ since tr.d.$(\overline{K}/\mathbb{Q}) = 1$ while tr.d.$(A/\mathbb{Q}) = 0$. Thus, Theorem 1.28D fails for countable algebraically closed fields. (See 1.28H below.)

1.28F COROLLARY. *If K is an uncountable algebraic closed field of characteristic 0, and if $L \supseteq K$ is a field countably generated over K, then the algebraic closure $\overline{L}$ of L is isomorphic to K.*

PROOF. $|\overline{L}| = |L| = |K|$, and char$\overline{L}$ = char$K$, hence Theorem 1.28D applies.

1.28G BIZARRE EXAMPLE. If $L = \mathbb{C}(X)$, then $\overline{L} \approx \mathbb{C}$, hence there is a subfield F of $\overline{L}$ such that $F \approx \mathbb{R}$, and $[\overline{L} : F] = 2$. (Cf. 1.30D below.) In this way one can construct a countable chain of fields $\{L_n\}_{n=1}^{\infty}$ with each $L_n \approx \mathbb{C}$. Furthermore, the field $K = \bigcup_{n=1}^{\infty} L_n$ is also algebraically closed with $|K| = |\mathbb{C}|$, hence $K \approx \mathbb{C}$. (Bizarre, no?)

The next theorem is a refinement of aspects of the three Steinitz theorems 1.28A,B and D.

1.28H THEOREM (STEINITZ. *op.cit.*). *If K and L are algebraically closed fields containing a field F, then $K \approx L$ over F iff tr.d.K/F = tr.d.L/F.*

PROOF. If X and Y are transcendence bases of K/F and L/F of the same cardinality, then $A = F(X) \approx B = F(Y)$, and hence $K = \overline{A} \approx L = \overline{B}$ are isomorphic algebraic closures of isomorphic fields. Conversely any isomorphism $K/F \approx L/F$ preserves tr.d. $\square$

1.29. Lüroth's Theorem.

1.29A LÜROTH'S THEOREM. *If $L = K(X)$ is a field of rational functions in a single variable over K, then any subfield F of L properly containing K is also, that is, $F = K(Y)$ for some transcendental element $Y \in L$.*

PROOF. See [VDW], p.198, or Cohn [77], p.221.

The proofs use the next lemma:

1.29B LEMMA. *In Lüroth's theorem, write $Y = g(x)/f(x)$, for $f(x), g(x) \in K[X]$, and let $n = max\{\deg f(x), g(x)\}$. Then L has dimension n over $K(Y)$.*

PROOF. *ibid.*

1.30. Artin-Schreier Theory of Formally Real Fields.

A field K is **formally real** if -1 is not sum of squares of elements of K, equivalently, a sum of nonzero squares is never 0, that is, for all $x_1, \dots, x_n \in K$,

$$x_1^2 + x_2^2 + \cdots + x_n^2 = 0 \Rightarrow x_1 = x_2 = \cdots = x_n = 0.$$

A field K is **real closed** if K is formally real but no proper algebraic field extension of K is formally real.

A field K is **ordered** provided that there is a set of elements P, called the **positive cone** of K, consisting of elements which are said to be **positive**, notation $a \in P$ iff $a > 0$, satisfying the rules of order:

(O1) (Trichotomy) If $a \in K$, then just one of the three possibilities hold:

$$a > 0, \qquad a = 0, \qquad \text{or} \qquad -a > 0$$

(O2) If $a > 0$ and $b > 0$, then $a + b > 0$ and $ab > 0$.

Furthermore, if $-a > 0$, we say that a is **negative**. The **order relation** in K is defined by:

$$a > b \Leftrightarrow a - b > 0$$

and

$$a < b \Leftrightarrow b - a > 0.$$

One checks that this is a linear order of K, namely given $a, b \in K$ just one of these possibilities hold:

$$a > b, \quad a = b, \quad \text{or} \quad a < b$$

The **absolute value** $|a|$ is defined by:

$$|a| = a \quad \text{if } a > 0 \text{ or } a = 0 \text{ and } |a| = -a \quad \text{if } a < 0.$$

One checks

$$|ab| = |a| \cdot |b|$$

$$|a + b| \leq |a + b| \qquad \textbf{triangle inequality}.$$

Furthermore, by (O1) or (O2), the characteristic of an ordered field K is zero: char$K = 0$. See, e.g. Artin-Schreier [26], or Van Der Waerden [48], vol. 1, pp.209-210. As before, we abbreviate this reference by [VDW].

1.30A THEOREM (ARTIN-SCHREIER [26]). *Every real closed field K has a unique order. A field K can be ordered iff K is formally real.*

PROOF. See [VDW], p.225. Also, Cohn [77],p.268.

1.30B THEOREM (*op.cit.*). *The field extension L of an ordered field K generated by all square roots of positive elements of K is formally real. Moreover, every ordered field has a real closure that is unique up to an order isomorphism.*

PROOF. See [VDW], p.228. Cf. Cohn *op.cit.*, p.268.

1.30C THEOREM (*op.cit.*). *If K is formally real, then K is real closed iff $K(i)$ is algebraically closed. In this case every nonconstant polynomial factors into linear and quadratic polynomials.*

PROOF. [VDW], p.228.

1.30D THEOREM (*op.cit.*). *If F is an algebraically closed field containing a formally real subfield K, then there exists a real closed field L between F and K such that $F = L(i)$, where $i^2 = -1$.*

PROOF. [VDW], p.230.

1.30E THEOREM (*op.cit.*). *Every formally real algebraic field extension of $\mathbb{Q}$ is isomorphic over $\mathbb{Q}$ to a subfield of the field A of real algebraic numbers.*

PROOF. [VDW], p.232.

1.31. Theorem of Castelnuovo-Zariski.

1.31A THEOREM (CASTELNUOVO-ZARISKI). *If K is algebraically closed of characteristic 0, and if $L = K(X,Y)$ is the field of rational functions in two variables, then every subfield F of L properly containing K is either isomorphic to $K(X,Y)$ or to $K(X)$ by a mapping that induces the identity on K.*

PROOF. Castelnuovo proved this theorem for the case $K = \mathbb{C}$, and Zariski [58] proved it for algebraically closed K of characteristic 0.

1.31B REMARK. Deligne [73] proves the failure of the Castelnuovo-Zariski Theorem for rational function fields in three variables over algebraically closed fields of characteristic 0.

1.32. Monotone Minimal Generator Functions.

Let $L \supset F \supset K$ be three fields, and let $n(F/K)$ denote the cardinal of any minimal set of generators of F/K. By Lüroth's theorem, $n(F/K) = 1$ for any proper intermediate field F of $K(X)$ and K, and by Castelnuovo-Zariski, $n(F/K) = 1$ or 2, for intermediate fields of $K(X,Y)$ when K is algebraically closed of characteristic 0.

1.32A THEOREM (FAITH [61D]). *Let L/K be a finitely generated field extension of tr.d.≤ 1. Then the minimal generator function $n(F/K)$, defined for any intermediate extension F/K, is monotone. Furthermore, the same is true for L/K of tr.d.≤ 2 when K is algebraically closed of characteristic 0.*

1.32B REMARKS. (1) A similar theorem holds for intermediate fields F of a finitely generated field extension for which either L/K or F/K is not separably generated: then $n(L/K) \geq n(F/K)$. (See *ibid.*, p.551); (2) The proofs depend fundamentally on a lemma proved by the author (*ibid.*, p.550) that the dimension function on the vector spaces $\mathcal{D}(F/K)$ of K-derivations is monotone, for any finitely generated field extension L/K, and intermediate fields F; (3) The proofs also require fundamental results of Jacobson [37] on abstract derivations, and other fundamental results on separably generated extensions, found e.g. in Zariski-Samuel [58], vol. 1.

1.33. Quigley's Theorem: Maximal Subfields without α.

Quigley [62] studied field extensions $K \supset F$. If $\alpha \in K\backslash F$, then by Zorn's Lemma, there exists a subfield $M \supseteq F$ that is **maximal without** α, that is, $M(\alpha)$ is the intersection of all subfields L of K properly containing M. We cite just Quigley's first two results:

1.33 QUIGLEY'S LEMMA 1. *If M is maximal without α, then K is an algebraic extension of M.*

PROOF. We observe first that α is algebraic over M; for if $\alpha^2 \notin M$, then $M(\alpha^2) \supset M(\alpha) \supset M(\alpha^2)$, so that α^2 is algebraic over M. Let $\overline{M}$ be the algebraic closure of M in K. If $\exists \eta \notin \overline{M}$, then η is transcendental over M, as is every element of $M(\eta)$ not in M. Since $M(\eta) \neq M$, it follows that $M(\eta) \supset M(\alpha)$; thus, since $\alpha \notin M$, it follows that α is transcendental over M; a contradiction. $\square$

1.34 QUIGLEY'S THEOREM 1. *If M is maximal without α, then there exists a prime number p such that $[N : M]$ is a power of p for every finite normal extension N of M. Either M is a perfect field, or else K is a purely inseparable extension of M. Furthermore $[M(\alpha) : M] = p$, and $M(\alpha)$ is a normal extension of M. All pth roots of unity lie in M, so that there exists $a \notin M$ such that $a^p \in M$ and $M(a) = M(\alpha)$, unless M is perfect and $p = q$, the characteristic of K.*

REMARK. The beautiful proofs require much of classical Galois Theory, field theory, group theory, and the Artin-Schreier Theorems. The complete structure theory of "maximal without α subfields" is quite elaborate (*ibid.*).

CHAPTER 2

Introduction to Ring Theory: Schur's Lemma and Semisimple Rings, Prime and Primitive Rings, Nil, Prime and Jacobson Radicals

If $M = M_1 \oplus \cdots \oplus M_n$, and if $N = M_1 \approx M_j, j = 1, \ldots, n$, then $M \approx N^n = N \times \cdots \times N$ (n factors) and End M_R is canonically isomorphic to the ring B_n of all $n \times n$ matrices over B, where B is a ring $\approx$ End N_R. (See, e.g., my Algebra I, pp.151-2, Prop. 3.33.) A right R-module is **simple** if M has precisely two submodules, namely M and 0. If N is a simple R-module, then by Schur's Lemma, B is a skew field (= sfield = division ring). Thus, any simple R-module V is a vector space over the skewfield End V_R.

REMARK. If I is a right ideal of R, then $V = R/I$ is simple iff I is a maximal right ideal.

Quaternions

The first noncommutative division algebra was discovered in 1843 by W. R. Hamilton after 12 years of research. (Now-a-days we expect our undergraduates in abstract algebra to discover them within the confines of a single lecture!) The algebra discovered by Hamilton is a 4-dimensional vector space over $\mathbb{R}$

$$\mathbb{H} = \mathbb{R} + \mathbb{R}i + \mathbb{R}j + \mathbb{R}k$$

where multiplication is given by:

$$i^2 = j^2 = k^2 = -1,$$

and

$$ij = k = -ji, \ jk = i = -kj, \ ki = j = -ik.$$

The multiplicative group generated by i, j, k has order 8 and is called the *Quaternion group*.

REMARK. The 8-dimensional non-associative Cayley-Dickson algebra over $\mathbb{R}$ is defined similarly.

2.0 THEOREM (FROBENIUS). *Every non-commutative division algebra A of finite dimension over the field $\mathbb{R}$ of real numbers is isomorphic to $\mathbb{H}$.*

Hilbert's Division Algebra

The first division algebra A of infinite dimension over its center k was discovered by Hilbert [03], namely, the field A of all skew Laurent power series

$$\sum_{i \geq n} \alpha_i x^i$$

where n is an integer (i.e., $n \in \mathbb{Z}$), α_i belongs to the rational function field $k = \mathbb{R}(t)$ in a variable t, and multiplication is "skewed" by the rule

$$x\alpha(t) = \alpha(2t)x$$

and its consequences, for any $\alpha(t) \in \mathbb{R}(t)$.

If $F = k\langle x \rangle$ denotes the power series ring over k in the variable x, then $A = F[x^{-1}]$ is the set of polynomials over F in the variable x^{-1}, with the stated skew multiplication. Hilbert [03] corrected his original construction in the first edition of his "Grundlagen der Geometrie" in 1899. (See Cohn [91], p.45 for a discussion.)

When Everybody Splits

A ring R is **simple** if R has no ideals except 0 and R. Any sfield D is simple, and so is the $n \times n$ matrix over D.

A right R-module M is **semisimple** if M has the equivalent two properties:
($SS1$) M is a direct sum of simple submodules,
($SS2$) Every submodule S of M is a direct summand of M: $M = S \oplus T$ for a submodule T (= every submodule of M splits).

Any vector space V over a skew field D is semisimple, in fact, V is the free D-module with a free basis of cardinality equal to the dimension of V over D. Over the full $n \times n$ matrix ring $R = D_n$ over a skew field D, every R-module is semisimple (see below), but no longer free if $n > 1$.

2.1 THEOREM (WEDDERBURN [08]-ARTIN [27]). *The following are equivalent conditions on a ring R:*

(1) *Every right R-module is semisimple,*
(2) *R is a finite direct product,*

$$R = R_1 \times \cdots \times R_t$$

where $R_i \approx (D_i)_{n_i}$ are full $n_i \times n_i$ matrix rings over skew fields $D_i, i = 1, \ldots, t$.

Then t is unique, and up to order so is the set $\{n_1, \ldots, n_t\}$. Furthermore, the skew fields D_i are unique up to order and isomorphism, $i = 1, \ldots, t$.

(3) *Every left R-module is semisimple.*

A ring R is **semisimple** if R satisfies the equivalent conditions of 2.1. Thus, a necessary and sufficient condition for a ring R to be semisimple is that R itself be a semisimple R-module, that is, a (necessarily) finite direct sum of minimal right ideals. Moreover R is a direct sum of the simple rings R_i.

2.1A COROLLARY. *The following are equivalent conditions on a ring R with center C:*

(1) *R is semisimple*

(2) *Every right ideal is generated by an idempotent.*

(3) *Every left ideal is generated by an idempotent.*

Moreover, then every (two-sided) ideal is generated by a central idempotent, that is, an idempotent of C.

PROOF. This follows from Theorem 2.1 and the results of 1.2 on the relationship between idempotents of $A = \text{End} M_R$ and direct summands of a right R-module M_R. In this case $M = R$, and $A \approx R$ canonically under the mapping sending $f \in A$ onto $f(1)$. □

2.1B COROLLARY. *Let R be a semisimple ring with center C. Then every two-sided ideal of the polynomial ring $R[x]$ is generated by a central idempotent, that is, an element of $C[x]$.*

PROOF. Cf. Goodearl-Warfield [89],p.263, Prop. 15.1 for a more general result. □

REMARKS. (1) By 2.1A, the proof reduces easily to the case when $R = D_n$ is a simple matrix ring over a sfield D, which is the case considered by Goodearl and Warfield, *ibid.*;

(2) In Corollary 2.1B, every simple $R[x]$-module has finite length as an R-module. Cf. 3.36D.

Artinian Rings and the Hopkins-Levitzki Theorem

The theorem of Artin [27] characterizes a semisimple ring R by (1) the descending chain condition (dcc) on right ideals (= R is *right Artinian*), and (2) R has no nilpotent ideals $\neq 0$ (Cf. 2.34As) (= R is **semiprime**). The original proof required the further assumption of the ascending chain condition (acc) on right ideals, that is, that R is *right Noetherian.* However, the theorems of Hopkins [39] and Levitzki [39] obviated this: every right Artinian ring **is** right Noetherian. Cf. *Noetherian and Artinian Modules and Composition Series*, **sup.** 2.17s8.

We next state the:

2.2 THEOREM (WEDDERBURN [08]—ARTIN [27]). *A ring R is a simple right Artinian ring iff $R \approx D_n$, a full ring of $n \times n$ matrices over a sfield D. In this case, R is left Artinian.*

Wedderburn proved this for simple algebras of finite dimension over a field. Another formulation of the theorem:

2.3 THEOREM (WEDDERBURN [08]—ARTIN [27]). *If R is a simple right Artinian ring, and if V is a simple right R-module, then (by Schur's lemma) $D = End\ V_R$ is a sfield, V has finite dimension n over D and*

$$R \approx D_n \approx End_D V.$$

2.4 COROLLARY. *If R is a finite dimensional algebra over a field k, and if k is algebraically closed, then the following are equivalent:*
(1) *R is a simple ring,*
(2) *$R \approx k_n$, a full $n \times n$ matrix ring over k, for some integer $n > 0$,*
(3) *R has a faithful simple right R-module V.*

A theorem of Burnside [11] states that $R \approx \text{End}_k V$ in (3). (Cf. Curtis-Reiner [62], p. 182, or Ribenboim [69], p. 78). Morita [58] proved very general theorems of this type for any ring R, namely that $R \approx \text{End}_A M$, where $A = \text{End} M_R$, whenever M is a right R-module such that a direct sum M^n of copies of M maps onto R (in Morita's terminology, M generates the category of right R-modules). (Cf. Faith [72a], Theorem 7.1, p. 326.)

Automorphisms of Simple Algebras: The Theorem of Skolem-Noether

If R is a ring, then the *center C of R* is a subring that consists of all $x \in R$ such that $xr = rx \quad \forall r \in R$. When R is simple, then C is a field, and when R is semisimple, C is a finite product of fields, i.e., semisimple.

An **automorphism** of an R-module M is a 1-1 onto R-homomorphism $M \to M$. An **automorphism of a ring R is a ring isomorphism $R \to R$.**

2.5A THEOREM (SKOLEM [27]-NOETHER [33]). *If R is a simple Artinian ring with center C, then any isomorphism $f: A \to B$ of two simple subalgebras over C that is C-linear in the sense that $f(ca) = cf(a) \quad \forall c \in C, a \in A$, is extendable to an inner automorphism of R, that is, there exists a unit $x \in R$ such that $f(a) = x^{-1}ax \quad \forall a \in A$.*

2.5B COROLLARY. *If the simple algebra R is finite dimensional over C, then every C-linear automorphism of R is inner.*

See, e.g. Herstein [68], pp.99-100. Also, see Jacobson's commentary in Noether [83] on p.19.

Similarly, any C-derivation D of a finite dimensional simple algebra R over C is inner: there exists $x \in R$ so that $D(r) = rx - xr \quad \forall r \in R$ (see, e.g. Jacobson [56,64], pp.101-2.).

Two subrings A and B of a ring R are **conjugate** if there is an inner automorphism f of R mapping A onto B.

The **centralizer** S' of a nonempty subset S of a ring R is a subring of R containing the center C of R, where

$$S' = \{x \in R \mid xs = sx \quad \forall s \in S\}.$$

If S is a subring then $S \cap S'$ is the center of S. Moreover, if x is a unit of R, then $(x^{-1}Sx)' = x^{-1}S'x$. Furthermore, setting $S'' = (S')'$, then $(x^{-1}Sx)'' = x^{-1}S''x$.

2.5C COROLLARY. *If A and B are isomorphic simple subrings of a simple Artinian ring R that contain the center C of R, then A and B, and their centralizers A' and B' are mutually conjugate:*

$$B = x^{-1}Ax \quad \textit{and} \quad B' = x^{-1}A'x$$

for a unit $x \in R$. Furthermore $B'' = x^{-1}A''x$.

2.5D COROLLARY. *If M and N are two sets of $n \times n$ matrix units of a simple Artinian ring $R = D_n$, where D is a sfield, then M and N and their centralizers M' and N' are mutually conjugate, and M' and N' are each conjugate to D.*

Wedderburn Theory of Simple Algebras

Let A be a simple central algebra A over the field C of finite dimension $d = \dim A$. If A^0 is the algebra opposite to A (i.e., anti-isomorphic to A), then $A \otimes_C A^0 \approx C_n$, the algebra of all $n \times n$ matrices over C. (Cf. the Brauer group $Br(C)$, sup. 4.16A, where $[A]^{-1} = [A^0]$). In particular, $d = n^2 = \dim_C C_n$ is a perfect square.

If A is a central division algebra over C, any maximal subfield F has $\dim_C F = n$, and $A \otimes_C F \approx F_n$, the $n \times n$ matrix algebra over F. (F is called a **splitting field** of A. Cf. Amitsur [56] on generic splitting fields. Also, see Herstein [68], pp.92–94, and Jacobson's discussion in Noether [83] on p.19.)

Crossed Products and Factor Sets

If A has a maximal subfield F that is a separable normal, or Galois, field extension of C then A is said to be a **crossed product** for reasons which we now describe. (Cf. also Jacobson [95], p.56ff.)

If G is the Galois group of F over C, then A has a free basis $\{u_g\}_{g \in G}$ of $n = (G:1)$ elements over F. Thus, by the Skolem-Noether theorem, any element $x \in A$ is uniquely expressible $x = \sum_{g \in G} a_g u_g$ with $a_g \in F$, where $u_g a u_g^{-1} = a^g \quad \forall a \in F$, and where a^g is the image of a under g, $u_g u_h = k_{g,h} u_{g,h}$, for elements $k_{gh} \in F$ satisfying these *associativity conditions*:

$$k_{g,h} k_{gh,f} = k_h^g k_{g,hf} \quad \forall g, h, f \in G.$$

(The set $\{k_{g,h}\}_{g,h \in G}$ is called a *factor set* of A. According to Jacobson in Noether [83], p.20, Dickson defined crossed products in 1926, except for the associativity conditions.)

If the Galois group G of F over C is Abelian (resp. cyclic) then A is said to be an *Abelian* (resp. *cyclic*) division algebra. Cyclic division algebras first were constructed by L. E. Dickson in 1906. See Dickson [23,76]. Also, see Amitsur and Saltman [78] for generic Abelian splitting fields.

According to Jacobson, *loc. cit.*, Dickson failed in his attempt to find noncylic division algebras, an achievement due to R. Brauer. Any division algebra finite dimensional over $\mathbb{Q}$ is a cyclic algebra over its center by a theorem of Brauer, Hasse and Noether [32]. (See Jacobson's discussion in Noether [83], p.21.) And many people attempted to find division algebras that were not crossed, and that honor belongs to Amitsur [72], who proved the existence over fields of characteristic 0 of "uncrossed" products of division algebras of degree n (= dimension n^2), for any n divisible by 8, or by a square of an odd prime. Cf. 15.16f.

Primitive Rings

Let M be a right R-module. The annihilator of M is

$$\text{ann}_R M = \{r \in R \mid xr = 0 \quad \forall x \in M\}$$

This is an ideal of R, and if $\text{ann}_R M = 0$, then M is called a *faithful* R-module. If R is a simple ring, then every right or left R-module $M \neq 0$ is faithful.

A ring R is **right primitive** if R has a faithful simple right R-module; equivalently, if there is a maximal right ideal I such that I contains no ideals $\neq 0$. (Then $V = R/I$ is simple and faithful.) Obviously, any simple ring is right and left primitive. In fact, any full (or dense) ring R of linear transformations of left a vector space V is right primitive (and V is a simple faithful right R-module N.B.). Not every right primitive ring is left primitive (Bergman [65]).

An ideal I of R is said to be **(right) primitive** (**qua** ideal) if R/I is a right primitive ring. A commutative ring R is primitive iff R is a field, and an ideal of R is primitive iff it is a maximal ideal.

Nil Ideals and the Jacobson Radical

An element x of a ring R is **nilpotent** (of index n) provided that some power $x^n = 0$ (for least exponent n). Thus, an integer $x \in \mathbb{Z}$ maps onto a nilpotent element of $\mathbb{Z}_{p^m}$, for a prime p, if $p \mid x$, i.e. x is in $p\mathbb{Z}_{p^m}$. Any matrix unit $e_{ij}, i \neq j$, is nilpotent of index 2: $e_{ij}^2 = 0$.

An ideal I is **nil** if every element of I is nilpotent. thus, $p\mathbb{Z}_{p^m}$ is a nil ideal of $\mathbb{Z}_{p^m}$. Furthermore, if $T_n(R)$ denotes the set of all lower triangular matrices over a ring R, then the set $LT_n(R)$ of all strictly lower triangular matrices is a nil ideal of T_n, and similarly for (strictly) upper triangular matrices (Cf. nilpotent ideals, 1.2s, 2.34As and 3.37s).

The **Jacobson radical**, rad$R$, of a ring R is the intersection of all primitive ideals. This coincides with the intersection of all left primitive ideals, and contains all **nil** one-sided ideals (= a right or left ideal consisting of nilpotent elements) (Jacobson [45A,55-64]). We shall come back to the Jacobson radical in several notes in §3.

The Chevalley-Jacobson Density Theorem

Below, we let l.t. abbreviate linear transformation.

2.6 Density Theorem (Chevalley-Jacobson [45a,b]). *If R is a right primitive ring with a faithful simple right R-module V, and $A = End\ V_R$, then R is canonically isomorphic to a dense ring of l.t.'s in V* **qua** *vector space over A, that is, given linearly independent elements $v_1, \dots, v_n$ of V, and corresponding arbitrary elements $w_1, \dots, w_n$ of V, there exists $r \in R$ so that $v_i r = w_i, i = 1, \dots, n$. Conversely, any dense subring R of l.t.'s of a vector space V over a sfield A is a right primitive ring (and V is a simple faithful right R-module).*

REMARK. The density theorem provides beautiful and beautifully easy proofs of the Wedderburn-Artin theorems since for any right Artinian primitive ring R necessarily $\dim_A V = n < \infty$, hence $R = \text{End}_A V \approx A_n$. It also shows that every nil one-sided ideal N is contained in every primitive ideal hence, as asserted **sup.** 2.6, $N \subseteq \text{rad} R$.

We give a proof of the density theorem in 3.8A. Also see §12.

Semiprimitive Rings

A ring R is a *subdirect product of a family* $\{R_i\}_{i\in I}$ of rings, if R embeds in the direct product $P = \prod_{i\in I} R_i$ in such a way that each projection $P_i : P \to R_i$ maps the image of R onto R_i $\forall i \in I$. An equivalent formulation is that R contains a family $\{K_i\}_{i\in I}$ of ideals so that $R/K_i \approx R_i$ $\forall i \in I$ and $\bigcap_{i\in I} K_i = 0$. It follows from the definitions that any ring R with zero Jacobson radical is isomorphic to a subdirect product of (right) primitive rings and is called a **semiprimitive** ring (called a semisimple ring by Jacobson [45a].) These rings are also subdirect products of left primitive rings since the radical is the intersection of left primitive ideals.

REMARKS. (1) A commutative ring R is semiprimitive (Cf. **sup.** 2.6) iff R is a subdirect product of fields (Example: $\mathbb{Z}$), and in this case the polynomial ring $R[X]$ is semiprimitive. Cf. 2.6A below.

(2) Similarly to the Remark following 2.6Af, if R is a right Artinian semiprimitive ring and P is a primitive ideal, then $\overline{R} = R/P \approx A_n$ (as before). Moreover, R has a direct factor $\approx \overline{R}$, and by the Artinian hypothesis and finite induction, R is a finite product of such rings. (Cf. 2.1.)

Semiprimitive Polynomial Rings

2.6A THEOREM (JACOBSON [45,55,64]). *If R is a ring with no nil ideals except 0, then the polynomial ring $R[x_1, \dots, x_n]$ in n variables is semiprimitive.*

PROOF. Jacobson [55,64], p.12, Theorem 4.

REMARK. It follows, e.g.: *the polynomial ring $R[x]$ may be semiprimitive even if R is not.* For example, if $R = k\langle x\rangle$ is the power series ring in the variable x, then $\text{rad} R = xR \neq 0$, but by 2.6A, $\text{rad}(R[x]) = 0$.

Matrix Algebraic Algebras

An algebra R over a field k is **matrix-algebraic** if every element a of every $n \times n$ matrix algebra R_n is algebraic over k, that is, $f(a) = 0$ for some nonzero $f(x) \in k[x]$.

2.6B REMARKS. (1) If R is any algebraic field extension of k, by the **Hamilton-Cayley Theorem** (see e.g. Herstein [75]), every $\alpha \in R_n$ satisfies its characteristic equation

$$f(x) = \det |\alpha - Ix| = 0$$

where I is the $n \times n$ identity matrix. The subalgebra L of R_n generated by α and the coefficients of $f(x)$ is thereby finite dimensional over k, hence α is algebraic over k. thus: *any algebraic field extension of k is matrix algebraic.*

(2) Similarly: *Any finite dimensional algebra A over k, hence any locally finite dimensional algebra over k, is matrix-algebraic over k.*

Furthermore:

2.6C THEOREM (AMITSUR [56]). *Every algebraic algebra R over an uncountable field k is matrix-algebraic.*

PROOF. See Jacobson [55,64], p.247, Theorem 2. □

2.6D COROLLARY. *If R is an algebraic algebra over an uncountable field k, and if S is a locally finite dimensional algebra over k, then the tensor product $R \otimes_k S$ is algebraic (hence matrix algebraic by 2.6B(2)).*

PROOF. *ibid.*,p.248, Corollary 1. □

2.6E THEOREM. *If R is an algebraic algebra over an uncountable field k, then $R \otimes_k F$ is an algebraic algebra over any field extension F of k.*

PROOF. *ibid.*, p.25.

Primitive Polynomial Rings

2.6F THEOREM (JACOBSON [45,55,64]). *Let R be a simple Artinian ring with center C. The following are equivalent right-left symmetric conditions:*
(1) *R is matrix-algebraic over C*
(2) *$R \otimes_C C(x)$ is a simple ring*
(3) *The polynomial ring $R[x]$ is* **right bounded** *(= every right ideal $\neq 0$ contains an ideal $\neq 0$)*
(4) *$R[x]$ is not a right primitive ring.*
(5) *Every $f(x) \in R[x]$ is a factor of a (central) polynomial, that is, of some $f_0(x) \in C[x]$.*

PROOF. See, e.g. Jacobson, *ibid.*, or Goodearl-Warfield [89], p.265, Theorem 15.2. □

A division ring D is transcendental if some $d \in D$ is not algebraic (= transcendental) over the center C.

2.6G COROLLARY (COZZENS-FAITH [75]). *Let D be a division algebra with center C. Then $R = D \otimes_C C(x)$ is a simple (right and left) principal ideal domain with (right and left) quotient field Q. Moreover, $R = Q$ iff $R[x]$ is not primitive. In particular, R is not a division ring whenever D is transcendental.*

PROOF. *Op.cit.*, p.62, Theorem 3.21. □

2.6H EXAMPLES. (1) The Hilbert division ring D (see 2.0f) is transcendental, hence $D[x]$ is primitive;

(2) In answer to a question of Cozzens-Faith [75], Resco [87] proved that all simple R-modules are injective over $R = D \otimes_C C(x)$ when D is an existentially closed field. Cf. 3.19-20 and 6.24-25.

The Structure of Division Algebras

Implicit in Corollary 2.4 is the fact that a division algebra D of finite dimension over an algebraically closed (= a.c.) field k is commutative. Another famed theorem of Wedderburn [05] states that any finite sfield D is commutative. (According to Parshall [83], there was an error in Wedderburn's proof that L. E. Dickson corrected.) Jacobson [74], vol. I, p.431 gives the celebrated short proof of E. Witt [31]. (Also see Cohn [77], vol.2, p.367 for a proof that invokes the Skolem-Noether theorem.) Wedderburn's theorem gave the first proof that Desargues' theorem implied Pappus' theorem for projective planes. (See, for instance, Blumenthal [80], Chapter V, esp. pp.96–98, or Parshall [83] and Schmidt [95].)

Tsen's Theorem

A student of Noether, Tsen [34,36] showed that if a division algebra D is finite-dimensional over an algebraic function field F in one variable (= F is a finite algebraic extension of the field $k(x)$ of rational functions in a variable x) over an algebraically closed field k, then D is commutative. (In the terminology of §4, *sup.* 4.16A, the Brauer group over F is trivial, i.e., $= 0$). Artin conjectured that this still held true when k is a finite field, and this was proved by Chevalley [36]. (See Artin [65], Preface, for additional comments by Lang-Tate.) Cohn [91], pp.278–9, gives a proof of Tsen's theorem using a result of Chevalley-Warning.

Tsen also proved that if D is a non-commutative division algebra over $K = k(x)$ where k is a real-closed field, then D is a 4-dimensional quaternion algebra over K.

These theorems of Tsen were extended to the situation where K is a subfield of D of finite relative dimension by Faith [61a], using a theorem of Jacobson [56,64]: *if D is finite dimensional over a subfield K, then D is finite dimensional over its center.*

Cartan-Jacobson Galois Theory of Division Rings

Let A be a division ring with center C, and G a group of automorphisms of A. The *algebra of the group* is the subalgebra $T(G)$ over C generated by all $0 \neq x \in A$ such that the inner automorphism I_x is an element of G (Recall $I_x(a) = x^{-1}ax \quad \forall a \in A$.) G is called an **N-group** (after Noether, or maybe Nakayama [50]) provided that $I_y \in G$ for all $0 \neq y \in T(G)$. A division subring F of A is said to be **Galois** provided that there is a group G of automorphisms of A so that F is the **fixring** of G, namely

$$F = \text{fix}G = \{a \in A \mid a^g = a \quad \forall g \in G\}.$$

Then the set $G(A/F)$ consisting of all automorphisms g of A with $\text{fix}(g) \supseteq F$ is a group called the **Galois group** of A/F (read A over F).

The **reduced order** of a group G of automorphisms of A is the product $(G : \mathcal{I}(G))[T(G) : C]$, where $\mathcal{I}(G)$ is the subgroup of inner automorphisms of G, and $[T(G) : C] = \dim_C T(G)$ **qua** vector space over C. If F is a non-commutative division subring of A, then in general the left vector space dimension $[A : F]_\ell$ is not equal to $[A : F]_r$, the right vector space dimension (N.B.).

Historical Note: Artin's Question

The equality, or inequality, of the two dimensions of a division ring A over a division subring F was raised by E. Artin at a Mathematical Congress at the Bicentennial of Princeton University in 1947. P. M. Cohn gave an example with $[A:F]_r = 2$ and $[A:F]_\ell = \infty$, and Schofield [85a,2.7] gave examples $[A:F]_r \neq [A:F]_\ell$ and both finite.

2.7. FUNDAMENTAL THEOREM OF GALOIS THEORY (CARTAN [47]-JACOBSON [47]). *Let $G = G(A/F)$ for a Galois subring F of a division ring A. Then $[A:F]_\ell < \infty$ iff G has finite reduced order n, and this case, $n = [A:F]_\ell = [A:F]_r$. Furthermore, the Galois correspondence*

$$K \longrightarrow G(A/K)$$

is $1-1$ between division subrings of K of A containing F and N-subgroups of G.

REMARKS. See Jacobson [56-64] for the Galois Theory of division rings, (Chap. VII), full rings of linear transformations and primitive rings (Chap. VI), including Nakayama's Galois Theory for simple Artinian rings (Nakayama [50], p.148ff *loc.cit.* Cf. Azumaya [49], Nakayama [49]). This theory includes the commutator (= centralizer) theory of simple algebras A, namely if $[A:C] < \infty$, and B is a simple central subalgebra, then $B = (B')'$, where B' is the centralizer of B in A. (Actually, $[A:C]$ need not be finite for this but $[B:C] < \infty$, for then B is an Azumaya algebra over C. See Theorem 4.15. Also see 6.30 and 12.A-C on Galois subrings of Ore domains and semiprime rings. (Cf. Kitamura [76,77] and Tominaga [73].)

See Tominaga and Nagahara [70] for a treatise on Galois Theory of Simple Rings, including that of G. Hochschild [50], T. Nakayama (*op.cit.*), G. Azumaya (*op.cit.*), J. A. Dieudonné, A. Rosenberg and D. Zelinsky, F. Kasch, S. A. Amitsur, C. Faith, H. Tominaga and T. Nagahara. Also Krull's Galois Theory of infinite dimensional algebraic field extensions, and the generalization to division rings of Jacobson and Nobusawa. Also see Chase, Harrison and Rosenberg on Galois theory of commutative rings.

Jacobson's $a^{n(a)} = a$ Theorems and Kaplansky's Generalization

Jacobson's generalization [45c], Theorem 8 of Wedderburn's theorem [05] states:

2.8A THEOREM (JACOBSON [45C]). *An algebraic division algebra R over a finite field is commutative.*

REMARK. These algebras have the property that some non-zero power of each element lies in the center, and also $a^{n(a)} = a$ for every element a of R.

Theorem 2.8A is used in the proof of:

2.8B THEOREM (JACOBSON, *op.cit.*). *If R is a ring, and if to each $a \in R$ there corresponds $n(a) \in \mathbb{N}$ so that $a^{n(a)} = a$, then R is commutative.*

The proof of 2.8B is a powerful application of Jacobson's density theorem and radical J of a ring. One first shows $J = 0$, then that R/P_α is commutative for every primitive ideal P_α. Since a subdirect product of commutative rings is commutative, *voilà*! See Jacobson [56,64], p.217, Theorem 1.

The next theorem generalizes Jacobson's theorem 2.8A by the remark following it.

2.9A THEOREM (KAPLANSKY [51,95]). *Any division ring, or, more generally, any semiprimitive ring, in which some power of each element lies in the center is commutative.*

Kaplansky's Characterization of Radical Field Extensions

A field extension F/H is **purely inseparable** iff $F \neq H$ and the following equivalent conditions hold.

(PI 1) No proper intermediate field L of F/H is separable over H.

(PI 2) F has characteristic $p > 0$ and some p-power of every $x \in F$ lies in H: that is, $x^{p^e} \in H$ for some $e > 0$.

CANONICAL EXAMPLE. If K is a field of characteristic $p > 0$, $F = k(x)$ the field of rational functions in the variable x, and $H = F(x^{p^e})$ for some $e > 0$, then F/H is purely inseparable.

Kaplansky's proof of Theorem 2.9A depended on his characterization of radical field extensions: F/H is a **radical extension** of fields if for every $x \in F$ some power $x^n \in H$.

2.9B THEOREM (KAPLANSKY *op.cit.*). *A field F is a radical extension of a subfield H iff F has characteristic $p > 0$ and either*

(KAP 1) *F is an algebraic extension of the prime subfield $P = GF(p)$; or,*

(KAP 2) *F is purely inseparable over H.*

REMARK. (Kap 2) is a radical extension by definition. Furthermore, if x is in F in (Kap 1), then $P(x)$ is a finite field, say $P = GF(p^m)$, so $x^{p^m} = x$. Then, $x \neq 0$ satisfies $x^{p^{m-1}} = 1 \in P \subseteq H$. Thus, *in* (Kap 1), *$F$ is radical over every subfield.*

Radical Extensions of Rings

Kaplansky's idea was generalized, and radical extensions of arbitrary subrings were studied, where the ring A is a *radical extension* of the ring B in case each $a \in A$ is radical over B in the sense that some power of a lies in B. In this connection Theorem A of Faith [60] states:

2.10 THEOREM (FAITH [60]). *If A is a simple ring with a minimal one-sided ideal, and radical over a subring $B \neq A$, then A is a field.*

This implies that the non-commutative simple ring A is generated by $\{a^{n(a)} \mid n(a) > 0,\ a \in A\}$. This is the best possible result of this type for which A/B is radical, and no restriction is placed on B ("best" in the sense that there exist non-commutative primitive rings with minimal one-sided ideals and radical over proper simple subrings (*loc.cit.*)).

2.11 THEOREM (FAITH [61B]). *If A is a ring with no nil ideals $\neq \{0\}$, and if A is radical over a division subring $B \neq A$, then A is a field.*

2.12 THEOREM (FAITH [61B]). *If A is semiprimitive, and A is radical over a commutative subring B, then A is commutative.*

Kaplansky's Theorem [51] is the special case when B is contained in the center of A, and the proof depends on this.

2.13 THEOREM (NAKAYAMA [55]–FAITH [60]). *If D is a division algebra over a field k, and Δ is a division subalgebra $\neq D$ such that to each $d \in D$ there corresponds elements $\alpha_1, \ldots, \alpha_r \in k$ such that for each $a \in k(d)$ there exists $p_a(x) \in [\alpha_1, \ldots, \alpha_r][x]$ (the polynomial ring over the subring $[\alpha_1, \ldots, \alpha_r]$ generated by $\alpha_1, \ldots, \alpha_r$) such that*

$$a^n - a^{n+1}p_a(a) \in \Delta$$

with $n = n(a)$ an integer > 0, then D is commutative.

REMARKS. (1) This was proved by Nakayama for the case Δ is the center of D (Cf. Jacobson [56,64], p.185. Theorem 3), and the general case by the author.

(2) This generalized certain theorems of Herstein [53] and others (see "Other commutativity theorems," 2.16Jf) and in turn was generalized by the author to simple rings with minimal one-sided ideals (*loc.cit.*)

(3) By Jacobson's theorem above, if D is algebraic over a finite field, or algebraically closed field k, then D is commutative, hence any non-commutative division algebra over k is necessarily transcendental.

We also cite a related theorem.

2.14A THEOREM (FAITH [62]). *If D is a non-commutative transcendental division algebra over a field k, and if A is a division subalgebra $\neq D$, then there exists a transcendental element u over k such that $k(u) \cap A = k$.*

2.14B COROLLARY. *If D is a transcendental (resp. non-commutative) division algebra over a field k, and if A is a division subalgebra $\neq D$ such that for every $a \in D$ there exists a non-constant $f_a(x) \in k(x)$ such that $f_a(a) \in A$ then D is commutative (resp. algebraic over k).*

The proof of this depends on a theorem on three fields of Herstein and the author (loc. cit.) If $L \supset F \supseteq k$ are three fields L/F is not purely inseparable and L/k is transcendental, there exists a transcendental element $u \in L$ so that $F \cap k(u) = k$. (Cf. the elementary proof of the author [61c].)

2.14C COROLLARY (FAITH [62]). *If D is a non-commutative transcendental division algebra over a field k and if to each $a \in D$ we pick a non-constant $f_a(x) \in k[x]$, then $S = \{f_a(a) \mid a \in D\}$ generates D, i.e., D is the smallest division subalgebra containing S.*

2.14D COROLLARY (*loc.cit.*). *If D is a division algebra over k, and if $F(x)$ is a non-constant polynomial over k, and if to each $a, b \in D$ there corresponds $f(x) \in k[x]$ so that $f(b)F(a) = F(a)f(b)$, then D is commutative.*

The Cartan-Brauer-Hua Theorem on Conjugates in Division Rings

Let D be a non-commutative division ring with center C, and let Δ be a proper division subring not contained in C. In [47] Cartan raised the question: is it possible for each inner automorphism of D to induce an automorphism of Δ? As is well-known, Cartan [47], Theorem 4, with the aid of his Galois Theory answered this negatively in case D is a finite dimensional division algebra. Later Brauer (47), and Hua [49], using elegant, elementary methods, extended Cartan's theorem to arbitrary division rings.

Hua's Identity

Hua [49] discovered and used the beautiful identity

$$x = [x^{-1} - (t-1)^{-1}x^{-1}(t-1)][t^{-1}x^{-1}t - (t-1)^{-1}x^{-1}(t-1)]^{-1}$$

Let $D^\star$ denote the group of all non-zero elements of D, and let $H(\Delta)$ be the subgroup of all elements of $D^\star$ which effect inner automorphisms of D that map Δ onto Δ. The following is an extension of the Cartan-Brauer-Hua theorem: $H(\Delta)$ *cannot have finite index in* $D^\star$. This theorem implies (and is implied by):

2.15A Theorem (Faith [58]). *Let D be a non-commutative division ring, and Δ a proper division subring not contained in the center. Then there exist infinitely many distinct subrings $x\Delta x^{-1}$.*

This result extended Herstein [56] to the effect that any non-central element has infinitely many conjugates.

Although this result implies that every finite division ring is commutative, its proof does not constitute a new proof of this theorem of Wedderburn. As a matter of fact, the proof requires not only Wedderburn's theorem but also Jacobson's theorem on algebraic division algebras over a finite field.

If D is any ring, and Δ is a subring, then the **centralizer** Δ' of Δ in D is a subring. Obviously $a^{-1} \in \Delta'$ for every $a \in \Delta'$ that is a unit of D, hence Δ' is a division subring of Δ when D is a division ring.

2.15A′ Theorem (Faith [58]). *Let D be a division algebra over an infinite field ϕ, and let Δ be any proper division subalgebra not contained in C. Then, card $\{(1+\alpha v)^{-1}\Delta(1+\alpha v) \mid \alpha \in \phi\} =$ card ϕ, for each $v \in D, v \notin \Delta$ where Δ' is the centralizer of Δ.*

2.15B Corollary. *Let D be a non-commutative division ring, and let Δ and A be division subrings such that the following conditions are satisfied:*

1. *Δ does not contain A.*
2. *The centralizer A' of A does not contain Δ.*

Then D contains infinitely many different subrings of the form $a\Delta a^{-1}$ with $a \in A$, provided any one of the following conditions are satisfied:

(I) *$\Delta \cap A$ is not contained in the center of A.*
(II) *$Z \cap A$ is infinite, where Z is the center of Δ.*
(III) *D has characteristic* 0.
(IV) *D is algebraic over the prime subfield.*

See *loc.cit.*, p.379, Corollary 1.

Amitsur's Theorem and Conjugates in Simple Rings

The following is an extension of the Cartan-Brauer-Hua theorem.

2.16A THEOREM (AMITSUR [56]). *If A is a simple algebra over a field k containing an idempotent $e \neq 0, 1$, and if A is not the ring of 2×2 matrices over a field of characteristic 2, then the only subalgebra $B \neq A$ invariant under every inner automorphism is contained in the center C.*

Theorem 2.16A depends on:

2.16B AMITSUR'S THEOREM [56]. *Let R be a simple ring not 4-dimensional over its center C when C has characteristic 2, and R has an idempotent $e \neq 0, 1$. Then for any C-subspace S invariant under all inner automorphisms effected by elements $1 + u$ for $u^2 = 0$, either $S \subseteq C$, or*

$$S \supseteq [R, R] = \{ab - ba \mid a, b \in R\}.$$

Cf. 2.42(3) and Herstein's review of *op.cit.*, #13.01.02 in Small [81],vol.1.

The next theorem, an application of 2.16A, is an extension of the author's theorem on division rings cited above, e.g. B has infinitely many conjugates when k is infinite.

2.16C THEOREM (FAITH [59]). *Under the condition of Amitsur's theorem, then for any non-central subalgebra $B \neq A$ the cardinal of the set*

$$\{(1 + \alpha u)^{-1} B (1 + \alpha u) \mid \alpha \in k\}$$

is the cardinal of k, for some element u with $u^2 = 0$.

Moreover, if char $k = 0$, *then different $\alpha \in k$ determine different subrings $(1 + \alpha u)^{-1} B (1 + \alpha u)$.*

2.16C′ COROLLARY. *Let A be a simple algebra as in Amitsur's Theorem. Then:*

(1) *Every element of A is a sum of units.*
(2) *A is not radical over any subring $B \neq A$. (Cf. 2.10 - 2.12).*
(3) *A contains no invariant right ideals.*
(4) *A is a sum of finitely many isomorphic principal right ideals generated by idempotents.*

PROOF. See *op.cit.* Theorem 5, and Corollaries 3 and 5. Cf. 2.16F below. □

REMARK. Regarding (1), see 2.16E(2), 2.16F,G,H and I.

Invariant Subrings of Matrix Rings

A subring B of R is said to be **invariant (characteristic)** if $f(B) \subseteq B$ for all inner automorphisms (resp. automorphisms).

The next result shows that a proper subring B of $R = A_n, n > 1$, is noninvariant if B contains any element belonging to a set $M(n)$ of matrix units in R. This improves on a result of G. Ehrlich [55] who proved the noninvariance of any proper subring B containing $M(n)$. She also proved the noninvariance of any noncentral subsfield B of $A_n, n > 1$ (*ibid.*). Her proof of the latter uses Hua's identity.

2.16D THEOREM (FAITH [59]). *Let A be a ring with identity 1, and let $M(n) = \{e_{ij} \mid i,j = 1,\dots,n, \sum_1^n e_{ii} = 1\}$ be a complete set of matrix units for $R = A_n, n > 1$. Then R is generated (as a ring) by the conjugates of any element $e = e_{ij} \in M(n)$.*

PROOF. Let B denote the subring generated by the conjugates of $e = e_{11}$. Then B contains every $e_{jj} = x_j^{-1} e_{11} x_j, j \neq 1$, where $x_j = x_j^{-1} = 1 - e_{11} - e_{jj} + e_{1j} + e_{j1}$. For each $a \in A$, and $i \neq j$, set

$$t_{ij}(a) = (1 + ae_{ij})e_{jj}(1 - ae_{ij}).$$

Then,

$$ae_{ij} = t_{ij}(a) - e_{jj} \in B,$$

for all $a \in A$, and all $i \neq j$. Then B contains every ae_{ii} as well, $i = 1,\dots,n$, so that $B \supseteq R = A_n$. The case $e = e_{jj}, j$ arbitrary, follows from this since $e_{11} = x_j e_{jj} x_j^{-1}$. Finally, if $e = e_{ij}, i \neq j$, note that $e_{jj} = e_{ij} - (1 - e_{ji})e_{ij}(1 + e_{ji})e_{ij} \in B$, so that $B = R$ in this case too. □

2.16E COROLLARY. *(1) The only ideal of $R = A_n$, $n > 1$, containing a matrix unit is R itself; (2) Moreover, every element of R is a sum of units.*

Shoda [32-33] proved (2) in the case A is a sfield. Cf. 2.16H.

Rings Generated by Units

If $S \neq \emptyset$ is a subset of a ring R, then

$$S^\perp = \{x \in R \mid sx = 0 \quad \forall s \in S\}$$

is a right ideal, called the **annihilator right ideal** of S. If $S = \{a\}$, let $a^\perp = S^\perp$. We let ${}^\perp S$ denote the left annihilator of S. An element $a \in R$ is **regular** in case $a^\perp = 0$ and ${}^\perp a = 0$.

2.16F REMARKS. (1) If R is a ring with center C with no invariant subrings $S \neq R$ except when $S \subseteq C$, then R is generated as a ring by the following:

(1a) Regular elements
(1b) Units

Furthermore, then every element of R is a finite sum of regular elements, and/or units. This follows since the subring S generated by regular elements (resp. units) consists of finite sums of (products of) regular elements (resp. units).

(2) Of course, R can be generated, e.g. by units, even when R has an invariant subring-1 $S \neq R$ and $S \not\subseteq C$. See 2.16G and H.

2.16G ZELINSKY'S THEOREM [54]. *Let $R = End\ V_D$ be a complete ring of linear transformations of a vector space V over a sfield D. Then every element of R is a sum of units.*

2.16H HENRIKSEN'S THEOREM [74]. *If $R = A_n$ is the $n \times n$ matrix ring, $n > 1$, over a ring A, every element of R is a sum of three units.*

2.16I REMARKS. (1) In general, "three" cannot be replaced by "two"; (2) Menal and Moncasi [81] pointed out in A_2 the following simple identity:

$$\begin{pmatrix} a & b \\ c & d \end{pmatrix} = \begin{pmatrix} a & 1 \\ -1 & 0 \end{pmatrix} + \begin{pmatrix} 0 & -1 \\ 1 & d \end{pmatrix} + \begin{pmatrix} 1 & 0 \\ c & 1 \end{pmatrix} + \begin{pmatrix} -1 & b \\ 0 & -1 \end{pmatrix}$$

i.e., every element of A_2 is a sum of 4 units.

Transvections and Invariance

Let A be a ring, and $R = A_n$ the ring of $n \times n$ matrices over A, $n \geq 2$. For each $a \in A, a_{pq} = ae_{pq}$ denotes the matrix with a in the (p, q) position and zeros elsewhere. An additive subgroup S of R is a TI-*subgroup* (*Transvectionally Invariant subgroup*) in case S is invariant under all inner automorphisms affected by the set of **transvections**, namely $\{1 + a_{ij}\}_{i \neq j, a \in A}$.

The additive **commutator** of two elements a, b of a ring R is

$$[a, b] = ab - ba.$$

If A and B are additive subgroups of R, then $[A, B]$ denotes the additive subgroup generated by all $[a, b]$, where $a \in A$ and $b \in B$.

2.16J THEOREM (ROSENBERG [56]). *The TI-subgroups of $A_n, n \geq 3$, are the subgroups of the center, or they have the form $[A_n, K_n] + D$, where K is a nonzero 2-sided ideal of A (consisting of the off-diagonal entries of the elements of the TI-subgroup), and D is an additive group of diagonal matrices.*

Other Commutativity Theorems

If R is radical over its center C, then for each pair $x, y \in R$, there exist positive integers $n = n(x, y)$ and $m = m(x, y)$ such that $(\star)$ $x^n y^m = y^m x^n$. The question raised by Faith and answered affirmatively in Anarín-Zjabko [74] and Herstein [76] is: does this condition $(\star)$ imply commutativity if R has no nil ideals $\neq 0$. Furthermore, it is proved (*loc.cit.*) that in any ring R that satisfies $(\star)$ the set I of nilpotent elements is an ideal such that R/I is commutative, equivalently, R has nil commutator ideal.

If R is a ring such that

$$(*) \qquad (xy)^n = x^n y^n \quad \forall x \in R$$

holds for 3 consecutive positive integers n, then Ligh and Richoux [77] prove by elementary methods that R is commutative. See Harmani [77] for the same theorem for two consecutive integers n such that $n(n!)^2$ is not divisible by the characteristic of R.

We now refer the reader to a paper of Kaplansky [95b] (in his *Selected Papers* [95a]) in which he totes up over 200 papers in Reviews in Ring Theory on commutativity, counting his own! He further adds Kaplansky [95b] to the total, by proving a theorem on ξ-rings defined **sup.** 4.19A. (Cf. Drazin [56], Utumi [57] and Nakayama [59]).

Noetherian and Artinian Modules

A set $C = \{M_i\}_{i\in I}$ of a set M is a **chain** (or **linearly ordered**) if for every pair $i, j \in I$ either $M_i \supseteq M_j$ or $M_i \subseteq M_j$. If the index set I is the set $\mathbb{N}$ of natural numbers, then C is said to be **ascending** if $i < j \Rightarrow M_i \subseteq M_j$; and **descending** if $i < j \Rightarrow M_i \supseteq M_j$, for all i and $j \in \mathbb{N}$. Then we say that M satisfies the **ascending chain condition (= acc)** if for every ascending chain C of submodules, there exists n so that $M_n = M_k \ \ \forall k \geq n$. In this case M is said to be **Noetherian**. The dual concept is the **descending chain condition (= dcc)** on submodules of M, in which case M is said to be **Artinian**.

EXAMPLES. A vector space V over a sfield k is seen to be a Noetherian (resp. Artinian) k-module iff V is finite dimensional. $\mathbb{Z}$ is Noetherian but not Artinian.

The Maximum and Minimum Conditions

A partially ordered set S (= poset) satisfies the maximum (minimum) condition provided that every nonempty subset contains a maximal (minimal) element. It is elementary to show that S satisfies the maximum (minimum) condition iff S satisfies the acc (dcc) on subsets. This equivalence is frequently applied in the text to the poset of submodules (right ideals) of a module (ring).

Inductive Sets and Zorn's Lemma

A set S of subsets of a set M is **inductive** if $S \neq \emptyset$, and the union $\cup C$ of every chain C of subsets of M belongs to S.

2.17A ZORN'S LEMMA. *If S is a nonempty set of subsets of a set M, and if S is inductive, then S contains a maximal element P, that is, $P \subseteq Q \Rightarrow P = Q \ \ \forall Q \in S$.*

Zorn's Lemma is stated more generally for partly ordered sets. For background and the connection of Zorn's Lemma with the Axiom of Choice (they are equivalent), see e.g. my Algebra [73], pp. 29–30, Also see Birkhoff [67], Chapter VIII, §7, p. 191ff.

2.17B Applications (1) For any $f \cdot g$ R-module M, and submodule $N \neq M$, the set of proper submodules of M containing N is inductive, hence by Zorn's lemma, N is contained in a maximal submodule. In particular, any proper (right) ideal of R is contained in a maximal (right) ideal. In particular, any ring R has maximal right ideals, and maximal ideals. Moreover, R is a sfield (resp. simple ring) iff 0 is a maximal right ideal (resp. two-sided ideal); (2) **Hausdorff's Maximal Principle:** every chain of a partly ordered set can be extended to a maximal chain. This is equivalent to Zorn's Lemma. See Birkhoff [67], p. 162. Obviously the Noetherian assumption on a set obviates the necessity of Zorn's Lemma or Hausdorff's maximal principle.

Subdirectly Irreducible Modules: Birkhoff's Theorem

An R-module M is **subdirectly irreducible** provided that the intersection V of all nonzero submodules of M is nonzero, that is, M has a unique minimal submodule V contained in every submodule $\neq 0$. If N is a submodule of M so that $N \neq M$ and M/N is a subdirectly irreducible module, then we say that N is a **subdirectly irreducible submodule** of M. By the straightforward application of Zorn's Lemma one proves:

2.17C Birkhoff's Theorem. *If M is an R-module, and N a submodule $\neq M$, then for any $x \in M \backslash N$ there is a submodule $N_x \supseteq N$ maximal with respect to excluding x. Furthermore, N_x is subdirectly irreducible, and N is the intersection of all such N_x.*

Proof. One easily checks that the set S of all submodules $\supseteq N$ that exclude x is inductive, hence S contains a maximal element N_x by Zorn's Lemma 2.17A. Furthermore, every submodule K of M that properly contains N also contains x, hence M/N_x is subdirectly irreducible. Obviously N is the intersection of the sets $\{N_x\}_{x \in M \backslash N}$.

Remark. Birkhoff's theorem holds more generally for lattices, and universal algebras. Cf. Birkhoff [67] Theorem 15, p. 193, and Theorem 16, p. 194.

A ring R is said to be **subdirectly irreducible (qua ring)** provided that the intersection of all nonzero ideals is a nonzero ideal S. In this case S is a minimal ideal contained in every ideal $\neq 0$. Obviously any simple ring R is a subdirectly irreducible ring. If R is commutative then R is subdirectly irreducible iff R is a subdirectly irreducible R-module, and in this case S is a simple R-module.

Example. If $R = \text{End}_D V$ where V is a left vector space over D, then R is subdirectly irreducible, and the set S of all $a \in R$ such that $\dim Va < \infty$ is an ideal contained in every ideal $I \neq 0$.

Below, the concept of subdirect product of R-modules is defined anologously to the definition of subdirect product of rings (2.6f): replace "ring" by "module" and "ideal" by submodule. For more general algebras, see Birkhoff [67], p. 140.

Here is another formulation of:

2.17D Birkhoff's Theorem. *Any R-module M is a subdirect product of subdirectly irreducible modules, and any ring R is a subdirect product of subdirectly irreducible rings.*

Proof. Apply the first formulation of Birkhoff's theorem for the case $N = 0$. Conversely, apply the second formulation to M/N to obtain the first.

2.17E Examples. (1) By the Fundamental Theorem (see 1.15A), an Abelian group C of finite order is irreducible iff C is cyclic of prime order, $C \approx Z_{p^n}$, for a prime p. In this case C is also subdirectly irreducible; (2) By Theorem 1.3, a divisible Abelian group D will be irreducible iff $D \approx \mathbb{Q}$, or $D \approx \mathbb{Z}_{p^\infty}$. In the latter case, D is subdirectly irreducible; (3) $\mathbb{Z}$ is irreducible but not subdirectly irreducible. However, $\mathbb{Z}$ is a subdirect product of the groups $\{\mathbb{Z}_p\}_{p \text{ prime}}$, that is, $\mathbb{Z}$ is a subdirect product of fields $\approx \mathbb{Z}/p\mathbb{Z}$, since $\cap p\mathbb{Z} = 0$. Likewise, since $\cap_{n=1}^{\infty} p^n\mathbb{Z} = 0$, for any prime number p, then by Birkhoff's Theorem, $\mathbb{Z}$ is also a subdirect product of the groups $\{\mathbb{Z}_{p^n}\}_{n=1}^{\infty}$ since $\mathbb{Z}_{p^n} \approx \mathbb{Z}/p^n\mathbb{Z}$. Thus $\mathbb{Z}$ has two subdirect product representations of sets of subdirectly irreducible groups, where the subdirect components are not isomorphic; (4) A ring R is a **Boolean ring** (after G. Boole, see Birkhoff [67], p. 44) provided that $a^2 = a$ for all $a \in R$. Such a ring is commutative (Cf. Jacobson's generalization, Theorem 2.8B), and satisfies $2a = 0$ $\forall a$. It follows that R is a subdirect product of copies of $\mathbb{Z}_2$.

Jordan-Hölder Theorem for Composition Series

A module M has finite (Jordan-Hölder) length if there is a finite series of submodules

$$M_0 = M \supset M_1 \supset \ldots \supset M_{n-1} \supset 0 = M_n \tag{3}$$

such that M_i/M_{i+1} is a simple module $\forall i$, called a *simple factor* of M. Then (3) is called a *Jordan-Hölder* or *Composition Series*, and any two composition series have the same length, denoted $|M|$, and the same simple factors up to order. Cf. Noether [26],p.56,§10.

2.17F APPLICATIONS. *(1) If $|M| = n < \infty$, then every strict chain $\{S_i\}_{i=1}^t$ of submodules can be "refined to a composition series," that is, by interposing additional modules if some S_i/S_{i+1} is not simple. Thus, no strict chain of submodule has more than n elements $\neq 0$; (2) It follows from (1) that $|M| = n$ implies that every submodule can be generated by $\leq n$ elements although, as in the case of $\mathbb{Z}_n = \mathbb{Z}/n\mathbb{Z}$, n elements may not be required: $|\mathbb{Z}_n| = n$ but every subgroup is cyclic.*

2.17G REMARKS. (1) $\mathbb{Z}_{p^e}$, for a prime p and $e \geq 1$, has a unique decomposition series (and, $|\mathbb{Z}_{p^e}| = e$). The proof of (2) of 2.17F then proceeds by induction. Cf. Uniserial modules, $5.1A'f$.

(2) *An R-module M has a composition series iff M has both dcc (= Artinian) and the acc (= Noetherian) on submodules.* Thus, by the Hopkins-Levitzki Theorem, any right Artinian ring R has a composition series as a right R-module. However, this condition is not right-left symmetric: There exist right Artinian rings that are not left Noetherian (see e.g. Faith [72a], p.337, 7.11′).

Two Noether Theorems

A submodule N of M is **irreducible** if any two submodules A and $B \supseteq N$ satisfy: $A \cap B = N \Leftrightarrow A = N$ or $B = N$. Before Emmy Noether [21], there were no theorems like the following:

2.18A NOETHER'S THEOREM. *A right R-module M is Noetherian iff each submodule is $f \cdot g$. In this case every submodule $N \neq M$ is the intersection of finitely many irreducible submodules.*

Cf. Birkhoff [67], p. 181 for a more general result.

Below, an ideal I of a ring R is **prime** if $I \neq R$, and if $A \supseteq I$ and $B \supseteq I$ are ideals, then $AB \subseteq I \Leftrightarrow A \subseteq I$ or $B \subseteq I$.

2.18B NOETHER'S THEOREM. *If R is a ring satisfying the acc on ideals, then every ideal I contains a product of prime ideals.*

THE FAMOUS PROOF. Let I be a maximal counterexample. Then I is not a prime ideal, hence I is properly contained in ideals A and B such that $AB \subseteq I$. But by the maximality condition on I, both A and B are products of primes, hence, so is AB. This proves Noether's theorem. □

Cf. Zariski-Samuel [58], p.200.

2.19A THEOREM (COHEN [50]). *A commutative ring R is Noetherian iff every prime ideal is finitely generated.*

Proof is similar to that of Noether's theorem. Cf. Nagata [62],p.8.

2.19B THEOREM (COHEN [50]-ORNSTEIN [68]). *If R is right Noetherian and R/P is Artinian for each prime ideal $P \neq 0$, then either R is prime or right Artinian.*

The proof uses 2.18 and the Chinese Remainder Theorem (Cohen's theorem was for commutative R).

Hilbert Basis Theorem

Hilbert's basis theorem below states that if every right ideal of R is $f \cdot g$ (= has a finite basis) then the same is true for $R[X_1, \dots, X_n]$.

2.20 HILBERT BASIS THEOREM [1888]. *If R is a (right) Noetherian ring, so is the polynomial ring $R[x_1, \dots, x_n]$ in a finite number of variables. Moreover, so is the power series ring $R[[x_1, \dots, x_n]]$.*

See e.g. Kaplansky [70], Theorems 70 and 71, or Zariski-Samuel [58],p.201, for proof. Also see the Hilbert Syzygy Theorem, 14.9, and the Hilbert basis theorem 7.15 for differential polynomials.

REMARKS. (1) If M is a Noetherian right R-module, then the polynomial module $M[x]$ (resp. power series module $M[[x]]$) is a right Noetherian module over $R[X]$ (resp. $R[[x]]$); (2) If M has finite length n, then every $R[x]$ (resp. $R[[x]]$) submodule of $M[x]$ (resp. $M[[x]]$) is generated by $\leq n$ elements.

See, e.g. my Algebra [73], p. 341, for (1); (2) is proved by induction: for $n = 1$, use the method used to prove the corresponding result for when $M = R$ is a sfield, in which case $R[X]$ and $R\langle X\rangle$ are principal right (and left) ideal domains.[1]

2.21A COROLLARY. *Any $f \cdot g$ commutative ring R is Noetherian, in fact, any commutative $f \cdot g$ algebra R over a Noetherian ring k is Noetherian.*

PROOF. If $r_1, \dots, r_n$ generate R **qua** ring, then R is an epimorphic image of $\mathbb{Z}[x_1, \dots, x_n]$ that sends $x_i \to r_i$, $i = 1, \dots, n$. The proof of the second statement follows by replacing $\mathbb{Z}$ by k. □

2.21B THEOREM. *If R is a right Noetherian (Artinian) ring, then any $f \cdot g$ right R-module M is Noetherian (Artinian).*

By induction the free R-module R^n is Noetherian (Artinian), hence so is any epic image, e.g., M.

[1] To emphasize, $R\langle X\rangle$ is a variant symbol that I used for $R[[X]]$, e.g. in my *Algebra*.

Hilbert's Fourteenth Problem: Nagata's Solution

(H 14): If $R = k[x_1, \ldots, x_n]$ is the polynomial ring in n variables, and K is a subfield of $Q(R) = k(x_1, \ldots, x_n)$, then is $K \cap R$ a $f \cdot g$ algebra over k?

Nagata [60] proved the answer negative. See Ulrich [97], p.180, for a discussion. The solution hinges on the fact that an affirmative answer implies that $K \cap R$ is Noetherian. The answer for $n = 1$ is affirmative. (See, for instance, the author's paper [61c].)

Noether's Problem in Galois Theory: Swan's Solution

(NP): Let G be a finite group acting faithfully as permutations on a finite set $x_1, \ldots, x_n$ of variables. Is the fixed field $F = \text{fix } G$ of the field $K = k(x, \ldots, x_n)$ of rational functions in $x_1, \ldots, x_n$ purely transcendental over k?

For the next theorem, see Swan's paper, in Srinivasan-Sally [82].

SWAN'S THEOREM. *If G is cyclic of order divisible by 8, then the fixed field F of G acting faithfully on $K = k(x_1, \ldots, x_n)$ is not purely transcendental over k.*

The proof depends on a theorem of Saltman on generic Galois extensions, and e.g. theorems of Kuyk and Wang.

REMARK. Swan [69] showed that (NP) had a negative answer for $\mathbb{Z}_{47}$, and Lenstra [74] for $\mathbb{Z}_8$. Cf. Vila [92], p.1055.

Realizing Groups as Galois Groups

Classically any finite group G can be realized as a Galois group of some Galois field extension, e.g. in the statement of Noether's problem, K/F is Galois with Galois group $\approx G$.

Hilbert [92] posed the *problem* $(= HP)$ *of realizing a prescribed finite group G as a Galois group over every algebraic number field K (= a finite extension of $\mathbb{Q}$).* His irreducibility criterion, *ibid.*, enabled him to solve (HP) for the symmetric group S_n and the alternating group A_n. (Cf. Vila [92].) Since every finite group G embeds in S_n, for $n = |G|$, this shows any finite group G is the Galois group of some finite field extension L of K, hence of *some* algebraic number field L.

Noether proved that the answer to (HP) would be affirmative if her problem (NP) had an affirmative answer. See Swan's papers in *op.cit.*, and in Brewer and Smith [81]. Noether's question and theorem appear in *Gleichungen mit vorgeschriebener Gruppe*, Math.Ann. **78** (1918), 221–229 (see Noether's *Collected Papers* [83]).

SHAFAREVITCH'S THEOREM [54,56][2]. *Any finite solvable group G can be the Galois group over any number field K, that is, $G \approx G(L/K)$ for a finite Galois field extension L of K, in particular for $K = \mathbb{Q}$.*

[2] Šafarevič is an alternative spelling of Shafarevitch. I followed Herstein [68, p.187ff] in the latter spelling.

See Vila [92] for a survey of realizable groups, and a number of related problems. Vila *iteralia* points out that the starting point of Shafarevitch's research is the theorem of Scholz (1937) and Reichardt (1937) who, independently proved:

SCHOLZ-REICHARDT THEOREM. *Every p-group, $p \neq 2$, is realizable as a Galois group over $\mathbb{Q}$.*

REMARK. Shafarevitch gave a new proof and extended the theorem to $p = 2$. See Vila [92], p. 1057.

Prime Rings and Ideals

A ring R is **prime** provided that R satisfies the equivalent conditions: (P1) every right ideal $I \neq 0$ is faithful; i.e., $I^{\perp} = 0$; (P2) every left ideal $L \neq 0$ is faithful; i.e., ${}^{\perp}L = 0$; (P3) 0 is a prime ideal (Cf. **sup.** 2.18Bs).

Examples: (1) any integral domain R; (2) any right primitive ring R is prime. (If V is a faithful simple R-module, and $I \neq 0$ in (P1), then $VI \neq 0$ so $VI = V$, hence every $a \in I^{\perp}$ annihilates V, so $a = 0$, hence $I^{\perp} = 0$); (3) any simple ring R is primitive, hence prime; (4) if R is prime (resp. right primitive), and if V_R is a faithful torsionless module (**sup.** 1.5), then End V_R is prime (resp. right primitive) by Zelmanowitz [67] (resp. Amitsur [71]). This holds, e.g. for V_R free. Cf. 13.32–33.

An ideal I is **prime** if $I \neq R$, and R/I is a prime ring .

REMARKS. (1) If R is prime, then R contains no nilpotent ideal $I \neq 0$, since if I has index n then $I^{\perp} \supseteq I^{n-1} \neq 0$; (2) It follows that a prime ideal P of R contains every nilpotent ideal.

Chains of Prime Ideals

A chain of distinct prime ideals $P = P_0 \supset P_1 \supset \cdots \supset P_n$ is said to be of *length* n, even though $n + 1$ prime ideals appear in the chain.

Following Kaplansky [70] P is said to have **rank** n (= *height* n in Zariski-Samuel [58]) if n is the maximal length of chains of prime ideals descending from P. The dual concept is called the **dimension of** P (= depth in [Z-S]). Thus a minimal prime ideal has rank 0. Since the intersection of prime ideals in a chain is a prime ideal for any proper ideal I there is at least one **minimal prime containing (or over)** I. (Cf. Kaplansky [70], Theorem 10.) We will come back to these concepts in §14.

The Principal Ideal Theorems and the DCC on Prime Ideals

The next theorem is of fundamental importance in ideal theory.

2.22 KRULL PRINCIPAL IDEAL THEOREM [28]. *If x is a non-unit of a Noetherian commutative ring, then a minimal prime ideal P over (x) has rank ≤ 1.*

See Kaplansky [70], Theorem 142. This appears in Krull [35–48], on p. 37, *Hauptidealsatz.*

2.23 GENERALIZED PRINCIPAL IDEAL THEOREM. *If $I = (a_1, \dots, a_n)$ is an proper ideal in a commutative Noetherian ring R, generated by n elements, then any prime ideal P over I has rank $\leq n$. Moreover, any prime ideal P of R of rank n is minimal over an ideal I generated by n but no fewer elements.*

See Krull [35–48], p.37, *Primidealkettensatz.* Also Kaplansky [70], Theorem 152.

2.23A COROLLARY. *A Noetherian commutative ring satisfies the dcc on prime ideals, that is, every prime ideal has finite rank.*

2.24 THEOREM (BASS [71]). *If R is a Noetherian commutative ring and M is a $f \cdot g$ R-module, then every chain of submodules is countable.*

Primary and Radical Ideals

An ideal of a commutative ring R is **irreducible** if it is not the intersection of two larger ideals. The **radical** $\sqrt{I} = \{a \in R | a^n \in I$ for some $n \geq 1\}$ is an ideal for any ideal I, and $\sqrt{I}/I$ is a nil ideal of R/I. (Cf. *sup* 14.1.) Any prime ideal is irreducible.

2.25 DEFINITION AND PROPOSITION. *(1) An ideal I of a commutative ring R is primary if for all $a, b \in R$ with $ab \in I$, it is true that either $a \in I$ or some power $b^m \in I$; (2) in this case $P = \sqrt{I}$ is a prime ideal called the* **associated prime** *of I, and I is said to be primary for P; (3) An ideal I is* **primary** *iff every zero divisor of R/I is nilpotent, and then the set of zero divisors of R/I is a prime ideal P/I; (4) If I is an irreducible ideal, and P/I is the set of zero divisors of R/I, then I is primary iff $P = \sqrt{I}$.*

See Zariski-Samuel (= [Z-S]) [58], p.152ff.

2.26 COROLLARY. *For any maximal ideal P, P^n is primary for P for all $n \geq 1$.*

Lasker-Noether Decomposition Theorem

2.27 THEOREM. *If R is Noetherian and commutative, then every ideal K is a finite irredundant intersection $\bigcap_{i=1}^n P_i$ of primary ideals, hence every irreducible ideal is primary. Moreover, K is a radical ideal iff P_i is prime, $i = 1, \dots, n$. If n is minimal in the sense that there is no such intersection with fewer terms, then each associated prime of R is equal to $\sqrt{P_i}$ for a unique index i.*

See Van der Waerden (= VDW) [48], Vol. II, p.31ff. For the last statement see Eisenbud [96], p.95, Theorem 3.10, which yields the primary decomposition for submodule K of a $f \cdot g$ R-module M.

2.28 EXAMPLES. (1) A primary ideal I need not be a power of its associated prime ideal P. Let $R = k[x, y]$ the polynomial ring in two variables over a field k. Then $P = (x, y)$ is a maximal ideal, and $I = (x, y^2)$ is primary for P, but of course not a power of P; (2) Powers of a prime ideal need not be primary. (See [Z-S],p.154.); (3) If $I = (x^2, 2x)$ in $R = k[x]$, then $P = (x)$ is maximal, and $P^2 \subseteq I$, hence $P = \sqrt{I}$ but I is not primary.

2.29 THEOREM. *The intersection K of a finite number of primary ideals of a commutative Noetherian ring R all having the same associated prime P is again primary for P, but the intersection of primary ideals with different radicals (= different associated prime ideals) is never primary.*

See [VW], Vol.II, p.32. For modules, see Eisenbud [96], p.94, Corollary 3.8. Also see Krull [58] for a study of Lasker rings (= rings in which every ideal is a finite intersection of primary ideals).

REMARKS. (1) If a commutative ring R has acc on irreducible ideals, every irreducible ideal is primary. Any irreducible ring R ($= 0$ is an irreducible ideal) with acc on (point) annihilators is primary ($= 0$ is a primary ideal). See the author's paper [98]. Cf. 8.5A,B,C and 16.40.

(2) Krull [28b] defined such concepts as the highest prime ideal dividing an ideal A, the prime ideals belonging to A, and isolated components, for noncommutative rings satisfying "finiteness conditions weaker than the Noetherian chain conditions."

(3) Fitting [35b] defined the prime and primary ideals of an ideal in a noncommutative ring "without finiteness conditions." However, assuming the ascending chain conditions each co-irreducible ideal, that is, an ideal not the intersection of two larger ideals is primary, and of course, every ideal is the intersection of primary ideals. Further properties of the prime ideals are developed, and the radical of an ideal A is defined as the set $\sqrt{A}$ of "properly nilpotent" elements modulo A; that is, elements c which generate a nilpotent ideal modulo A. Fitting applies his results to characterize when an order R in a simple algebra (of finite dimensions) is a product of primary rings. This happens, iff the proper prime ideals are comaximal and commutative (compare with 5.1A').

Hilbert Nullstellensatz

Let k be a field, let K be an algebraically closed field, containing k, and let I be an ideal in the polynomial ring $R = k[X_1, \dots, X_n]$ in n variables over K.

The **variety** $\mathcal{V}(I)$ of I is defined by:

$$\mathcal{V}(I) = \{(\alpha) = (\alpha_1, \dots, \alpha_n) \in K^n \mid f(\alpha) = f(\alpha_1, \dots \alpha_n) = 0 \ \forall f \in I\}.$$

A point $(\alpha) \in \mathcal{V}(I)$ is called a **zero of the ideal** I.

Conversely for any nonempty subset $W \subseteq K^n$ let

$$\mathcal{I}(W) = \{f \in R \mid f(\alpha) = 0 \quad \forall \alpha \in W\}.$$

Then $\mathcal{I}(W) = \mathcal{I}(\overline{W})$, where $\overline{W} = \mathcal{V}(\mathcal{I}(W)$ is the variety of $\mathcal{I}(W)$.

A variety $V = \mathcal{V}(I)$ is **irreducible** provided that V is not the union of two varieties $V_1 \cup V_2$ which are proper subsets of V.

2.30A THEOREM. *A variety V is irreducible iff its ideal $\mathcal{I}(V)$ is a prime ideal.*

PROOF. See Zariski-Samuel [58], p.162, denoted below by [Z-S].

2.30B THEOREM. *Every variety is V can be represented as a finite irredundant union $V = \bigcup_{i=1}^n V_i$ of irreducible varieties, which is unique up to order.*

PROOF. [Z-S], p.162. The varieties $V_1, \dots, V_n$ are called the **irreducible components** of V.

2.30C HILBERT NULLSTELLENSATZ (1893). *The ideal $\mathcal{I}(\mathcal{V}(I))$ of the variety of an ideal I of $R = k[X_1, \dots, X_n]$ is the radical $\sqrt{I}$ of I.*

PROOF. [Z-W], p.164, or Van Der Waerden [50], §79.

Cf. The Lasker-Noether Theorem 2.27, and also 2.30A,B and 3.36ff.

2.30D COROLLARY. *If $f, f_1, \dots, f_t$ are polynomials in $R = k[X_1, \dots, X_n]$, and if $f(\alpha) = 0$ whenever $f_i(\alpha) = 0$, $i = 1, \dots, t$, for any α in the algebraic closure of k, then there exist polynomials $p_1, \dots, p_t$ such that $f^m = \Sigma_{i=1}^t p_i f_i$ for some $m \geq 1$.*

2.30E COROLLARY. *The correspondences*

$$I \mapsto \mathcal{V}(I) \qquad \text{and } V \mapsto \mathcal{I}(V)$$

induce inverse bijections between the sets of radical ideals of $k[X_1, \dots, X_n]$ and varieties of K^n, where K is the algebraic closure of k.

2.30F THEOREM. *Every maximal ideal m of $R = K[X_1, \dots, X_n]$, when K is algebraically closed, is of the form*

$$m_p = (X - \alpha_1, \dots, X - \alpha_n)$$

for some point $p = (\alpha_1, \dots, \alpha_n)$ in $\mathcal{V}(m)$.

In particular, the maximal ideals of R are in 1-1 correspondence with the points of K^n.

PROOF. Follows easily from above and the Nullstellensatz.

Prime Radical

If K and Q are ideals of R, and if $K \supseteq Q$, then there is a ring isomorphism $R/K \approx (R/Q)/(K/Q)$. This proves the following proposition.

2.31 PROPOSITION. *If $K \supseteq Q$ are ideals of R, then K is a prime ideal of R if and only if K/Q is a prime ideal of R/Q.*

The **prime radical** of R is defined to be the intersection of the prime ideals of R. In view of 2.31, we have the following corollary.

2.32 COROLLARY. *Prime rad $(R/\text{prime rad}R) = 0$.*

An element $a \in R$ is **strongly nilpotent** if, for each infinite sequence $\{a_n \mid n \geq 0\}$ such that $a_0 = a$, and $a_{n-1} \in a_n R a_n, n = 0, 1 \dots$ there exists an integer k such that $a_n = 0 \quad \forall n \geq k$. If a is strongly nilpotent, and if $\{a_n \mid n = 0, 1, \dots\}$ is the sequence $a_0 = a, a_1 = a^2, \dots, a_n = a^{2n}$, then $a_{n-1} = a^{2n-1} = a^{2^n} \cdot a^{2^n} a^{2^n} \cdot a^{2^n} = a_n^2 \in a_n R a_n \quad \forall n$. Thus, $a_k = a^{2^k} = 0$ for some k, so each strongly nilpotent element is nilpotent. If R is commutative, then conversely, each nilpotent element is strongly nilpotent.

2.33 PROPOSITION (LEVITZKI [51]). *The prime radical is the set of all strongly nilpotent elements of R.*

PROOF. Let a be an element of R not in prime rad R. Then a lies outside of some prime ideal P and $aRa \not\subseteq P$, so there is an $a_1 \in aRa$ and $a_1 \notin P$. Assuming $a_n \notin P$, then $a_nRa_n \not\subseteq P$, so there is an $a_{n-1} \in a_nRa_n$ and $a_{n-1} \notin P$. Since $a_n \notin P\ \forall n, a_n \neq 0\ \forall n$, so a is not strongly nilpotent.

Conversely, assume a is not strongly nilpotent, and let $\{a_n\}_{n=1}^{\infty}$ be a sequence of elements in R such that $a_0 = a$, and $a_{n-1} \in a_nRa_n\ \ \forall n$. Let

$$T = \{a_n \mid n = 0, 1, \dots\}.$$

Then $0 \notin T$, and by Zorn's lemma, there exists an ideal P that is maximal in the set of ideals not containing an element of T.

Next let A, B be right ideals of R such that $A \supsetneq P, B \supsetneq P$. Since $A + P \neq P$, $B + P \neq P$, both $A + P$ and $B + P$ meet T, say $a_i \in A + P$, $a_j \in B + P$. If $m = \max\{i, j\}$, then

$$a_{m-1} \in a_mRa_m \subseteq (A + P)(B + P) \subseteq AB + P.$$

But $a_{m-1} \notin P$, hence $AB \not\subseteq P$. Thus, P is prime and $a_0 = a \notin P$. Thus $a \notin$ prime rad R. □

2.34 REMARKS. (1) For a ring R, the following are equivalent:

1. R is semiprime.
2. Prime rad $R = 0$.
3. R is a subdirect product of prime rings.
4. For any pair A, B of ideals, $AB = 0$ if and only if $A \cap B = 0$.

(2) (McCoy [49]) For any $n \times n$ matrix ring R_n,

$$\text{prime rad } (R_n) = (\text{prime rad } R)_n$$

Nil and Nilpotent Ideals

A subring-1 S of R is **nil** if every element of S is nilpotent. Since a nil subring-1 S cannot have unit, we refer to nil subrings dropping the "minus." The subring S is **nilpotent of index** k if all products of k elements of S (in whatever order) is zero.

The following are easily verified properties of nil (nilpotent) ideals

2.34A PROPOSITION. *Let R be a ring.*

(1) *The sum of two nil (nilpotent) ideals is nil (nilpotent).*

(2) *The sum N of all the nil ideals of R is a nil ideal containing all nilpotent right (or left) ideals.*

PROOF. (1) If A and B are nil, then

$$\overline{A} = (A + B)/B \approx A/(A \cap B)$$

is a homomorph of A, hence nil, so if $x \in A + B$, then $x^n \in B$ for $n \geq 1$, and then $x^{nm} = 0$ for some $m \geq 1$. Thus, $A + B$ is nil;

(2) By induction, any finite sum of nil ideals is nil, hence if $x \in N$, then x belongs to a finite sum of ideals, hence is nilpotent. Thus N is nil.

If I is a nilpotent right ideal of index k, then $(RI)^k = RI^k = 0$. Since I is contained in the nilpotent ideal RI, then $I \subseteq N$. □

2.34B THEOREM. *Let R be a ring.*

(1) *If I_1 and I_2 are nilpotent right ideals of indices n_1 and n_2, then $I_1 + I_2$ is nilpotent of index $\leq n_1 + n_2 - 1$.*
(2) *If I is a nilpotent right ideal of index n, I generates a nilpotent ideal of index n.*
(3) *A sum of finitely many nilpotent right ideals is nilpotent.*
(4) *The sum N of all nilpotent right ideals is the sum of all nilpotent left ideals. Furthermore, N is* **locally nilpotent** *in the sense that every finite subset generates a nilpotent subring.*
(5) *The sum L of all the locally nilpotent ideals is a locally nilpotent ideal containing every locally nilpotent right (left) ideal.*

PROOF. (1) is an easy consequence of the binomial theorem. See, e.g. Van Der Waerden [50], pp.139–140;

(2) was proved in (2) of 2.34A above;

(3) follows from (1) by induction, and (4) follows from (2) and (3). The first part of (5) is proved similarly to (1) of Proposition 2.34A. See Jacobson [55,64], p.197, Props. 1 and 3 for the second part of (5). □

2.34C REMARKS. (1) If R is commutative, then every nilpotent element x generates a nilpotent ideal (x), hence by (1) of Proposition 2.34A, every $f \cdot g$ nil ideal of R is nilpotent. Consequently if R is Noetherian, every nil ideal is nilpotent, a result that holds for non-commutative Noetherian rings by a theorem of Levitzki below. (Cf. 3.39–3.42.);

(2) Regarding (4) and (5) of 2.34B, compare the Köthe radical **sup.** 3.50, and theorems of Amitsur and Klein 3.50A,B.

2.34D THEOREM (LEVITZKI). *If R is a ring satisfying the acc on right annihilators of nil subsemigroups generated by finitely many non-nilpotent elements, then any $f \cdot g$ nil multiplicative subsemigroup of R is nilpotent.*

PROOF. Jacobson, *ibid.*, p.199

REMARK. See Shock's Theorem 3.39 for a more general result. (Cf. his Corollary 3.40.)

2.34E THEOREM (LEVITZKI [45]). *If R is right Noetherian, then every nil right (left) ideal of R is nilpotent.*

PROOF. Jacobson, *ibid.*, p.199, Theorem 1.

REMARK. See the more general theorem of Levitzki and Herstein-Small 3.41 on rings satisfying acc$\perp$ and dcc$\perp$.

Nil Radicals

As stated, an ideal A of a ring R is **nil** provided that every element of A is nilpotent. A union of a chain of nil ideals of R is again a nil ideal, hence by Zorn's lemma there exists a maximal nil ideal N of a ring R. By use of the binomial theorem if I is a nil ideal of R, then $I + N$ is a nil ideal $\supseteq N$, hence $I \subseteq N$ by maximality. Thus, R has a largest nil ideal N denoted nil rad R. Furthermore,

$$\text{rad } R \supseteq \text{nil rad } R \supseteq \text{prime rad } R$$

the right inclusion by **sup**. 2.33, but each inclusion may be proper. However, if R is commutative, each nilpotent element is strongly nilpotent, so 2.33 implies:

2.35A PROPOSITION. *If R is commutative then nil rad $R =$ prime rad R.*

We now define the **nil radical of an ideal** A of R to be η^{-1}(nil rad R/A), when η is the canonical map $\eta : R \to R/A$; that is, nil rad A is the largest ideal of R which is nil modulo A. Prime rad A is defined similarly.

2.35B COROLLARY. *If R is commutative and A is an ideal, then nil rad A is the intersection of the prime ideals containing A.*

The prime ideals containing A are called the **prime ideals of (belonging to)** A, and a **minimal prime ideal of** A is just one that is minimal in the set of prime ideals of A ordered by inclusion. Since the intersection of a chain of prime ideals is prime, then any prime ideal $P \supseteq A$ contains a minimal prime ideal $\supseteq A$.

2.36 PROPOSITION (MCCOY [49]). *If A is an ideal of a ring R, then every prime ideal belonging to A contains a minimal prime ideal belonging to A, and prime rad A is the intersection of the minimal prime ideals of A.*

PROOF. Immediate from the proof of 2.33. □

Semiprime Ideals and Unions of Prime Ideals

An ideal I of a ring R is **semiprime** in case $\forall$ right ideals K, if $K^n \subseteq I$ for some n, then $K \subseteq I$. Expressed otherwise, I is semiprime if and only if the factor ring R/I is semiprime. A combination of 2.32 and 2.33 establishes the next proposition.

2.37A PROPOSITION. *An ideal I of R is semiprime if and only if I is the intersection of the prime ideals of R containing it. Therefore every semiprime ideal of R contains prime rad R.*

2.37B THEOREM (MCCOY [57]). *If $I, I_1, \dots, I_n$ are finitely many ideals of a ring R such that I is contained in the union of the others, but not in the union of any $n-1$ of the others, then some power I^k is contained in their intersection.*

PROOF. See *op.cit.* Theorem 1.

2.37C COROLLARY. *If $I, I_1, \dots, I_n$ are finitely many ideals of R such that I is contained in the union of the others, and if at least $n-2$ of the others are semiprime, then $I \subseteq I_j$ for some j.*

2.37D REMARKS. (1) Similar theorems hold for groups (*ibid.*); (2) Under the same assumptions as 2.37C, then $K = I_1 \cup \dots \cup I_n$ is an ideal iff $K \subseteq I_j$ for some j (in which case $K = I_j$); (3) See "prime avoidance," 16.5–16.8C.

Maximal Annihilator Ideals Are Prime

An ideal is I of R is an **annihilator ideal** (= **annulet**) if either (1) $({}^{\perp}I)^{\perp} = I$, or (2) ${}^{\perp}(I^{\perp}) = I$. In case (1) I is **right annihilator ideal** (= right **annulet**) to be distinguished between an **annihilator right ideal**, which satisfies $({}^{\perp}K)^{\perp} = K$ but need not be an ideal. A **maximal right annihilator ideal** P (= **right maxulet**) if P is maximal in the set of right annihilator ideals $\neq R$.

2.37E THEOREM. *Any maximal right annihilator ideal P is a prime ideal.*

PROOF. Suppose $A \supset P$ and $B \supset P$ are ideals properly containing P, and let $K = {}^{\perp}P$. If $AB \subseteq P$, then $KAB = 0$, hence $KA \subseteq {}^{\perp}B$. But ${}^{\perp}B = 0$ by maximality of P, so $K \subseteq {}^{\perp}A = 0$ so $K = 0$ too. This contradicts maximality of P. □

2.37F THEOREM. *If I is a right annihilator ideal of R, and if R satisfies the acc on right annihilator ideals (resp. acc $\perp$), then so does the factor ring R/I.*

PROOF. Let $A \supset I$ and $B \supseteq I$ be ideals so that $\overline{B} = B/I$ is a right annihilator ideal of $\overline{R} = R/I$ and $\overline{A}^{\perp} = \overline{B}$. If $K = {}^{\perp}I$ in R, then $KAB = 0$ since $AB \subseteq I$, hence $(KA)^{\perp} \supseteq B$ in R. But if $(KA)r = 0$ in R for $r \in R$, then $Ar \subseteq I$, hence $r \in B$, that is, $(KA)^{\perp} = P$, so B is a right annihilator ideal of R. (The proof for acc $\perp$ is similar. See Herstein [69], Lemma 5.3.) □

Rings with Acc on Annihilator Ideals

2.37G THEOREM. *Let R be a ring satisfying the acc on right annihilator ideals (=* **accra***).*

(1) *R has just finitely many maximal right annihilator ideals.*
(2) *If I is a right annulet of R, then R/I also satisfies accra.*
(3) *If R is semiprime, and if I is a right annulet, then I has just finitely many minimal primes and is semiprime.*

PROOF. (1) is proved in 16.31 for commutative R, and a similar proof (using 2.37C instead of 16.1) suffices here; (2) is immediate from 2.37F; (3) First note that I is a semiprime ideal. For if $X^{\perp} = I$, and if $A \supseteq I$ is nilpotent modulo I say $A^n \subseteq I$, then

$$(XA)^n = (XA) \cdots (XA) \subseteq XA^n = 0$$

hence $A^n \subseteq X^{\perp} = I$. Thus, it suffices to prove (3) assuming $I = 0$. Hence suppose R is a semiprime accra ring. If A and B are ideals, and if $AB = 0$, then $(BA)^2 = 0$ so $BA = 0$. It follows that (*) every right annihilator ideal (= annulet) is a left annulet. Since accra implies the dcc on left annulets, then (*) implies that R satisfies **dccra**, the dcc on right annulets.

If $P_1, P_2, \ldots, P_n$ are right maxulets, then each P_i is prime by Theorem 2.37E. By dccra, then the descending chain $A_n = P_1 \cap \cdots \cap P_n$ is stationary for some n, and then $P_k \supseteq A_n$ $\forall k \geq n$. By Theorem 2.37C, then $P_k = P_j$ for some $j \leq n$, hence $P_1, \ldots, P_n$ constitute all of the right maxulets.

We next show that (**) each P_i is a minimal prime; for if P is a prime ideal properly contained in P_i, the fact that for $A = {}^{\perp}P_i$ we have $0 = AP_i \subseteq P$, so $A \subseteq P$. But then, $A^2 = 0$, hence $A = 0$, a contradiction.

It follows that $\{P_1, \ldots, P_n\}$ is the totality of minimal primes of R, hence $\bigcap_{i=1}^n P_i = 0$ by Theorem 2.36. This completes the proof of (3), and the proof of the theorem. □

REMARKS. (1) Cf. Aldosray [96]. Also see 14.34, 16.25, and Rowen [88],pp.364–5; (2) If a prime ideal P is not an essential right ideal, then P is a right annulet. Moreover, if $({}^{\perp}P)^2 \neq 0$, then P is a minimal prime ideal.

2.37H COROLLARY. *A ring R is a semiprime accra ring iff there exists a finite set of prime ideals $P_1, \ldots, P_n$ with zero intersection. In this case:*

(1) *Every right annulet is the intersection of P_i's, hence every chain of annihilator ideals has length $\leq n$.*
(2) *The prime ideals P_i are the right maxulets, and the minimal prime ideals.*
(3) *R satisfies dccra, and conversely, dccra implies accra.*

PROOF. The proof follows as the proof of the theorem 2.37G. □

2.37I THEOREM (FAITH [91B]). *If R is a ring, and if $\mathcal{A} = \{P_i\}_{i=1}^n$ is a set of right maxulets such that the corresponding left minulets $\{{}^{\perp}P\}_{i=1}^n$ is a maximal independent set of minulets, then $\mathcal{A}$ is the set of all maxulets.*

PROOF. The proof is the same *mutatis mutandis* as Proposition 16.15. □

2.37J THEOREM. *If R satisfies dccra (= the dcc on right annulets), then R has just finitely many right maxulets (= right maximal annihilator ideals).*

PROOF. Same proof as the proof of (3) of Theorem 2.37G, and also Theorem 16.25. □

REMARK. If I is an annilator right ideal, say $I = L^{\perp}$, for a left ideal L, and if I is an ideal, then $LI = (LRI) = 0$, so $I = (LR)^{\perp}$ is a right annihilator ideal (= annulet). Thus:

$$acc \perp \Rightarrow accra$$

and

$$dcc \perp \Rightarrow dccra.$$

The converse implications fail for, e.g., for a non-Artinian simple von Neumann regular ring. (It is known that if A is an integral domain not a right Ore domain, then its injective hull R is such a ring. See Theorem 6.28.)

The Baer Lower Nil Radical

2.38A THEOREM (LEVITZKI [51]). *Let $N(\alpha)$ be the ideal of R defined inductively for any ordinal α by setting:*

$N(0) =$ *the sum of all nilpotent ideals of R;*

$N(\alpha + 1) =$ *the inverse image in R of the ideal $N(0)$ defined for $R/N(\alpha)$;* $N(\alpha) = \bigcup_{\beta<\alpha} N(\beta)$ *when α is a limit ordinal.*

Then, there is a least ordinal α such that $N(\alpha) = N(\alpha + 1)$, and $N(\alpha)$ is then called the **Baer lower nil radical**. *Moreover, $N(\alpha) =$ prime rad R.*

PROOF. Clearly $M = \text{prime rad } R \supseteq N(0)$, and moreover, assuming there exist an ordinal α_0 so that $M \supseteq N(\beta)\ \forall \beta \leq \alpha_0$, one sees that $M \supseteq N(\alpha_0)$. (This follows since M contains any ideal which is nilpotent modulo $M(\beta)$.) Thus, by transfinite induction, $M \supseteq N(\alpha)$. However, $N(\alpha) \supseteq M$ by 2.37A. □

REMARK. Baer [43b] first defined the lower nil radical.

2.38B THEOREM. *The Baer lower nil (or prime) radical $N(\alpha)$ of a ring R is locally nilpotent, that is, every finite set of elements generates a nilpotent subring.*

PROOF. See, e.g. Kharchenko [96], p.783.

2.38C THEOREM (HERSTEIN-SMALL). *If R is an acc$\perp$ ring, then any nil subring-1 N, is locally nilpotent.*

PROOF. See Herstein [69], p.88, Corollary. □

2.38D THEOREM (KAPLANSKY [48,95]). *If R is a PI-algebra (see §15), then any nil subring-1 is locally nilpotent.*

PROOF. See Herstein, *ibid.*, p.91,Corollary 1. □

2.38E REMARKS. (1) Herstein attributes 2.38D theorem to Kaplansky (Cf. 15.9), and 2.38C jointly with Small (*ibid.*, p.87);

(2) Cf. Nil $\Rightarrow$ nilpotent theorems 3.37–3.43.

(3) If R satisfies the acc on principal right annihilators, that is, on $\{x^\perp \mid x \in R\}$, then every nil right ideal $\neq 0$ contains a nilpotent right ideal $\neq 0$. Furthermore, if R is semiprime, then R has no nil one-sided ideals $\neq 0$. (See Herstein [69], or Rowen [88], p.205.)

(4) By a theorem of Kaplansky [46] and Levitzki [46] algebraic algebras over a field satisfying a PI, or of bounded degree, are locally finite dimensional (see Theorems 15.11–12. Cf. Kurosch's Problem, Nagata-Higman Theorem and Golod-Shafarevitch Theorem 3.43f–3.43As.

Group Algebras over Formally Real Fields

A field F is **formally real** provided that

$$x_1^2 + x_2^2 + \cdots + x_n^2 = 0 \Rightarrow x_1 = x_2 = \cdots = x_n = 0$$

for any elements $x_1, \ldots, x_n \in F$.

2.39 LEMMA. *If G is a group, and if F is a formally real field, then the group algebra FG is semiprime.*

PROOF. The involution of G sending $g \mapsto g^{-1}$ extends to an involution of FG; namely if $a = \sum_{g\in G} a_g g \in FG$, where $a_g \in F\ \forall g \in G$, let $a^\star = \sum_{g\in G} a_g g^{-1}$. The mapping $t : FG \to F$ such that $t(a) = a_1$, where 1 is the group identity, is a linear transformation over F, and $t(ab) = t(ba)\ \forall a, b \in F(G)$. Moreover, $t(aa^\star) = \sum_{g\in G} a_g^2$. Thus

$$aa^\star = 0 \Rightarrow a = 0$$

since F is formally real. Now if b is an element in a nil ideal of FG, and then $a = bb^\star \neq 0$ is a nilpotent element of FG such that $a^\star = a$. Suppose $a^t \neq 0$, and $a^{t-1} = 0$. Then $c = a^t$ satisfies $c^2 = 0$. But $c^\star = c$, and so $cc^\star = 0$, whence $c = 0$ by $(\star)$. This contradicts $c = a^t \neq 0$. □

Jacobson's Conjecture for Group Algebras

If F is a field of characteristic 0, then a conjecture of Jacobson is that any group algebra FG is semiprimitive, that is, $\text{rad}(FG) = 0$. The next theorem and corollary verifies this conjecture for the case F is uncountable. See Passman [97,98] for a survey of progress on the conjecture and recent results. Also see Snider's theorem (§12) that shows *inter alia* that FG embeds in a semiprimitive ring, in fact, in a von Neumann regular ring (§4).

A field F is **absolutely algebraic** if F is algebraic over the prime subfield.

2.40 PROPOSITION (AMITSUR [59]). *If F is a field of characteristic 0, and if F is not absolutely algebraic, then every group algebra FG is semiprimitive for any group G.*

PROOF. (See *op.cit.*, or the author's Algebra II, pp. 258–260).

2.41 COROLLARY. *Every group algebra FG over an uncountable field of characteristic 0 is semiprimitive.*

PROOF. An absolutely algebraic field is countable. □

Cf. 3.43ff.

Simplicity of the Lie and Jordan Rings of Associative Rings: Herstein's Theorems

If A is a subset of an associative ring R that is an additive subgroup, then A is a **Lie** (resp.**Jordan**) **subring** provided that $[a, b] = ab - ba \in A$ (resp. $a \circ b = ab + ba \in A$) $\forall a, b \in A$. A **Lie** (resp.**Jordan**) ideal of A is a subgroup U of A so that $[u, a] \in U$ (resp. $u \circ a \in U$) $\forall u \in U, a \in A$. Then A is said to be **Jordan simple** if A has no proper Jordan ideals $\neq 0$.

2.42 THEOREM (HERSTEIN [55B]). *Let R be a simple ring of characteristic $\neq 2$, then 1) R is Jordan simple; 2) Any Lie ideal U is either contained in the center of R or else contains $[R, R] = \{ab - ba \mid a, b \in R\}$; 3) Any Lie ideal U of R that is a subring of R is either R or is contained in the center of R, except when R is 4-dimensional over C.*

REMARK. 2) is often expressed by saying R is *Lie simple*; in fact:

$$[R, R]/([R, R] \cap C)$$

is a simple Lie ring.

Simple Rings with Involution

An involution $\star$ is an anti-isomorphism of a ring R with itself, that is, $\star$ is a 1-1 map R onto R so that

$$a^{\star\star} = a, \quad (ab)^\star = b^\star a^\star, \quad \text{and } (a+b)^\star = a^\star + b^\star.$$

(A ring R has an involution iff R is isomorphic to its opposite ring R^0, say by a map f. Then defining $a^\star = f^{-1}(a^0)$ defines an involution, etc.)

The set $S = \{x \in R \mid x^\star = x\}$ is the set of **symmetric elements** of R and $K = \{x \mid x^\star = -x\}$ is the set of **skew-symmetric elements.**

2.43 REMARKS. (1) If $R = A_n$ is the $n \times n$ matrix ring over a ring A, then every non-trivial involution $\star$ of R is composed of the involution taking a matrix into its transpose

$$(a_{ij})^\star = (a_{ij}^\star)^T.$$

And conversely, any involution $\star$ of A extends to an involution of A_n via this formula. A similar result holds for an automorphism of A_n with T replaced by an inner automorphism by an invertible matrix X of A_n (Cf., e.g. Jacobson [56, 64], p.45, for a theorem on isomorphisms of vector spaces.)
(2) If G is a group, then $(ab)^{-1} = b^{-1}a^{-1}$ $\forall a, b \in G$, and the map $a \to a^{-1}$ defines an involution of G. In this case, for any commutative ring A the group ring $R = AG$ has an involution $\star$ such that

$$\left(\sum a_g g\right)^\star = \sum a_g g^{-1}.$$

In fact, if A is any ring, with involution $\star$, then $\star$ can be extended to AG via the formula

$$\left(\sum a_g g\right)^\star = \sum a_g^\star g^{-1}.$$

DEFINITION. Let R be a simple ring of characteristic $\neq 2$ with center C, and let S be the Jordan ring of symmetric elements of R. The involution is of the **first kind** if $C \subset S$; the involution is of the **second kind** if $C \not\subset S$.

2.44 THEOREM (HERSTEIN [56,69]). *Let R be a simple ring with center C having an involution of the second kind. Then*

(1) *If the characteristic of R is not 2 then S is a simple Jordan ring.*
(2) *Any Lie ideal of $[K, K]$ is either in C or is $[K, K]$ except if R is of characteristic 2 and is 4-dimensional over its center.*
(3) *Any Lie ideal of K either is in C or contains $[K, K]$ except if R is of characteristic 2 and is 4-dimensional over C.*

See Herstein [69], p.27.

2.45 THEOREM (HERSTEIN [69]). *If R is a simple ring of characteristic not 2 and if dim $R/Z > 16$, then any proper Lie ideal of $[K, K]$ is in C.*

See *loc.cit.*, p.46.

2.46 THEOREM. *The only Jordan ideals of S are (0) and S; that is, S is a simple Jordan ring.*

Loc.cit., p.32.

Symmetric Elements Satisfying Polynomial Identities

For the background on polynomial identities in the following, see Chapter 15.

2.47 THEOREM (HERSTEIN [67]). *Let R be a simple ring of characteristic $\neq 2$, with center C. If R has an involution $\star$, and the symmetric (or skew-symmetric) elements satisfy a polynomial identity over the centroid of R, of degree n, then* $\dim_C R = n^2/4$ *if $\star$ is of the first kind, and* $\dim_C R \leq n^2$ *if $\star$ is of the second kind.*

2.48 REMARK. A number of improvements on Herstein's theorems have been made by W. E. Baxter, W. D. Burgess, C. Lanski, P.H. Lee, S. Montgomery, among others, e.g. Montgomery's paper [70] extends the Lie simplicity of simple rings to characteristic 2.

2.49 THEOREM (AMITSUR [68]). *If R is a ring with involution, and if the set S of the symmetric elements satisfy a polynomial identity of degree d, then R satisfies a power $S_{2d}[X]^m = 0$ of the standard identity $S_{2d}[X] = 0$ of degree $2d$.*

2.50 REMARK. If R is a simple ring (or a semiprime algebra of characteristic $\neq 2$), then R satisfies an identity of degree $\leq 4d$ (resp. $\leq 2d$). This is a result of Martindale [69], Herstein [67,69] and Amitsur [68] (Martindale's paper surveys the status of this genre of theorems up to about 1967).

There are also a number of papers that place various conditions on the set of symmetric elements, e.g. a paper by Herstein and Montgomery [71] extends Jacobson's theorem to division algebras with involution with the condition $a^{n(a)} = a$ required for just the symmetric elements. This was generalized by various authors including the junior author (*Vide*, Small [81,85], especially, §11 (Rings with Polynomial Identity) and §12 (Rings with Involution)).

Historical Notes

(1) One day at tea at the Institute, Freeman Dyson asked me if such a theorem as 2.49 held true for a certain class of rings with involution. Just about this time I had become aware of Amitsur's theorem, and I told Freeman. He was a bit taken back, I think, by the extent of generality of Amitsur's theorem!

(2) According to Kaplansky [95], "Afterthought: Rings with a polynomial identity" p.65: "Marshall Hall [characterized] a quaternion division algebra by the identity that says that the square of any commutator is central." This is Wagner's Identity (see 15.1(4))

$$(xy - yx)^2 z = z(xy - yx)^2$$

which Hall [43] proved holds for any field k or generalized quaternion algebra over k. Kaplansky [48] (also [95]) showed that any primitive algebra, e.g., any division algebra D over a field k that satisfies *any* polynomial identity is necessarily finite dimensional.

(3) Wagner's main theorem [37] states that any ordered PI-algebra over $\mathbb{R}$ is commutative.

(4) Albert [40] proved that any ordered finite dimensional algebra over a field is a field.

(5) (3) and (4) are commented on in Kaplansky [48], pp. 579–580 (in Kaplansky [95],pp.63–64).

Separable Fields and Algebras

An algebraic field extension L of k is **separable** if the minimal polynomial over k of every element of L has distinct roots in the algebraic closure of k. If k has characteristic 0, then every algebraic extension of k is separable. (See, e.g. Van der Waerden [48], Artin [55], or Jacobson [51].

2.51 SEPARABILITY CRITERION. *Let A be a finite dimensional algebra over a field k. Then, A is said to be* **separable over** *k, or A/k is separable, provided that the following conditions, which are equivalent, hold:*

(1) *$L \otimes_k A$ is semisimple for all field extensions L of k.*

(2) *For some algebraically closed field $L \supseteq k$, the L-algebra $L \otimes_k A$ is a product of total matrix algebras over L.*

(3) *A is semisimple, and center A is a product of finitely many separable field extensions of k.*

In this case, (2) holds for $L = \overline{k}$, the algebraic closure of k, and also for some finite separable field extension L contained in $\overline{k}$. (Cf. Splitting field, p.19.)

Wedderburn's Principal or Factor Theorem

The next theorem is also called the **Wedderburn Principal Theorem**.

2.52 WEDDERBURN FACTOR THEOREM. *If A is an algebra of finite dimension over a field k, and if $A/\text{rad } A$ is separable, then there is a semisimple subalgebra S of A canonically isomorphic to $A/\text{rad } A$ such that $A = S + \text{rad } A$ and $S \cap \text{rad } A = 0$.* **The algebra S is the semisimple factor, or part, of A.**

PROOF. See Jacobson [42], p.117, or the author's **Algebra I**, p. 371, Theorem 13.15. □

REMARK. See Wehlen [94] for various generalizations, e.g. to Hochschild's separable algebras defined **sup.** 4.15B, and also generalizations of Mal'cev's conjugacy theorem stated in the next paragraph. Wehlen, *ibid.*, §5 also discusses work of G. Azumaya, W. C. Brown, G. Hochschild, R. Reisel, A. Rosenberg and D. Zelinsky and I. Stewart.

Invariant Wedderburn Factors

Assume that an algebra A of finite dimension over a field k is separable modulo radical. If S_1 and S_2 are two semisimple factors of A, then there exists an element $x \in \text{rad } R$ such that $S_1 = (1-x)^{-1}S_2(1-x)$ (Mal'cev [42]). (This remains true, assuming only that $A/\text{rad } A$ is finite dimensional (Eckstein [69]; also see Curtis (1954), as cited by Eckstein). Also see Wehlen [94] for additional generalizations.)

A semisimple factor S (also called a **Wedderburn factor**) is said to be G-invariant relative to a finite group G of automorphisms and anti-automorphisms of A provided that $g(S) \subseteq S \ \forall g \in G$. If A is finite dimensional, and G is finite, then A has G-invariant Wedderburn factors provided that the characteristic of k does not divide $|G|$ (Taft [57]), or when G is a completely reducible group acting on A and k has characteristic 0 (Mostow (1956), Cf. Taft [68]). Taft [64] established a strong uniqueness of G-invariant Wedderburn factors for a completely reducible group G.

CHAPTER 3

Direct Sum Decompositions of Projective and Injective Modules

A right R-module M is *free* if M is isomorphic to a direct sum of copies of R. E.g. if R is a skew field, then every R-module is free, but this fails for $n \times n$ matrix rings over a skew field D if $n > 1$. However, by the Wedderburn-Artin theorems, D_n is a semisimple ring, and every R-module over a semisimple ring R is isomorphic to a direct summand of a free R-module.

Recall that an *epimorphism* is an onto homomorphism. An R-module M is said to be *projective* if M is isomorphic to a direct summand of a free R-module, equivalently, every R-epimorphism $f : N \longrightarrow M$ splits in the sense that the kernel

$$K = \ker f = \{x \in N \mid f(x) = 0\}$$

of f is a direct summand of N. Thus, if M is projective in this sense, and $f : F \longrightarrow M$ is a *presentation* of M (= F is a free R-module and f is an epimorphism), then $F = K \oplus N$, where $K = \ker f$, and $M \approx N \approx F/K$ is isomorphic to a direct summand of F. Similarly, when R is a semisimple ring, then K is a direct summand of F for every submodule K of F by a basic property of semisimple modules, so every R-module is projective when R is semisimple. Cf. 3.4A.

Direct Sums of Countably Generated Modules

Kaplansky gave the first general theorem on the structure of projective modules over an arbitrary ring.

3.1A THEOREM (KAPLANSKY [58A]). *Over any ring R, any projective module is a direct sum of countably generated modules.*

3.1A is a consequence of a more general:

3.1B THEOREM (KAPLANSKY [58A]). *If a right R-module M is a direct summand of a direct sum of countably generated modules, then M is itself a direct sum of countably generated modules.*

3.1C THEOREM. *Any direct summand M of a direct sum of $\aleph$-generated modules, for an infinite cardinal $\aleph$, is a direct sum of $\aleph$-generated modules.*

PROOF. Same proof *mutatis mutandis* as Theorem 3.1A. Cf. Osofsky [78], Lemma 3.8.

A number of consequences of these theorems that appear in the same paper are given later in this chapter.

Injective Modules and the Injective Hull

Dual to the concept of projective is that of an injective R-module. This concept, due to Baer [40], is a far reaching generalization of divisible Abelian groups, and provided the setting for the development of homological algebra and category theory (see e.g. Cartan-Eilenberg [56], MacLane [50,63], Freyd [64], Gabriel [62], Faith [73,76,81] and Mitchell [65]).

Injective Hulls: Baer's and Eckmann-Schöpf's Theorems

An R-module E is *injective* iff every embedding $E \to M$ of E in an R-module, splits, equivalently, E is a direct summand of every R-module M containing E as a submodule. If R is an injective right R-module, then R is **right self-injective**; e.g. any semisimple ring R is both right and left injective.

3.2 THEOREM (BAER [40]). *An Abelian group A is an injective $\mathbb{Z}$-module iff A is divisible.*

3.2A BAER'S CRITERION [40]. *A right R-module E is injective iff for every mapping $f : I \to E$ of a right ideal I, there exists $m \in E$ so that $f(x) = mx\ \forall x \in I$.*

Essentially Baer's criterion states that any mapping $f : I \to E$ extends to a mapping $R \to E$. This readily implies:

3.2B THEOREM. *Any direct product of injective modules is injective. (Cf. 3.4B)*

3.2C BAER'S THEOREM ([40]). *Every R-module M has an embedding into an injective module E. In fact M can be embedded into a minimal injective R-module $E(M)$, called the* **injective hull** *of M; and $E(M)$ is unique up to an isomorphism (see 3.2D).*

An overmodule E of M is *essential* provided $S \cap M \neq 0$ for any submodule $S \neq 0$ of E; then S is called an **essential submodule**.

3.2D THEOREM (ECKMANN-SCHÖPF [53]). *The injective hull $E(M)$ of a right R-module M is the maximal essential extension of M.*

Complement Submodules and Maximal Essential Extensions

If M is an R-module, and S is a submodule, then by Zorn's Lemma there exists a submodule maximal in the set of those submodules T such that $S \cap T = 0$. Then T is a **complement** to S, and any submodule complement to some submodule is called a **complement submodule**. Again by Zorn's lemma if T_0 is complement to S, then there is a complement submodule $S_0 \supseteq S$ complement to T_0. Furthermore S is an essential submodule of S_0 since if N is a submodule of S_0, and if $N \neq 0$, then

$$S \cap N = S \cap (T_0 + N) \neq 0$$

by maximality of T_0 w.r.t. $T_0 \cap S = 0$.

3.2E THEOREM. *(1) Every submodule S of an R-module M is an essential submodule of a complement submodule S_0 that is a maximal essential extension of S in M. Furthermore,*

(2) Every maximal essential extension of S in M is a complement submodule;

(3) If M is injective, then in (1), S_0 is an injective hull of S.

PROOF. Straightforward application of the concepts and remarks preceding, and the fact that the injective hull $E(S)$ is the maximal essential extension of S.

□

The Cantor-Bernstein Theorem for Injectives

$A \hookrightarrow B$ denotes a 1-1 homomorphism of R-modules, and then one says that A *embeds* in B.

3.3 THEOREM (BUMBY [65] AND OSOFSKY). *The Cantor-Bernstein theorem for sets holds for injective R-modules A and B; namely, $A \approx B$ iff $A \hookrightarrow B$ and $B \hookrightarrow A$ as R-modules.*

See also Faith ([76], p.171n). This theorem also holds for quasi-injective modules (Bumby [65]), and implies that the minimal injective cogenerator is unique up to isomorphism. See below.

Generators and Cogenerators of Mod-R

A right R-module G is said to be a **generator** of mod-R, or **generates** mod-R if G satisfies the equivalent conditions:

For every right R-module M:
(G1) There is an epimorphism $G^{(I)} \to M$ of a direct sum of copies of G.
(G2) There is an epimorphism $G^n \to M$ for an integer $n > 0$.
(G3) Some finite product G^m contains R as a direct summand as a right R-module, that is, $G^m \approx R \oplus X$ in mod-R for some $m > 0$ and right R-module X.

PROOF. See my **Algebra I**, p. 144, 3.25-6. □

Dually, a **cogenerator** C of R is defined by the property that every right R-module M embeds in a direct product C^α of copies of C, for some index α depending on M.

3.3′ REMARKS. (1) A cogenerator C of mod-R has the property for each right ideal I of R, there is a subset $X \subseteq C$ such that I is the annihilator $\text{ann}_R X$ in R of X. This follows since if $h : R/I \hookrightarrow C^\alpha$ is an embedding, one verifies that $I = \text{ann}_R X$, for the set $X = \{x_i\}_{i \in \alpha}$ such that $h(1) = (\dots, x_i, \dots) \in C^\alpha$.

(2) If C is a cogenerator of mod-R, and V is a simple right R-module, then since $V \hookrightarrow C^\alpha$ for some α, then $V \hookrightarrow C$, since the projection $V = p_\alpha(V) \neq 0$ for at least one projection map $p_\alpha : C^\alpha \to C$.

(3) If C is any cogenerator of mod-R, and V is a simple right R-module, along the lines of (1) above, one sees that the injective hull $E(V)$ of V embeds in C. Furthermore, if $\{V_i\}_{i \in I}$ is an isomorphy class of simple right R-modules, then the set $\{E(V_i)\}_{i \in I}$ of all such injective hulls embedded in C is independent (Cf. 1.4), hence $C_0 = \oplus_{i \in I} E(V_i)$ embeds in C. Moreover, C_0 itself is a cogenerator of mod-R. (See, e.g. my Algebra I, p. 167, proof of 3.55(1). See below.)

(4) A ring R is a **right cogenerator** ring if R_R is a cogenerator of R, equivalently, by 1.5, every right R-module is torsionless. Cf. $3.5B'$.

Minimal Cogenerators

It follows from (3) in Remark 3.3′ that the direct sum C_0 of the injective hulls $E(V_i)$ of the set $\{V_i\}_{i\in I}$ of non-isomorphic simple right R-modules is a minimal right cogenerator for R. While the injective hull $E(C_0)$ is, by the Bumby-Osofsky Theorem 3.3, the unique (up to isomorphism) minimal injective right cogenerator, Osofsky [91] showed a minimal cogenerator is not necessarily unique even for commutative R, but that it is when R is either right Noetherian, semilocal (defined following 3.10A), or C_0 is quasi-injective, e.g., as discussed in 3.19A, any right V-ring. In Faith [97b], we call a ring R a **right Osofsky ring** when C_0 is the unique minimal right cogenerator, and show that rings studied by Camillo [78] with the property that $\mathrm{Hom}_R(E(V_i), E(V_j)) = 0$ for $i \neq j$, are right Osofsky. We call these **right Camillo rings**, and show that commutative SISI rings of Vámos [75] (see §9), and locally perfect commutative rings; in fact, any 0-dimensional ring, among others, are Camillo, hence Osofsky rings.

Cartan-Eilenberg, Bass, and Matlis-Papp Theorems

The terminology injective originated in Eilenberg-Steenrod's book [52].

3.4A THEOREM (CARTAN-EILENBERG [56]). *The following are equivalent conditions on a ring R:*
(1) *R is semisimple Artinian,*
(2) *Every right R-module is projective,*
(3) *Every right R-module is injective.*
(4) *(i') = the left-right symmetry of (i), $i = 2, 3$.*

3.4B THEOREM (CARTAN-EILENBERG [56], BASS [62], PAPP [59]). *A ring R is right Noetherian iff every (countable) direct sum of injective right R-modules is injective.*

In this case every injective is a direct sum of indecomposable modules, i.e., modules which have only trivial direct summands.

3.4C THEOREM OF MATLIS [58] AND PAPP [59]. *One may decompose all injective right modules over R into a direct sum of indecomposable modules iff R is right Noetherian.*

Two Theorems of Chase

The following theorems are relatives of Chase's Theorem 1.17A.

3.4D CHASE'S FIRST THEOREM. *If every right R-module is a direct sum of indecomposable modules, then R is right Artinian.*

PROOF. R is right Noetherian by Theorem 3.4C, and R satisfies the dcc on principal left ideals by Theorem 1.17A, and this implies that the radical J of R is a nil ideal (see e.g. Bass' Theorem 3.31). Then by Levitzki's Theorem 2.34E, R right Noetherian implies that J is nilpotent, and the method of proof of the Hopkins-Levitzki Theorem yields R is right Artinian. □

3.4E CHASE'S SECOND THEOREM. *If every right R-module is a direct sum of $f \cdot g$ modules, then R is right Artinian.*

PROOF. Similar to the theorem above. Cf. Theorem 3.5A below. □

Sets vs. Classes of Modules: The Faith-Walker Theorems

An indecomposable injective R-module F is the injective hull $E(S)$ of any nonzero submodule S (since $E(S)$ is a direct summand $\subseteq F$). In particular, $F = E(S)$ for a cyclic submodule S. It follows then that the isomorphism class of indecomposable injective R-modules form a set, since the isomorphism class of all cyclic modules is a set. In this connection a theorem:

3.5A FAITH-WALKER THEOREM ([67]). *A ring R is right Noetherian iff there is a set S of right R-modules such that every module embeds in a direct sum of modules in that set. Furthermore, if either every module is contained in a direct sum of $f \cdot g$ right R-modules, or if every module is isomorphic to a direct sum of modules from a given set, then R must be right Artinian.*

The last statement follows easily from the first as was observed by Faith [71], Griffith [70], and Vámos [71]. A *quasi-Frobenius* ($= QF$) *ring* is an Artinian ring in which every left and right ideal is an annihilator ideal (Nakayama [39]). Ikeda [51,52] characterized these as Artinian rings which are right (or left) self-injective.

3.5B FAITH-WALKER THEOREM ([67]). *The following conditions on a ring R are equivalent:*

(1) *R is QF,*
(2) *Every injective right R-module is projective,*
(3) *Every injective left R-module is projective.*

3.5B' REMARKS. (1) By Baer's Theorem 3.2C, every right R-module M embeds in an injective module, hence if R is QF, then M embeds in a free R-module. By 3.5B, it follows from this, that any QF ring R is a right cogenerator of R, and by symmetry in 3.5B, a left cogenerator as well.

(2) There exist cogenerator rings that are not QF by an example of Osofsky considered in §4, **sup.** 4.25.

(3) By a theorem of Kato, also taken up in §4 (see 4.23) any ring R that is a right and left cogenerator of R is right and left injective over R. By theorem 4.20, this happens iff every faithful R-module (either side) is a cogenerator.

3.5C FAITH'S THEOREM ([66]). *The following are equivalent conditions on a ring R:*

(1) *R is QF,*
(2) *Every projective right R-module is injective,*
(3) *Every projective left R-module is injective,*
(4) *R is right self-injective with the acc on annihilator right ideals,*
(5) *R is left self-injective with the acc on annihilator left ideals.*

As an application of 3.5C, every right module can be isomorphic to a direct sum of right ideals only if R is QF, since the condition implies that every injective module is projective.

3.5D THEOREM (FAITH-WALKER [67]). *A ring R is QF iff every cyclic right, and every cyclic left R-module embeds in a free R-module.*

The QF rings are the Artinian rings with a duality between finitely generated right and left modules induced by $\mathrm{Hom}_A(\ ,A)$ for the ring A and as noted can be characterized as right self-injective rings with the a.c.c. on left (resp. right) annihilators, a fact used in the proof of:

3.5E LAWRENCE'S THEOREM ([77]). *Any countable right self-injective ring R is QF*

(Cf. Megibben's Theorem 3.8).

3.5F BJÖRK [70]-VINSONHALER'S THEOREM [73]. *If R is a ring such that the injective hull $E(R_R)$ of R is Noetherian, then R is right Artinian.*

Polynomial Rings over Self-injective or QF Rings

In this section we consider the polynomial ring $R[X]$ over a self-injective or QF ring R.

A ring R has a *classical right quotient ring* $Q = Q^r_{\mathrm{cl}}(R)$ provided that (1) every **regular element** (= not a zero divisor) $b \in R$ has an inverse $b^{-1} \in Q$, and (2) $Q = \{ab^{-1} \mid a$, regular $b \in R$. $Q^r_{\mathrm{cl}}(R)$ is unique up to isomorphism; and, moreover, exists for any commutative ring R. (Cf. 3.12Bf and Ore domains, 6.26ff.)

3.6A PILLAY'S THEOREM [84]. *A ring R has QF classical right quotient ring $Q^r_{c\ell}(R)$ iff the polynomial ring $R[X]$ has the same property. In this case $Q^r_{cl}(R[X])$ is QF for any set X of variables.*

Herbera and Pillay generalized this:

3.6B HERBERA-PILLAY THEOREM [93]. *If $Q^r_{c\ell}(R[X])$ exists and is right and left self-injective then $Q^r_{cl}(R)$ exists and is QF.*

There are many generalizations of this theorem in their paper.

For example:

3.6C COROLLARY. *If $Q^r_{cl}(R[X])$ exists and is right self-injective, and if R is VNR (see §4), then R is semisimple.*

3.6D HERBERA-PILLAY PROPOSITION [93]. *If $Q^r_{cl}(R[X])$ exists and is right self-injective for an infinite set X of variables, then R satisfies acc$\perp$.*

An immediate corollary of this and 3.5C:

3.6E COROLLARY. *If R is right self-injective, then $Q^r_{cl}(R[X])$ is right self-injective for an infinite set X, only if R is QF.*

Σ-Injective Modules

An injective right R-module E is *Σ-injective* provided that every direct sum of copies of E is injective.

3.7A THEOREM (FAITH [66A]). *An injective right R-module is Σ-injective iff R satisfies the acc on the set $\mathcal{A}_r(E,R)$ of right ideals I of R that are annihilators of subsets of E.*

Cf. Cailleau's theorem, 3.14.

3.7B THEOREM (*Ibid.*). *For any right R-module E, the set $\mathcal{A}_r(A,R)$ satisfies acc iff for each right ideal I there is a $f{\cdot}g$ right ideal $I' \subseteq I$ so that $ann_E I = ann_E I'$.*

3.7C MEGIBBEN'S THEOREM ([82]). *A countable injective right R-module is Σ-injective.*

Quasi-injective Modules and the Johnson-Wong Theorem

A right R-module M is *quasi-injective* if every R-homomorphism $M \to M$ extends to an endomorphism of $E(M)$, equivalently, by a theorem of R. E. Johnson and E. T. Wong [61], M is a fully invariant submodule of $E(M)$. Any semisimple module is quasi-injective.

The proposition below comes from Wedderburn-Artin and Artin-Tate [51] for (semi) simple modules, and Jacobson [56,64] (see the lemma on p.27 *ibid.*), and Johnson and Wong [61].

3.8 PROPOSITION. *Let M be a quasi-injective R-module with endomorphism ring A.*

(a) Every finitely generated A-submodule F of M satisfies the double annihilator condition

$$F = ann_M\, ann_R F$$

(b) If F is a finitely generated A-submodule, and if N is an A-submodule that satisfies the double annihilator condition, then so does $N+F$.

PROOF. (a) is the $N=0$ case of (b); (b) is proved by induction on the number of generators of F. It suffices to prove (b) for the case $F = Ax$. We let $X^\perp$ denote annihilator in R of a subset X of M, and ${}^\perp Y$ annihilation in M of a subset Y of R. For any subset X, ${}^\perp(X^\perp \supseteq X$, so we must show that ${}^\perp((N+Ax)^\perp) \subseteq N+Ax$. Now

$$(N+Ax)^\perp = N^\perp \cap (Ax)^\perp = N^\perp \cap x^\perp.$$

Let $y \in^\perp ((N+Ax)^\perp) = {}^\perp(N^\perp \cap x^\perp)$, so $y(N^\perp \cap x^\perp) = 0$.

Consider the correspondence:

$$\theta : xa \mapsto ya, \qquad a \in N^\perp.$$

If $a, b \in N^\perp$ are such that $xa = xb$, then $a-b \in x^\perp = (Ax)^\perp$, hence $(a-b) \in N^\perp \cap x^\perp$, and therefore $y(a-b) = 0$; that is, $ya = yb$. This shows that θ is a mapping $xN^\perp \to yN^\perp$. Since $\theta(xar) = \theta(xa)r$ $\forall r \in R$, θ is a map of the R-submodules $xN^\perp$ and $yN^\perp$. Since M is quasi-injective, θ is induced by an element of A, which we also designate θ. Since $(\theta x - y)N^\perp = 0$, then $z = \theta x - y$ is an element of $N = {}^\perp N^\perp$, and hence $y = -z + \theta x \in N + Ax$, proving

$${}^\perp((N+Ax)^\perp) = N + Ax.$$

□

R is right **self-injective** provided that R is injective in mod-R.

COROLLARY. *Any finitely generated left ideal F of a right self-injective ring is an annihilator left ideal:* $F = {}^{\perp}(F^{\perp})$

Dense Rings of Linear Transformations and Primitive Rings Revisited

If M is a left vector space over a field D, then by our convention of writing homomorphisms opposite scalars, if a is any element of $L = \text{End}_D M$, and $x \in M$, then we let $xa = (x)a$. A subring A of L is said to be **dense** provided that for each finite subset $y_1, \dots, y_n$ of elements of M, with $n \leq \dim_p M$, and any set $x_1, \dots, x_n$ of n linearly independent vectors, there is an element a of A such that $x_i a = y_i, i = 1, \dots, n$. (Intuitively, this means that A has "enough" l.t.'s.) We also say that A is **dense** in L, or a **dense ring of l.t.'s in** (or **on**) M.

As defined in Chapter 2, **sup.** 2.6, a ring R is **right primitive** if R has a faithful simple right R-module. Any dense ring A of l.t.'s in a left vector space $M \neq 0$ is right primitive since M is faithful over A, and simple since given any nonzero A-submodule M', we must have $M' = M$ by density of A. (To wit: if $y \in M$, and if $x \neq 0$ in M', then $y = xa \in M'A = M'$ for some $a \in A$.) The converse is the:

3.8A CHEVALLEY-JACOBSON DENSITY THEOREM. *(Cf. 2.6) Let M be a simple right R-module with endomorphism ring A. If $x_1, \dots, x_n$ are finitely many elements of M that are linearly independent over A, and if $y_1, \dots, y_n$ are corresponding arbitrary elements of M, then there exists $r \in R$ such that $x_i r = y_i, i = 1, \dots, n$.*

PROOF. Let F denote the A-submodule of M generated by $x_1, \dots, x_n$, and, for each i between 1 and n, let $N_i = F - Sx_i$. By 3.8, ${}^{\perp}(N_i^{\perp}) = N_i, i = 1, \dots, n$. Since $x_i \notin N_i$, then $x_i N_i^{\perp} \neq 0$, and since M_R is simple, then $x_i N_i^{\perp} = M, i = 1, \dots, n$. Then $y_i = x_i r_i$, with $r_i \in N_i^{\perp}, i = 1, \dots, n$, and $r = r_1 + \dots + r_n$ has the desired property. □

COROLLARY. *A ring R is right primitive if and only if R is isomorphic to a dense ring of linear transformations in a left vector space.*

Harada-Ishii Double Annihilator Theorem

3.8B THEOREM (HARADA-ISHII [72]). *Let M_R be quasi-injective, and $A =$ End M_R. If L is a $f \cdot g$ left ideal of A, then*

$$L = ann_A ann_M L.$$

Double Annihilator Conditions for Cogenerators

It is known that any cogenerator R_R satisfies the double annihilator condition (DAC): For any right ideal I of R

$$I = \text{ann}_R \text{ann}_F I.$$

We also note another DAC for F.

3.8C THEOREM (KURATA [91], FAITH [95B]). *If F is any right cogenerator of R, and I and M are submodules of R_R and F_R respectively, then they satisfy the DAC's:*

(a) $$I = ann_R ann_F I$$

(b) $$M = ann_F ann_A M$$

where $A = End\ F_R$.[1]

PROOF. (1) Since F is a cogenerator then $R/I \hookrightarrow F^\alpha$ for some cardinal α, and if (x_i) is the image in F of the coset $1 + I$ in R/I, one sees that

$$I = \text{ann}_R\{x_i\},$$

so (a) follows.

(2) F/M embeds in a direct product F^α of copies of F, and hence there is a map $h : F \to F^\alpha$ that has ker $h = M$. Then, if $p_\alpha : F^\alpha \to F$ is the α-th projection, it follows that $\omega_\alpha = p_\alpha \circ h \in A$ and that

(3) $$M = \bigcap_\alpha \ker \omega_\alpha.$$

Then,

(4) $$M = \text{ann}_F L,$$

where $L = \sum_\alpha A\omega_\alpha$.

Since (4) $\Rightarrow$ (b), the proof is complete. □

Koehler's and Boyle's Theorems

3.9 REMARKS. (1) An R-module M is faithful iff $R \hookrightarrow M^\alpha$ for some product of α copies of M (Cf. my Algebra I, **sup.** 3.24, p.143).

(2) M is said to be **compactly faithful** (also **cofaithful**) provided that $R \hookrightarrow M^n$ for a finite cardinal n. Any quasi-injective cofaithful module is injective by Baer's criterion.

(3) Any injective R-module E is $\prod$-injective in the sense that any product E^α is injective for any cardinal number α. A quasi-injective module M is $\prod$-quasi-injective iff M is an injective $R/\text{ann}_R M$ module (Fuller [69b]).

(4) If $N = R \oplus M$ is a quasi-injective right R-module, then by (2), N is injective, hence so are both R_R and M_R.

3.9A KOEHLER'S THEOREM ([70]). *If every quasi-injective right R-module is injective (= R is a* **right** *QI* **ring***), then R is right Noetherian.*

A ring R is **right (semi) hereditary** if every $(f \cdot g)$ right ideal is projective. A ring R is a right **V-ring** if every simple right R-module is injective. (See 3.19A below.) Trivially, any right QI-ring is a right V-ring.

[1]After this was written (in 1995), I found Kurata's report *ibid.*, where (b) is stated without proof.

3.9B BOYLE'S THEOREM ([73]). *A ring R is a right and left QI ring iff R is right and left hereditary Noetherian V-ring.*

3.9C BOYLE'S CONJECTURE. *Every right QI ring R is right hereditary.*

Boyle's conjecture was verified for Noetherian rings of Krull dimension 1 by Michler and Villamayor [73] (Cf. Faith [76a] and [86a] and Kosler [82]). See 7.40ff.

Quasi-injective Hulls

See Theorems 3.2D and E for the background to the next theorem.

3.9D THEOREM (FAITH-UTUMI [64], HARADA [65]). *If M is a quasi-injective R-module, then*

(1) *Any complement submodule S of M is a direct summand, and is quasi-injective;*

(2) *Any maximal essential extension S_0 in M of a submodule S of M is a minimal quasi-injective submodule of M containing S;*

(3) *Any minimal quasi-injective submodule S_0 of M containing S is an essential extension, and any two such minimal quasi-injective submodules of $M \supseteq S$ are isomorphic.*

PROOF. See *op.cit.* □

Cf. Miyashita [64,65]

The Teply-Miller Theorem

3.10 TEPLY-MILLER THEOREM ([79]). *If E is an injective R-module, then the dcc in $\mathcal{A}_r(E,R)$ implies the acc in $\mathcal{A}_r(E,R)$, hence E is then $\sum$-injective.*

This theorem was motivated by the Hopkins-Levitzki theorem (see §2).

3.10A REMARK. The condition on E in 3.10 is called **Δ-injective** by the author in [82a], Part I. Thus, Δ-injective modules are Σ-injective. Cf. Nastasescu [80].

Semilocal and Semiprimary Rings

A ring R is *semilocal* if $R/\text{rad } R$ is semisimple. A ring R is *semiprimary* if R is semilocal, and rad R is nilpotent. Any semiprimary ring satisfies the *dcc* on $f \cdot g$ one-sided ideals (left or right) (cf. e.g. Bass[60]). (See 3.31 and Björk's theorem following.)

Let E be a right R-module. By writing endomorphisms $a \in A = \text{End}E_R$ on the left, say $a(x) = ax \quad \forall x \in E$, then E is a canonical left A-module, and *mutatis mutandis*, E is canonically a right module over the biendomorphism (= bicentralizer) ring $R'' = \text{End}_A E$. See my Algebra [72a], pp. 119–120 for a discussion of the "**homomorphisms opposite scalars**" convention.

3.11A HANSEN'S THEOREM ([74]). *In theorem 3.10, the double centralizer (= biendomorphism) ring R'' of E is a semiprimary ring.*

3.11B CONVERSE OF THE TEPLY-MILLER-HANSEN THEOREM (FAITH [67,82A]). *If E is a $\sum$-injective right R-module with semiprimary biendomorphism ring R'', then E is Δ-injective.*

3.12A REMARK. For sufficient conditions for when Δ-injective implies that R'' is right Artinian, see Theorem 7.45.

3.12B REMARK. The Cartan-Eilenberg-Bass theorem can be obtained as a consequence of 3.7 since, by 3.8C, any right ideal of R is the annihilator of any injective cogenerator E of mod-R, so $\sum$-injectivity of E implies R is right Noetherian. Moreover, if E is any faithful module, then any right annihilator I of R is the right annihilator of EL, where L is the left annihilator in R of I, hence $\sum$-injectivity of a faithful injective module E implies acc on right annihilators of $R(=$ acc $\perp)$. Cf. Goursaud and Valette [75].

Regular Elements and Ore Rings

An element x of a ring R is **regular** if $x^{\perp} = 0$ and ${}^{\perp}x = 0$. We let $R^{\star}$ denote the set of regular elements of R. If R is commutative then R has a classical quotient ring

$$Q = Q_{c\ell}(R) = \{a/b \mid a \in R, b \in R^{\star}\}$$

in which every $b \in R$ has an inverse $b^{-1} = 1/b$ (see below). When R is a domain, then Q is the quotient field of R. A ring R satisfies the **right Ore condition** if, for each $a, b \in R$, with b regular, there exists $c, d \in R$ with d regular so that $ad = bc$ Mnemonic: $b^{-1}a = cd^{-1}$. In this case we say R is a **right Ore ring**.

Let Q denote the set of equivalence classes a/b with equality $a/b = c/d$ holding iff the implication below holds: if x, y belongs to the set $R^{\star}$ of regular elements, then

$$bx = dy \Rightarrow ax = cy.$$

Addition and multiplication in Q are defined in the standard manner. See my *Algebra I*, p.390–91 for details, or Jacobson [42].

In this case, Q is a ring, called the **classical right quotient ring** $Q^r_{c\ell}(R)$ of R. Q contains R as subring via the embedding $r \mapsto r/1$ $\forall r \in R$, every regular element of R is a unit of Q, and

$$Q = \{ab^{-1} \mid a \in R, b \in R^{\star}\}$$

Finite Goldie Dimension

A right R-module M has infinite Goldie dimension if there is an infinite independent set $\{M_i\}_{i=1}^{\infty}$ of nonzero submodules. Otherwise, M has *finite Goldie dimension*, equivalently, M satisfies the acc on direct sums of submodules ($=$ acc $\oplus$). An R-module has finite Goldie dimension iff $E(M)$ is a finite direct sum of indecomposable modules (cf. §8). A module is **quotient finite dimensional** (q.f.d.) if every factor module is finite dimensional.

REMARKS. (1) Shock [72] proved that a polynomial ring $R[x]$ in a variable x over R has the same finite right Goldie dimension as R;

(2) Wilkerson [75] proved the same result for the twisted polynomial ring $R[x, \alpha]$ for an automorphism α of R;

(3) Wilkerson [73] proved for a finite, or free Abelian, or $f \cdot g$ abelian group G that the group ring RG has finite right Goldie dimension iff R does.

A ring R is *right Goldie* provided that R satisfies both acc$\perp$ and acc$\oplus$.

3.13 THEOREM (GOLDIE [58,60] AND LESIEUR-CROISOT [59]). *A ring R has a semisimple right quotient ring $Q = Q^r_{c\ell}(R)$ iff R is semiprime right Goldie.*

Also see 3.55.

Small [66] generalized Goldie's theorem to Artinian Q (see 3.55B), and in [68] showed then the polynomial ring and power series rings have right Artinian quotient rings. Pillay [84] extended Small's theorem to rings with perfect or QF quotient rings and to infinite polynomial rings (see 3.6A). Lesieur-Croisot [59] proved 3.13 independently.

Cailleau's Theorem

3.14 THEOREM (CAILLEAU [69]). *An injective module E is $\sum$-injective iff E is a direct sum of indecomposable $\sum$-injective modules (Cf. 3.15C).*

It follows that any $f \cdot g$ submodule S of E has injective hull $E(S)$ which is a finite direct sum of indecomposable injectives, and hence that S contains an essential finite direct sum $V_1 \oplus \cdots \oplus V_n$ of uniform submodules $\{V_i\}_{i=1}^n$, i.e., S has finite Goldie dimension.

Moreover, from the stated results on $\sum$-injective modules, it follows that any ring R with $\sum$-injective injective hull $E(R)$ is a Goldie ring (Goursaud and Valette [75]). Moreover, if, in addition, R is semiprime then $Q = Q^r_{c\ell}(R)$ exists, is semisimple, and furthermore $E(R) = Q$. As an example, when R is a domain, then $E(R)$ is $\sum$-injective iff R is a right Ore domain (= has a field of right quotients $Q = E(R)$. Cf. 6.26ff and 6.36B(2).

Local Rings and Chain Rings

If the set of non-units of a ring R is an ideal, equivalently R has a unique maximal right ideal m, then R is said to be a *local ring.* In this case, R/m is a sfield and m is the radical of R. A local ring R has no idempotents except 0 and 1. Moreover, for any prime ideal P of a commutative ring R, the complement $S = R\backslash P$ is multiplicatively closed, and there is a local ring R_P consisting of all symbols a/s with $a \in R$, $s \in S$ and equality defined by: $a/s = b/t \Leftrightarrow (at - bs)s' = 0$ for some $s' \in S$. Addition and multiplication are defined in the same way as ordinary rational numbers. Thus, R_P is the *local ring at* P, and the unique maximal ideal is the set $\{p/s | p \in P, s \in S\}$, denoted PR_P. Moreover, R_P is a flat R-module. (See *sup.* 4.A.)

A ring R is a *right chain ring* if the lattice of right ideals is linearly ordered. A right chain ring is a local ring, and a commutative chain ring is called a *valuation ring.* For example, for any prime number p we have a prime ideal $P = (p)$ of $\mathbb{Z}$, and the local ring $\mathbb{Z}_{(p)}$ is a valuation ring. Similarly, if k is a field, and $R = k[x]$ the polynomial ring, any irreducible polynomial $p(x)$ generates a prime ideal $(p(x))$ and the local ring $R_{(p(x))}$ at $(p(x))$ is a valuation ring.

Uniform Submodules and Maximal Complements

3.14A PROPOSITION AND DEFINITION. *An R-module M is* **uniform** *if $M \neq 0$ and satisfies the equivalent conditions:*

(U1) *M has Goldie dimension $= 1$.*

(U2) *$E(M)$ is indecomposable.*

(U3) *$A \cap B \neq 0$ for any two submodules $A \neq 0$ and $B \neq 0$.*

(U4) *M has no complement submodules except 0 and M.*

(U5) *End $E(M)_R$ is a local ring.*

PROOF. This goes back to Matlis [58], Papp [59], and Goldie [58,60]. See 8.A and 16.9B. □

3.14B COROLLARY. *If U is a uniform submodule of M, then any complement K of U in M is a maximal complement; and conversely, if K is a maximal proper complement submodule of M, then any complement U of K is uniform.*

PROOF. Straightforward application of 13.14A and 3.2E. □

3.14C REMARK. Any proper maximal complement submodule K of M is an **irreducible submodule** in the sense that M/K is an irreducible module.

Beck's Theorems

Let R be a commutative ring, and $\mathcal{P}$ a nonempty set of prime ideals. Then R is said to be *$\mathcal{P}$-Noetherian* provided that for every ascending chain of ideals $I_1 \subseteq I_2 \subseteq \cdots \subseteq I_n \subseteq \cdots$ there is an integer k so that $\forall P \in \mathcal{P}$

$$I_n R_P = I_k R_P \quad \forall n \geq k$$

where R_P is the local ring at P.

3.15A BECK'S THEOREM [72A]. *A ring R is $\mathcal{P}$-Noetherian iff*

$$E = \oplus_{P \in \mathcal{P}} E(R/P)$$

is $\sum$-injective.

A prime ideal P is said to be *Noetherian* if R_P is Noetherian.

This is Theorem 1.11 of Beck [72a].

3.15B COROLLARY. *If P is a prime ideal, then $E(R/P)$ is $\sum$-injective iff P is Noetherian. In this case $E(R/P)$ is an injective R_P-module.*

Regarding 3.15B, see Dade's Theorems 3.17A,B below.

We also note a corollary of 3.15B and Cailleau's theorem [69]:

3.15C COROLLARY. *An R-module E is $\sum$-injective iff there exists a set of Noetherian primes $\mathcal{P}$ so that $E = \oplus_{P\in\mathcal{P}} E(R/P)$.*

3.16A THEOREM (BECK [72A]). *If M is a flat R-module and if E is a $\sum$-injective R-module then $E \otimes_R M$ is $\sum$-injective.*

The proof uses the Govorov-Lazard Theorem 4A.

3.16B COROLLARY. *If E is a $\sum$-injective R-module, then for any multiplicative closed subset S of R, ES^{-1} is an injective RS^{-1} module.*

Corollary 3.16B follows, since, as Beck and also Dade [81] observed, an RS^{-1}-module, e.g. $E \otimes RS^{-1} \approx ES^{-1}$, is injective as an RS^{-1}-module iff it is injective as an R-module. Thus, injectives localize to injectives for Noetherian rings. (This is in Kaplansky's book [74],pp.162–163.) However, Dade showed that in general an injective R-module E does not "localize" to an injective module. (See 3.17B.) *However: $E(R/P)$ is an injective R_P-module.* (See 16.12.)

See Faith [72b], Facchini-Puninski [95], and Puninski-Wisbauer [96] for additional results on Σ-injective modules. Cf.4.6D.

A prime P of a commutative ring R is an **associated** prime provided that $P = \chi^{\perp}$ for some $0 \neq x \in R$. See 16.11(2)ff. Then Ass R denotes the set of all such prime ideals. (See §16.)

3.16C THEOREM (BECK [72A]). *The following are equivalent conditions:*

1. *There exists a finite family $\mathcal{P}$ of Noetherian prime ideals so that the canonical map $R \to \prod_{P\in\mathcal{P}} R_P$ is an embedding.*
2. *The set AssR is a finite set of Noetherian primes whose union is the set of zero divisors of R.*
3. *There is a flat embedding of R into a Noetherian ring T.*

Dade's Theorem

3.17A DADE'S THEOREM. *If (1) R is a coherent ring (see 6.6 below), (2) every ideal of R is countably generated, and (3) S is a multiplicatively closed subset such that for all finitely generated ideal I any chain of ideals*

$$I \subseteq (I : s) \subseteq (I : s^2) \subseteq \cdots \subseteq (I : s^n) \subseteq \cdots$$

terminates for any $s \in S$, where $(I : s) = \{r \in R \mid sr \in I\}$, then ES^{-1} is injective (both as R and RS^{-1} modules) for any injective R-module E.

3.17B THEOREM (DADE [81]). *If A is a commutative Noetherian ring, then the polynomial ring $R = A[X]$ satisfies*

> *(LI) ES^{-1} is injective for any multiplicative subset S and for any injective R-module E,*

when X is a countable set of commuting variables but not if X is uncountable. Moreover, (LI) fails for a factor ring of $A[X]$ when X is countable.

When Cyclic Modules Are Injective

A ring R is *right PCI* if all proper cyclic modules (= cyclic modules $\not\approx R$) are injective. If all simple right modules are injective then R is a *right* V-ring, Cf. 3.19A. A right PCI ring is a *right* V*-ring.*

3.18A OSOFSKY'S THEOREM ([64]). *If all cyclic right* R*-modules are injective, then* R *is semisimple Artinian.*

A ring R is *right (semi-) hereditary* if every $(f \cdot g)$ right ideal is *projective*, e.g. semisimple rings, and PID's are hereditary, and valuation domains are semi-hereditary.

Osofsky's theorem was the inspiration for the following:

3.18B BOYLE'S THEOREM [74]. *A right and left Noetherian ring* R *is right PCI iff* R *is semisimple or a right hereditary* V*-domain. Moreover,* R *is then left PCI.*

And Boyle's Theorem inspired:

3.18C DAMIANO [79]-FAITH [73] THEOREM. *Every right PCI ring* R *is either semisimple, or a right Noetherian hereditary simple domain.*

The author (*loc.cit.*) proved R is either semisimple or a right semihereditary simple domain. Cf. Yue Chi Ming [81], who proves this assuming proper cyclic right modules are "p-injective."

When Simple Modules Are Injective: V-Rings

The *radical of a module* M is the intersection rad M of its maximal submodules.

3.19A VILLAMAYOR'S THEOREM. *A ring* R *is called a* **right** V**-ring** *provided that the following equivalent conditions hold:*

1. *Every simple right* R*-module is injective,*
2. *Every right ideal* I *is an intersection of maximal right ideals, equivalent,* $\text{rad}\,(R/I) = 0$,
3. $\text{rad}\, M = 0$ *for every right* R*-module* M.

These rings were named in honor of Villamayor who introduced the concept (Cf. Faith [72], p.356 and Michler-Villamayor [73]). Since every simple module is quasi-injective, then every right QI-ring is a right V-ring. Menal and Faith [95] showed that R is a right V-ring iff there is a semisimple module W so that annihilation in W and back in R is 1-1 on the set of right ideals.

3.19B KAPLANSKY'S THEOREM. *A commutative ring* R *is a* V*-ring iff* R *is a von Neumann regular ring* (see §4)*; equivalently* R_m *is a field for every maximal ideal* m.

For a generalization, see **Max Rings Theorems** following 3.33. In response to my inquiry, Professor Kaplansky [94] wrote that 3.19B "fell into the public domain", i.e., unpublished.

3.20 FAITH [67,72B]-ORNSTEIN [68] THEOREM. *Every right Goldie right* V*-ring* R *is a finite product of simple right* V*-rings.*

See e.g. Faith [81], p.357.

Cozzens' V-Domains

Cozzens [70] supplied the first known examples of right V-domains that were not fields, namely, the ring $R = k[y, D]$ of differential polynomials over a (Kolchin) universal differential field k with respect to a derivation D. This ring has a unique simple R-module, namely k, and it is injective. Other examples are localizations of the ring of twisted polynomials over an algebraic closed field (Cozzens [70], cf. Osofsky [71], Faith [72,81], pp.361–362, Faith [86a], and Resco [87].

REMARK. Osofsky [72] gave the first examples of V-domains with infinitely many simple modules over fields of characteristic $p > 0$, and Cozzens, using different methods, extended the results to characteristic 0.

Projective Modules over Local or Semilocal Rings, or Semihereditary Rings

3.21 THEOREM OF KAPLANSKY ([58A]). *Over a local ring R, every projective R-module is free.*

The proof uses Kaplansky's theorem 3.1A.

3.22A CARTAN-EILENBERG THEOREM [58]. *R is right hereditary iff every factor module of every injective right R-module is injective and iff every submodule of a projective right R-module is projective.*

3.22B THEOREM (*loc.cit.*). *A ring R is right semihereditary iff each $f \cdot g$ submodule of a projective right R-module is projective.*

Theorem 3.22B does not match up with 3.22A: what about factor modules of injectives? Gupta's Theorem 4.2B supplies the missing link: R is right semihereditary iff every factor of every injective is weak $\aleph_0$-injective (see 4.2B).

3.22C THEOREM (KAPLANSKY [52]). *If R is right hereditary, then each projective module, hence each submodule of a free module, is isomorphic to a direct sum of right ideals.*

3.23A THEOREM (ALBRECHT [61]). *Any projective right or left R-module over a right semihereditary ring is isomorphic to a direct sum of $f \cdot g$ one-sided ideals.*

Albrecht's theorem was proved by Kaplansky [58a] for commutative (or hereditary) R (and conjectured by him). A ring is a right *(semi)fir* if every ($f \cdot g$) right ideal is free of unique rank. The free algebra in n non-commuting variables over a field is a *fir* (cf. Cohn [71]), hence a hereditary ring.

3.23B COROLLARY (ALBRECHT [61], BASS [64A]). *Over a right semifir R, every projective R-module is free.*

3.23C HINOHARA'S THEOREM [62]. *Over a commutative semilocal ring with no nontrivial idempotents, every projective module is free.*

3.24A THEOREM (BASS [64A]). *All $f \cdot g$ projective modules over the free monoid or free group Π over a PID are free.*

3.24B THEOREM (DICKS AND MENAL [79]). *A group ring RG is a semifir iff R is a sfield and G is a locally free group (i.e., any $f \cdot g$ subgroup is free).*

Serre's Conjecture, the Quillen-Suslin Solution and Seshadri's Theorem

3.25. Serre's conjecture on the freedom of any $f \cdot g$ projective module P over the polynomial ring $k[x_1, \dots, x_n]$ in n commuting variables over a field k was proved by Quillen [76], using a lemma of Horrocks, and independently by Suslin [76]. The conjecture previously was verified for $n \leq 2$ by Seshadri [68].

Bass' Theorem on When Big Projectives Are Free

If P_R is projective then P is *uniformly* big if for every ideal A of R, then P/PA requires as many generators as P, i.e., if P is generated by α elements then P/PA cannot be generated by fewer than α elements. Any free R-module is uniformly big.

3.26 BASS' THEOREM ([63]). *If R is right Noetherian modulo its Jacobson radical, then all uniformly big projective right R-modules are free.*

A right R-module G is a *generator* if there exists an onto R-homomorphism $G^n \to R$, equivalently $G^n \approx R \oplus X$ for some R-module X, where $n \in \omega$.

3.27 THEOREM (BASS [63]). *If R is a Noetherian commutative ring with no non-trivial idempotents, then all non-$f \cdot g$ projectives are free.*

REMARK. O'Neill [92] (Prop. 0.1, p.272) corrected an example in Bass [63].

3.28 THEOREM (BECK [72B]). *If P is a projective right R-module such that $P/(P \cdot rad\ R)$ is a free right $R/rad\ R$ module, then P is free.*

3.29. THEOREM (AKASAKI [70]). *If $R/rad\ R$ is a finite product of skew fields, and if every projective right R-module is a generator, then every projective right R-module is free.*

These all generalize Kaplansky's theorem 3.21.

Projective Modules over Semiperfect Rings

R is *semiperfect* if every $f \cdot g$ right R-module M has a **projective cover** $P(M)$ in a sense dual to $E(M)$. The concept is left-right symmetric:

3.30 THEOREM (BASS [60]). *The following are equivalent conditions:*

(SP1) *R is semiperfect,*

(SP2) *$\overline{R} = R/rad\ R$ is semisimple and idempotents of $\overline{R}$ lift (to idempotents of R),*

(SP3) *$R = \oplus_{i=1}^n e_i R$, where $1 = e_1 + \dots + e_n$, $e_i e_j = 0 \quad \forall i \neq j$, e_i is an idempotent and $e_i R e_i$ is a local ring, $i = 1, \dots, n$,*

(SP4) *The left right-symmetry of (SP3).*

Note: Let $J =$ rad R (=the Jacobson radical). Then in (SP3) $V_i = e_i R / e_i J$ is a simple R-module, and the canonical map $e_i R \to V_i$ is the projective cover of $V_i, i = 1, \dots, n$.

DEFINITION. A principal indecomposable right R-module is one $\approx e_i R$ for some i in 3.30, a **right prindec**, or **principal cyclic module**.

REMARK. R is called a **lift/ring** if idempotents of $R/\text{rad } R$ lift. See Jacobson [56,64], p.53, where these are called SBI-rings. Any ring R with nil radical is classically known to be a lift/ring (*loc.cit.*). Cf. 3.54f.

Bass' Perfect Rings

A ring R is *left perfect* if every left R-module has a projective cover $P(M)$.

3.31 THEOREM (BASS [60]). *A left perfect ring R is characterized by the equivalent conditions:*

($P1$) *R has the dcc on principal right ideals.*

($P2$) *$R/\text{rad } R$ is semisimple and rad R is* **right vanishing** *(=*__left T-nilpotent__ *in [60]) in the sense that any infinite sequence $\{x_1x_2\dots x_n\}_{n=1}^{\infty}$ of products of elements of rad R terminates in zeros.*

($P3$) *Every flat left R-modules is projective.*

($P4$) *R is semilocal, and rad R is right vanishing.*

($P5$) *R is semilocal, and every nonzero left module has a maximal submodule (= R is* **left max***).*

($P6$) *Every left R-module has a projective cover.*

Regarding ($P3$), we have anticipated the definition of a flat module introduced in §4. By Govorov-Lazard Theorem 4.A:

($P3$) $\Leftrightarrow$ ($P3'$) Every direct limit of projective left modules is projective.

3.32 COROLLARY (BASS [60]). *When R is (semi)perfect, every ($f{\cdot}g$) projective right (left) R-module P is isomorphic to a direct sum of principal indecomposable R-modules each of which is isomorphic to e_iR (resp. Re_i) for some $i, i = 1, \dots, n$.*

REMARKS. (1) Mueller [70b] showed that any projective module P over a semiperfect ring has the structure stated in 3.32; (2) Brauer's theory of blocks can be generalized to perfect rings (from Artinian). See Faith [76], p.171, 22.34; (3) Also see Mares [63] and Kasch-Mares [66] on semiperfect (resp. perfect) modules, i.e., modules such that every factor module of M (resp. of a direct sum of copies of M) has a projective cover. Cf. Nicholson [75b]; (4) See Oberst and Schneider [71] for a characterization of when every $f \cdot p$ module has a projective cover ($R/\text{rad } R$ is VNR and idempotents can be lifted). Cf. Meyberg and Zimmermann-Huisgen [77]. Also see F-semiperfect rings, 6.52s.

Theorems of Björk and Jonah

Björk [69] showed that (P1) is equivalent to the dcc on $f \cdot g$ right ideals, and Jonah [70] showed this is equivalent to the condition that all left R-modules satisfy the acc on cyclic submodules! Moreover Björk proved that all right R-modules over a left perfect ring have the dcc on $f \cdot g$ submodules; more precisely, if an R-module M over any ring R satisfies the dcc on cyclic submodules, then M satisfies the dcc on $f \cdot g$ submodules.

Max Ring Theorems of Hamsher, Koifman, and Renault

A ring R is a *right max* ring if every right R-module M has a maximal submodule. Any right perfect ring is right max (Bass [60]) (and so is any right V-ring).

Hamsher [67], Koifman [70] and Renault [67] prove that a commutative ring R is max iff R has T-nilpotent Jacobson radical J and R/J is VNR. Moreover, Faith [95a] showed that R is max iff R_m is perfect for each maximal ideal m (equivalently, iff R is locally a max ring. Cf.3.19B). Any right max ring has left vanishing Jacobson radical J (see 3.31); and conversely if R/J is right max (see Hamsher et al, *op.cit.*).

REMARK. Hamsher *etal* thus answered a question of Bass for commutative rings: R is perfect iff R is a max ring with no infinite set of orthogonal idempotents. Koifman [70] and Cozzens [70] disproved Bass' question in general: V-domains not fields are counterexamples (Cf. **sup.** 3.21A).

The Socle Series of a Module and Loewy Modules

The **socle** of a right R-module, denoted $\mathrm{soc}(M)$, is either 0, or the sum of all simple (or minimal) submodules. (The terminology "socle" owes to the French, probably Dieudonné [42].)

The **socle series** of a module M is defined inductively. The **second socle** is the submodule $\mathrm{soc}_2(M) \supseteq \mathrm{soc}(M)$ such that

$$\mathrm{soc}(M/\mathrm{soc}(M)) = \mathrm{soc}_2(M)/\mathrm{soc}(M).$$

(Possibly $\mathrm{soc}_2(M) = \mathrm{soc}(M)$.) By transfinite induction one may define $\mathrm{soc}_\beta(M)$ for a limit ordinal β as the union $\bigcup_{\alpha<\beta} \mathrm{soc}_\alpha(M)$, and

$$\mathrm{soc}_{\alpha+1}(M)/\mathrm{soc}_\alpha(M) = \mathrm{soc}(M/\mathrm{soc}_\alpha(M))$$

for every ordinal α. The least ordinal α such that

$$\mathrm{soc}_{\alpha+1}(M) = \mathrm{soc}_\alpha(M)$$

is called the **socle length** of M. If M has socle length α, and if $M = \mathrm{soc}_\alpha(M)$, then M is said to be a **Loewy module of Loewy length** α.

REMARK. The concept of a Loewy module dates back to Loewy [05] and [17]. See Fuchs [69b] and Shores [74] for a bit of history. (I summarized a bit of this on p.176 of my *Algebra II.*)

Semi-Artinian Rings and Modules

A right R-module M is **semi-Artinian** provided that for every submodule $N \neq M$, $\mathrm{soc}(M/N) \neq 0$. A ring R is **right semi-Artinian** provided that R is a semi-Artinian right R-module.

3.33A THEOREM. *(1) A right R-module M is semi-Artinian iff M is a Loewy module; (2) A ring R is right semi-Artinian iff every right R-module is Loewy.*

3.33B COROLLARY. *A ring R is right semi-Artinian iff every right R-module M has essential socle.*

REMARK. Right semi-Artinian is called "right socular" in my *Algebra II*, p.156, Prop.22.10.

3.33C THEOREM (BASS [60]). *A ring R is left perfect iff R is semiperfect and right semi-Artinian.*

3.33D THEOREM (NASTASESCU-POPESCU [68]). *A ring R is right semi-Artinian iff R/J is right semi-Artinian and J is right vanishing, where $J =$ rad R.*

3.33E COROLLARY. *A commutative ring R is semi-Artinian iff J is vanishing and R/J is VNR and semi-Artinian.*

REMARK. If R is commutative semi-Artinian then R is a max ring by 3.33E and the Max Ring Theorems stated above. See Remark 3.33G below.

3.33F THEOREM (CAMILLO-FULLER [74]). *If R is right Loewy of Loewy length $n < \infty$ then R is left Loewy of Loewy length $\leq 2^n$.*

3.33G REMARKS. (1) von Neumann regularity (= VNR) is defined in §4: Semi-Artinian VNR rings have equal right and left Loewy length, *ibid.*

(2) Osofsky [74] constructs two-sided perfect rings whose left and right Loewy lengths are two arbitrary cardinals.

(3) Fuchs (1970) constructed commutative VNR Loewy rings of length $\beta + 1$, for any ordinal β. See Camillo-Fuller [74], Theorem 2.2.

3.33H THEOREM (CAMILLO-FULLER [79]). *(1) Any right Loewy ring with acc on (left or right) primitive ideals is a left max ring.*

(2) Any right Loewy PI-ring is a left max ring.

REMARK. There exist two-sided Loewy rings that are not max rings, *ibid.* p.78. By 3.33F, R is also right max in 3.33H.

The Perlis Radical and the Jacobson Radical

Call a one-sided ideal I of a ring R *quasi-regular* if $(1+x)^{-1}$ exists for all $x \in I$ (e.g. any nil ideal is quasi-regular). The maximal quasi-regular (q.r.) ideal $P(R)$ exists, contains all q.r. one-sided ideals and equals $J(R)$, the Jacobson radical of R (Jacobson [45a]). Originally $P(R)$ was called the Perlis-Jacobson radical, after Sam Perlis (my teacher and Ph.D. advisor at Purdue University, circa 1951–5) who showed that $P(R)$ is the maximal nilpotent ideal of a finite dimensional algebra R over a field k (Perlis [42]). Jacobson carried the theory through for arbitrary rings, and showed that $P(R) = J(R)$ is the intersection of all primitive ideals of R.

The Frattini Subgroup of a Group

The concept of the radical $\Phi(G)$ of a group G is a much older concept than that of a module. Indeed in groups, the intersection $\Phi(G)$ of the maximal subgroups of a group G was introduced by Frattini in 1885! (*Vide*). Of course, for Abelian $G, \Phi(G) = \text{rad}G$. Frattini proved that for a finite group $G, \Phi(G)$ is a nilpotent group. Moreover, a theorem of Wielandt states that G is nilpotent iff the derived group $[G,G] \subseteq \Phi(G)$ (see, for example, Huppert [67], pp.268–271, esp. Sätzes 3.6 and 3.11).

Krull's Intersection Theorem and Jacobson's Conjecture

3.34 THEOREM (GENERALIZED KRULL INTERSECTION THEOREM). *If R is a commutative Noetherian ring, and I is an ideal $\neq R$, then there exists $a \in I$ so that*

$$I^\omega = \bigcap_{n\in\omega} I^n = (1-a)I^\omega.$$

Furthermore, if R is a domain, or if I is contained in the Jacobson radical J of R, then $I^\omega = 0$.

For proofs see [Z-S], vol. I, pp.215–216. Or Kaplansky [70], Theorem 79.

Jacobson's Conjecture. If R is a 2-sided Noetherian ring, does the Krull Intersection Theorem still hold, i.e., is $J^\omega = 0$?

3.34A REMARKS. (1) Jategaonkar [74b] proved the conjecture for two-sided fully bounded Noetherian (= FBN) rings, and Cauchon [74] for left Noetherian right FBN rings (Cf. Krause [72]. Also see Jans' Theorem 7.49;

(2) Gordon [74] proved *inter alia* that any FBN ring embeds in an Artinian ring;

(3) The answer was shown to be negative for one-sided Noetherian R by Herstein [65], and Jategaonkar [68] for a right PID;

(4) It is known for any left or right Noetherian ring that J is transfinitely nilpotent in the sense that $J^\alpha = 0$ for some ordinal power α (Jacobson [45a], Theorem 11). (Cf. Jategaonkar [69], who shows that α may be of arbitrarily large cardinal. Also see the Remark and Note preceding 13.17.)

Nakayama's Lemma

A famous result first due to Azumaya [51] in a special case is:

3.35 NAKAYAMA'S LEMMA [51]. *If M is finitely generated R-module over any ring R then $MI = M$ implies $M = 0$ for any right ideal $I \neq 0$ such that $1-a$ is regular for some $a \in I$ (e.g. any ideal $I \neq 0$ contained in the Jacobson radical or any ideal $I \neq 0$ contained in an integral domain).*

Azumaya's and Nakayama's papers were reviewed by I. Kaplansky in Math. Reviews 12, 669g and 13,313f (reprinted in Reviews in Ring Theory, vol. 1, #8.01.001 and #2.01.007). He also reviewed Azumaya [48] and [50].

Nakayama's lemma is ubiquitous in module and ring theory, attaining a status something like Zorn's lemma. (Seriously.) It is at the basis of parts of many proofs in ring theory, e.g. it can be used in the proof of the Krull Intersection Theorem.

The Jacobson Radical and Jacobson-Hilbert Rings

Krull [50] pointed out the relationship between Hilbert's Nullstellensatz and the Jacobson radical. It hinges on the question: when is the Jacobson radical of a finitely generated algebra over a field a nil ideal? This holds true for commutative algebras (Krull [51], Goldman [51]), algebras over nondenumerable fields (Amitsur [56]), and algebras satisfying a polynomial identity (Amitsur [57]). The latter theorem is related to a non-commutative Hilbert Nullstellensatz, and many of the foregoing results on Jacobson and Hilbert rings are generalized by Amitsur and Procesi [66] and Procesi [67].

A commutative ring R is a *Jacobson-Hilbert* ring if R satisfies the equivalent conditions:
$(H1)$ Every factor ring R/I has nil Jacobson radical.
$(H2)$ Every prime ideal is an intersection of maximal ideals.
$(H3)$ Every maximal ideal of the polynomial ring $R[x]$ contracts to a maximal ideal of R.
$(H4)$ Every maximal ideal of $R[x]$ contains a monic polynomial.
$(H5)$ If the quotient field $Q(R/P)$ of a prime ideal is finitely generated over R/P, then P is maximal.
Note: Krull [51] called these rings Jacobson rings, and Goldman [51] called them Hilbert rings. (H4) is a combination of an exercise in Kaplansky [74], and an observation of Faith [89a], i.e., a maximal ideal M of $R[x]$ is monic iff M contracts to a maximal ideal of R.

3.36 THEOREM (GOLDMAN [51]-KRULL [51]). *A commutative ring R is Jacobson-Hilbert iff the polynomial ring $R[x]$ is Jacobson-Hilbert.*

Note: The power series ring $R\langle x\rangle$ over **any** Noetherian commutative ring is Hilbert (Brewer-Heinzer [80]), i.e., R **need not be Hilbert for** $R\langle x\rangle$ **to be.**

A ring R is **Jacobson-Hilbert** if for every prime ideal P, R/P is semiprimitive. This agrees with the above definition for commutative R.

3.36B GENERALIZED OR WEAK NULLSTELLENSATZ (GOLDMAN [51], KRULL [51]). *If R is a Jacobson-Hilbert ring, then any $f \cdot g$ R-algebra A is also, and the contraction $P = M \cap R$ of a maximal ideal M of A is a maximal ideal of R. Moreover, A/M is a finite dimensional field extension of R/P.*

PROOF. See Eisenbud [96], p.132, Theorem 4.19.

3.36C REMARK. See 11.13 and the following 3.36D for another "weak Nullstellensatz."

3.36D THEOREM (AMITSUR-SMALL [78]). *If R is a simple Artinian ring, any simple module V over the polynomial ring $R[x_1, \dots, x_n]$ in n variables has finite length over R.*

PROOF. See Goodearl and Warfield [89], p.270, Theorem 5.6.

A ring R is **right fully bounded** (= **right FB**) for every prime ideal P, every nonzero right ideal of R/P contains an nonzero ideal of R/P. A ring R is **right FBN** if R is right FB and right Noetherian. An **FBN ring** is one which is right and left FBN.

REMARK. A right primitive ring R is right FB iff R is a field.

3.36E. THEOREM (RESCO-STAFFORD-WARFIELD). *If R is a fully bounded Noetherian Jacobson ring, then any simple right module M over the polynomial ring $R[x_1, \dots, x_n]$ in n variables is annihilated by a maximal ideal of R, and M is semisimple of finite length as an R-module.*

PROOF. See Goodearl-Warfield [89], p.276, Corollary 15.11. □

3.36F THEOREM (VÁMOS [77]). *Let K and L be field extensions of a field F and suppose*

$$tr.d.\ K/F \geq tr.d. L/F = n < \infty.$$

Then $R = K \otimes_F L$ is a Hilbert ring in which every maximal ideal has rank n.

3.36G REMARK. The proof of this depends on Vámos' Proposition 4, *ibid.*, p.276, which implies that $K[X_1, \dots, X_n]S^{-1}$ for $S = F[X_1, \dots, X_n]\backslash\{0\}$ is a Hilbert ring in which every maximal ideal has rank n.

When Nil Implies Nilpotency

If I is a nil ideal, when is I **nilpotent** in the sense that $I^n = 0$ for some $n \geq 1$, that is, all products $x_1 \dots x_n$ of n elements of I are $= 0$? The **index of nilpotency** is the least exponent n such that $I^n = 0$.

3.37 THEOREM (KÖTHE [30]-LEVITZKI [31]). *If k is a field, then any multiplicative closed nil submonoid of $k_n \approx End_k k^n$ is nilpotent of index $\leq n$.*

This follows from the next theorem.

3.38 THEOREM (LEVITZKI [31], FITTING [33]). *If M is an R-module of length n, then any multiplicative nil submonoid of End M_R is nilpotent of index $\leq n$.*

PROOF. See 3.69 below. □

Shock's Theorem

Recall for any subset A of a semigroup R with 0, the left annihilator ${}^\perp A = \{r \in R | ra = 0 \ \ \forall a \in A\}$. Similarly for $A^\perp$. And, as is the case for rings, $\perp$acc (resp. acc$\perp$) denotes the acc on left (right) annihilators.

3.39 SHOCK'S THEOREM [71B]. *Let R be a semigroup with 0 and acc$\perp$. If R has a non-nilpotent nil submonoid, N, then (1) there is a set $\{a_i\}_{i\in\omega}$ of elements of R so that*

$${}^\perp A_1 \subset \cdots \subset {}^\perp A_n \subset \cdots$$

where $A_n = \{a_i\}_{i\geq n}$; (2) If R is a ring, then $\{a_i\}_{i\in\omega}$ can be chosen so that $\{a_iR\}_{i\in\omega}$ are independent right ideals, i.e., the sum $\oplus_{i\in\omega} a_iR$ is direct.

Shock's theorem implies the next corollary and the following theorems of Levitzki, Herstein-Small and Lanski.

3.40 COROLLARY. *If R is a semigroup with 0, and R satisfies the acc on right and left annihilators (= acc$\perp$ and dcc$\perp$), then every nil multiplicative submonoid of R is nilpotent.*

3.41 THEOREM (LEVITZKI [63], HERSTEIN AND SMALL [64]). *If R is a ring with acc$\perp$ and dcc$\perp$, then any nil subring is nilpotent.*

3.42 THEOREM (HERSTEIN-SMALL [64] AND LANSKI [69]). *If R is a right Goldie ring, then every multiplicative nil submonoid S of R is nilpotent.*

3.43 REMARKS. 1) Herstein and Small proved this for nil subrings S (*loc.cit.*), and Levitzki proved it in [45] for right and left Noetherian rings in answer to a problem of Köthe. Finally Goldie [60] proved that nil subrings are nilpotent in one-sided Noetherian rings. (See Goodearl and Warfield [89], p.98 for additional references); 2) In their Addendum to [64], Herstein and Small [66] point out that an example of Sasiada shows that nil does not imply nilpotency in one-sided acc$\perp$ rings; 3) Lenagan [73] showed that nil implies nilpotency for subrings of rings with Krull dimension (see 14.29D). Moreover, Gordon, Lenagan and Robson [73] announced that the prime radical of a ring R with Krull dimensions is nilpotent, a result (14.30A) that appears in R. Gordon and Robson [73]. (Cf. Goldie and Small [72].)

Kurosch's Problem

Kurosch's Problem (=KP): If A is an algebraic algebra over a field k, is every finitely generated subalgebra finitely dimensional? The answer is negative (see below) but theorems of Kaplansky [46] and Levitzki [46] answered (KP) positively for algebraic algebras of bounded degree (nil algebras of bounded index of nilpotency): thus every finite subset generates a finite dimensional (resp. nilpotent) subalgebra. In particular, if a finitely generated ring satisfies the identity $x^n = 0$, for a fixed integer $n \geq 1$, then the ring is nilpotent.

The Nagata-Higman Theorem

A related theorem of Nagata-Higman states that in any associative algebra A over a field k of characteristic p that any subalgebra B satisfying the identity $x^n = 0$ for an integer $n < p$ when $p \neq 0$ is nilpotent of index $\leq 2^n - 1$. A short proof of this by P. J. Higgins appeared in Jacobson [64, p.274]. (The theorem of Nagata [52] is for nilpotency of B over characteristic 0.)

Note if N is the commutative algebra over a field k of characteristic p with an infinite basis $\{x_i\}_{i=1}^{\infty}$ such that $x_i^p = 0 \ \forall i$, then N is nil but not nilpotent.

$\aleph_0$-Categorical Nil Rings Are Nilpotent

A ring R is **$\aleph_0$-categorical** if there exists a unique countable model for R.

THEOREM (CHERLIN [80]). *Any $\aleph_0$-categorical associative nilring N is nilpotent.*

REMARK. This answered a question of Baldwin and Rose who proved:

THEOREM (BALDWIN AND ROSE [77]). *Any $\aleph_0$-categorical ring R has nil Jacobson radical.*

COROLLARY. *Any $\aleph_0$-categorical ring R has nilpotent Jacobson radical.*

The Golod-Shafarevitch Theorem

Jacobson [64], p.260 also cites the example of Golod of a finitely generated nonnilpotent nil algebra B over a field F. Thus, B is not finite dimensional (the main point of the example). Any such example has to be non-commutative: that is, nil $\Longrightarrow$ nilpotency for finitely generated commutative nil algebras. A theorem of Golod-Shafarevitch enables one to find over any countable field an infinite dimensional nil algebra generated by three elements, thus answering (KP) in the negative. For an exposition, See Herstein [68], p.192.

Some Amitsur Theorems on the Jacobson Radical

We now cite some fundamental theorems of Amitsur on the Jacobson radical.

3.43A THEOREM (AMITSUR [56]). *If R is an algebra over a field k of cardinal $|k| > dim_k R$, then the Jacobson radical rad R is a nil ideal.*

3.43B THEOREM (AMITSUR-SMALL [96]). *Under the same conditions as 3.43A, R is an algebraic algebra, iff every right or left regular element (i.e., non zero-divisor) is a unit.*

3.44 COROLLARY. *If R is a $f \cdot g$ (**qua ring**) algebra over an uncountable field k, then rad R is nil.*

3.45 COROLLARY. *If N is a nil algebra over an uncountable field k, then the $n \times n$ matrix algebra N_n is a nilalgebra.*

Cf. Jacobson [65], p.20ff. Also see 3.50 below.

3.46 THEOREM (AMITSUR [59]). *If R is an algebraic algebra over an uncountable field k, then the $n \times n$ matrix algebra R_n is algebraic.*

We remark on another theorem of Amitsur; a consequence of 3.43, namely:

3.47 THEOREM (AMITSUR [56]). *If k is a field of characteristic 0 not algebraic over $\mathbb{Q}$, e.g., if k is uncountable, then the group algebra kG of any group G is semiprimitive.*

See Jacobson [64], p.254, Theorem 3 and Corollary.

3.48 THEOREM (AMITSUR). *If R is a ring, then rad $R[x] = N[x]$ for the polynomial ring $R[x]$ and a nil ideal N of R.*

Cf. 3.50 below.

3.49A COROLLARY. *If R has no nil ideals $\neq 0$, then $R[x]$ is semiprimitive.*

3.49B COROLLARY (SNAPPER [51]). *If R is commutative, then rad $R[x] = N[x]$ where N is the maximal nil ideal of R.*

3.49C REMARK. By 3.49A, $R[x]$ is semiprimitive for any commutative domain, i.e., R need not be semiprimitive for $R[x]$ to be, e.g. consider any local domain R not a field.

Köthe's Radical and Conjecture

Let $K_\ell(R)$ denote the sum of all nil left ideals of a ring R. If $x \in K_\ell(R)$ and $r \in R$, then ra is nilpotent, say $(ra)^n = 0$, hence $(ar)^{n+1} = 0$, that is, ar is nilpotent. This shows the well-known fact that $K_\ell(R)$ is an ideal $= K_r(R)$. If $K_\ell(R)$ is nil, then it is called the **Köthe nil radical**, denoted $K(R)$.

Let $N(R)$ denote the sum of all nil ideals of R, and consider the property:

(K1) The sum of two nil left ideals of a ring R is nil.

The truth of $(K1)$ for all rings R is called **Köthe's conjecture** and is equivalent to:

(K2) $K(R) = N(R)$ for all rings R.

(K3) $K(R) = 0$ iff $N(R) = 0$ for all rings R.

(K4) $N(R_n) = N(R)_n$, for all $n \times n$ matrix rings R_n over R.

Then $(K1) \Leftrightarrow (K4)$ are clearly equivalent. Cf. Rowen [89], 2.6.35.

3.50 AMITSUR'S THEOREM[56,59,71]. *If R is an algebra over an uncountable field k, then $(K1)$ – $(K4)$ hold, and moreover, the Jacobson radical $J(R[X]) = N(R)[X]$, for the polynomial ring $R[X]$.*

PROOF. See 3.43, 3.45, and 3.48.

Cf. Rowen *loc. cit.*

3.50A REMARK. Kegel [63] proved that the sum of two nilpotent rings is nilpotent, and Kegel [64] asked the same question for nil or locally nil subrings. The latter question (for locally nil) was given a counter-example by Kelarev [93].

Cf. Kelarev [97] where this question and related ones are discussed. Also see:

3.50B KLEIN'S THEOREM [94]. *The sum of finitely many nil ideals of bounded index is nil of bounded index.*

A General Wedderburn Theorem

A right R-module M is **balanced** if the canonical homomorphism of R into the biendomorphism (bicentralizer, bicommutator) ring of M is surjective. The classical Wedderburn-Artin Theorem states that every minimal right ideal M of a simple right Artinian ring R is balanced, in which case R has a representation as a full ring of linear transformations on a finite dimensional vector space M over the (generally noncommuntative) field $D = \text{End}M_R$. A theorem of Morita [58] applies to any ring R: *any generator M of the category* mod -R *is balanced, and M is finitely generated projective as a canonical left module over $B = EndM_R$.* (See the author's paper [66b] for an elementary proof employing just linear and matrix algebra. Also see "Snapshots" ("Rieffel, Lang, Smale and Me").) Now, if R is any simple ring, then it is a triviality to verify that any nonzero right ideal M is a generator of mod -R, so $R \to \text{End}_B M$ is surjective. Moreover, simplicity of R implies the kernel is zero, $R \approx \text{End}_B M$, and so Morita's theorem implies the General Wedderburn Theorem of Rieffel [65]. Then, finite projectivity of M over B yields an idempotent e in the full $n \times n$ matrix ring B_n, for an appropriate integer n, a ring isomorphism $R \approx eB_n e$, and $B_n e B_n = B_n$ so Morita's theorem implies that of Hart [67]. In the case M is a minimal right ideal of R, then B is a sfield by Schur's lemma, so then M is a finite dimensional (by the finite generation of M over B) vector space over B. In this case, M is free over B, say $M \approx B^n$, and then R is isomorphic to the $n \times n$ matrix ring B_n so Morita's theorem implies the

Principal Wedderburn [08]-Artin [27] Theorem. Moreover, if M is a uniform right ideal, then B is a right Ore domain with a right quotient sfield D (see 6.26f), and R has a full right quotient ring $Q(R) \approx D_m$, for some integer m (Theorem of Goldie [58] and Lesieur-Croisot [59], see 3.13).

REMARK. For more on balanced modules, see 13.29–30 (Theorems of Camillo and Fuller).

Koh's Schur Lemma

In this connection, a theorem of Koh [66] is of interest: If a ring R has a maximal annihilator right ideal I, say $I = x^{\perp}$, then $B = \mathrm{End}(R/I)_R$ is an integral domain $\approx \mathrm{End}(xR)_R$. If in addition, R is a simple ring, then Morita's theorem on balanced modules applies, namely, $U = xR$ is $f \cdot g$ projective over an integral domain B, and $R \approx \mathrm{End}_B U$ canonically. (See, e.g. Faith [72], p.350, 724-25.)

Categories

We shall not give here the somewhat tedious definitions of a category C, the morphisms of C, the functors $T : C \rightsquigarrow D$ between categories, the natural equivalence of two functors

$$T : C \rightsquigarrow D \quad \text{and} \quad S : C \rightsquigarrow D,$$

the equivalence $C \approx D$ of categories, all of which are due to Eilenberg and MacLane [45], and are to be found in many places in the literature (e.g. my *Algebra* [72], p. 72ff (vol. I), Freyd [64], Mitchell [65], among others).

Morita's Theorem

The categories mod-R and mod-S over two rings R and S consist of the right modules over these rings, and the morphisms of mod-R are the homomorphisms $f : M \to N$ between them.

3.51 MORITA'S THEOREM [58]. *The categories mod-R and mod-S over two rings R and S are equivalent iff there exists a $f \cdot g$ projective generator P of mod-R and a ring isomorphism $S \approx \mathrm{End}P_R$. In this case we write mod-$R \approx$ mod-S.*

In this case, $\mathrm{Hom}_R(P, -)$ induces the category equivalence mod-$R \rightsquigarrow$ mod-S, and $\mathrm{Hom}(P^\star, -)$ is the inverse equivalence, where $P^\star = \mathrm{Hom}_R(P, R)$. Cf. my Algebra [72,81], p.450ff. Also see Arhangel'skii, Goodearl, and Huisgen-Zimmermann [97].

The condition mod-$R \approx$ mod-S is reflexive, symmetric and transitive; R and S are then said to be **Morita Equivalent or Similar Notation:** $R \sim S$. If R is a sfield, then $S \sim R$ iff $S \approx R_n$, the full $n \times n$ matrix ring over R. (Thus two sfields R and S are similar iff they are isomorphic **qua** rings. The same is true of two commutative rings, or two integral domains.)

Theorems of Camillo and Stephenson

The next theorem answered a question of W. Stephenson.

3.51′ Camillo's Theorem [84]. *Two rings R and S are Morita equivalent iff there exists a ring isomorphism $End_R R^{(\omega)} \approx End_S S^{(\omega)}$ between their respective rings of column finite matrices.*

Stephenson [69] generalized the fundamental theorem of projective geometry by showing that a lattice isomorphism $L(F) \approx L(G)$ between the lattices of subspaces of free modules F and G of ranks ≥ 3 over rings A and B induces a Morita equivalence $A \sim B$. See end-of-chapter 17 notes (Excerpts of Letters of Victor Camillo).

The Basic Ring and Module of a Semiperfect Ring

The following folkloric results are fundamental in the study of semiperfect rings.

3.52 Definition and Proposition. *Let R be a semiperfect ring with Jacobson radical J and let $\{e_i\}_{i=1}^n$ be orthogonal indecomposable idempotents such that $1 = e_1 + \cdots + e_n$. Then R is said to be a* **basic ring** *iff the equivalent conditions hold:*

(1) $e_i R \approx e_j R \Leftrightarrow i = j, \quad \forall i,j = 1, \ldots, n$
(2) $e_i R/e_i J \approx e_j R/e_j J \Leftrightarrow i = j, \quad \forall i,j = 1, \ldots, n.$
(3) *R/J is a finite product of sfields.*
(4) *$(3+i)$ Right left symmetry of $(i), i = 1,2$.*

Note, by (SP3) of 3.30, $e_i R e_i$ is a local ring $\forall i$.

3.53 Definition and Theorem. *Let R be a semiperfect ring with radical J, as in the first statement of the last Definition, and renumber $\{e_i\}_{i=1}^n$ so that $\{e_i\}_{i=1}^t$, for maximal $t \leq n$ is such that the equivalent conditions (1) and (2) above hold. Then $e_0 = e_1 + \cdots + e_t$ is called a* **basic idempotent** *and $e_0 R$ is called a (right)* **basic module** *of R. Furthermore: (I) $R_0 = e_0 R e_0$ is a basic ring, called the* **basic ring of** *R; (II) R_0 is Morita equivalent to R; (III) If e_0' is another basic idempotent of R, there exists a unit $x \in R$ so that $e_0' = x e_0 x^{-1}$, and*

$$e_0' R e_0' = x(e_0 R e_0) x^{-1},$$

that is, any two basic rings of R are isomorphic by a map extendible to an inner automorphism of R; (IV) The basic module $e_0 R$ is a minimal generator of mod-R, that is, $e_0 R$ is isomorphic to a direct summand of every generator of mod-R. Any two basic modules are isomorphic; (V) Two semiperfect rings are Morita equivalent iff they have isomorphic basic rings, hence two basic rings are Morita equivalent iff they are isomorphic.

For proofs, see my Algebra II, [76], p.44,18.24, or Morita [58]. These results were known and used in finite dimensional algebras and Artinian rings by myriad researchers, e.g. by J.H.M. Wedderburn, M. Hall, Jr., R. Brauer and C. Nesbitt, T. Nakayama and others; in particular by Brauer in his theory of blocks for group algebras and Artinian rings, generalized by the author to perfect rings in *op.cit.*,p.171f., 22.34ff.

We also note:

3.54 THEOREM. *A ring R is semiperfect iff R is a semilocal ring in which idempotents modulo the radical J lift.*

See *op.cit.*,p.45, where the appellation **lift/rad ring** is applied for rings with the stated property on idempotents, and called SBI rings elsewhere (see 8.4Aff).

The Regularity Condition and Small's Theorem

If I is an ideal of a ring R then $\mathcal{C}(I)$ denotes the set of all $x \in R$ such that $\overline{x} = x + I$ is a regular element of $\overline{R} = R/I$, that is, $\overline{x}^{\perp} = {}^{\perp}\overline{x} = 0$. Thus $\mathcal{C}(0)$ is the set of regular elements of R. R is said to satisfy the **regularity condition** if $\mathcal{C}(0) = \mathcal{C}(N)$, where N =prime rad R.

3.55A THEOREM (SMALL [66], P.23, 2.13). *A commutative Noetherian ring R satisfies the regularity condition iff the associated primes of R are the minimal primes of R. In this case $Q(R)$ is Artinian.*

Cf. Theorem 16.31 on commutative acc$\perp$ rings and the Remark.

3.55B THEOREM (SMALL [68A,B], TALINTYRE [63]). *A right Noetherian ring R with prime radical N has right Artinian classical right quotient ring $Q = Q^r_{c\ell}(R)$ iff R satisfies the regularity condition. In this case $Q = Q^{\ell}_{c\ell}(R)$.*

3.55C THEOREM (SMALL [66A]). *A right Noetherian right hereditary ring R has right Artinian $Q^r_{c\ell}(R)$.*

Reduced Rank

If R is a semiprime Goldie ring with right quotient ring $Q = Q^r_{c\ell}(R)$, and M is a right R-module, the **reduced rank** of M is defined as the Jordan-Hölder length $\rho(M)$ of $M \otimes_R Q$ as a right Q-module. The concept of reduced rank was originated by Goldie [64].

3.56 REMARKS. (1) Any $f \cdot g$ right R-module M over a semiprime Goldie ring R has finite reduced rank $\rho(M)$; (2) Reduced rank plays an important part in Warfield's characterization [79b] of rings R for which $Q^r_{c\ell}(R)$ exists and is Artinian (see 3.57); (3) The **reduced rank** of M over a right Noetherian ring R with prime radical N is defined to be

$$\rho_R(M) = \sum_{i=0}^{n-1} \rho_{R/N}(M_i/M_{i+1})$$

where $\{M_i\}_{i=0}^{n}$ is a descending chain of submodules such that $(M_i/M_{i+1})N = 0$, $i = 0, \ldots, n-1$; (4) Any $f \cdot g$ right R-module over a Noetherian ring R has finite reduced rank; (5) The concept of reduced rank of a module extends to a ring R such that R/N is right Goldie and N nilpotent.

Using the above concepts and results one can prove:

3.57 THEOREM (SMALL [66], WARFIELD [79B]). *A ring R with prime radical N has right Artinian $Q = Q^r_{c\ell}(R)$ iff R/N is right Goldie, N is nilpotent, R has finite right reduced rank, and satisfies the regularity condition.*

See Goodearl and Warfield [89], p. 175 for a discussion of this and the previous results, including Talintyre's contributions in 3.55, the results of Warfield [79b], and

Stafford [82]. Robson [67] characterized rings with Artinian Q somewhat differently. Cf. Theorem 7.6B and the remark following.

Finitely Embedded Rings and Modules
Theorems of Vámos and Beachy

A module M is **finite embedded** (= f.e.) if soc(M) is $f \cdot g$ and essential.

EXAMPLE. Any Artinian module M is f.e. Cf. 3.61.

See Vámos [68], Beachy [71] and my Algebra II, pp.67–69, 19.13A–19.16B, for the background and proofs of the following.

3.58 THEOREM. *A right R-module M is f.e. iff the submodules of M have the* **finite intersection** *property (= fip); namely, if* $S = \{S_i\}_{i \in I}$ *is a set of submodules of M, then the intersection is zero iff a finite subset of S has zero intersection.*

3.59 COROLLARY. *If a f.e. right R-module M embeds in a product* $\prod_{i \in I} N_i$ *of modules* $\{N_i\}_{i \in I}$*, then M embeds in a finite product* $\prod_{a \in A} N_a$ *where A is finite subset of I.*

3.60 THEOREM (BEACHY [71]-KAMIL [76]). *A ring R is right finitely embedded iff R embeds in a finite product of copies of any faithful right R-module N.*

Cf. my Algebra II, 19.13A.

3.61 THEOREM (BEACHY [71], VÁMOS [68]). *A right R-module is Artinian iff every factor module is finitely embedded.*

Cf. my Algebra II, 19.16B. Also see Shock's Theorem 7.28.

REMARKS. (1) Vámos introduced f.e. modules as the dual of $f \cdot g$ modules; hence an f.e. module is also called **cofinitely generated**; (2) An equivalent condition to that of 3.61: M is quotient finite dimensional and semiArtinian (= Loewy). Cf. 7.49-50.

3.61B THEOREM (BEACHY [71]). *A ring R is right Artinian iff every factor ring is right finitely embedded.*

3.61C COROLLARY. *The following are equivalent conditions on a simple ring R:*

(1) *R is right f.e.*
(2) *R is left f.e.*
(3) *R is right Artinian*
(4) *R is left Artinian.*

PROOF. (Trivial) If S is the right socle of R, then $RS = R$, hence R is a sum, hence a direct sum, of simple R-modules. This implies that R is semisimple. See Theorem 2.1. □

The Endomorphism Ring of Noetherian and Artinian Modules

If A is an endomorphism of M, we investigate the effect on A of various chain conditions on M.

3.62 REMARKS. Let A be an endomorphism of a module M.

(1) If k is a natural number such that $\ker A^k = \ker A^{k+1}$, then $A^kM \cap \ker A^k = 0$, A induces a monomorphism in A^kM, and A induces a nilpotent endomorphism in $\ker A^k$.

(2) If t is a natural number such that $A^tM = A^{t+1}M$, then A induces an epimorphism in A^tM, A induces a nilpotent endomorphism in $\ker A^t$, and $M = A^tM + \ker A^t$.

3.63 REMARKS. If M is any R-module, then an element $f \in S = \text{End } M_R$ has a (two-sided) inverse in S if and only if f is an automorphism of M, that is, if and only if f is an isomorphism of M.

Let M be an R-module, let $S = \text{End } M_R$, and let $A \in S$. Then:

(1) If M is Artinian, A is an automorphism if and only if A is a monomorphism.

(2) If M is Noetherian, A is an automorphism if and only if A is an epimorphism.

(3) **Uniqueness of the basis number**. Let R be a right Noetherian ring, and let M be a free module with a finite basis $x_1, \ldots, x_n$.

(a) If $y_1, \ldots, y_m$ is any basis of M, then $m = n$, and there exists an automorphism f of M such that $f(x_i) = y_i$, $i = 1, \ldots, n$;

(b) If $y_1, \ldots, y_m$ generates M, then $m \geq n$. Furthermore, $y_1, \ldots, y_m$ is a basis if and only if $m = n$.

(4) (Vasconcelos) Any epimorphism $f : M \to M$ of a $f \cdot g$ module M over a commutative ring R is an automorphism. Conclude that (3) also holds for commutative R.

Fitting's Lemma

Applying 3.63 to the situation of 3.62 (1) and (2), if M is Artinian, then so is the submodule A^kM of (1), and A induces an automorphism a of A^kM. If M is Noetherian, then so is the submodule A^tM of (2), and A induces an automorphism b of A^tM. Now let M be both Artinian and Noetherian, and assume A is not an automorphism. Then there exist integers k and t such that

$$M \supset AM \supset \cdots \supset A^tM = A^{t+1}M = \ldots$$
$$0 \subset \ker A \subset \cdots \subset \ker A^k = \ker A^{k+1} = \cdots .$$

Since A induces an automorphism of A^kM, necessarily $A^{k+1}M = A^kM$, and therefore $k \geq t$. On the other hand, A induces an automorphism b of A^tM. Hence, if $x \in \ker A^{t+1}$, then $0 = A^{t+1}x = bA^tx$, and $A^tx = 0$, so $x \in \ker A^t$. Then $M = A^sM \oplus \ker A^s$ by 3.62 (1) and (2). This proves the next result.

3.64 THEOREM (FITTING'S LEMMA). *If M is a module of finite length s, then any endomorphism A induces an automorphism in the submodule A^sM, induces a nilpotent endomorphism in $\ker A^s$, and $M = A^sM \oplus \ker A^s$.*

3.65 COROLLARY. *Let M be an indecomposable R-module of finite length s, and let $S = \text{End } M_R$.*

(1) *Every element of S is either a unit, or is nilpotent.*

(2) *The set of N of nilpotent endomorphisms of M is a nilpotent ideal of S of index $\leq s$* (Cf. 3.68).

PROOF. (1) If $A \in S$, then by Fitting's lemma, $M = A^s M \oplus \ker A^s$. Since M is indecomposable, either $M = a^s M$, or else $M = \ker A^s$. By Fitting's lemma, A is an automorphism in the former case, and A is nilpotent in the latter case.

(2) Now let $A, B \in S$ be such that AB is an automorphism, that is, such that AB has an inverse $C \in S$. Then $A(BC) = (CA)B = 1$, and consequently neither A nor B is nilpotent; that is, both A and B are automorphisms. Thus, if $B \in N$, then $AB \in N$ and $BA \in N$. Hence, in order to show that N is an ideal of S it remains only to show that N is an additive subgroup of S. Let $A, B \in N$, and suppose for the moment that $A - B = C \notin N$. Then C is an automorphism, so $A_1 - B_1 = 1$, where $A_1 = AC^{-1}, B_1 = BC^{-1}$. We already know that $A_1, B_1 \in N$. Thus, $A_1^s = 0$, and, by the binomial theorem,

$$0 = A_1^s = 1 + sB_1 + \cdots + B_1^s = 0.$$

Since $B_1 \neq 0$, there exists a natural number k such that $B_1^{k-1} \neq 0$, and $B_1^k = 0$. Then

$$0 = 0 \cdot B_1^{k-1} = B_1^{k-1} + sB^k + \cdots + B_1^{s+k} = B_1^{k-1}$$

a contradiction, which proves that $A - B = c \in N$ for all $A, B \in N$. Therefore N is an ideal of S, and the following Proposition 3.68 proves that N is nilpotent of index $\leq s$. □

3.66 COROLLARY (FITTING[33, SATZ 3 AND 8]). *The endomorphism ring of an indecomposable module of finite length s is a local ring with nilpotent radical of index $\leq s$.*

Köthe-Levitzki Theorem

Let K be a ring, and let S be a multiplicative submonoid of the $n \times n$ matrix ring K_n. Then, S is said to be in **(strict) upper triangular form**, provided that for some set $\{e_{ij}\}_{i,j=1}^n$ of matrix units of K_n, and unit some $x \in K_n$ every element $s \in x^{-1}Sx$ has the canonical form

$$s = (s_{ij}) = \sum_{i,j=1}^n s_{ij}e_{ij}$$

with $s_{ij} = 0$ whenever $i \geq j$ (resp. whenever $i > j$). then, the elements of S are said to be placed **simultaneously** into (strict) upper triangular form.

3.67 REMARKS. (1) If K is a field, and if $T_n(M)$ is the set of (strict) upper triangular matrices of K_n relative to some set M of $n \times n$ matrix units, then a submonoid S can be placed into (strict) upper triangular form if and only if there is a unit $x \in K_n$, and a set M such that $xSx^{-1} \subseteq T_n(M)$.

(2) S can be put into (strict) upper if and only if S can be placed in (strict) lower triangular form.

3.68 PROPOSITION (KÖTHE [30B], LEVITZKI [31]). *If K is a sfield, then any multiplicative nil submonoid S of $K_n \approx End_K K^n$ can be placed simultaneously into strict triangular form, and $S^n = 0$.*

PROOF. For $n = 1$, there is nothing to prove. Assume the proposition for vector spaces K^m of dimension $m < n$. Let $V = K^n$. then, there is a vector subspace U of V of dimension $< n$ such that $US \subseteq U$. Otherwise, $VS = V$, and then there exist elements $s_1, \dots s_k \in S$ such that

$$V = \sum_{i=1}^{k} V s_i.$$

Then, for any integer $j > 0$,

$$V = \sum_{1 \le i_1, \dots, i_j \le k} V s_{i_1} \dots s_{i_j}.$$

This implies the existence of a sequence $\{s_{i_t}\}_{t>0}$ such that $s_{i_t} \in \{s_1, \dots, s_k\}$, and

$$s_{i_1} \dots s_{i_t} \neq 0, \qquad \forall t \in \mathbb{Z}^+.$$

Now one of the elements $s_1, \dots, s_t$, say $a = s_1$, occurs infinitely often in the sequence $\{s_{i_t}\}_{t \in \mathbb{Z}^+}$. Then, there is a sequence $\{s'_{i_t}\}_{t \in \mathbb{Z}^+}$ of elements of S such that

$$s'_1 a s'_2 a \dots s'_t a \neq 0, \qquad \forall t > 0.$$

Put $W = Va$. Since a is nilpotent, dim $W <$ dim V. Furthermore,

$$WSa \subseteq Va = W$$

so that Sa induces a nil submonoid of $\text{End}_K W$. By the induction hypothesis,

$$s'_1 a s'_2 a \dots s'_{n-1} a s'_n a = 0$$

which is a contradiction.

Since, therefore, there exists $t > 0$ such that

$$V \subset VS \supset VS^2 \supset \dots \supset VS^{t-1} \supset VS^t = 0$$

is strictly decreasing, then there is a basis $x_1, \dots, x_n$ of $V = K^n$, such that $x_1, \dots, x_{n_i}$ is a basis for $VS^{t-i}, i = 1, \dots, t$, and the matrix representations of the elements of S relative to this basis all have the strict lower triangular form.

□

Levitzki-Fitting Theorem

3.69 THEOREM (LEVITZKI [31], FITTING [33]). *If M is an R-module of length n, then any nil submonoid of $A = EndM_R$ is nilpotent of index $\le n$.*

PROOF. We prove it first under the assumption that M is semisimple. If M is homogeneous, that is, if M is a direct sum of isomorphic simple modules, then $A \approx K_n$, where $K =$ End V_R is the endomorphism sfield of a simple submodule V. In this case, the theorem follows from 3.68. Otherwise, $M = H \oplus G$ is a direct sum of two fully invariant submodules of lengths h and $n - h$ respectively. If S is a nil submonoid of A, then S induces a nil submonoid $\overline{S}$ of End H_R, and a nil submonoid S' of End G_R. Then $\overline{S}^h = 0, S'^{n-h} = 0$, and $S^n = 0$.

In the general case, we may assume that the socle H of M has length $m < n$. Since H is fully invariant, an element $s \in S$ induces $\overline{s} \in \text{End}(M/H)_R$, and

$s' \in \text{End } H_R$, both of which elements are nilpotent. By the induction hypothesis, $\overline{S} = \{\overline{s} \mid s \in S\}$ and $S' = \{s' \mid s \in S\}$ are nilpotent of indices $n - m$, and m respectively. Thus, given a seqeuence $\{s_i\}_{i=1}^n$ of elements of S, then

$$s_n s_{n-1} \dots s_{n-m+1} \dots s_2 s_1 M \subseteq s_n s_{n-1} \dots S_{n-m+1} H = 0.$$

Thus $S^n = 0$, completing the induction and proof. □

Kolchin's Theorem

A matrix A is **unipotent** if $A = 1 + B$, where B is nilpotent. (An equivalent formulation: the characteristic roots of A are all $= 1$.) Kolchin's theorem states that any multiplicative semigroup S of unipotent matrices over a commutative field k can be placed simultaneously in triangular form. It is tempting to derive Kolchin's theorem from 3.69; however, the B's do not form a multiplicative submonoid in general. (They do, obviously, when S consists of commuting matrices, but this is an unnecessary restriction. Moreover, there is a theorem which permits the diagonalization of commuting matrices over an algebraically closed field. See Jacobson [53, p.134].)

3.70 Burnside's Theorem. *Any irreducible semigroup S of linear transformations on a vector space V of dimension n over an algebraically closed field k contains n^2 linearly independent transformations.*

Proof. Let R be a subalgebra of $L = \text{End}_k$ spanned by S. Then, V is a simple R-module, so (2.4 and 2.4f) applies, so that $R = L$, so S has $n^2 = \dim_k L$ linearly independent transformations. □

3.71 Corollary. *If g is the number of distinct traces of elements of S in 3.70, then S has at most t^{n^2} elements.*

Proof. On the algebra of all linear transformations on V, introduce the inner product $(A, B) = Tr(AB)$. (This is easily seen to be nonsingular.) Let $c_1, \dots, , c_t$ be the distinct traces that occur. Let $A_j (i = 1, \dots, n^2)$ be n^2 linearly independent elements in S (Theorem 3.70). Each X in S satisfies equations $Tr(A_i X) = b_i$ where $b_1, \dots, b_{n^2}$ are chosen from the c's.

These equations determine X uniquely, so there are at most t^{n^2} choices for X. □

3.72 Theorem (Kolchin [48]). *Let S be a multiplicative semigroup of unipotent matrices. Then the elements of S can be put simultaneously into triangular form.*

Proof. Let n be the size of the matrices. We argue by induction on n. The case $n = 1$ is trivial.

Case I. The scalar field is algebraicaly closed. If S is irreducible, then by 3.71, S has only one element, and one matrix is always reducible (here $n > 1$). Hence S is reducible. Then by choosing a basis for the invariant subspace of S and extending it to a complete basis, all the elements of S will have matrices of the block form:

$$\begin{pmatrix} B & C \\ O & D \end{pmatrix}.$$

Now the sets S_L, of the upper left corners B, and S_R, of the lower right corners D, form multiplicative semigroups of unipotent matrices of dimension less than n.

One can then use the induction hypothesis to triangulate simultaneously these matrices, and all elements of S will then have been put in triangular form.

Case II. An arbitrary scalar field k. Form the algebraic closure of k and triangulate the elements of S simultaneously as matrices over the extension field. Then any product of n matrices $(T-1)$, where T is in S and 1 is the identity matrix, must be zero. Let r be the smallest integer such that the product of any r elements $(T-1)$ is zero. Then there exist elements $T_1, \dots, T_{r-1}$ in S such that

$$(T_1 - 1)(T_2 - 1)\dots(T_{r-1} - 1) \neq 0.$$

Find a vector x such that

$$x(T_1 - 1)\dots(T_{r-1} - 1) = y \neq 0.$$

Then for any T in $S, y(T-1) = 0$, or $yT - y$. This shows that S is reducible. The argument can now proceed as in Case I. □

3.73 REMARK. The group T of nonsingular upper triangular matrices of degree n over the field k is a solvable group, and in fact, T is an extension by an abelian group of a nilpotent group.

[**Hint:** Let $U = 1 + N$, consisting of all unipotent matrices, where 1 is the identity matrix, and N is the ideal of strictly upper triangular matrices. Then, $1 + N^i, (1 \leq i \leq n)$ is a normal subgroup of U, and the commutator formula $[1 + N^i, 1 + N^j] \subseteq 1 + N^{i+j}$ for the case of $i = 1$, and variable j, shows that U is a nilpotent group. Moreover, U is a normal subgroup of T, and T/U is abelian.]

Historical Notes on Local and Semilocal Rings

The importance of semilocal rings stems from a vast number of applications from such diverse fields as algebraic geometry, commutative and noncommutative algebra, group theory, module theory, and category theory. In algebraic geometry, or commutative algebra, for example, one can consider the local ring at a point on an algebraic variety, or at a prime ideal of a ring.

According to Bourbaki [65, *Note Historique*, p.131] the general idea of a local ring developed very slowly: Grell (1926), and Krull (1938), for domains, Chevalley (1944) for Noetherian rings, and the general case by Uzkov (1948).

Beginning about 1940, the local ring R_P at a prime ideal P of a domain R was consistently used by Krull (and his school), and in algebraic geometry by Chevalley and Zariski. Krull's term *Stellenring* was superseded by Chevalley's terminology *local ring*, a ring associated with a point of a variety which gives "local properties" of the variety, for example, the ring of all functions regular at that point (also see Nagata [62, p.xi]). This chronology omits the important work of Hensel at the turn of the century on p-adic numbers; however, Hensel considered not R_P but the completion of R_P, that is, the p-adic completion (see Bourbaki, *loc.cit.* and also Matlis' Theorem 5.4B).

Köthe [30a] studied noncommutative semilocal rings with a **Köthe radical**, that is, a nil ideal K containing every nil onesided ideal. (It is an open question if every ring has a Köthe radical, but see Theorem 3.50.) Köthe (*loc.cit.*) proved that a semilocal ring R with Köthe radical K is isomorphic to a full matrix A_n over a local ring A iff R/K is simple. Moreover $R \approx \text{End}_A E$, where E is one of the rows of A_n considered as a module over A. This theorem generalizes the Wedderburn-Artin theorem, and the theorem of Noether [29] (for simple semisimple R). Moreover, Köthe [30a, p.182,Satz 13] proved that in a semilocal lift/rad ring with radical J, and **right prindecs** (=principal indecomposable right ideals) eR and fR, that $eR \approx fR$ iff $eR/eJ \approx fR/fJ$. (Köthe's proof of this is for J =Köthe radical.)

REMARK. Lambek [66] extended Köthe's theorem to a semiperfect ring R having R/J simple.

Köthe [30b] generalized two theorems of Shoda to a semilocal ring R with nil radical K:

(1) *The intersection of all the nilsubrings of R is K.*
(2) *In case R is Artinian, any two maximal nilpotent subrings of R are conjugate, and every nilpotent subring is contained in a maximal nilpotent subring.*

(Shoda proved (1) for Artinian, and (2) for finite rings.) Also (1) has been generalized by Michler [66] to right Noetherian R;

(2) implies that *any nilpotent subring S of an $n \times n$ matrix ring k_n over a field* k can be placed simultaneously into triangular form, inasmuch as the ring $T_n(k)$ of upper triangular matrices is maximal nilpotent. In answering a conjecture of Köthe, Levitzki [31] (also [45a]) showed that every nilsubring of an Artinian (also Noetherian) ring is nilpotent, and thereby sharpened Köthe's theorem, that is, then nil subrings of k_n can be placed simultaneously in $\triangle$-form (see 3.68). Levitzki [50] generalized nil $\Rightarrow$ nilpotence to multiplicatively closed systems (M-systems) of left and right Noetherian rings. (See 2.33 and 2.38 A–B.)

Fitting [33] established the relationship between the direct decomposition of a module M of finite length and endomorphism ring A (*loc.cit.*, p.528, Satz 4). In particular, M is indecomposable iff A is local (*loc.cit.* p.533, Satz 8). This is called Fitting's "lemma" (see 3.65 and 8.8f).

Kaplansky [68,p.4] expressed his doubt that Artinian rings are the natural generalization of finite dimensional algebras because "natural examples are not common," and suggests, as an alternative, rings which are finitely generated modules over Noetherian subrings of their centers (that is, PI-algebras). Be that as it may, semiprimary rings certainly do arise naturally as endomorphism rings of (Jordan-Hölder) modules of finite length.

Levitzki [44] characterized semiprimary rings as semilocal rings modulo the ideal N generated by all nilpotent one-sided ideals, and satisfying the dcc on products of ideals in N; and Björk (or who?) characterizes semiprimary rings as follows: there exists an integer n such that R does not contain a strictly decreasing sequence of n principal left ideals.

Chase anticipated Björk's theorem (3.32f) stating that right modules satisfy the dcc on finitely generated submodules whenever R satisfies the dcc on principal right ideals. Chase proved this holds (on both sides) for a semiprimary ring. See the Appendix of my paper [66a].

Theorem 3.68, the placing nil submonoids of K_n simultaneously into triangular form is attributed to Levitzki by Jacobson [43], but the theorem can be deduced from a theorem of Köthe [30].

A theorem generalizing both Kolchin's Theorem 3.72 and the commutative field case of 3.68 is proved by Kaplansky [69b], p.137.

Further Notes for Chapter 3

The purely algebraic theorem of Kolchin [48b], given in 3.72, was preceded by the theorem of Lie-Kolchin which states that any connected solvable algebraic matrix group over an algebraically closed field can be placed simultaneously into triangular form (Kolchin [48a]). This has been proved by Borel [69, p.243] in a way which yields as an immediate corollary Mal'cev's theorem [49,51] which states that if M is any solvable subgroup of the group $GL(n,k)$ [the general linear group of nonsingular $n \times n$ matrices over a field k] over an algebraically closed field k, then M is an extension by a finite group of a group which can be placed simultaneously into triangular form. Moreover, a theorem of Zassenhaus [38] states that the derived series of any solvable subgroup of $GL(n,F)$, for any field F, is bounded by a number $z(n)$ independent of F.

Free Subgroups of $GL(n,F)$

A theorem of Tits [72] on a conjecture of Bass and Serre states that any finitely generated subgroup G of $GL(n,F)$, for any field F, and $n > 1$, contains either a free group of rank > 1, or else is an extension by a finite group of a solvable group. (Over characteristic 0, the subgroup G is not required to be finitely generated.) A number of theorems on the structure of solvable subgroups of $GL(n,F)$ of Zassenhaus, (e.g. a maximal irreducible subgroup has a unique maximal Abelian normal subgroup when F is infinite), Mal'cev and others, are presented in Suprunenko [63].

Sanov's Theorem

On the subject of free subgroups of $GL(n,F)$ is the following explicit example.

3.74 THEOREM (SANOV). *If $c_2, c_2 \in \mathbb{C}$ and if $|c_1| = |c_2| \geq 2$, then $\begin{pmatrix} 1 & c_1 \\ 0 & 1 \end{pmatrix}$ and $\begin{pmatrix} 1 & 0 \\ c_2 & 1 \end{pmatrix}$ generate a non-Abelian free subgroup of $GL(2,\mathbb{C})$.*

PROOF. This theorem is quoted in Jespers' survey [98], p.146, Prop. 2.1, which refers to the book by Kargapolov and Merzljakov [79].

Recall, G is **Hamiltonian** if every subgroup is normal.

A consequence to Sanov's theorem is a corollary to the Hartley-Pickel theorem (see below).

3.75 COROLLARY. *If G is a non-Abelian and non-Hamiltonian finite group, then the units group $\mathbf{U}(\mathbb{Z}G)$ of the integral group ring $\mathbb{Z}G$ contains a non-Abelian free subgroup.*

PROOF. The Wedderburn-Artin decomposition of the rational group algebra $\mathbb{Q}G$ contains an $n \times n$ matrix ring D_n for some skewfield D and $n \geq 2$, so the corollary follows easily from Sanov's theorem. Cf. Jespers, *ibid.*, p.146. □

Hartley-Pickel Theorem

3.76 HARTLEY-PICKEL THEOREM [80]). *If G is a non-Abelian finite group, then $\mathbf{U}(\mathbb{Z}G)$ contains a finite non-Abelian free group iff G is not Hamiltonian.*

PROOF. See Jespers, *ibid.*, p.147, Corollary 2.6.

REMARK. Also refer to Jespers, *ibid*, for related results of Berman, Ritter and Seghal, and Bass and Milnor on $\boldsymbol{U}(\mathbb{Z}G)$.

Steinitz Rings

A free right R-module F is an (f)-**Steinitz module** if every (finite) independent subset of elements extends to a free basis. A right (f)-**Steinitz ring** is one in which every $(f \cdot g)$ free right R-module is (f)-Steinitz.

3.77 THEOREM (BRODSKII [72]). *A free right R-module of rank $n \geq 2$ is f-Steinitz iff R is a local ring and $L^{\perp} \neq 0$ for every n-generated left ideal $L \neq R$.*

3.78 THEOREM (BRODSKII [72])-LENZING [71]). *A ring R is right Steinitz iff R is a right perfect local ring.*

Free Direct Summands

3.79 THEOREM (BECK AND TROSBORG [78]). *If F is a free right R-module, and J is the Jacobson radical of R, and if G is a submodule of F such that $F = G + FJ$, then G has a direct summand $\approx F$.*

Essentially Nilpotent Ideals

A right ideal I of a ring R is said to be **essentially nilpotent** if I contains a nilpotent right ideal that is essential in I.

3.80 SHOCK'S THEOREM [71B]. *Any right vanishing (= left T-nilpotent) right ideal I of a ring R is essentially nilpotent. Furthermore, so is any right ideal I of prime rad R in an acc$\oplus$ or acc $\perp$ ring.*

CHAPTER 4

Direct Product Decompositions of von Neumann Regular Rings and Self-injective Rings

A ring R is **von Neumann regular (= VNR)** if it satisfies the equivalent conditions:

(VNR) For every $a \in R$ there exists $x \in R$ so that $axa = a$.

(VNR 1) Every principal right ideal pR is generated by an idempotent, i.e., $pR = eR$ for some $e = e^2 \in R$.

(VNR 2) Every $f \cdot g$ right ideal is generated by an idempotent.

(VNR 3) Every $f \cdot g$ right ideal is a direct summand of R.

(VNRi)$'$ Left-right symmetry of $1+i$, $i = 0, 1, 2, 3$.

These rings were introduced by von Neumann [36] as co-ordinate rings of infinite dimensional projective and continuous geometries. Cf. Kaplansky [69b], p.111. (A **projective geometry** in the classical sense can be viewed as a lattice of submodules of a vector space, equivalently, as a lattice of right ideals of a full matrix ring over a sfield. Cf. 12.4.)

A VNR ring is **unit-regular** is x in (VNR) can be chosen to be a unit.

Remarks and Results. (1) Any semisimple ring R is *a fortiori* VNR since by Theorem 2.1, every submodule of every module is a direct summand.

(2) Any union of any chain of VNR rings is VNR.

(3) Any $n \times n$ matrix ring R_n over a VNR ring R is VNR. (see von Neumann, *ibid.*)

(4) One checks that every product of (unit) regular VNR's is (unit) regular.

(5) Every full linear ring $R = \text{End}_D V$ of a vector space V over a sfield D is VNR. (See Kaplansky [69b], p.111.)

(6) Ehrlich [76] shows that R in (5) is unit-regular iff $n = \dim_D V < \infty$, in which case $R \approx D_n$.

(7) Ehrlich, *ibid.*, shows that if 2 is a unit of a unit regular ring R, then every element of R is a sum of two units. Cf. her characterization of unit regular rings in Chapter 6, Theorem 6.3B.

(8) Any right and left self-injective VNR ring R is unit regular. (This is a theorem of Utumi—see 4.7B.)

Flat Modules

A right R-module M is *flat* provided that the functor

$$M_R \otimes (\ ) : R - \text{mod} \rightsquigarrow Ab$$

is exact from the category R-mod of left R-modules to the category Ab of Abelian additive groups.

4.A THEOREM (GOVOROV [65]-LAZARD [64]). *An R-module M is flat iff M is a direct limit of projectives.*[1]

Thus: any projective R-module is flat. A consequence of Schanuel's lemma is that any finitely presented flat R-module is projective (Cf. Faith [72a,81], pp.436–7, esp.11.30). Furthermore, any $f \cdot g$ flat module over a commutative integral domain is projective hence finitely presented. See Endo's theorem below, and also see Sandomierski [68], where this is proved also for finite dimensional nonsingular rings: $f \cdot g$ flat modules are projective. Also see Sahaev [69,77], and Jøndrup's [76]. Cf. 12.9.

4.A1 THEOREM (ENDO [62]). *If R is commutative, and $Q = Q_{c\ell}(R)$ is semilocal, then every $f \cdot g$ flat R-module is projective.*

REMARKS. (1) Examples of commutative rings R with semilocal Q are:

(a) Rings with acc$\perp$, e.g. whenever Q has acc$\perp$ (see, e.g. 16.31–32; for example whenever Q is Noetherian).

(b) When R is semilocal.

4.A2 THEOREM (SIMSON [72]). *If a ring R is embeddable in a right Noetherian or right perfect ring R, then every flat right R-module is an $\aleph_0$-directed union of countably generated flat right R-modules, and every $f \cdot g$ submodule of a flat right R-module embeds in a free module.*

Below, let $R \in P$ denote that $f \cdot g$ flat R-modules are projective.

4.A3 JØNDRUP'S THEOREM [70]. *If R is commutative, then $R \in P$ iff the power series ring $R\langle x\rangle \in P$. Moreover, then $A \in P$ for any R-algebra A when $R \subseteq$ center of A.*

Character Modules and the Bourbaki-Lambek Theorem

If M is a left R-module, then $M' = \text{Hom}(M, \mathbb{Q}/\mathbb{Z})$ is a canonical right R-module called the *character module of M*, where

$$(fr)(x) = f(rx) \quad \forall x \in M, r \in R, f \in M'.$$

4.B BOURBAKI [61A]-LAMBEK [64] THEOREM. *A left R-module M is flat iff its character module M' is an injective right R-module.*

Cf. Cheatham and Stone [81].

---

[1] This theorem has long been ascribed to Lazard, and I wish to thank J-L. Gómez-Pardo for the reference to Govorov [65]. See P. M. Cohn's review of the latter, e.g. on p.324 of Small [81].

4.C PROPOSITION.

(1) *If M is a (B, A)-bimodule, then the following are equivalent:*

(a) *$Hom_A(M,\ )$: mod-$A \rightsquigarrow$ mod-B preserves injectives.*

(b) *If Q is the smallest injective cogenerator of mod-A, then $Hom_A(M, Q)$ is injective in mod-B.*

(c) *The left adjoint $\otimes_B M$ of the functor (a) is exact.*

(d) *M is a flat B-module.*

(2) *If I is an ideal of a ring A, then every injective right A/I module is injective in mod-A if and only if A/I is flat in A-mod.*

(3) *If B is a subring of A, then every injective right A-module is an injective right (canonical) B-module iff A is flat as a (canonical) left B-module.*

4.D REMARKS. (1) This is a corollary to a more general theorem (due to Gabriel) on injective-preserving functors of Abelian categories. See Theorem 6.28 of my **Algebra I**.

(2) Theorem 4.B is a corollary of 4.C: see my **Algebra I**, p.440, Prop. 11.35 (and the remark following).

(3) (Faith [72B]). If R is VNR, then an R-module E is Σ-injective iff $R/(\text{ann}_R E)$ is semisimple Artinian.

This follows from 3.7 and 4.C.

When Everybody Is Flat

Semisimple rings are the rings over which every module is projective. We now consider when every R-module is flat:

4.1A HARADA [56] AND AUSLANDER [57] THEOREM. *All right R-modules over a ring R are flat iff R is von Neumann regular (= VNR) iff all left R-modules are flat.*

4.1B CAMILLO [74]-PILLAY [80] THEOREM. *If the polynomial ring $R[x]$ is left or right semihereditary, then R is VNR.*

Pillay extended Camillo's theorem to non-commutative R. According to C. V. Jensen in his review of Pillay's theorem, which is the case R is non-commutative, the converse does not hold, by an unpublished example of Jøndrup. However, Fieldhouse's theorem [78] states that the converse holds assuming that $[x]$ is, e.g., left coherent (see 6.6): R then VNR implies that $R[x]$ is left semihereditary. Moreover:

4.1C MCCARTHY'S THEOREM [73]. *If R is a commutative VNR then $R[x]$ is semihereditary.*

REMARK. Power series rings over a VNR ring are considered in §6, 6.12ff.

4.1D KAPLANSKY'S THEOREM ON TORSION SPLITTING [60]. *A commutative integral domain is Prüfer (= semihereditary) iff the torsion submodule of every $f \cdot g$ module splits off.*

4.1D′ CHASE'S THEOREM [60]. *A commutative domain is Dedekind (= Noetherian Prüfer) iff every torsion module of bounded order splits off.*

Singular Splitting

The **singular submodule** *sing* M of an R-module M consists of all $m \in M$, whose annihilators are essential right ideals of R.

One verifies that $S = \text{sing } M$ is a **fully invariant** submodule in the sense tht $f(S) \subseteq S \ \forall f \in \text{End } M_R$. Thus sing R_R is an ideal called the **right singular ideal** of R. M is **nonsingular** if *sing* $M = 0$. (For a domain R, the torsion submodule of $M = \text{sing } M$.) R has **singular splitting** if sing M splits in M for all right R-modules M.

R is **right nonsingular** provided that sing $R_R = 0$.

Examples of Nonsingular Rings. (1) Any domain; (2) Any simple ring R. (Note: sing R_R is an ideal, hence necessarily $= 0$ in (2).)

4.1E THEOREM (CATEFORIS AND SANDOMIERSKI [68]). *The following are equivalent conditions on a commutative ring R:*
(1) *R has singular splitting,*
(2) *Sing M is injective for all M in mod-R,*
(3) *Sing $R_R = 0$ and R/I is semisimple for every essential right ideal I.*
(4) *R is VNR and every bounded singular submodule splits.*
Then, R is hereditary.

Cf. 4.1G and H(2).

REMARKS. (1) Huynh [98] pointed out that in general (2) does not imply (4). Huynh also pointed out that a right PCI domain not a field is a counter-example to this implication (see 4.1K below); (2) Since $E(M)/M$ is singular for any R-module M, then injectivity of singular modules implies R is hereditary. (In the terminology of §14, R has right global dimension ≤ 1.) Any semihereditary ring is right nonsingular. Cf. 4.2B′.

4.1E′ COROLLARY. *If R is commutative self-injective ring with singular splitting, then R is semisimple.*

4.1F FUELBERTH AND TEPLY THEOREM [72]. *Over a commutative ring R, the singular submodule splits off in every $f \cdot g$ module iff R is semihereditary and every non-singular $f \cdot g$ module is "almost finitely presented."*

4.1G THEOREM (GOODEARL [72]). *Every singular right R-module is injective iff R/I is semisimple for all essential right ideals I. Then R is right hereditary, $(\text{rad } R)^2 = 0$, and $R/(\text{soc } R_R)$ is Noetherian.*

Cf. 12.4C′-H and 12.8. also see Osofsky-Smith [91].

4.1H REMARKS. (1) The conditions of 4.1G are not left-right symmetric: see 4.1K below for characterizations of these rings: (2) Under the conditions of 4.1G, then R/I is singular hence injective. Thus, e.g. if R is right uniform, then R is right PCI, hence by 3.18C, R is either semisimple or a right hereditary Noetherian domain. Any right PCI domain has singular splitting (*op.cit.*).

For the next theorems, consult Chapter 14 for the concept of global dimension.

4.1I THEOREM (TEPLY [70]). *If R has singular splitting, then R has global dimension ≤ 2.*

4.1J THEOREM (FUELBERTH-TEPLY [72B]). *There exists a splitting ring R of global dimension 2.*

4.1K THEOREM (GOODEARL [72], KOIFMAN [71,A,B]). *Every singular right R module is injective ($= R$ is right SI) iff R is a finite product $R = K \times R_1 \times \cdots \times R_n$ such that $K/soc(K_K)$ is semisimple, and each R_i is Morita equivalent to a right SI-domain.*

PROOF. See Goodearl, *op.cit.*, p.56.

Utumi's Theorems

4.2 THEOREM OF UTUMI ([56,67]). *If R is a right self-injective ring, then the Jacobson radical rad R is the right singular ideal J, and $\overline{R} = R/J$ is a right self-injective VNR ring.*

4.2A THEOREM (FAITH-UTUMI [64], OSOFSKY [68A]). *If A is the endomorphism ring of a quasi-injective right R-module M, then $J =$ rad A is the set of all $a \in A$ whose kernels are essential submodules, A/J is a right self-injective VNR ring, and idempotents lift module any subset I containing J.*

Wong-Johnson [59] is the case $J = 0$ of Utumi's theorem.

REMARK. Osofsky proved (*inter alia*) that A/J is right self-injective.

Weak or F·G Injectivity

A right module M over a ring R is *weak $\aleph_0$-injective* (or *$f \cdot g$-injective*) provided for any homomorphism $f : I \to M$ of a $f \cdot g$ right ideal there exists $m \in M$ so that $f(x) = mx \ \ \forall x \in I$. In contradistinction to injective modules, any direct sum of weak $\aleph_0$-injective modules is weak $\aleph_0$-injective. (Cf. 3.4B and 3.7) *$\aleph_0$-injectivity* denotes the same property whenever I is countably generated. (Cf. **sup.** 6.1)

REMARK. A $f \cdot g$ or weak $\aleph_0$-injective module is called **f-injective** by Gupta [69] and **semi-injective** by Matlis [85].

The next theorem indicates a plentitude of $f \cdot g$-injective modules that are not necessarily injective (Cf. 3.22).

4.2B GUPTA'S THEOREM [69]. *A ring R is right semihereditary iff every factor module of an injective right module is $f \cdot g$ injective.*

4.2B′ (KOBAYASHI [82]). *A ring R is semihereditary iff $E(M)/M$ is $f \cdot g$ injective for each R-module M.*

4.2C THEOREM (GOODEARL [79], PROP. 9.31). *If R is a right $\aleph_0$-injective VNR ring and I is an ideal such that R/I has no uncountable independent set of right ideals, then R/I is right self-injective.*

Abelian VNR Rings

A von Neumann regular (= VNR) ring R is **Abelian**, or **strongly regular**, if R satisfies the following equivalent conditions.

(1) All idempotents are central.
(2) R has no nonzero nilpotent elements (= R is *reduced*).
(3) All one-sided ideals are ideals.
(4) For each $a \in R$ there exists $x \in R$ so that $a^2x = a$.
(5) Every one-sided ideal $\neq 0$ contains a central idempotent $\neq 0$.
(6) R/P is a skew field for each prime ideal P.

REMARKS. (1) These equivalences are due to Arens and Kaplansky [48], and Forsythe and McCoy [46]. See, e.g. Goodearl [79], p.35. Also see theorems 3.2 and 3.5, pp. 26–28 of *loc.cit.*; (2) Forsythe and McCoy proved more generally that all idempotents are central in *any* reduced ring; (3) (Ehrlich [76]) If R is Abelian VNR, then all $n \times n$ matrix rings R_n are unit-regular (Cf. Goodearl *loc.cit.*).

If e is an idempotent of a VNR ring R, then eRe is a VNR ring, and e is said to **Abelian** provided that eRe is an Abelian VNR ring.

4.3A THEOREM (UTUMI [60-61]). *If R is a right self-injective VNR ring, then $R = R_1 \times R_2$ where R_1 is Abelian, and R_2 has no Abelian idempotents $\neq 0$. Moreover, R_2 is left self-injective.*

COROLLARY. *Any right self-injective Abelian VNR ring is left self-injective.*

REMARK. See **sup.** 12.4C for definition of right continuous used in the next theorem.

4.3B THEOREM (UTUMI [60]). *Any right continuous VNR ring R is a product $R_1 \times R_2$ where R_1 is Abelian and R_2 is right self-injective.*

The Maximal Regular Ideal

The existence of the maximal regular (=VNR) ideal $M(R)$ of a ring has been demonstrated by Brown-McCoy [50], and that $M(R/M(R)) = 0$.

4.4 THEOREM (FAITH [59,61,85]). *The maximal VNR ideal $M(R)$ of a ring R splits off as a ring direct factor iff the image of $M(R)$ splits off in $R/radR$ as a right ideal. Sufficient conditions for this are:*

(1) *The image $\overline{M(R)}$ of $M(R)$ in $\overline{R} = R/radR$ splits off as a right or left ideal.*
(2) *R is right and left continuous (e.g. self-injective).*
(3) *$\overline{R}$ is a finite product of simple rings (e.g. $\overline{R}$ is semisimple).*

This type of theorem was originated by M. Hall, Jr. [40] and Mal'cev [42] who show essentially that the maximal semisimple ideal $M(R)$ splits off of any finite dimensional algebra R, a result that Brown-McCoy extended to right Artinian rings. For an application of Theorem 4.4 to VNR rings, see Faith [96]. Cf. Birkenmeier-Kim-Park [97].

Products of Matrix Rings over Abelian VNR Rings

The next theorem generalizes a theorem of Kaplansky [50].

4.5 KAPLANSKY [50] AND ARMENDARIZ-STEINBERG THEOREM ([74]). *If R is a right self-injective VNR ring, then the following conditions are equivalent:*

(1) *R is isomorphic to a finite product of full matrix rings over Abelian (= idempotents are central) VNR rings.*

(2) *R/P is Artinian for all primitive ideals P.*

(3) *R/M is Artinian for all maximal ideals M.*

(4) *R has bounded index (of nilpotency).*

This theorem is due to Kaplansky [50] (all R/M have the same index n and $R \approx A_n$ for some Abelian VNR ring A); and Armendariz-Steinberg [74] in the general case. See Goodearl [79, p.79] for other credits, including Goodearl and Utumi.

Cf. Π-Regular rings **sup.** 8.4F. Cf. also 4.18.

Products of Full Linear Rings

A full right linear ring is defined to be $\text{Hom}_D(V, V) = \text{End}_D V$, for a right vector space V over a skew field D.

4.6 THEOREM (CHASE-FAITH [65]). *A ring R is isomorphic to a direct product of full right linear rings iff R is VNR, right self-injective and has essential right socle (= every nonzero right ideal contains a minimal right ideal).*

The special case when R is a prime ring characterizes a full right linear ring and is due to Utumi (Cf. Utumi [63d]). Actually, Chase-Faith ([65]) characterize when the maximal right quotient ring of a ring R is a direct product of full linear rings. Cf. Lambek-Michler [76]. Hudry [75] and O'Meara [75] independently characterize when R has classical right quotient ring Q that is a product of linear rings.

Dedekind Finite

4.6A DEFINITION AND PROPOSITION. *R is* **Dedekind Finite** *(***Directly Finite** *in Goodearl [79]) provided that $xy = 1 \Longrightarrow yx = 1$ for each $x, y \in R$. A necessary and sufficient condition for R to be Dedekind Finite (= DF) is that $R/\text{rad}R$ be DF. R is* **stably finite** *provided that every matrix ring R_n is Dedekind Finite. Noetherian, commutative and semilocal rings are examples of stably finite rings.*

4.6A′ HISTORICAL NOTES. (1) In his 1887 classic "Was Sind und Was Sollen Die Zahlen," Dedekind introduced the concept of finiteness of sets, namely: if $F : S \to T$ is a 1-1 mapping of a set S onto a subset T, then $S = T$; (2) Leary [92] generalized the concept to rings R, namely: if whenever $f : R \to R$ is a monomorphism of R-modules, then f is an isomorphism. In this case R is *right Dedekind finite*, equivalently, if $x \in R$ and if $x^\perp = 0$ then x is a unit. (This follows by considering $x = f(1)$.) (3) For a VNR ring R, $x^\perp = 0$ is equivalent to the existence of an element $y \in R$ with $yx = 1$. Thus, the definition in 4.6A for a VNR ring R is equivalent to that given in 4.6A'(2); (4) Since $xy = 1 \Rightarrow e = yx$ is idempotent, any Abelian VNR is DF.

Jacobson's Theorem

4.6B THEOREM (JACOBSON [50]). *A Dedekind Infinite ring R possesses an infinite set $\{e_{ij}\}_{i,j=1}^{\infty}$ of matrix units, hence an infinite set $\{e_{ii}\}_{i=1}^{\infty}$ of orthogonal idempotents.*

PROOF. If $xy = 1$, then $g = yx$ is an idempotent such that

$$x(1-g) = (1-g)y = 0$$

and the elements

$$e_{ij} = y^i(1-g)x^i$$

are the desired matrix units, since

$$e_{ij}e_{pq} = \delta_{jp}e_{iq}$$

where δ_{jp} is the Kronecker delta ($= 1$ if $j = p$, and 0 if $j \neq p$). □

As defined **sup.** 3.13, R has acc$\oplus$ denotes finite Goldie dimension, that is, R contains no infinite direct sums of right ideals.

4.6C COROLLARY. *Any ring R with no infinite set of orthogonal idempotents, in particular, any acc$\perp$ or acc$\oplus$ ring R, is Dedekind Finite.*

REMARK. (1) This also shows any semilocal ring R, or any ring R such that $R/\text{rad}R$ has acc$\perp$, *or* acc$\oplus$, is DF; (2) Any subring of a DF ring is DF.

Shepherdson's and Montgomery's Examples

(1) Shepherdson [51] shows there exists a domain R such that the 2×2 matrix ring R_2 is not DF.

(2) Montgomery [83] shows that if L is an algebraic extension field of a field k, $L \neq k$, then there exists an algebra A over k which is a domain, hence DF, such that $A \otimes_R L$ is not DF.

Group Algebras in Characteristic 0 Are Dedekind Finite

4.6D KAPLANSKY'S THEOREM [69]. *If K is a field of characteristic 0, then for any group G, the group algebra KG is Dedekind Finite.*

PROOF. The proof uses a non-trivial result of Kaplansky (*op.cit.*) and Zalesski [72] which states that the "trace" $a = tr\ e$ any idempotent $e \neq 0, 1$ is a rational number such that $0 < a < 1$. If $xy = 1$ in KG, then $e = yx$ is an idempotent which has trace $= 1$, hence $e = 1$. (See e.g. Passman [77], p.38.)

4.6E REMARK. Zalesski [72] proved for any idempotent e of a group algebra KG, that $tr\ e$ lies in the prime subfield of K. See Passman [77], p.48, Theorem 3.5.

Prime Right Self-injective VNR Rings

4.7A UTUMI'S THEOREM [65]. *A right self-injective VNR prime ring R is simple under the following assumption: R is Dedekind Finite, in particular, if R is also left self-injective, or left $\aleph_0$-injective.*

Cf. Goodearl [79], p.106, Theorem 9.30 and also references on p. 109.

4.7B THEOREM (UTUMI [65]). *Any right and left self-injective VNR ring R is Dedekind Finite and unit regular.*

PROOF. See Goodearl, *loc.cit.*, p.105 Theorem 9.29. □

4.7C COROLLARY. *Any right and left self-injective ring is Dedekind Finite.*

PROOF. By Utumi's theorems 4.2 and 4.7B, $R/\text{rad } R$ is DF, hence Prop. 4.6A applies. □

4.7D THEOREM (HENRIKSEN [73]). *If R is unit-regular, then every $n \times n$ full matrix ring R_n is Dedekind Finite.*

PROOF. See Goodearl, *loc.cit.*, p.50, Prop. 5.2. □

Goodearl-Handelman Characterization of Purely Infinite Rings

A ring R is *purely infinite* if R has no Dedekind Finite central idempotents $e \neq 0$ (= such that eRe Dedekind Finite).

Let ${}_RR^\omega$ denote a direct product of $\aleph_0$ copies of R as a left R-module.

4.8 THEOREM (GOODEARL [79] AND GOODEARL-HANDELMAN [75]). *A VNR right self-injective ring R is purely infinite iff ${}_RR \approx {}_RR^\omega$ (equivalently, ${}_RR^\omega \hookrightarrow {}_RR$).*

Goodearl and Handelman in *op.cit.* classified simple right self-injective rings. Cf. Hannah [80] on countability in VNR self-injective rings, and also 4.9 and 4.12.

4.9 COROLLARY (GOODEARL [79]. *If R is a prime VNR right self-injective ring, then R is Dedekind Finite iff R has at most countable left Goldie dimension (=R has no uncountable independent sets of left ideals).*

For proofs, see Goodearl [79], Theorem 10.19 and Corollary 10.20, pp.118–119.

Kaplansky's Direct Product Decompositions of VNR Rings

A VNR ring R has Type I: if R has a faithful Abelian idempotent e (that is, no central idempotent except 0 is orthogonal to e). I_f denotes Directly Dedekind and I_∞ Dedekind Infinite of Type I.

R has Type II, if R has a faithful Dedekind Finite idempotent $e(= eRe$ is Dedekind Finite) but no Abelian idempotents $\neq 0$. Types II_f and II_∞ are defined similarly as I_f and I_∞.

R has Type III if R contains no nonzero Dedekind Finite idempotent.

4.10 KAPLANSKY'S THEOREM (SPECIAL CASE). *Every VNR right self-injective ring R is uniquely a product of rings of Types I, II, and III. Furthermore, every VNR right self-injective ring is a product of a Dedekind Finite ring and a Dedekind infinite ring, hence R decomposes uniquely into products of rings of Types* $I_f, I_\infty, II_f, II_\infty$, *and III.*

This theorem is a special case of Kaplansky's more general theorems [68]. See Goodearl [79], Theorem 10.13 and 10.22, pp. 115 and 120ff.

Kaplansky's Conjecture on VNR Rings: Domanov's Counterexample and Goodearl's and Fisher-Snider's Theorems

The conjecture of Kaplansky: Are prime VNR rings primitive, equivalently, do they have a faithful simple module?

Domanov [77,78] constructed non-primitive prime VNR group algebras over arbitrary fields. See Theorem 11.9′.

Goodearl [73b] verified this conjecture for a right self-injective VNR ring.

4.11 GOODEARL'S THEOREM [73B]. *If R is a right self-injective VNR ring, then:*

(1) *The ideals of R that contain a prime ideal P are linearly ordered, under inclusion, and are all prime ideals.*
(2) *If P is a closed prime ideal of* $R(= P_R$ *has no essential extension in* $R_R)$, *then the ideals of R containing P are well-ordered.*
(3) *If R is a prime ring, then R is primitive.*
(4) *If R is directly indecomposable as a ring then R is prime, hence primitive.*

A set B of nonzero ideals of R is a *base* if every nonzero ideal contains an ideal in B.

4.12 FISHER-SNIDER THEOREM [74]. *If R is a prime VNR ring with a countable base of ideals* $\{I_i\}_{i=1}^{\infty}$, *then R is primitive. In particular, any countable prime VNR ring is primitive.*

REMARK. Domanov's counterexample and Theorem 4.12 owe to some pioneering work of Formanek and Snider [72]. (See the introduction to Domanov [77].)

Azumaya Algebras

In considering products of rings we have thus far failed to consider tensor products, which leads inevitably to the consideration of Azumaya algebras, and the Brauer group.

4.13 AZUMAYA ALGEBRA DEFINITION AND THEOREM. *An algebra A over a commutative ring k is called an* **Azumaya algebra** *if A satisfies the equivalent conditions.*

(*Az*1) *A is a projective module over the enveloping algebra* $A^e = A \otimes_k A^0$, *where* A^0 *is the opposite algebra.*
(*Az*2) *A is a finitely generated projective module over k, and* $A^e = \mathit{End}_k A$ *canonically.*
(*Az*3) A^e *is Morita equivalent to k.*

(Az4) *A is a finitely generated module over k, and for all maximal ideals m of k, the factor algebra A/mA is a central simple k/m-algebra.*
(Az5) *A is finitely generated projective and central over k and every ideal I of A is of the form $I = I_0 A$, where $I_0 = I \cap k$.*
(Az6) *A generates mod-A^e and $k = End_{A^e} A$.*

When this is so, then k is the center of A.

Most of this is due to Azumaya [51] over local k, and the carry-over to general k is due to Auslander-Goldman [60]. (Az 5) is due to Rao [72], and this condition was axiomatized to rings by Azumaya [80]. M. Artin [69] characterized Azumaya algebras containing a field in the center via polynomial identities and Procesi [71] removed the field requirement. See 15.8. Also see Faith [92], p. 550, for additional references and the following:

4.14 PROPOSITION. *If A is an Azumaya algebra over a commutative ring k, then the f.a.e.*

A) *A is right self-injective.*
B) *A is left self-injective.*
C) *k is self-injective.*
D) *A^e is right self-injective.*
E) *A^e is left self-injective.*

In regard to 4.14, see 4.16.

4.15 AZUMAYA'S THEOREM [51]. *If A is an Azumaya algebra over a commutative ring k, and if A is a subalgebra of an algebra K over k, then $K \approx A \otimes_k A'$, where A' is the centralizer of A in K.*

A proof and partial converse is given by Bass [68].

4.15A REMARK. A finite dimensional simple central algebra A over k is characterized by the property that $K = A \otimes_k A'$ for any algebra K over k. See Jacobson [55,64], p.118. The same is true for Azumaya Algebras over k assuming that K is $f \cdot g$ projective over k. (See Bass, *ibid.*)

Hochschild's Theorem on Separable Algebras

DEFINITION. (1) An algebra A over a commutative ring k is **separable** provided that A is a projective module over its **enveloping algebra** $A^e = A \otimes_k A^0$. (This agrees with the definition given in 2.51. Cf. Hochschild [56], DeMeyer and Ingraham [71].);

(2) An algebra A over a commutative ring k is a k-**split-split algebra**, provided that an exact sequence $0 \to A \to B \to C \to 0$ of A-modules splits whenever it is split as k-modules.

4.15B THEOREM (HOCHSCHILD [56]). *An algebra A over a commutative ring k is separable iff its enveloping algebra A^e is a k-split-split algebra.*

4.15C COROLLARY. *An algebra A over k is an Azumaya algebra iff A is a central separable module-finite algebra over k.*

The Auslander-Goldman-Brauer Group of a Commutative Ring

If k is a commutative ring, then the set k-ALG of all k-algebras are a semigroup with respect to the tensor product $A \otimes_k B$ of algebras A and B over k. Let $[A] = [A']$ denote that A and A' are Morita equivalent k-algebras, which by Morita's theorem (see 3.51) means that $A' \approx \text{End}\, P_A$ for a $f \cdot g$ projective generator P for right A-modules. (A generator P for A means that every right A-module is an epimorphic image of copies of P, equivalently that $P^n \approx A \oplus X$ for an integer $n \geq 1$ and a right A-module X.)

Then the set of classes $[A]$ of algebras is a semigroup under the product $[A][B] = [A \otimes_k B]$. Furthermore, the set $Br(k)$ of all $[A]$ such that A is an Azumaya algebra over k is a group such that $[A]^{-1} = [A^0]$, called the *Brauer Group* (Auslander-Goldman [60]). This is named in honor of Richard Brauer [29] who discovered the Brauer group of a field k. In this case every $[A] = [D]$ for a unique finite dimensional division algebra D over k depending on A. The theorems of Wedderburn, Tsen and Chevalley, discussed in §2, are statements about the triviality of the Brauer group over the indicated fields (e.g. finite fields). Moreover, Gröthendieck [65] showed that $Br(k)$ is a torsion group for any commutative ring k.

Menal's Theorem on Tensor Products of SI or VNR Algebras

4.16A Menal's Theorem [81]. *If A and B are algebras over a field k, and if $A \otimes_k B$ is either right self-injective, or VNR, then both A and B enjoy the same property, and furthermore, either A or B is an algebraic algebra over k.*

See Menal's General Theorem in Faith [92] and also Menal [82] on the radical and semi-primitivity of $A \otimes_R B$ for VNR rings A and B.

4.16A is a grand generalization of a lemma used in the proof of Hilbert's Null-stellensatz, namely that a commutative ring extension $K = k[a_1, \ldots, a_n]$ of a field k can be a field only if K is algebraic over k.

Lawrence's Theorem on Tensor Products of Semilocal Algebras

4.16B Lawrence's Theorem [75]. *If A and B are algebras over a field k, and if $A \otimes_k B$ is semilocal, then A and B are algebraic algebras.*

4.16C Remark. If A is a two-sided self-injective algebra over a field k, then every non-zero divisor is a unit. Thus, by the Amitsur-Small theorem [96] (see 3.43B), then A is an algebraic algebra when $|k| > \dim_R A$.

One consequence of Menal's theorem is:

4.17A Busqué-Herbera Theorem. *If in Menal's theorem the product $A \otimes_k B$ is QF, then either A or B is QF.*

See Busqué [93], Prop. 2.1. The proof uses Theorem 3.7.

Another theorem of Menal is of interest to the study of self-injective VNR rings.

4.17B Theorem (Menal). *Every right self-injective VNR ring is isomorphic to a product of algebras over fields.*

See Busqué [93], Lemma 2.4.

Armendariz-Steinberg Theorem

It has been noted in several places in the literature that the center of a self-injective ring need not be self-injective. (See, e.g. Pascaud-Valette [79], Herbera-Menal [89] or Clark [89]). In this connection, one notes:

4.18 ARMENDARIZ-STEINBERG THEOREM [74]. *If R is a right self-injective VNR ring, then the center of R is also.*

Strongly Regular Extensions of Rings

As defined by Arens and Kaplansky [48] a ring A is *strongly regular* (s.r.) in case to each $\alpha \in A$ there corresponds $x = x_\alpha \in A$ depending on α such that $\alpha^2 x = \alpha$. As stated following 4.26C, Abelian VNR is a variant term.

A ring A is defined to be a *s.r. extension* of a subring B in case each $\alpha \in A$ satisfies $\alpha^2 x - \alpha \in B$ with $x = x_\alpha \in A$. S.r. rings are, then, s.r. extensions of each subring. A ring A which is a s.r. extension of the center has been called a ξ-ring (see Utumi [57], Drazin [57], and Martindale [58].

4.19A THEOREM (ARENS AND KAPLANSKY [48]). *A s.r. ring is a subdirect product of division rings.*

Since any s.r. ring is semiprimitive, a later result stating that any semiprimitive ε-ring is a subdirect product of division rings (see (Martindale [58], *op.cit.*) contains this result.

4.19B THEOREM (FAITH [62B]). *If a semiprimitive ring A is a s.r. extension of a commutative subring B, then A is a subdirect product of division rings.*

The proof depends on the following:

4.19C THEOREM (*op.cit.*). *If A is a primitive ring, not a division ring, and if A/B is s.r., then B is dense in the finite topology on A.*

4.19D COROLLARY (*op.cit.*). *In order that a semiprimitive ring A be a s.r. extension of a subring B, it is necessary that B be a subdirect product of primitive rings and integral domains.*

Pseudo-Frobenius (PF) Rings

Right PF rings, generalizing QF rings, were introduced by Azumaya [66] as rings R such that every faithful right R-module is a generator in the category of right R-modules.

4.20 THEOREM (AZUMAYA [66]-OSOFSKY [66]-UTUMI [67]). *A ring R is right* **Pseudo-Frobenius** *(= PF) iff R satisfies the equivalent conditions:*

($PF1$) *R is right self-injective semiperfect and has essential right socle (= ES).*

($PF2$) *R is right self-injective and has finite essential right socle (= right finitely embedded).*

($PF3$) *R is a finite direct sum*

$$R = \bigoplus_{i=1}^{n} e_i R$$

where $e_i^2 = e_i \in R$, *and* e_iR *is indecomposable injective with simple socle,* $i = 1, \dots, n$.

($PF4$) R *is an injective cogenerator for mod-*R.

($PF5$) R *is right self-injective and every simple right module embeds in* R.

($PF6$) *Every faithful right* R*-module* M *generates mod-*R*, that is, there exists an epimorphism* $M^n \to R$ *for some integer* n*, depending on* M.

Cf. Faith [76a], 24.32. Every 2-sided PF ring has a Morita self-duality induced by $\mathrm{Hom}_R(\ ,R)$. See 13.7.

Remark. In connection with (PF2), see the theorem of Beachy-Kamil 3.60. Thus, (PF2) via 3.60 implies that R embeds in a finite product M^n of any faithful module, so injectivity implies that R splits in M^n, hence (PF6) holds.

Note, 3.33B and C imply that over a left perfect ring R, every right R-module M has ES. Thus by (PF1):

4.21 Corollary. *Every left perfect right self-injective ring* R *is right PF. Consequently, a semiprimary right self-injective ring is right PF.*

4.22 Theorem (Osofsky [66]). *The following are equivalent conditions:*

(1) R *is* QF.

(2) R *is left perfect, right and left* PF.

(3) R *is right self-injective, left perfect, and* J/J^2 *is finitely generated in mod-*R.

Kasch Rings

4.22A Proposition and Definition. *A ring* R *is* **right Kasch** *if* R *satisfies the equivalent properties:*

(1) *Every maximal right ideal of* R *is a right annihilator;*

(2) *Every simple right* R*-module embeds in* R*;*

(3) *The injective hull* $E(R_R)$ *is a cogenerator of mod-*R.

Proof. (1) $\Rightarrow$ (2) since if M is a maximal right R-module, and if (1) holds then $M = x^\perp$ for some $0 \neq x \in R$, hence

$$(*) \qquad V = R/M = R/x^\perp \approx R \subset R.$$

Since every simple right module $V \approx R/M$ for some such M, then (1) $\Rightarrow$ (2).

Conversely, (2) $\Rightarrow$ (1) since if $f : R/M \hookrightarrow R$, then $x^\perp = M$ where $x = f(1)$. (3) is an immediate consequence, since R is essential in $E(R)$, hence a simple module $V \hookrightarrow E(R)$ iff $V \hookrightarrow R$. □

4.22B Theorem. *A ring* R *is right PF iff* R *is right Kasch and right self-injective.*

Proof. This follows immediately from (PF5) of Theorem 4.20. □

4.22C Theorem (Kato [67]). *Any right PF ring* R *is right and left Kasch. (Cf. 4.23A.)*

4.23A THEOREM (KATO [67]).

1. *A right PF ring R is left PF iff R is left self-injective.*
2. *A ring R is left and right PF iff R is a left and right cogenerator.*
3. *Every factor module of R^2 (both in mod-R and R-mod) is torsionless iff R is left and right PF.*

4.23B COROLLARY (OSOFSKY-KATO). *A left perfect right and left self-injective ring is QF.*

4.24 OSOFSKY'S EXAMPLE. **The Split-Null or Trivial Extension** $B \ltimes E$ of a (B,B)-bimodule E (i.e., $(ax)b = a(xb)$ $\forall a, b \in B, x \in E$) is a ring whose additive group is the product $B \ltimes E$ and whose multiplication is defined by $(a,x)(b,y) = (ab, xb + ax)$. $B \times E$ is isomorphic to a ring 2×2 matrices under a mapping

$$(a,x) \mapsto \begin{pmatrix} a, x \\ 0, a \end{pmatrix} \quad \text{under the usual definition of matrix operations.}$$

Osofsky [66] gave the first non-Artinian, i.e., non-QF example of a PF ring, namely, the split-null or trivial extension $\mathbb{Z}_{(p)} \ltimes \mathbb{Z}_{p^\infty}$ where $Z_{(p)} \approx \text{End}\mathbb{Z}_{p^\infty}$ is the **ring of p-adic integers**, i.e., the completion of the local ring $\mathbb{Z}_p$.

Moreover:

4.25 THEOREM (FAITH [79], FOSSUM-GRIFFITH-REITEN [75]). *If E is a (B,B)-bimodule such that ${}_BE$ is faithful, then the split-null extension $R = B \ltimes E$ is right self-injective (resp. right PF) iff E_B is injective (resp. injective cogenerator) and $B = \text{End}E_B$.*

Using this theorem, Dischinger and Müller [86] proved that a right PF ring need not be left PF.

FPF Rings

A ring R is **right FPF** if every $f \cdot g$ faithful right R-module generates mod-R. Any right PF ring is trivially FPF; also any Prüfer domain is FPF. See Faith-Pillay [90] for additional examples. Endo [67] originated the concept of FPF rings. See 4.27ff.

4.26 CAMILLO-FULLER [76]. *A ring R is right (F)PF iff every $(f \cdot g)$ faithful right R-module is (left) flat over its endomorphism ring.*

See, e.g. Reiner [75], for the concept of A-order in the next two theorems.

4.27 ENDO'S THEOREM [67]. *A Noetherian commutative ring R is FPF iff R is a finite product of Dedekind domains and QF local rings.*

4.28 THEOREM (ENDO [67]). *If A is a Noetherian commutative domain with quotient field K, then a projective A-order R of a semisimple K-algebra S is right FPF iff R is a hereditary maximal order in S.*

The next theorem generalizes Endo's theorem for Noetherian R. Cf. 5.42.

4.29 SPLITTING THEOREM (FAITH [79A], FAITH-PILLAY [90]). *Any commutative FPF ring R splits: $R = R_1 \times R_2$ where R_1 is semihereditary and R_2 has essential nil radical.*

See, e.g. Faith-Pillay, p.57. An ideal I of R is a **waist** if every ideal of R either contains I or is contained in I.

4.30 LOCAL FPF RING THEOREM (OP.CIT.). *If R is a commutative local ring then R is FPF iff $Q(R)$ is injective, the singular ideal $Z(R)$ is a* **waist** *of R, and $R/Z(R)$ is a valuation domain.*

See, e.g., Faith-Pillay, p.56, Theorem 3.9. The proof uses the characterization of commutative FPF rings—see 5.42.

4.31 THEOREM (FAITH [76-77,I]). *A one-sided perfect two-sided FPF ring is QF.*

This generalizes Kato-Osofsky's theorem for one-sided perfect two-sided PF (see 4.24).

REMARK. For additional results on FPF rings, see the lecture notes of Faith-Page [84e] and Faith-Pillay [90]; and the papers of Clark [89], Herbera [91], Herbera and Menal [89], Herbera and Pillay [93],Faticoni [84,87,88], Kobayashi [85,88], Yoshimura [91],[94], and Yousif [91]. See the author's papers [77b], [79a,b,,c], 82b] and [84b,c,d]. Also see 5.23ff for when products of FPF rings are FPF and 5.40ff. for some FPF structure theorems. For when cyclic faithful modules are generators, see Birkenmeier [87,89] and Yoshimura [95], [98].

CHAPTER 5

Direct Sums of Cyclic Modules

A semisimple ring R has the following property: *right Σ-cyclic*: every right R-module M is a direct sum of cyclic modules.

The corresponding property for finitely generated right R-modules is called *right σ-cyclic* or *FGC* ring.

5.1A THEOREM (KÖTHE [35]). *Artinian commutative Σ-cyclic rings are the finite product of Artinian chain rings.*

Cohen-Kaplansky [51] proved the Artinian hypothesis redundant.

To proceed, we pause for additional definitions.

A ring R is *completely primary* in case it is a local ring with nilpotent radical. A *primary* ring is a ring which is isomorphic to a full ring of $n \times n$ matrices over a completely primary ring. As stated previously, R is *semiprimary* in case rad R is nilpotent, and $R/\text{rad}R$ is semisimple. Thus, by 3.31 a semiprimary ring is left and right perfect. A semiprimary ring, then, may not be a direct sum of primary rings. We cite Asano's criterion:

5.1A′ ASANO'S CRITERION [49]. *For a semiprimary ring R with radical J, the conditions are equivalent:*

(*a*) *R is primary-decomposable,*
(*b*) *R/J^2 is primary-decomposable,*
(*c*) *the prime ideals of R are commutative,*
(*d*) *the prime ideals are comaximal in pairs.*

PROOF. The proof requires the Chinese Remainder Theorem. See for example, the author's Algebra II, p.50, 18.37. □

REMARK. Fitting [35b] proved this for an order R in a finite dimensional simple algebra.

Uniserial and Serial Rings

A *generalized left uniserial* or *left serial ring* is a left Artinian ring R in which each principal indecomposable left ideal has a unique composition series as a left R-module. A *left uniserial ring* is a left serial ring which is a direct product of finitely many primary rings. A *left serial* ring R is a ring (not necessarily Artinian) which is a finite direct sum of *uniserial* (= chain) left modules (e.g. a finite direct sum of valuation rings is serial).

An *Artinian principal ideal ring* is a ring which is both left and right Artinian and a left and right principal ideal ring.

5.1B Classical PIR Theorem. *Any commutative PIR is FGC.*

This follows classically since PIR's are elementary divisor rings (see 6.3), or one can apply 7.5A and reduce to Artinian PIR's (which are uniserial, hence $\sum$-cyclic) and PID's which are FGC by the Fundamental Theorem of Abelian groups.

Uzkov put on the finishing touches:

5.1C Theorem (Uzkov [63]). *A commutative Noetherian ring R is FGC iff R is a PIR.*

5.2A Theorem (Köthe [35], Nakayama [39,40,41]). *Artinian serial rings are Σ-cyclic.*

Below, $J(R) = \text{rad } R$.

5.2B (Nakayama Theorem [40a]). *A semiprimary ring R is Artinian serial if $R/J(R)^2$ is QF.*

See Faith [76], p.23F, Cor.25.4.3. Also see 5.2E below, 113.7A, and 13.15.

5.2C Theorem (Murase [63,64], Amdal-Ringdal [68], and Eisenbud-Griffith [71b]). *Any Artinian serial ring A decomposes into a product $A_0 \times A_1 \times A_2 \times A_3$ where $J(R)$ denotes the radical of any ring R, and:*
1. *A_0 is semisimple and A_i has no semisimple direct factors, $i > 0$.*
2. *A_1 is Artinian PIR.*
3. *A_2 has finite global dimension modulo $J(A_2)^2$ (= the square of its radical).*
4. *A_3 is QF modulo $J(A_3)^2$.*

5.2D Theorem (Skornyakov, cited by Eisenbud-Griffith [71b], p.120). *A ring R is a serial Artinian ring iff every left R-module is a direct sum of uniserial modules.*

Cf. Skornyakov [69].

5.2E Theorem (Fuller [69a,b]). *For a left Artinian ring R the following are equivalent:*
1. *R is serial,*
2. *every 2-generated module is a direct sum of uniserial modules,*
3. *every indecomposable (right and left) module is quasi-injective (= QI),*
4. *every indecomposable module is quasi-projective (= QP),*
5. *every indecomposable left R-module is both QI and QP,*
6. *$R/J(R)^2$ is serial,*
7. *every indecomposable injective, or projective, left R-module is uniserial.*

Cf. Xue [90], p.136.

5.3A Theorem (Eisenbud-Griffith [71a], Eisenbud-Robson [70b]). *A hereditary Noetherian prime (= HNP) ring R is an Artinian serial ring modulo any ideal $I \neq 0$* (Cf. 7.1).

5.3B Theorem. *Over a hereditary Noetherian prime (HNP) ring R: every finitely generated module M is a direct sum of uniserial (= has a unique decomposition series) modules and right ideals.*

This follows since Kaplansky's theorem for modules over hereditary rings implies that the torsion submodule $t(M)$ splits off, and that $M/t(M)$ is isomorphic to a direct sum of right ideals; then the theorem of Eisenbud, Griffith and Robson above applies: modulo any nonzero ideal R is a (generalized uni) serial ring, so by Nakayama's theorem just cited $M \approx t(M) \oplus M/t(M)$ is isomorphic to a direct sum of uniserial modules and right ideals.

5.3C THEOREM (WARFIELD [75]). *A Noetherian serial ring is a direct product of an Artinian serial ring and a finite product of hereditary prime rings.*

Cf. Chatter's Theorem 7.2; also Robson's Theorem 7.4.

Nonsingular Rings

A ring R is **left nonsingular** if $x = 0$ is the only element with essential left annihilator ${}^{\perp}x$, that is, R has zero left singular ideal. (See §12 for an elaboration and an equivalent formulation.)

Thus, R is a left nonsingular ring if ${}_RR$ is a nonsingular module (**sup.** 4.1E).

For the concept $\oplus$acc, see **sup.** 3.13.

5.3D THEOREM (WARFIELD [75]). *For a semiperfect ring R the following are equivalent:*

1. *R is left nonsingular (= n.s.) and serial.*
2. *R is left semihereditary (= s.h.), $\oplus$acc, and $Q^{\ell}_{\max}(R)$ is left flat over R.*
3. *R is left n.s., has finite left Goldie dimension (= $\oplus$acc) and all $f \cdot g$ nonsingular left modules are projective.*
4. *R is serial and s.h. (both sides).*

Singh proved that the converse to 5.3A holds for **(essentially)right bounded** (= every nonzero (essential) right ideal contains an ideal $\neq 0$) Noetherian prime rings. (Cf. **sup.** 5.44.) The two concepts coincide for prime rings. More generally:

5.3E THEOREM (SINGH [76]). *If R is a prime right Goldie right bounded ring such that R/I is an Artinian serial ring for every ideal $I \neq 0$, then R is right hereditary.*

5.3F THEOREM (LEVY AND SMITH [82]). *If R is an essentially right bounded, right Noetherian semiprime ring such that R/I is a serial ring for each right essential two-sided ideal, then R is right hereditary.*

5.3G THEOREM (SINGH [84]). *If R is a right Noetherian semiprime ring whose proper factor rings are all (left and right) serial rings, then either R is a serial Noetherian ring, or a prime ring.*

FGC Rings

All rings in this section are commutative.

Kaplansky [49,52] (cf. [69], p.80) initiated the problem of determining all FGC commutative rings; when all $f \cdot g$ modules decompose into a direct sum of cyclic modules. In 1952, Kaplansky proved that almost valuation domains are FGC domains, and for several decades, these and PID's were the only known FGC domains.

A domain R is (Matlis) *h-local* provided that: (1) every prime ideal $P \neq 0$ is contained in exactly one maximal ideal (= R/P is a local ring); 2) every ideal $I \neq 0$ is contained in just finitely many maximal ideals (= R/I is semilocal).

5.4A MATLIS' THEOREM [64]. *A domain R is h-local iff*

$$Q(R)/R \approx \bigoplus_{M \in \text{mspec} R} (Q/R)_M.$$

The "h" stands for "homological" (Cf. Matlis [66] and [72], p.27), and mspec$R$ is the set of all maximal ideals.

Linearly and Semicompact Modules

A right R-module M is *linearly compact* (=l.c.) in the discrete topology if ($\star$) any finitely solvable system $\{x \equiv x_a (\text{mod } I_a) \mid a \in A\}$ of congruences is solvable for any index set A, where $x_a \in M$ and I_a is a submodule. Any Artinian module is l.c.; any right l.c. ring R is semiperfect (Sandomierski [72]).

A ring R is **semicompact** (Matlis [59]) if ($\star$) holds true assuming just that each $I = \text{ann}_M X$ for a subset $X \neq \phi$ of R (i.e., I is an **annihilator submodule**).

SC 1 THEOREM (MATLIS [59]). *Any injective module over a commutative ring R is semicompact.*

SC 2 THEOREM (FLEISCHER [58]). *If R is a Prüfer domain, any divisible semi-compact module is injective.*

SC 3 THEOREM (MATLIS [85]). *For an $f \cdot g$-injective module M over a commutative ring, the following are equivalent:*

1. *M is injective,*
2. *M is semi-compact,*
3. *M is "principally" semicompact.*

Cf. Theorem 6.B.

Maximal Rings

R is a *maximal* ring if it is commutative and linearly compact. Then R is a finite product of local rings, hence semilocal (Zelinsky [53]). Furthermore, then and only then R has a Morita duality (Ánh [90]). Cf. Morita [58] and Mueller [70]). If R is l.c. with Jacobson radical J and $J^\omega = \bigcap_{n \in \omega} J^n$ then R/J^ω is Noetherian (Mueller [70]).

See Zariski-Samuel [60], Nagata [62] and Cohen [46] for the completion $\hat{R}_p$ of a local ring R_p. Any complete local Noetherian ring, e.g. any Artinian ring, is maximal (see 5.4B).

Almost Maximal Valuation, and Arithmetic Rings

A ring R is *almost maximal* if R/I is maximal for all ideals $I \neq 0$. R is *locally almost maximal* if R_M is almost maximal for each maximal ideal M. An *(almost) maximal valuation ring* is a valuation ring that is (almost) maximal. A ring R is *(right) Bezout* if all *$f \cdot g$ (right) ideals* are principal. Any valuation ring is Bezout, and conversely for a local ring. R is an **Arithmetic ring** if R is locally a valuation ring. Any semilocal Arithmetic ring is Bezout (Hinohara [62]). Cf. Kaplansky [70-74], p.37, Theorem 60.

5.4B THEOREM (MATLIS [58]). *If R is commutative and Noetherian, and P is a prime ideal, then the completion $\hat{R}_p$ is a commutative Noetherian complete local ring, hence maximal (i.e., linearly compact), and $\approx$ $End_R E(R/P)$. Moreover, $E(R/P)$ is an Artinian $\hat{R}_p$-module* (cf. 5.4E).

5.4C THEOREM (MATLIS [59]). *A commutative valuation domain R with quotient field Q is almost maximal iff Q/R is injective.*

Note, by Gupta's theorem 4.2B, for any valuation domain, Q/R is $f \cdot g$-injective.

5.4D THEOREM (GILL [71]). *A valuation ring R with maximal ideal M is almost maximal iff $E(R/M)$ is a uniserial module.*

5.4E THEOREM (FACCHINI [81]). *If R is commutative and M is an Artinian right R-module with simple socle, then $A = End_R M$ is a complete local Noetherian commutative ring and M_A is the minimal injective cogenerator.*

5.5 THEOREM (MATLIS [66]). *A locally almost maximal h-local Bezout domain is FGC.*

5.6 THEOREM (BRANDAL [73], TH. 2.9). *A commutative domain is almost maximal iff it is h-local and locally almost maximal* (cf. S. Wiegand [75]).

5.7 THEOREM (KAPLANSKY [52], BRANDAL [73] AND SHORES-WIEGAND [74]). *An almost maximal commutative Bezout domain is FGC.*

Kaplansky [52] proved this for Almost Maximal Valuation Domains (= AMVD), whereas Gill [71] and Lafon [69] extended it to general almost maximal valuation rings (= AMVR).

Torch Rings

R is a *torch ring* (Shores and Wiegand [74]; also called an *umbrella* ring elsewhere) provided:

1. R is not local,
2. R has a unique minimal prime ideal P,
3. The ideals of R contained in P form a chain under inclusion,
4. R/P is h-local, and
5. R is Bezout and locally almost maximal.

REMARK. See Facchini [83] for examples of torch rings that are not trivial extensions of h-local domains.

5.8 THEOREM (SHORES AND WIEGAND [74]). *A torch ring is FGC.*

Fractionally Self-injective Rings

A commutative ring R is *fractionally self-injective* (= FSI) if for every ideal I the factor ring R/I has self-injective classical quotient ring $Q(R/I)$. In this case $Q(R/I)$ is FSI for every ideal I.

5.9 THEOREM (VÁMOS [77,79]). *Every FSI ring R is a finite product of indecomposable FSI rings. The indecomposable FSI rings are either (I) AMVR's; (II) h-local domains which are locally almost maximal; or (III) torch rings.*

5.10 THEOREM (W. BRANDAL AND R. WIEGAND [76]). *Every FGC ring has only finitely many minimal prime ideals.*

This not only classified reduced FGC rings, but actually in a footnote Vámos [76] pointed out that this theorem together with his results completed the classification of FGC rings as follows (cf. Vámos [79]).

FGC Classification Theorem

5.11 FGC CLASSIFICATION THEOREM. *The following conditions are equivalent conditions on a commutative ring R.*

(FGC 1) *R is a FGC ring (= $f \cdot g$ modules are direct sum of cyclic modules),*
(FGC 2) *R is FSI and Bezout,*
(FGC 3) *R is Bezout and a finite product of the three types of rings (I), (II), and (III) of 5.9.*

Vámos [75] also studied commutative rings, called **SISI rings**, over which every subdirectly irreducible factor ring is self-injective. Morita rings, locally Noetherian rings, and FSI rings are important examples (see **sup.** 9.1).

REMARKS. G. Köthe proved that an Artinian commutative ring R with the property

(right Σ-cyclic): every right module is a direct sum of cyclics

is an uniserial (einreihig) ring (see 5.1A). Cohen and Kaplansky [51] countered with the observation that it was redundant to assume that R is Artinian. S.U. Chase [60] proved commutativity is not necessary to assert the ring is right Artinian, and moreover, that finitely generated modules can replace the cyclics in the statement. (But, then, the ring is no longer necessarily serial, of course.) Finally Faith and Walker [67], Faith [71], and Vámos [71] noticed that finite cardinality of the modules in the direct summands played no role; if there exists a *set* of modules such that every right module decomposes into a direct sum of modules isomorphic to modules in that set, then the ring is right Artinian. Cf. Griffith [70]. The proof of this makes heavy use of another theorem of Chase [60] (see 1.17A) on direct sum decompositions of modules: If there is a cardinal number c not less than the cardinal of R such that the product R^c is a pure submodule (for example, a direct summand) of a direct sum of right R-modules having cardinal not exceeding c, then R satisfies the dcc on principal left ideals (= right perfect rings).

5.12 REMARK. As defined **sup** 4.26, a ring is *FPF* if every finitely generated faithful module generates mod-R, and CFPF if all factor rings of R are FPF. A theorem of Faith ([82a]) shows that any FPF commutative ring R has self-injective quotient ring, hence any CFPF ring is FSI and conversely (cf. Faith and Pillay [90], pp. 67–68). Also see 5.31ff, esp. 5.42.

5.13 REMARK. A reduced commutative ring R is FPF iff R is a semihereditary ring with self-injective quotient ring Q (Faith [79], cf. Faith-Pillay [90]). This is an easy consequence of 5.42.

Maximal Completions of Valuation Rings

The principal ideals of a valuation ring R with maximal ideal m is a totally ordered monoid under reverse order by inclusion, called the **valuemonoid** M (Shores [74]). Let (R, m, M) denote this situation. Then (R', m', M') is said to be an **extension** of (R, m, M) if $R' \supseteq R$ amd $m' \cap R = m$. The extension is said to be **immediate** provided that $k' = R'/m' \approx k = R/m$ and $M' \approx M$ canonically. Then R is **maximally complete** if (R, m, M) has no proper immediate extension.

R' is called a **maximal completion** of R if (R', m', M') is an immediate extension and R' is maximally complete. See Kaplansky [42], p.303, Schilling [50], p.39; Ostrowski [35], p.368 gives a slightly different definition. Also see Vicknair [87], p.56, for this and the following:

5.14A PROPOSITION. *The cardinality of a valuation ring* $[R, m, M]$ *is bounded by that of* $k[[M]]$, *where* $k = R/m$.

5.14B COROLLARY. *Every valuation ring* R *has a maximal completion.*

The next theorem is due to Kaplansky [42], Theorem 4 when R is a domain, and Vicknair [87], Theorem 1.2 for general R.

5.15A THEOREM. *A maximal valuation ring* R *is maximally complete, and conversely for a valuation domain* R.

REMARK. (1) By Gill [70], any almost maximal valuation ring R with zero divisors is maximal; (2) It is easy to see that **any FGC ring is FPF**: if M is finitely generated, then $M = R/I_1 \oplus \cdots \oplus R/I_n$ for ideals $I_1 \subset I_2 \subset \cdots \subset I_n$ (i.e. R is an elementary divisor ring (see **sup** 6.3) in accordance with the structure theory of FGC rings (see e.g. Brandal [79] or Vámos [77]), so M is faithful only if $I_1 = 0$, so $M \approx R \oplus X$ in mod-R as required. Thus, **any FGC ring is CFPF**.

MacLane's and Vámos' Theorems

5.15B THEOREM (MACLANE [39],[24] IN MACLANE [79]). *If* K *is a field with a maximal valuation, then* K *is algebraically closed iff the value group is divisible and the residue field of the valuation ring is algebraically closed.*

Cf. Kaplansky's article in *op.cit.*, esp.p.523, §6. Also Vámos [75b], p.107,Prop.9, and Theorem A, p.108. Ánh [97] has a new approach and discusses the relation with Newton's method of approximating roots.

5.15C THEOREM (VÁMOS [75B]). *A field* K *is multiply maximally complete with respect to two inequivalent non-trivial valuations iff* K *is algebraically closed and* $|K|^{\aleph_0} = |K|$.

5.15D COROLLARY. *A maximally complete field* K *is multiply maximally complete iff* K *is algebraically closed.*

This follows from Vámos' theorem because any maximal valuation domain R not a field satisfies $|R|^{\aleph_0} = |R|$ by Proposition 4, *op.cit.*

Gill's Theorem

Let (R, m) denote a local ring with maximal ideal m.

5.16 THEOREM (GILL [71]). *A commutative local ring (R, m) is an almost maximal valuation ring (= AMVR) iff $E = E(R/m)$ is uniserial. (Thus, if R is not a domain, then R is an MVR iff E is uniserial.)*

5.17 MUELLER-ÁNH THEOREM. *A commutative local ring (R, m) is maximal, equivalently, linearly compact iff $E = E(R/m)$ is linearly compact over R and $R \approx End_R E$ canonically.*

This follows from Ánh's Theorem 13.6, and Mueller's Theorems 13.5 and 13.1.

5.18 COROLLARY. *A commutative local ring (R, m) is a MVR iff R is linearly compact and $E(R/m)$ is uniserial.*

Vamosian Rings

A commutative ring R is **Vamosian** (after the author's paper [86b]) provided $E(R/m)$ is linearly compact (= l.c.) for all maximal ideals m. A ring R is SISI provided that subdirectly irreducible factor rings are self-injective, equivalently (Vámos [75]), $E(R/m)$ is fully invariant for all maximal ideals m (equivalently, submodules of $E(R/m)$ are quasi-injective).

EXAMPLE (VÁMOS [75]). Any commutative Morita ring R, (hence by Ánh's theorem 13.6 any l.c. R), is Vamosian, and any Vamosian ring is SISI.

Another (Vámos) example: If R_m is Noetherian for all maximal ideals m (= R is locally Noetherian), then R is SISI. In this case, $R[x]$ is locally Noetherian, hence SISI, for any variable x (see, e.g. Faith [86b]).

5.19 THEOREM (VÁMOS [75], FAITH [86B]). *A commutative valuation ring R is an AMVR iff R is SISI. (Thus, a SISI valuation ring with zero divisors is maximal.)*

REMARK. Vámos' Theorem [75], p.126, was stated for R Vamosian (= **classical** in *op.cit.*).

Quotient Finite Dimensional Modules

An R-module M is **quotient finite dimensional** (= q.f.d.) provided that every factor module M/N of M has finite Goldie dimension.

5.20A THEOREM (KURSHAN [70]). *A right R-module M is q.f.d. iff every factor module has $f \cdot g$ socle.*

5.20B THEOREM (CAMILLO [77]). *An R-module M is q.f.d. iff for every submodule N there is a $f \cdot g$ submodule S such that rad$(N/S) = N/S$, that is, N/S has no maximal submodules.*

Camillo's proof requires a theorem of Shock [74]. See 7.21ff.

It is known that any linearly compact module M is q.f.d. See Herbera-Shamsuddin [95] for references. Also see 6.37.

REMARK. A module M is Noetherian iff M is q.f.d. and satisfies the acc on subdirectly irreducible submodules (Faith [98]). See 16.50.

5.21 THEOREM (VÁMOS [75]). *Over a semilocal commutative Vamosian ring R, every $f \cdot g$ module M is q.f.d.*

5.22 EXAMPLE (*loc.cit.*). If $R = k\langle x_1, \ldots, x_n, \ldots \rangle$ is the power series ring in infinitely many variables, and if A is the factor ring module the ideal generated by all products $\{x_i x_j\}_{i,j=1}^{\infty}$, then A is SISI but not Vamosian. Every subdirectly irreducible factor ring of A has Jordan-Hölder length ≤ 2, hence is QF, hence self-injective, so A is SISI, but not q.f.d., hence not Vamosian.

The Genus of a Module and Generic Families of Rings

A right module M over a ring R is said to have a **unimodular element** (UME) if there exists $u \in M$ such that uR is a direct summand of M canonically isomorphic to R. Thus, M has a UME iff there is an epimorphism $M \twoheadrightarrow R$.

Thus, M generates mod-R if and only if $M^n \to R$ is onto for some integer $n > 0$; equivalently, M^n has a UME. In this case, we let $\gamma(M)$ denote the infimum of all such integers n, and call this the **genus** of M. If M does not generate mod-R, we set $\gamma(M) = \infty$. The **(little) right genus** of a ring R will be denoted by $g_r(R)$ and is defined to be the supremum of $\gamma(M) < \infty$ for M finitely generated in mod-R. The **big right genus** $G_r(R)$ is defined similarly without restriction on finite generation of M. Clearly, $g_r(R) \leq G_r(R)$, and equality holds when R is a right Noetherian ring (R. Wiegand, 5.25). Also by a theorem of Vasconcelos and Wiegand, if $\dim R = n < \infty$, then $G(R) \leq n+1$ (see 5.25).

A family $F = \{R_i\}_{i \in I}$ of rings is **generic of (with) bound** B or **right** B-**generic** if there exists a function $B : \mathbb{Z}^+ \to \mathbb{Z}^+$ such that for all modules M if $\gamma(M) < \infty$ and $\nu(M) < \infty$ is the minimal number of elements in any set of generators of M, then $\gamma(M) \leq B(\nu(M))$. The product theorem states that any product of a generic family of rings of bound B is a ring which is generic of bound B (considering a ring as a family with one member) (see Theorem 5.28).

For example, a family of rings each of genus $\leq g$ is generic with bound $\leq g$, where g also denotes the constant function. Moreover, any family of commutative rings is generic of bound $1_{\mathbb{Z}^+}$. A corollary of the product theorem is that any product $R = \Pi_{i \in I} R_i$ of a generic family of right FPF rings is right FPF. (In particular, the product any family of commutative FPF rings is FPF.) This implies that any product of self-basic right FPF rings, in particular, any product of self-basic right PF rings is right FPF.

Another corollary to the product theorem states that if $\{R_k\}_{i \in I}$ is any family of commutative rings each having the property $P(n,g)$: there exist integers $n \geq 0$ and $g > 0$ with the property that for all $i \in I$ every finitely generated R_i-module of free rank $\geq n+1$ has genus $\leq g$, then their product R also has property $P(n,g)$ (see Corollary 5.34). The FPF theorem is the case $P(0,1)$.

That a commutative ring is generic with bound the identity is given by:

5.23A THEOREM (W. VASCONCELOS). *If R is a commutative ring, then $\gamma(M) \leq \nu(M)$ for any $f \cdot g$ generator M.*

PROOF. Let $M^n \twoheadrightarrow R$. Then there exist elements $x_1, \dots, x_n \in M$, $f_n \dots, f_n \in M^\star$ such that $\sum_{i=1} f_i(x_i) = 1$. If $t = \nu(M)$, and if $m_1, \dots, m_t$ generate M, then $x_i = \sum_{j=1}^t m_j a_{ij}$ for some $a_{ij} \in R$, $i = 1, \dots, n$. However, $f'_j = \sum_{i=1}^n f_j a_{ij} \in M^\star$, $j = 1, \dots, t$ is such that $\sum_{j=1}^t f'_j(m_j) = 1$, so that $M^t \twoheadrightarrow R$ holds, that is, $\gamma(M) \leq t = \nu(M)$.

5.23B COROLLARY. *If M is a $f \cdot g$ faithful projective over a commutative ring R, then (M generates mod-R) and $\gamma(M) = \gamma(M^\star) \leq \nu(M)$.*

PROOF. M generates mod-R by a theorem of Azumaya [66].

5.24 THEOREM (R. WIEGAND AND W. VASCONCELOS [78]). *If R is commutative of dimension n, then $G(R) \leq n+1$, hence $\gamma(M) \leq n+1$ for any generator M.*

5.25 THEOREM (WIEGAND). *If R is a right Noetherian ring, then $G(R) = g(R)$.*

For proofs, see the author's paper [79c], pp. 621-622.

5.26 REMARKS. (1) An unpublished result of D. Eisenbud states for $R = k[x,y]$, the polynomial ring in 2 variables over a field k, that $G(R) = 1$; (2) R. Wiegand has asked which commutative rings have the property that every generator has a faithful direct summand.

The Product Theorem

The proof of the product theorem requires:

5.27 LEMMA. *The only $f \cdot g$ right ideal H of a product $\Pi_{i \in I} R_i$ of rings which contains the direct sum $\oplus_{i \in I} R_i$ is the unit ideal.*

5.28 THE PRODUCT THEOREM (FAITH [79C]). *A family $\{R_i\}_{i \in I}$ of rings is right B-generic iff the product $R = \Pi_{i \in I} R_i$ is B-generic. Thus, for every generator of mod-R with $\nu(M) = n < \infty$ we have:*

$$\gamma(M) = \sup\{\gamma(M_i)\}_{i \in I} \leq B(n)$$

where $M_i = Me_i$, and $e_i \in R_i$ is the identity element, $\forall i \in I$.

5.29 COROLLARY. *If M is an $f \cdot g$ module over a product of rings $R = \Pi_{i \in I} R_i$, if $M_i = Me_i$ generates mod-R_i where $e_i : R \to R_i$ is the projection idempotent, and if sup$\{\gamma(M)_i)\}_{i \in I} = \gamma < \infty$, then M generates mod-R and $\gamma(M) = \gamma$. Thus*

$$\gamma(M) = \sup\{\gamma_{R_i}(M_i)\}_{i \in I}.$$

5.30 COROLLARY. *Let $R = \Pi_{i \in I} R_i$. Then*

$$g_r(R) = \sup\{g_r(R_i)\}_{i \in I}.$$

This follows from the corollary and the proof of the product theorem.

5.31 COROLLARY. *Any product of commutative FPF rings is FPF. Similarly for products of right FPF self-basic rings.*

PROOF. Both are generic families.

5.32 COROLLARY. *Any right generic product of right $(F)PF$ rings is right FPF.*

5.33 EXAMPLES. (1) If F_n is the $n \times n$ matrix ring over a local ring R, then the product $R = \prod_{n \in Z^+} F_n$ is not generic, since $\gamma(M) = \infty$ for the cyclic module $M = eR$, where $e = e^2$ is the idempotent the j-th component of which is the e_{11}-matrix in F_n; (2) An infinite product of right PF rings is never right PF since a semiperfect ring contains no infinite sets of orthogonal idempotents. Furthermore, a product of PF rings is not necessarily FPF, e.g. $R = \prod_{n \in Z^+} F_n$ is not. Nevertheless, any product of right PF rings of right genus $\leq g$ is right FPF of genus $\leq g$, according to Corollary 1.22E; (3) Let $R = \prod_{i \in I} M_n(F_i)$, where F_i is a self-basic right $(F)PF$ ring. Then R is right FPF of genus n, according to Cor. 5.32 since $G_r(M_n(F_i)) = n$, $\forall i \in I$, by Example (1) 5.33.

Serre's Condition

DEFINITION. The **free rank** of a module M, denoted *frk(M)* is the smallest integer t so that for every maximal ideal P, the local module M_P has a free direct summand $\approx R_p^t$.

Below $frk =$ **free rank**.

5.34 COROLLARY. *Let $R = \prod_{i \in I} R_i$ be a product of commutative rings such that there exists an integer $n > 0$ such that each R_i satisfies Serre's condition $P(n, g)$: that is, any finitely generated R_i-module of $frk > n+1$ has an unimodular element. Then R satisfies $P(n, g)$.*

PROOF. Let M be any finitely generated R-module of $frk \geq n+1$. If P_i is any maximal ideal of R_i, then $P = P_i \oplus R_i$ where $R_i = \prod_{j \neq i} R_j$, is maximal in R, and $(M_i)_p = M_p$ has $rk \geq n + 1$, so M_i has unimodular element, that is, $\gamma(M_i) = 1$; hence $\gamma(M) = 1$ by Corollary 5.29.

5.35 THEOREM. *Let $\{R_i\}_{i \in I}$ be a family of rings such that R_i is a commutative ring of one of the following types:*

(i) *a Bezout domain,*
(ii) *a local FPF ring (e.g. any AMVR, or any self-injective local ring),*
(iii) *an FPF ring of genus 1,*
(iv) *any product of rings $\{R_i\}$ where R_i has type (i)–(iv).*

Then: $R = \prod_{i \in I} R_i$ is FPF of genus 1.

PROOF. The rings (i)–(iii) are all FPF of genus 1; hence by Corollary 5.29 so are the rings in (iv); hence so is $R = \prod_{i \in I} R_i$.

A commutative local ring R is FPF iff every faithful module M with $\nu(M) = 2$ is a direct sum of cyclics [79a]. This is generalized to arbitrary products of commutative rings of genus 1 in Theorem 5.38 and Corollary 5.39. Any self-injective commutative ring is FPF [79a].

5.36 PROPOSITION. *If R is any ring and M is a generator such that $\gamma(M) = 1$ and $2 \leq \nu(M) = n < \infty$, then*

$$M \approx R \oplus B/K$$

where B is an $f \cdot g$ projective such that

$$R^n \approx R \oplus B.$$

See the author's paper [79c], Prop.13.

5.37 COROLLARY. *If R is commutative, then in the proposition, B is a progenerator ($= f \cdot g$ projective generator).*

PROOF. By Azumaya's theorem, all that is required is that B be faithful. But $R^n = R_1 \oplus B \Rightarrow R^n a = (Ra)^n \approx R_1 a$ for all $a \in R$ which annihilates B, and this implies $n = 1$ since $R_1 a$ is cyclic, contrary to the assumption.

5.38A THE 2×2 THEOREM. *If R is FPF and commutative of genus 1, then every faithful module M with $\nu(M) = 2$ is a direct sum of two cyclics: $M \approx R \oplus R/K$.*

PROOF. $\gamma(M) = 1$ so the corollary applies: $M = R \oplus B/K$, where $R^2 \approx R \oplus B$, and B generates mod-R. Then $B \approx R \oplus Y$ so $R^2 \approx R^2 \oplus Y$ which means that $Y_m = 0$, $\forall$ maximal ideals m; hence $Y = 0$, and $B \approx R$, so $M = R \oplus R/K$ is a direct sum of cyclics.

We shall abbreviate the conclusion of the 2×2 Theorem by the terminology: **Every faithful 2-generated module is 2-cyclic.** In this case, **we say the 2×2 Theorem holds**.

5.38B COROLLARY. *Any product of commutative FPF rings of genus 1 is FPF, and hence the 2×2 Theorem holds.*

PROOF. Apply 5.35(iii) to the 2×2 theorem 5.38A.

5.39 REMARK. The FGC Theorem 5.11 shows that no infinite product of rings can be CFPF, that is, that product theorem for FPF rings fails for CFPF rings. (Finite products of CFPF rings are CFPF however.)

In [67], Endo initiated the study of FPF rings.

5.40 ENDO'S THEOREM [67]. *A commutative Noetherian ring R is FPF iff R is a finite product of Dedekind domains and QF rings.*

FPF Split Null Extensions

5.41 THEOREM. *Let $R = B \ltimes E$ be the split-null or trivial extension of a faithful module E over a commutative ring B. R is an FPF ring iff the partial quotient ring BS^{-1} with respect to the set S of elements of B with zero annihilator in E is canonically the endomorphism ring of E, that is $BS^{-1} = \mathrm{End}_B ES^{-1}$.*

Every finitely generated ideal with zero annihilator in E is invertible in BS^{-1}, and $E = ES^{-1}$ is an injective module over B.

The proof uses the author's following characterization of commutative FPF rings, (5.25), and also the characterization of self-injectivity of a split-null extension (4.25).

Characterization of Commutative FPF Rings

5.42 FPF RING THEOREM (FAITH [82A]). *A commutative ring is FPF iff R has the following two properties:*

(FPF 1) Every finitely generated faithful ideal is projective.
(FPF 2) R has injective quotient ring $Q_c(R)$.

When (FPF 2) holds, R is said to be **quotient-injective**. This theorem is illustrated by the result that a domain R is FPF iff R is Prüfer. (In view of the fact that by the theorem a commutative self-injective ring R is FPF, the theorem indicates that finitely generated faithful ideals of a self-injective ring R are projective, but the fact that $f \cdot g$ ideals in a self-injective, or $f \cdot g$-injective ring are annihilators implies that R is the only one!) Theorem 5.42 is contained in the lecture notes of Pillay and the author.

Semiperfect FPF Rings

5.43 THEOREM (YOUSIF [91]). *If R is a semiperfect right FPF ring, then R is right self-injective iff the Jacobson radical $J(R)$ is the right singular ideal.*

5.44A COROLLARY (FAITH [76,I]). *(1) If R is a semiperfect right FPF ring with nil Jacobson radical, then R is right self-injective; (2) Moreover, a local right FPF ring R is self-injective iff rad R consists of zero divisors.*

As Yousif remarks: To deduce (1) of the Corollary from the theorem requires a result of Faticoni [87] stating that nil ideals are right singular in semiperfect right FPF rings.

A ring R is **right bounded** if every essential right ideal contains an ideal $\neq 0$. Then R is **strongly right bounded** if every nonzero right ideal contains a nonzero ideal. The two concepts coincide for prime rings.

5.44B THEOREM (FAITH [76,77B]). *Any right FPF ring R is right bounded.*

5.45 THEOREM (FAITH [77B]). *(1) Any right self-injective semiperfect ring R with strongly right bounded basic ring R_0 is necessarily right FPF; (2) Semiprime semiperfect right FPF rings are semihereditary and finite products of full matrix rings of finite rank over right bounded local Ore domains which are right and left valuation rings; (3) The basic ring of a semiperfect right CFPF ring is right duo (= right ideals are two-sided), right FGC, and are finite products of right valuation rings; (4) Conversely, any ring Morita equivalent to a finite product of rings that are right FGC, right VR and right duo is a semiperfect right CFPF ring.*

5.46 REMARKS. Faticoni [84,87] made significant additions to the theory. (Cf. 5.46A,B and 5.48.) Endo [67] also studied the situation where every finitely generated projective faithful R-module generates mod-R.

Faticoni's Theorem

We next cite two theorems of Faticoni. Also see Theorem 5.48.

5.46A THEOREM (FATICONI [87,88]). *Let R be semiperfect in which* $(*)$ *every right regular element is regular. Then: R is right FPF iff (i) the basic ring R_0 is strongly right bounded; (ii) every $f \cdot g$ faithful right ideal is a generator, and (iii) R has a semiperfect right self-injective classical right quotient ring $Q = Q^r_{\text{cl}}(R)$.*

5.46B THEOREM (*Op.cit.*). *If R is right FPF, then $Q^r_{\max}(R)$ is semiperfect right self-injective iff R has finite right Goldie dimension and* $(*)$ *holds.*

Kaplansky's and Levy's Maximal Valuation Rings

5.47 EXAMPLES. (1) Kaplansky [42] constructed rings of formal power series $\Sigma_{\gamma\in\Gamma}\alpha_\gamma x^\gamma$ in a variable x, with coefficients α_γ in a field, and exponents γ coming from a totally ordered group Γ, and showed these rings are MVR's i.e., there exist MVR's with arbitrary value group Γ; (2) By the FGC Classification Theorem 5.11, a ring R is CFPF iff FSI, and a local ring R is CFPF iff R is AMVR. Cf. Faith-Pillay [90], p.66, Theorem 3.17; (3) Levy [66] gave an example of a non-Noetherian commutative ring R of which all factor rings modulo nonzero ideals are self-injective rings, and some of the factor rings are PF. The ring exhibited is the ring R of all formal power series in a variable x indexed by the family W of all well-ordered sets of nonnegative real numbers. Thus, an element r of R has the form $r = \Sigma_{i\in W} a_i x^i$, with $a_i \in R$, and unique $i \in W$. The only nonzero ideals of R are: the principal ideals (x^b), and those ideals of the form

$$(x^{>b}) = \{x^c u \mid c > b, \quad \text{and } u \text{ a unit of } R\}.$$

Thus, if I is any nonzero ideal, then $\overline{R} = R/I$ is completely self-injective (and non-Noetherian). By 5.9, R is an AMVR, and FGC by 5.11.

The next theorem generalizes Endo's theorem 5.40.

5.48 THEOREM (FAITH-PAGE [81], FATICONI [85]). *A Noetherian ring R is FPF iff R is the finite product of bounded Dedekind Prime rings and a QF ring.*

REMARK. Faith-Page *op.cit.* further assumed that R is semiperfect.

Page's Theorems

5.49 PAGE'S THEOREM [78]. *A VNR ring R is right FPF iff R is a right self-injective of bounded index. In this case R is left self-injective and left FPF.*

5.50 COROLLARY (*Loc.Cit.*). *A right self-injective VNR ring R is right FPF iff R is Morita equivalent to a strongly regular (= Abelian) right and left self-injective ring.*

REMARK. Any strongly regular right self-injective ring R is left self-injective by Utumi's Theorem 4.3A.

The next result belongs in §12 where the concepts maximal right (left) quotient rings are defined.

5.51 THEOREM (PAGE [83]). *If R is a right nonsingular right FPF ring R, then $Q = Q^r_{\max}(R) = Q^\ell_{\max}(R)$, and is FPF.*

See our Lecture Notes Faith-Page [81], for additional results on nonsingular FPF rings.

Further Examples of Valuation Rings and PF Rings

A ring R is a **right chain ring**, or **right valuation ring (= VR)** provided the set of right ideals are linearly ordered. As before (see Remark 4.84f), $R = B \ltimes E$ denotes the split-null extension of a B-bimodule E. A module E_B is **uniserial** if its submodules are linearly ordered. We next cull some examples from the author's paper [79b].

5.52 PROPOSITION. *A split-null extension $R = B \ltimes e$ is a right VR iff B is a right VR, E is uniserial, and $bE = E$, $\forall 0 \neq b \in B$.*

PROOF. If R is a right VR, then $B \approx R/(0,E)$ is right VR, and $E \approx (0,E)$ is uniserial. If $b \neq 0 \in B$, then $(b,0)R \not\subset (0,E)$; hence

$$(b,0)R = (bB,0) + (0,bE) \supseteq (0,E),$$

so $bE = E$. The converse follows by reading up. □

A VD is a domain which is a VR. *For simplicity, from here on we shall assume that B whence R is commutative.*

5.53 COROLLARY. *Let E be a faithful B-module over a commutative ring B. Then $R = B \ltimes E$ is a VR iff B is a VD and E is an uniserial divisible B-module.*

PROOF. Immediate. □

5.54 COROLLARY. *Let E be a torsion free module over a commutative domain B. Then $R = B \ltimes E$ is a VR iff B is a VD and E is a uniserial injective B-module. In this case R is injective iff E is* **strongly balanced**, *i.e., $B = End_B E$.*

PROOF. Any torsion free divisible module over a domain is injective, so apply the corollary. (Conversely, any injective module is divisible.) The last sentence follows from Theorem 4.25. □

5.55 THEOREM. *Let $R = B \ltimes E$ be a split-null extension where B is commutative. Then, the following are equivalent:*

1. *R is a $PFVR$ (= a VR which is PF).*
2. *B is an almost maximal valuation domain ($AMVD$), $E = E(B/rad\ B)$ is the injective hull of $B/rad\ B$, and $B = End_B E$.*
3. *B is a local domain such that $E = E(B/rad\ B)$ is uniserial and strongly balanced.*
4. *B is an MVD and $E = E(B/rad\ B)$ is strongly balanced.*

PROOF. By Gill's theorem 5.4D, a local ring B is $AMVR$ iff $E(B/J)$ is uniserial, where $J =$ rad B. Thus, using Theorems 4.25 and 5A, (3) $\Leftrightarrow$ (3) follows. Moreover, (1) $\Leftrightarrow$ (3) by 5.54 and Corollary 4A of the author [79b]; and (2) $\Leftrightarrow$ (4) by a theorem of Vámos [75]. □

5.56 COROLLARY. *If B is a Noetherian commutative local domain, and $E = E(B/J)$, then the ring $R = B \ltimes E$ is an injective VR iff B is a complete discrete valuation domain. In this case R is PF.*

PROOF. Follows from 5.55 and Matlis' theorem 5.4B (since B is a Noetherian VD). □

Historical Note

The proof of the characterization of commutative FPF rings (see Theorem 5.42) is nontrivial and requires the concept of a maximal quotient ring $Q_{\max}(R)$ (see 9.27s and §12).

REMARK. Yoshimura [98] characterizes an FPF ring R by a 1-1 correspondence between invertible ideals and $f \cdot g$ overmodules (in which case they are projective by Theorem 5.42), where "invertibility" and "overmodule" are defined with respect to the injective hull $E(R)$ of R. Cf. Faith-Pillay [90], p.45, Theorem 2.21 which implies that $E(R) = Q_{\max}(R)$ in this case. Also see 12.14.

CHAPTER 6

When Injectives Are Flat: Coherent FP-Injective Rings

A right R-module M is **finitely presented** ($= f \cdot p$) provided that there is an exact sequence

$$R^m \to R^n \to M \to 0$$

where m and n are integers ≥ 1. Any $f \cdot p$ R-module is $f \cdot g$, and any $f \cdot g$ right R-module over a right Noetherian ring is $f \cdot p$. Moreover, any $f \cdot g$ projective R-module is $f \cdot p$.

A right R-module M is **FP-injective** provided that $\mathrm{Ext}^1_R(F, M) = 0$ for every finitely presented R-module F, equivalently, every homomorphism $S \to M$ of a $f \cdot g$ submodule S of a free module F extends to a homomorphism $F \to M$.

A submodule A of a right R-module B is (Cohn) **pure** in B provided that the induced homomorphism $\mathrm{Hom}_R(M, B) \to \mathrm{Hom}_R(M, B/A)$ is an epimorphism for all $f \cdot p$ R-modules M, or equivalently, for any $f \cdot p$ left R-module F, the canonical homomorphism $A \otimes_R F \to B \otimes_R F$ is an embedding. (See Cohn [59], or Warfield [69].)

6.A THEOREM. *A right R module M is FP-injective iff M is a pure submodule of an injective module.*

6.A was cited in Menal-Vámos [84] as "probably folklore, and the proof is left to the reader."

Pure Injective Modules

As defined informally, **sup.** 1.26, a right R-module M is **algebraically compact**, or **pure-injective** provided that every system of linear equations for arbitrary index sets I and J of the form

$$L(I) \qquad \Sigma_{j\in J} x_j r_{ij} = m_i \in M \qquad (i \in I, r_{ij} \in R)$$

in the unknowns $(x_j)_{j\in J}$, and $r_{ij} \in R$ are almost all $= 0$ for each $i \in I$, can be solved provided that $L(I')$ can be solved for each finite subset I' of I.

Thus: **linear equations are simultaneously solvable iff finitely solvable** (Mycielski [64]), cited by Warfield [69a], p.707, who establishes that pure-injective envelopes exist (see 6.46 below), and determines the algebraically compact modules over Prüfer rings, extending Kaplansky's classification over PID's.

REMARK. The terminology for pure-injective derives from the property that a right R-module M is pure-injective iff M is a direct summand of any module containing M as a pure submodule (See, e.g. Fuchs [69], Warfield [69a], or Kaplansky [69], Proposition, p.84) The next result is a consequence of this and 6.A.

6.B THEOREM. *A right R-module M is injective iff M is FP-injective and pure-injective.*

6.C THEOREM. *(See Theorem 6.52) If R is commutative and M a pure-injective R-module, then End M_R is an "F-semiperfect" ring.*

Cf. Jensen-Lenzing [89], p. 180, Corollary 8.27.

6.D THEOREM (WARFIELD [69A]). *If R is commutative, then any linearly compact R-module is pure-injective.*

6.D′ THEOREM (ZIMMERMANN [77]). *If $_RM_S$ is a bimodule, and if $_RM$ is linearly compact, then M_S is pure-injective.*

REMARKS. (1) Zimmermann [82] gives an example of a right Artinian, hence linearly compact, ring not right pure-injective; (2) Also see Onodera [81] for a simple proof of 6.D'; (3) A left linearly compact bimodule $_AM_B$ is right pure-injective by Zimmermann's Theorem [72,77]. See Jensen-Lenzing [89], p.289, Theorem 11.18. (The converse does not hold (*loc.cit.*) p.303.); (4) For pure-injective group rings, see 11.8.

A ring R is right (self) FP-injective if R_R is an FP-injective R-module. A ring R is *right $\aleph_0$-injective* if $\text{Ext}^1_R(R/I, R) = 0$ for all countably generated right ideals I, equivalently for every mapping $f : I \to R$ there exists $a \in R$ so that $f(x) = ax \quad \forall x \in I$. R is right (self) **p-injective (weak-$\aleph_0$ or $f \cdot g$-injective)** if this holds true for all principal (resp. $f \cdot g$) right ideals I. A right self-FP-injective ring R is weak $\aleph_0$-injective. A VNR ring is right and left FP-injective (cf. 6.2).

6.E THEOREM (PUNINSKI, CITED BY NICHOLSON-YOUSIF [95]). *A ring R is right FP-injective iff R_n is right p-injective for all $n \geq 1$.*

6.1 THEOREM OF MENAL-VÁMOS [89]. *Every ring A is embedded in a ring R that is right and left (self) FP-injective. (Cf. 6.21.)*

6.2A REMARK. In certain cases, e.g. when R contains a field, or has torsionfree additive group, or is commutative and coherent, then R is FP-injective iff R is pure in all its ring extensions (Menal-Vámos [89]). It is known that not every ring A embeds in a self-injective ring (*loc.cit.), and Mal'cev [39] showed that not every integral domain embeds in a sfield. Cf. 6.26ff.*

6.2B THEOREM (STENSTRÖM [70]-S. JAIN [73]). *A ring R is left FP-injective iff every finitely presented right R-module is torsionless.*

From 6.2B one can show that R is left FP-injective iff every $n \times n$ matrix ring R_n is left $f \cdot g$-injective and iff every $f \cdot g$ right ideal of R_n is a right annihilator $\forall n$.

6.2C THEOREM (MEGIBBEN [70]). *If R is a Prüfer domain, then an R-module M is FP-injective iff M is divisible: $Mr = M \quad \forall 0 \neq r \in R$.*

6.2D COROLLARY. *Every factor module of an injective module over a Prüfer domain is FP-injective* (cf. 4.2B).

Elementary Divisor Rings

A ring R is an **elementary divisor ring** (=EDR) if every $n \times n$ matrix over R is equivalent to a diagonal matrix. Prominent examples are VR's, VNR rings, and PID's. That $\mathbb{Z}$ is an EDR is due to Smith in 1870, and 50 years later Steinitz [11,12] studied matrices and modules over the ring A of algebraic integers. See O'Neill [96b] where I found this reference.

6.3A THEOREM (KAPLANSKY [49], LARSEN, LEWIS AND SHORES [74]). *A commutative ring R is an elementary divisor ring iff every finitely presented module is a direct sum of cyclic R-modules.*

In [69], Kaplansky states (p.80) that Theorem 6.3 for a domain R "ought to have been the main theorem of [49]" but it was never clearly stated! He also asks whether every Bezout domain is an EDR.

If an ideal I has the property that for every $0 \neq a \in I$ there exists $b \in I$ such that $I = aR + bR$, then I is said to be **generated by $1\frac{1}{2}$ elements**.

REMARK. (1) Any EDR is Arithmetic (see 6.4 and 6.5A below), in fact, Bezout (*loc.cit.*). A semilocal Arithmetic ring is Bezout by the theorem of Hinohara [62], **sup** 5.4B; (2) Heitman and Levy [H-L] discovered interesting examples of Prüfer domains in which every $f \cdot g$ ideal is generated by 1-1/2 elements, e.g. every Prüfer domain R of dimension 1 has this property, and if R has Jacobson radical *not* $= 0$, then R is Bezout ($= f \cdot g$ ideals are principal).

A ring R is right(left) **Hermite** if every 1×2 (resp. 2×1) matrix is equivalent to a diagonal matrix. Any right Hermite ring is left Hermite (Menal-Moncasi [82]). A ring R is **unit-regular** if for every $a \in R$ there exists a unit $u \in R$ so that $aua = a$. A VNR ring R is unit-regular iff R cancels from direct sums, i.e., $R \oplus A \approx R \oplus B \Rightarrow A \approx B$. More generally, one has the following:

6.3B THEOREM (EHRLICH [76]). *Let M be a right R-module such that $A = \mathrm{End} M_R$ is VNR. The following conditions are equivalent:*
(1) *A is unit-regular.*
(2) *If $M = M_1 \oplus M_2 = N_1 \oplus N_2$ and if $M_1 \approx N_1$, then $M_2 \approx N_2$.*
(3) *If e and f are idempotents of A such that $eA \approx fA$, then $(1-e)A \approx (1-f)A$.*

6.3C COROLLARY. *Any Abelian VNR is unit-regular.*

PROOF. See Goodearl [79], p.38.

DEFINITION. An R-module M has the **cancellation property** if (2) of 6.3B holds.

6.3D THEOREM (HENDRIKSEN [73]). *Every unit-regular ring R is an elementary divisor ring; indeed, any rectangular matrix is equivalent to a diagonal matrix, and the $n \times n$ matrix ring R_n is unit regular.*

This answered a question of Handelman and Vasershtein, and the proof required a theorem of Vasershtein [71].

6.3E THEOREM (MENAL-MONCASI [82]). *Over a VNR ring R every rectangular matrix is equivalent to a diagonal matrix iff $R^2 \oplus A \approx R \oplus B$ implies $R \oplus A \approx B$ for all right R-modules A and B.*

REMARK. It is known that over any ring R, that any Artinian module cancels from direct sums (Camps-Dicks [93]; see 8.D). Also see 6.3G below.

Stable Range and the Cancellation Property

We say that 1 is in the **stable range** for a ring E if whenever $xa + b = 1$ in E, there is an element $y \in E$ such that $a + yb$ is a unit. Bass proved (see Swan [68], 11.8), that a semisimple artinian ring has 1 in the stable range. Since E clearly has 1 in the stable range if $E/J(E)$ does, it follows in particular that any semilocal ring has 1 in the stable range.

6.3F THEOREM (EVANS [73B]). *If M is a right R-module over any ring R, and if 1 is in the stable range for End M_R, then M has the cancellation property.*

6.3G COROLLARY. *Any R-module with semilocal endomorphism ring has the cancellation property.*

PROOF. Apply 6.3F to Bass' theorem stated **sup.** 6.3F.

6.3H THEOREM (GOODEARL-WARFIELD [76]). *Let R be a locally module-finite algebra over a commutative ring S such that $S/J(S)$ is von Neumann regular. Then:*

(1) *Finitely generated R-modules have the cancellation property*
(2) *Let A, B be finitely generated R-modules. Then:*
  (a) *If A^n is isomorphic to a direct summand of B^n, then A is isomorphic to a direct summand of B.*
  (b) *If $A^n \approx B^n$, then $A \approx B$. (Cancellation of powers.)*

PROOF. The proof uses Evans' Theorem 6.3F. □

6.3I THEOREM (GOODEARL-MONCASI[89]). *If R is a VNR ring which is left $\aleph_0$-injective (or right or left $\aleph_0$-continuous,* Cf. 12.4Cs*), then the following are equivalent:*
(a) R has bounded index (of nilpotence)
(b) All primitive factor rings are Artinian
(c) All $f \cdot g$ R-modules have 1 in the stable range of End M_R.
(d) All $f \cdot g$ R-modules have the cancellation property.

6.3J THEOREM (MENAL [81]). *If R is a VNR ring with Artinian primitive factor rings, then every $f \cdot g$ R-module M has 1 in the stable range of End M_R.*

REMARK. This is not true for an Arbitrary VNR ring, as Menal [88] pointed out. Cf. Goodearl-Moncasi [89].

Fractionally Self FP-Injective Rings

If P is a property of rings, then a ring R is *fractionally* P if $Q(R/I)$ has property P for all ideals I, where $Q(A)$ denotes the classical quotient ring of a ring A (cf. FSI rings in 5.9). A ring R is (right) *Kasch* if every maximal (right) ideal m is an annihilator right ideal, equivalently, ${}^{\perp}m \neq 0$, equivalently, every simple (right) module embeds in R. Any right PF ring is right Kasch by 4.20, and any acc$\perp$ commutative ring R has Kasch quotient ring (Faith [91b]).

6.4 THEOREM (FACCHINI-FAITH [96]). *A ring R is fractionally (self) FP-injective (=FSFPI) iff R is fractionally (self) p-injective. These rings are Arithmetic (= R_M is a valuation ring for every maximal ideal M). Conversely, an Arithmetic ring R is FSFPI under any of the following conditions:*
(a) *semilocal;*
(b) *fractionally semilocal;*
(c) *a 1-dimensional domain; or*
(d) *fractionally Kasch.*

REMARK. Any Prüfer domain, in fact any semihereditary ring, is locally a chain domain, hence Arithmetic.

Cf. the remark following 6.3A.

The proofs rely on Warfield's characterization of arithmetical rings:

6.5A WARFIELD'S THEOREM[1] ([70]). *A commutative ring R is Arithmetical iff every finitely presented R-module M is a direct summand of a direct sum of cyclicly presented (=CP) modules.*

The key ingredient in the proof is the following

6.5B THEOREM (WARFIELD [70]). *If R is a local ring, and if there is a bound on the minimal number $g(M)$ of generators of indecomposable finitely presented modules M, then R is a valuation ring.*

Coherent Rings: Theorems of Chase, Matlis and Couchot

6.6 CHASE'S THEOREM [60]. *A ring R is said to be* **left coherent** *provided R satisfies the equivalent conditions:*
(a) *any product of copies of R_R is a flat right R-module,*
(b) *any product of flat right R-modules is a flat right R-module,*
(c) *any $f \cdot g$ left ideal of R is finitely presented (= $f \cdot p$),*
(d) *any $f \cdot g$ submodule of a free left R-module is $f \cdot p$.*

A ring R is *coherent* provided R is both left and right coherent. A left Noetherian ring is left coherent, and a VNR ring is coherent. Cf. 6.60ff.

6.7A THEOREM (MATLIS [82]). *A commutative ring R is coherent iff Hom_R (A, B) is a flat R-module for all injective R-modules A and B.*

6.7B THEOREM (COUCHOT [82]). *A commutative ring R is coherent iff for every FP-injective R-module M and every maximal ideal P, M_P is an FP-injective R_P-module and R_P is coherent.*

6.7C THEOREM (COUCHOT [77]). *(1) If every finitely embedded injective right R-module is flat, then R is left FP-injective; (2) The converse holds if R is a right linearly compact ring, or if the ideals of R are linearly ordered.*

Cf. 6.10A and B and 6.11.

---

[1] As I reported in my *Algebra II* ([76], p. 129_n) in a letter of June 1974, Kaplansky attributed this theorem in the case R is a valuation ring to W. Krull. In this case M is a direct sum of CP modules.

When Injective Modules Are Flat: IF Rings

While earlier we considered when all injective right R-modules are projective (see 3.5B), we now consider when they are all flat. A ring R is (*right*) IF provided that all injective (right) R-modules are flat. By Bass' theorem 3.31, over a right perfect ring, any flat right R-module is projective, hence right perfect right IF rings are QF by 3.5B.

6.8 THEOREM (COLBY [75], WÜRFEL [73]). *A ring R is right IF if and only if all finitely presented right R-modules embed in a free right R-module.*

6.9 THEOREM (COLBY [75],GOMEZ-PARDO AND GONZALEZ [83],JAIN [73], MATLIS [85] AND WÜRFEL [73]). *The following conditions on a ring R are equivalent:*

(IF1) *R is IF.*
(IF2) *R is coherent and every finitely generated ideal (either side) is an annihilator ideal.*
(IF3) *R is coherent and flat modules (either side) are FP-injective.*
(IF4) *The classes of weak $\aleph_0$-injective modules and FP-injective modules are equal.*
(IF5) *Annihilation induces a duality on the finitely generated one-sided ideals.*
(IF6) *R is coherent and self FP-injective (both sides).*

REMARK. Matlis' theorems are for commutative R, and his term for "IF" is "semiregular".

6.10A THEOREM (MATLIS [85]). *(1) A commutative ring R is IF if and only if R is coherent and locally IF; (2) A domain R is Prüfer (i.e., arithmetical) if and only if R/I is IF for all finitely generated ideals $I \neq 0$.*

6.10B THEOREM (MATLIS [85]). *A commutative ring is IF iff R is coherent and every $f \cdot p$ module is R-reflexive.*

The next Corollary follows from Theorems 6.9 and 6.10 (Cf. 6.7B):

6.11 COROLLARY. *A commutative coherent ring is self FP-injective if and only if it is locally self FP-injective.*

Power Series over VNR and Linear Compact Rings

$R\langle x\rangle$ denotes the ring of formal power series over R (also denoted $R[[x]]$ in the literature.)

6.12 THEOREM (BREWER, RUTTER AND WATKINS [77]. *Let R be a VNR commutative ring. Then the power series ring $R\langle x\rangle$ is Bezout iff R is $\aleph_0$-injective.*

Note: By Theorem 6.9 (IF4), R is then FP-injective.

This is Theorem 42, p. 54 and the following is Theorem 43, p. 61, of Brewer [81].

6.13 THEOREM (BREWER, RUTTER AND WATKINS [77]. *Let R be a VNR commutative ring. Then the following are equivalent conditions:*

(1) *$R\langle x\rangle$ is semihereditary.*
(2) *$R\langle x\rangle$ is a coherent ring.*

(3) *R is $\aleph_0$-injective, and $I^\perp$ is countably generated for every countably generated ideal I.*
(4) *R is an $\aleph_0$-injective pp (= principal ideals are projective) ring.*
(5) *$R\langle x\rangle$ is a Bezout pp ring.*

Herbera [91], Lemma 5.16 and Proposition 5.17, generalized these theorems to non-commutative rings.

6.14 THEOREM (HERBERA [91]). *If $R\langle x\rangle$ is right semihereditary, then R is a left $\aleph_0$-complete VNR. Moreover, if R is a left $\aleph_0$-complete and left $\aleph_0$-injective VNR, then $R\langle x\rangle$ is right semihereditary and right Bezout.*

See Lemma 5.16 and Proposition 5.17 (also see Lemma 5.18 and Theorem 5.19) of Herbera [91]. (Note right semihereditary and right Bezout = right Bezout and right *pp*.)

REMARK. Ribenboim [97] considers a generalized power series ring $A = [[R^{S,\leq}]]$, where $(S, +, \leq)$ is a strictly ordered monoid, and characterizes when A is VNR (resp. skew field). Cf. Elliott-Ribenboim [90].

For a good background on linearly compact rings, consult Ánh [90] and Xue [92].

6.15 THEOREM (HERBERA AND XUE). *If R is a linearly compact commutative ring, then $R\langle x\rangle$ is linearly compact iff R is Noetherian (and iff R is locally Noetherian).*

For proof, see Xue [96b]. By Brewer-Heinzer [80], $R\langle x\rangle$ is necessarily Hilbert for $R\langle x\rangle$ to be linearly compact.

Historical Note

Herbera proved 6.15 while a Fulbright Postdoctoral Fellow at Rutgers in Fall '92. She later told me that 6.15 was a "bit folkloric," e.g. "known to Menini and Ánh." So perhaps more names ought to be attached to 6.15. I am very happy that Xue solved this problem, which I asked him (in a letter), and that he published his results.

Locally Split Submodules

A submodule A of an R-module B is said to be **locally split** if for every $f \cdot g$ submodule A' of A there exist a map $f : B \to A$ such that f restricted to A' is the identity. (This is equivalent to the same requirement for any cyclic submodule $A' = aR$.)

6.16 THEOREM (RAMAMURTHI-RANGASWAMY [73]). *The following are equivalent conditions on a right R-module M:*

1. *For each right R-module B, and each $f \cdot g$ submodule A of B, the canonical map*

$$Hom(B, M) \to Hom(A, M)$$

   is onto.
2. *M is locally split in each overmodule.*
3. *For every finite subset X of M there is an injective module E such that $X \subseteq E \subseteq M$.*

See Ramamurthi-Rangaswamy [73], where M with the above properties is said to be **finitely injective**, or **strongly absolutely pure**. Also see Facchini [95], who discusses aspects of strongly absolute pure modules relative to the question of whether R is necessarily Noetherian under the condition: (P) every $f \cdot p$ injective (or absolutely pure) module is strongly absolutely pure (see 6.18).

Below, see §14 for the concept of projective dimension.

6.17 THEOREM (FACCHINI [95]). *An almost maximal valuation domain $R \neq Q$ satisfies (P) iff Q has projective dimension $= 1$.*

A module M over a commutative domain R is *h-divisible* (after Matlis [72]) if M is a factor module of an injective module, equivalently, is an epimorphic image of a vector space over $Q = Q(R)$.

6.18 THEOREM (FACCHINI'S [95]). *Any absolutely strongly pure R-module over a domain is h-divisible, and conversely if R is an almost maximal valuation domain.*

The proof uses the theorem:

6.19 THEOREM (MATLIS [59], SALCE AND ZANARDO [81]). *A valuation domain R with quotient field Q is almost maximal iff every epimorphism from Q^n is injective over R, $n \geq 1$.* (Cf. 5.4C.)

REMARK. See Facchini [94], §4. for a generalization.

Existentially Closed Rings

A ring R is right *weakly* linearly existentially closed (= **Welex**) if every system of linear equations (LI) and a single linear inequation (LIE)

$$(LE) \qquad \begin{cases} x_1 a_{11} + \cdots + x_n a_{in} & = b_1 \\ \qquad \cdot & \quad \cdot \\ \qquad \cdot & \quad \cdot \\ \qquad \cdot & \quad \cdot \\ x_1 a_{mi} + \cdots + x_n a_{mn} & = b_m \end{cases}$$

$$(LIE) \qquad \left\{ x_1 a_{m+1,1} + \cdots + x_n a_{m+1,n} \neq b_{m+1} \right.$$

which has a solution in some ring extension of R already has a solution in R.

A ring R is *right linearly existentially closed* (= LEX) if R is WELEX with finitely many additional linear inequalities (FLIE) replacing the single (LIE).

A subring R of a ring A is *existentially closed* in A if every LE with FLIE with coefficients in R which has a solution in A already has a solution in A. R is *existentially closed* (EC) if R is EC in every ring extension A.

6.20 THEOREM (EKLOF AND SABBAGH [71]). *Any ring R can be embedded in an EC ring A.*

This is Theorem 7.2 of *loc.cit.*

6.21 COROLLARY. *Any ring R can be embedded in a right and left FP-injective ring S.*

This is a corollary of the following:

6.22 THEOREM (MENAL AND VÁMOS [89]). *A ring R is right and left FP-injective iff R is right and left WELEX.*

6.23 THEOREM (MENAL AND VÁMOS [89]). *A ring R is right and left LEX iff R is right and left FP-injective and 0 is the only finite one-sided ideal.*

Existentially Closed Fields

Existential closed skew fields are defined similarly, (see, e.g. Cohn [95], §6.5). According to Cohn, p.329, Notes and Comments, the concept EC was developed by "A. Robinson [63]; [and] the applications to sfields in §6.5 are taken from Cohn [75]."

6.24 THEOREM (COHN [95],P.311,6.5.3). *If D is a sfield, then there exists an EC sfield L containing D, in which every finite consistent set of equations over D has a solution. When D is infinite, then L can be chosen to have the same cardinal as D, while if D is finite, then L can be chosen to be countable.*

REMARK. A commutative field k is EC over k precisely when k is algebraically closed. Moreover, the center of any EC sfield D is algebraically closed.

6.25 THEOREM (ZIG-ZAG LEMMA). *Two EC sfields K and L that are countably generated over a field k are isomorphic iff they have the same families of subfields that are $f \cdot g$ over k.*

See Cohn [95], p. 312, 6.5.4.

Other Embeddings in Skew Fields

6.26 ORE'S THEOREM [31]. *An integral domain R has a right quotient field Q of the form $\{ab^{-1}$ for all $a, 0 \neq b \in R\}$ iff $aR \cap bR \neq 0$ for $a \neq 0, b \neq 0$,*

In this case R is a **right Ore domain**. It follows that a domain R is right Ore iff R is a uniform right R-module. (See 6.29 following.)

As mentioned **sup.** 6.2B:

6.27 THEOREM (MAL'CEV [36]). *Not every domain R can be embedded as a subring in a sfield.*

A similar theorem holds for semigroups with cancellation (cf. Lambek [51]).

Note that the theorem of Menal and Vámos (Corollary 6.21) implies that a **Mal'cev domain** R (= one not embeddable in a field) embeds in an FP-injective ring. However, anticipating a concept introduced in §12, **sup.** 12A, a better result holds:

6.28 THEOREM. *The maximal right quotient ring $Q = Q^r_{\max}(R)$ of an integral domain R is a right self-injective VNR. Moreover, Q is a simple ring, and is a sfield iff R is a right Ore domain.*

Cf. the lectures of the author [67], or Goodearl [79]. The point here is that Q being VNR is right and left FP-injective, and right self-injective as a bonus.

(Simplicity of Q follows from the fact that if I is a nonzero ideal of Q, then $\exists 0 \neq x \in I \cap R$, and since $x^\perp = 0$, there exists $y \in Q$ so that $yx = 1 \in I$. Thus, $I = Q$.)

6.29 GOLDIE'S THEOREM [58]. *A domain R is right Ore iff R has acc$\oplus$. In particular, any right Noetherian domain R is right Ore.*

Cf. 3.13. Also see 15.7B: PI-domains are Ore.

Galois Subrings of Ore Domains Are Ore

6.30 THEOREM (FAITH [72C]). *If G is a finite group of automorphisms of a right Ore domain R, then the Galois subring R^G is a right Ore domain.*

REMARK. This answered a question of Bergman. Also see Bergman and Isaacs [73], Cohn [73], Cohn [75], Harčenko [74,75], Kitamura [76,77], and Tominaga [73]. See 12.A and B.

6.31 COHN'S THEOREM [71]. *Any fir can be embedded in a universal sfield of fractions.*

This is done by inverting all full matrices (see Cohn [95], p.95) a method that extends to semifirs, in fact, to Sylvester domains presently defined (Cohn [95], Theorem 4.58 and Corollary 4.59, pp.181–182.)

Sylvester domains were introduced by Dicks and Sontag [78] as domains over which any two matrices A and B with the number n of columns of A equal to the number of rows of B satisfy the nullity condition:

$$AB = 0 \Longrightarrow r(A) + r(B) \leq n$$

on their "inner ranks" $r(A)$ and $r(B)$. Then *Sylvester's law of nullity* (dating to 1884) holds:

$$r(A) + r(B) \leq n + r(AB)$$

(see *loc. cit.*).

Rings with Zero Intersection Property on Annihilators: Zip Rings

Zelmanowitz [76b] introduced the ring concept, which we call *right zip rings*, with the defining properties below, which are equivalent:

(ZIP 1) If the right annihilator $X^\perp$ of a subset X of R is zero, then $X_1^\perp = 0$ for a finite subset $X_1 \subseteq X$.

(ZIP 2) If L is a left ideal and if $L^\perp = 0$, then $L_1^\perp = 0$ for a finitely generated left ideal $L_1 \subseteq L$.

6.32 REMARKS.

(1) Trivially, any left Kasch ring is right zip. In [76b] Zelmanowitz noted that any ring R satisfying the d.c.c. on annihilator right ideals (=dcc $\perp$) is a right zip ring, and hence, so is any subring of R. He also showed by example that there exist zip rings which do not have dcc $\perp$.

(2) In [91b] the author characterized a right zip ring by the property that every injective right module E is divisible by every left ideal L such that $L^\perp = 0$. Thus, $E = EL$. (It suffices for this to hold for the injective hull of R.)

(3) We also showed that a left and right self-injective ring R is zip iff R is pseudo-Frobenius (= PF), and that a semiprime commutative ring R is zip iff R is Goldie.

Beachy and Blair [75] studied rings that satisfy the condition that every faithful right ideal I is *co-faithful* in the sense that $I_1^\perp = 0$ for a finite subset $I_1 \subseteq I$, equivalently, $R \hookrightarrow I^n$ for $n < \infty$. Right zip rings have this property, and conversely for commutative R.

6.32A THEOREM (BEACHY-BLAIR [75]). *If faithful ideals of R are co-faithful then the same is true of $R[X]$, for any commutative ring R, and any set X of variables.*

6.32B COROLLARY. *If R is a commutative zip ring, then any polynomial ring $R[X]$ over R is a zip ring, for any set X of variables.*

In [91b] the author raised the questions that resulted in the next theorem.

6.33 THEOREM (CEDÓ [91]). *There exist right zip rings R so that (1) the polynomial ring $R[x]$ is not right zip, (2) the $n \times n$ matrix ring R_n is not right zip and (3) the group ring RG of a finite group G is not right zip.*

In fact, the example for (2) is an integral domain, hence a Mal'cev domain. The proof relied on coproduct contructions due to G. Bergman [74b]: Let k be a field, and A an algebra over k. Then the free product

$$M_n(k) \star_k A \approx M_n(S)$$

where S is an $(n-1)$-fir.

6.34 COROLLARY. *There exist integral domains not embeddable in left Noetherian nor right Artinian (nor right dcc$\perp$) rings.*

This follows from Corollary 3 of Cedó [91], stating that a free product $A \star_k B$ of algebras over k can be right zip only if both A and B are, that S is a domain hence zip while $M_n(S)$ is not zip. Thus, S is a Mal'cev domain with the property stated in the Corollary. Since k can be any field, it follows that the situation corresponding to the Camillo-Guralnick-Roitman Theorem 9.3 cannot transpire.

On a Question of Mal'cev: Klein's Theorem

Can the multiplicative semigroup $D^\star$ of a Mal'cev domain D be embedded in a group?

6.35 THEOREM (KLEIN [69]). *If an integral domain D has the property that any nilpotent $n \times n$ matrix over D has index of nilpotency $\leq n$, then the multiplicative semigroup $D^\star$ of D can be embedded in a group.*

6.36 REMARK. The examples of Mal'cev domains of Bokut' [67], Bowtell [67], Johnson [69] and Klein [69] are such that $D^\star$ can be embedded in a group. This gave an affirmative answer to Mal'cev's question. (A typo nullified this statement in my Algebra II [76], p.140ff. Related questions are treated in papers of Klein (1967 and 1972 (cited *loc.cit.*), and papers of Bokut' listed in Small [80] (Reviews in Ring Theory (1940–1979).)

Weakly Injective Modules

M is **weakly R_n-injective** provided every n-generated submodule S of $E(M)$ is contained in a submodule X of $E(M)$ with $X \approx M$. M is **weakly-injective (= WI)** if this holds for every n, that is, for every $f \cdot g$ submodule S.

REMARK. (1) Tuganbaev [77,82] uses the term (also poorly injective) quite differently: every endomorphism of any submodule S is induced by one of M. If R is semi-Artinian, these modules are all quasi-injective [77]; (2) A module M is called **pseudo-injective** if every monomorphism of a submodule S is induced by an endomorphism of M. These are not necessarily injective (Teply [75], Jain-Singh [75]) but they are when R is a serial ring (*ibid.*).

Gauss Elimination and Weak Injectivity

6.36A REMARK. At the Special Session in Ring Theory at Ohio State University in August 1990 organized by S. K. Jain and the Ohio group (López-Permouth, T. Rizvi, M. Yousif), Sergio López-Permouth pointed out that the ring $R = T(k_n)$ of $n \times n$ upper triangular matrices over a field k is left weakly self-injective, and that Gauss' elimination method of reducing an $n \times n$ matrix $q \in Q = k_n$ to upper triangular form by row operations yields a nonsingular matrix $x \in Q$ so that $xq = r \in R$. Since $Q = E({}_RR)$, then R is weakly R_1-injective as asserted. (I am indebted to S. López-Permouth [98] for refreshing my memory).

6.36B EXAMPLES (JAIN, LÓPEZ-PERMOUTH, SALEH [72]). (1) R is QF iff R is right Artinian and right WI; (2) A domain R is a right Ore domain iff R is right weakly R_1-injective; (3) A domain R is left and right Ore iff R is right WI; (4) A right WI semiprime right Goldie ring is left Goldie; (5) A right WI right continuous ring is right self-injective.

6.37 THEOREM (AL-HUZALI, JAIN, LÓPEZ-PERMOUTH [92]). *A ring R is right quotient finite dimensional iff every direct sum of (weakly) injective right R-modules is weakly injective.*

Zip McCoy Rings

The topics that follow in 6.38 are taken up in more detail in §16, 16.27ff.

6.38 REMARKS. (1) If $R \subseteq S$ are rings such that S_R is an essential extension of R, and if S is right zip, then so is R (Faith [76b]);

(2) On the other hand, if S_R is a free left R-module, and if S is right zip, then so is R (Cedó [91], Lemma 2]);

(3) Any left Kasch ring S trivially is right zip, hence any ring R satisfying conditions of Remark 1 or 2 is right zip;

(4) If R is right zip, then every annihilator right ideal $\neq$ contains a minimal right annihilator $\neq 0$ (Faith [91b], Theorem 3.1), hence every left annihilator ideal $L \neq R$ is contained in a maximal left annihilator $\neq R$ (= R has **left** $\mathcal{E}_{\max}$);

(5) A commutative ring R is **McCoy** in case every $f \cdot g$ faithful ideal I contains a regular element. Furthermore, R is McCoy iff $Q(R)$ is McCoy. Huckaba-Keller [79]: Any polynomial ring $R[X]$ over any ring

R is McCoy (= property A in Huckaba [88]); See Huckaba [88], p.9, Theorem 2.7;

(6) A commutative ring R has Kasch quotient ring $Q = Q(R)$ iff R is zip McCoy by a result of the author [91b]. It follows from the above Remark and 6.32B that if R is zip then $Q(R[X])$ is Kasch. Also, $Q(R)$ is semilocal Kasch iff $Q(R[X]$ is semilocal Kasch. Any commutative acc$\perp$ ring R has semilocal Kasch $Q(R)$, hence is zip McCoy (*ibid.*);

(7) If a ring R has the *finite left intersection property (= flip)* for any family of left ideals $\{I_\alpha\}_{\alpha \in A}$ the intersection is zero iff a finite subfamily has zero intersection, then ${}_RR$ is finitely embedded (f.e. = has $f \cdot g$ essential socle. Cf. **sup.** 7.8). This is "the dual of $f \cdot g$" according to Vámos [68]. (Cf. Faith [76], pp.67–70, esp. 19.13A–19.16B, which equates flip with the property that every faithful left ideal is cofaithful. Moreover, R is left Artinian iff every cyclic left module is *f.e.* The same result holds for modules, i.e., M is Artinian iff every factor module of M is *f.e.* (*Vide* Vámos, *loc.cit.* and Onodera [73]);

(8) In [91b] the author showed that any group ring RG of a finite Abelian group G over a commutative zip ring R is zip;

(9) Cedó [91] *interalia* showed the $n \times n$ matrix ring over a commutative zip ring is zip, and extended the above theorem on Abelian group rings to a non-commutative algebra R over an algebraically closed field k of characteristic not dividing $|G|$;

(10) Regarding (4), any ring R with the acc on left point annihilators has left $\mathcal{E}_{\max}$.

$\mathcal{E}_{\max}$ is an acronym for **enough maximal annihilators**. See 6.38 (4).

6.39 THEOREM (FAITH [91B]). *For commutative R, the restriction map $Ass^\star R[X] \to Ass^\star R$ is a bijection under any of the following assumptions:*

1. *$R[X]$ has $\mathcal{E}_{\max}$ (op.cit. 7.1),*
2. *R/P is Noetherian $\forall P \in Ass^\star(R)$ (op.cit. 7.2),*
3. *R is a* **zip ring**, *i.e., has a finite intersection property for zero intersections of annihilators (op.cit. 3.1),*
4. *A reduced ring R with $\mathcal{E}_{\max}$ (op.cit. 7.7).*

REMARK. By (*op.cit.*) [7.6], (4) holds iff the maximal quotient ring $Q_{\max}(R)$, defined in 9.27s and in §12, is a direct product of fields.

In his "Mathematical Notes" paper, McCoy [57] discussed annihilator ideals in polynomial rings: "if f is a zero divisor in the polynomial ring $R[X]$, where R is a commutative ring, there exists a nonzero element c in R such that $cf = 0$...it is clear that the theorem as stated does not generalize to more than one indeterminate. Moreover, it has been pointed out in Problem 4419 of this Monthly (1950, p.692 and 1952, p.336) that the theorem is not necessarily true for noncommutative rings...".

McCoy went on to consider annihilator ideals of the polynomial ring $R[X]$ in a finite set X of variables. If S is any subset $\neq \emptyset$ of a ring A, let

$$S^\perp = \{a \in A \mid sa = 0 \quad \forall s \in S\}$$

be the right ideal of A annihilated by S. After R. Baer, we will call this a **right annulet**. If S is a right ideal, then $S^\perp$ is an ideal of A, called an **annulet**. In this terminology, McCoy proved:

6.40 THEOREM (McCOY [57]). *Any nonzero annulet ideal I of a polynomial ring $A = R[X]$ in any set of variables has nonzero intersection with R.*

REMARK. In a seminar talk that I gave in March 1989, my colleague, Professor W. Vasconcelos, pointed out that McCoy's proof and theorem for finite X could be extended to arbitrary X.

6.41 EXAMPLES. (1) See Remarks 6.38, (5) and (6) for the definition, and examples of McCoy rings, e.g. any commutative ring R with Kasch $Q(R)$ is McCoy;
(2) Any commutative ring R such that $f \cdot g$ ideals of $Q(R)$ are annihilators, e.g. when $Q(R)$ is VNR, or FP-injective, is a McCoy ring. See. 6.42 below.
(3) Any valuation ring R is McCoy, hence so any ring R with $Q(R)$ a VR. However, Huckaba [88], p.191, Example 17, constructs a ring R such that R_m is a VD for all maximal ideals m, and R is not McCoy, hence $Q(R)$ is not McCoy. This example defeated the "FP^2F conjecture" of the author [92b]. (Cf. Prüfer rings 9.24(3).)

6.42 HUCKABA-KELLER THEOREM [79]. *A reduced coherent commutative ring R is McCoy iff $Q(R)$ is VNR.*

Cf. Huckaba's book [88], p.20, Theorem 4.7.

Elementary Equivalence

6.43 DEFINITION. Two rings (resp. fields, modules, ...) R and S are called *elementarily equivalent* (notation: $R \equiv S$) if R and S satisfy the same first order sentences in the corresponding language.

By the upper Löwenheim-Skolem theorem one gets an explicit criterion for elementary equivalence of algebraically closed fields:

6.44 THEOREM. *Two algebraically closed fields K and L are elementarily equivalent if and only if $char(K) = char(L)$.*

PROOF. See, e.g. Jensen-Lenzing [89], Theorem 1.13, p. 5.

A field K is **real closed** if K is formally real and $K[\sqrt{-1}]$ is algebraically closed. See 1.29Bf.

6.45A THEOREM (TARSKI [49,56]). *Any two real closed fields are elementarily equivalent.*

PROOF. Jensen-Lenzing *ibid.* p.9.

6.45B THEOREM (TARSKI [49]). *The theory of algebraically closed fields is not finitely axiomatizable.*

PROOF. See Bell and Slomson [71], p.101 (and p.106) for proofs of more general statements (and remarks). □

Pure-Injective Envelopes

6.46 THEOREM (KIEŁPIŃSKI [67]-WARFIELD [69]). *Each R-module M has a pure-injective envelope*

$$M \hookrightarrow A(M),$$

i.e., M embeds as a pure submodule into a pure-injective module R-module $A(M)$ such that—for each R-modules X—an R-linear map $f : A(M) \to X$ is a pure monomorphism if and only if its restriction $f|_M : M \to X$ to M is a pure monomorphism. $A(M)$ is the smallest pure submodule of $E(M)$ which is pure-injective and contains M.

PROOF. See Jensen-Lenzing [89], p. 128.

6.46A REMARKS. (1) Fuchs, Salce and Zanardo [98] point out that pure-essential extensions over a Prüfer domain R are transitive iff R is a discrete valuation domain (= DVD).

(2) Gómez Pardo and Guil Asensio [98] point out that Stenström's proof of Theorem 6.46 cannot work for the reason given in (1), and that the cardinality argument by transfinite induction given in Kiełpiński [67] suffices, as in the case of injective hulls.

(3) Gómez Pardo and Guil Asensio also point out in [97c] that a finitely presented pure-injective module M has an indecomposable decomposition iff it is **quotient pure-injective** (= all its pure factor modules are pure-injective). Furthermore, then $\text{End}M_R$ is semiperfect. Cf. 6.49–59 below.

6.47 EXAMPLE (JENSEN-LENZING, *Ibid.*). (1) If R is a von Neumann regular ring, the pure-injective envelope $A(M)$ of an R-module M coincides with its injective envelope $E(M)$. More generally, if R denotes an arbitrary ring, we have $A(M) = E(M)$ exactly for the FP-injective R-modules (Cf. 6.B.);

(2) If R is a commutative local Noetherian ring with maximal ideal m then the m-adic completion $\widetilde{R}_m$ of R, viewed as an R-module, serves as the pure-injective envelope of R with respect to the natural embedding $R \to \widetilde{R}_m$;

(3) The pure-injective envelope $A(\mathbb{Z})$ of the $\mathbb{Z}$-module $\mathbb{Z}$ can be obtained as the direct product

$$A(\mathbb{Z}) = \prod_{p \in P} J_p,$$

(with respect to the diagonal embedding of $\mathbb{Z}$ into $\prod_{p\in P} J_p, x \mapsto (x/1)$). Here P is the set of prime numbers and, for each prime number p, $J_p = \mathbb{Z}_{(p)}$ is the ring of p-adic integers viewed as a $\mathbb{Z}$-module.

6.48 THEOREM (SABBAGH [70]). *Every R-module M is elementarily equivalent to a pure-injective R-module, actually to its pure-injective envelope $A(M)$.*

PROOF. *Ibid.* Theorem 7.51, p. 158.

Ziegler's Theorem

6.49 THEOREM (ZIEGLER [84]). *Let R be any ring. Each left R-module M is elementarily equivalent to a direct sum of indecomposable pure-injective modules.*

6.50 THEOREM. *Let R be an Abelian VNR. The following conditions on an R-module M are equivalent:*

1. *M is simple.*
2. *M is indecomposable Σ-injective.*

3. *M is indecomposable pure-injective.*
4. *M is indecomposable.*

6.51 COROLLARY. *If R is an Abelian VNR, then each R-module M is elementarily equivalent to a semisimple R-module.*

For proofs, see *op. cit.*, p.181. Regarding 6.50, Cf. also the author's paper [72b]. Also see Trilfaj [96] where a problem of Ziegler is reduced to this question: **Does there exist a simple non-Artinian VNR ring R having a unique indecomposable injective R-module?**

A ring R is **F-semiperfect** if $\overline{R} = R/\text{rad}R$ is VNR and idempotents lift.

6.52 THEOREM. *For each pure-injective R-module M the factor ring*

$$End_R(M)/rad\ End_R(M)$$

is von Neumann regular and right self-injective. Moreover, idempotents can be lifted modulo rad $End_R(M)$. In particular $End_R(M)$ is F-semiperfect.

REMARK. This theorem generalizes Utumi's Theorem 4.2, and the Faith-Utumi-Osofsky Theorem 4.2A.

6.53 THEOREM. *Each pure-injective R-module M admits a decomposition*

$$M = M_d \oplus M_c, \quad M_d = \oplus_{a \in I} M_\alpha$$

where each M_α is an indecomposable direct factor of M and M_c does not have any indecomposable direct factor. Moreover, M_d is uniquely determined by M, while M_c and the M_α are uniquely determined up to isomorphism (and ordering).

PROOF. See Jensen-Lenzing [89], p.180 for proofs.

Noetherian Pure-Injective Rings

By Warfield's Theorem 6.D, every linearly compact commutative ring is pure-injective. The next theorem establishes the converse for Noetherian rings. (Cf. Matlis' theorem 5.4B.)

6.54 THEOREM. *A commutative Noetherian ring R is pure-injective (**qua** R-module) iff R is a finite product of complete local rings.*

PROOF. See *op.cit.* 11.3, p.283.

Σ-Pure-Injective Modules

A pure-injective module M is **Σ-pure-injective** if every direct sum of copies of M is pure-injective. By Zimmermannn's theorem 1.25, M is Σ-pure-injective iff every direct sum of copies of M splits off in the direct product.

6.55 THEOREM (LENZING [76], ZIMMERMANNN [76], FAITH [66A]). *If R is left perfect and right coherent, or pure-injective of cardinality $< 2^{\aleph_0}$, then R is Σ-pure-injective. Any Σ-pure-injective ring is semiprimary.*

PROOF. See Jensen-Lenzing, 11.1, p.281, for proof and attributions.

Cf. Zimmermann's Theorem 1.25, Lenzing's Theorem 1.24A and Corollary 1.24B.

Pure-Semisimple Rings

A ring R is right **pure-semisimple** provided that every right R-module is pure-injective.

6.56 THEOREM (ZIMMERMANN-HUISGEN [79]). *A ring R is right pure-semisimple iff every right R-module is a direct sum of $f \cdot g$ modules.*

REMARK. By Chase's Theorem 3.4E, then R is right Artinian.
Fuller's Theorem [76] can thus be expressed:

6.57 THEOREM (FULLER [76]). *A ring R is right pure-semisimple iff every right R-module has a decomposition that complements direct summands.*

6.58 THEOREM (HERZOG [94A]). *Every $f \cdot p$ left module over a right pure-semisimple ring has finite length over its endomorphism ring.*

6.59 REMARK. See Rings of Finite and Bounded Module Type, end of §8, for some connections with pure semisimple rings.

Π-Coherent Rings

R is **right Π-coherent** if an arbitrary product R^α of copies of R is coherent, i.e., every $f \cdot g$ submodule of R^α is $f \cdot p$. A ring R is a left $\star$-ring if the R-dual functor $\mathrm{Hom}_R(\ ,R)$ takes $f \cdot g$ left modules M into $f \cdot g$ right modules $M^\star$.

6.60 CAMILLO'S THEOREM [90]. *For a ring R, the following are equivalent: (1) R is right Π-coherent; (2) R is left $\star$-ring; (3) For each matrix ring R_n, all right annihilators are $f \cdot g$.*

6.61 THEOREM (*ibid.*). *If R is a two-sided Noetherian ring, and if R is either semiprime, or is an algebra over a non-denumerable field, then $R[X]$ is right Π-coherent for any set X of variables.*

Cf. Finkel Jones [82], Anderson and Dobbs [88], and the author's paper [90a].

CHAPTER 7

Direct Decompositions and Dual Generalizations of Noetherian Rings

If every factor M/S of a module M modulo an essential submodule S has property P, then M is said to satisfy "restricted P."

7.1 THEOREM (WEBBER [70], CHATTERS [71]). *Over a hereditary Noetherian ring, every $f \cdot g$ module is restricted Artinian.* (Cf. 5.3A.)

Cf. Ornstein [68] for related results for restricted Artinian.

7.2 THEOREM (CHATTERS ([72]). *A Noetherian hereditary ring is a finite product of rings that are either Artinian or prime rings.*

PP Rings and Finitely Generated Flat Ideals

A ring R *is right PP (or pp)* if every principal right ideal pR is projective, equivalently, the canonical map $R \to pR$ splits, equivalently $p^{\perp} = eR$ for an idempotent e. Integral domains and right semihereditary rings are trivially right PP.

7.3A THEOREM (JØNDRUP [71]).
1. *If R is commutative, then R is pp iff the polynomial ring $R[x]$ is pp. Moreover, all ideals are flat iff the 2-generated ideals are,*
2. *"n-generated left ideals are flat" is a right left symmetric property.*

7.3B THEOREM (LEVY[63]). *A right Goldie right PP ring is a finite product of prime right Goldie PP rings. Moreover, the prime direct factors are the minimal annihilator ideals of R.*

Cf. Koifman [71a].

7.4 THEOREM (ROBSON ([74]). *A Noetherian ring R with prime radical N (=the maximal nilpotent ideal) is a product of an Artinian ring and a semiprime ring iff*

$$Nc = cN = N$$

for every $c \in R$ such that $\bar{c}$ is a regular element of $\overline{R} = R/N$.

This can be used to give another proof of Chatters' theorem. Cf. Asano's theorem 3.12A.

7.5A THEOREM (KRULL [24], ASANO [38,49], GOLDIE [62]). *A left Noetherian principal right ideal ring R is a finite product of prime rings and primary Artinian rings.*

Right Bezout Rings

A ring R is **right Bezout** if all $f \cdot g$ right ideals are principal, e.g. any VNR is right and left Bezout. Any semilocal Arithmetic ring is Bezout (Cf. **sup.** 5.4B).

7.5B THEOREM (GOLDIE [62] AND ROBSON [67B]). *A semiprime right Bezout ring R is right semi-hereditary and a finite product of prime right Bezout rings each isomorphic to a full $n \times n$ matrix ring F_n over a right Ore domain F, for various integers $n \geq 1$.*

Goldie's Theorem [62] is for R a principal right ideal ring (PRIR). Swan [62] showed that F need not be right Bezout in 7.5B. For the next theorem, cf. Ore domains 6.26f.

7.5C THEOREM (WARFIELD [79]). *If R is right Bezout (= $f \cdot g$ right ideals are principal) and R/N is right Goldie, where $N = N(R)$ is the prime radical (= the intersection of all prime ideals), then $R = R_1 \times R_2 \times \cdots \times R_t$, where $R_i/N(R_i)$ is prime, and R_i is a full matrix ring over a ring D_i such that $D_i/N(D_i)$ is an Ore domain $\forall i$.*

7.5D COROLLARY (OP.CIT.). *If R is semilocal and $R/N(R)$ is right Goldie, then R is right Bezout iff $R = R_1 \times R_2 \times \ldots \times R_t$ is a finite product of full matrix rings over semilocal right Bezout rings D_i such that $D_i/N(D_i)$ is an Ore domain, $i = 1, \ldots, t$.*

Faith-Utumi Theorem

7.6A THEOREM (FAITH-UTUMI STRUCTURE TH. [65]). *A semiprime ring R is right Goldie iff R contains a finite product P of full $n_i \times n_i$ matrix rings $(F_i)_{n_i}$ over right Ore domains F_i (not necessarily having unit element but) having right quotient sfields $D_i, i = 1, \ldots, n$ such that $Q^r_{c\ell}(R) = \prod_{i=1}^n (D_i)_{n_i}$.*

Expressed otherwise, R is an essential extension of P as a right P-module.

7.6B FAITH-UTUMI THEOREM FOR SEMILOCAL RINGS. *If $Q = Q^r_{c\ell}(R)$ exists and $Q = D_n$ is a full $n \times n$ matrix semilocal ring over a ring D, then (after a possible change of matrix units) R contains the ring F_n of $n \times n$ matrixes over a subring F of D (which may not have a unit element) such that $Q^r_{c\ell}(F) = D$, hence $Q^r_{c\ell}(F_n) = D_n$.*

The caveat about the change of matrices does not apply if Q is also the left quotient ring of R. The bit about not having a unit may be illustrated: let R be the ring of all 2×2 integer-valued matrices whose diagonal entries are divisible by 2. Thus, $R \supseteq (2\mathbb{Z})_2$ and $Q_{c\ell}(R) = \mathbb{Q}_2$.

REMARK. As remarked in [72a] (p.411,Prop.10.19) the original proof of 7.6A suffices for 7.6B and obviated the case of Artinian Q of Robson [67] (p.607,3.3).

7.7 SANDOMERSKI'S THEOREM [68]. *A right hereditary ring R of finite right Goldie dimension is Noetherian. A right semihereditary ring of finite Goldie dimension is left semihereditary.*

Finitely Embedded Rings

A ring R is *right finitely embedded* (=f.e.) if R has finite essential right socle (= an essential $f \cdot g$ semisimple right ideal).

7.8 THEOREM (GINN AND MOSS [75]). *A two-sided Noetherian ring is two-sided Artinian iff R is right f.e.*

7.9 THEOREM (BEACHY [71]-VÁMOS [68]). *A ring R is right Artinian iff every cyclic right module is f.e.*

Cf. 6.38(7).

7.9A THEOREM (LENAGAN'S [75]). *If ${}_RM_S$ is a bimodule such that ${}_RM$ has finite length, and M_S is Noetherian, then M_S has finite length.*

The next theorem was inspired by Shizhong's theorem which follows.

7.10 THEOREM (FAITH [91A]). *Any f.e. commutative acc$\perp$ ring is Artinian.*

7.11 COROLLARY (SHIZHONG [91]). *Any commutative subdirectly irreducible acc$\perp$ ring is QF.*

REMARK. Regarding 7.11, see 13.26 for a generalization to uniform acc$\perp$ rings.

Simple Noetherian Rings

A problem raised by the author in [64], [71b], and by Cozzens-Faith [75] is: 1) is every simple Noetherian ring R Morita equivalent to a domain; and a related problem: 2) is R a domain or does it have nontrivial idempotents; indeed 3) is it a full matrix ring over a domain? 4) if R has finite global dimension, does 1) hold? (Global dimension is taken up in Chapter 14.)

Zalesski and Neroslavski [77] gave negative answers to 2) and 3). Both Goodearl [78] and Stafford [78] showed that the same example also gave a negative answer to 1). However, this left 4) still open.

Cozzens and the author proved

7.12 THEOREM (COZZENS AND FAITH [75]). *If R is a simple right Noetherian ring of global dimension ≤ 2, then R is Morita equivalent to a domain A.*

Thus, the domain A is also simple, and gl.dim$A \leq 2$.

7.13 THEOREM (STAFFORD [79]). *If R is a simple Noetherian ring of finite left global dimension, and if R has left Krull dimension $n < \infty$, then R has Goldie dimension $\leq n$. Furthermore if $n = 1$, then R is Morita equivalent to a domain, and has an idempotent $e \neq 0, 1$ if R is not a domain.*

REMARKS. (1) For the concept of left Krull dimension, see 14.26ff; (2) Question of Camillo-Krause: If R is a simple ring of left Krull dimension 1, is R left Noetherian? Cf. Shamsuddin [98]; (3) In [71b], the author characterized the situation that a simple Noetherian ring R is isomorphic to the endomorphism ring of a $f \cdot g$ projective module over a simple domain A. See also Cozzens and Faith [75], p.2.29ff., Theorem 2.20.

Simple Differential Polynomial Rings

Let A be a ring, and let $\varphi : A \to A$ be a ring monomorphism. Then a φ-*derivation* of A is a mapping $D : A \to A$ such that

$$D(a+b) = D(a) + D(b)$$
$$D(ab) = \varphi(a)D(b) + D(a)b$$

$\forall a, b \in A$. If φ =identity, then D is called an **(ordinary) derivation**. For each element $x \in A$, there is a derivation

$$D_x : a \mapsto ax - xa,$$

$\forall a \in A$, called the **inner derivation** defined by x. A derivation is **outer** if it is not inner.

7.14 EXAMPLES. 1) If R is a ring, and if $A = R[x]$ is the polynomial ring, then the usual derivation D, defined by

$$D(\sum_{i=0}^{n} a_i x^i) = \sum_{i=0}^{n} i a_i x^{i-1},$$

where $a_i \in R, i = 1, \dots, n$, is a derivation of A.

2) Let $A \supseteq k$ be sfields such that $\dim_k A = 2$. (We do not require that $\dim A_k = 2$.) Then $A = k + kx$ for any $x \in A, x \notin k$. If $a \in k$, then there exist unique elements $\varphi(a)$ and $D(a) \in k$ such that

$$xa = D(a) + \varphi(a)x.$$

This defines a ring monic φ of A and a φ-derivation D.

Let D be a φ-derivation of a ring A, and let $A[y]$ be the polynomial ring. Then $A[y]$ is a ring under addition of polynomials and multiplication defined by the rule

$$ya = D(a) + \varphi(a)y \tag{1}$$

and its consequences; this ring is called the **ring of φ-differential polynomials** in D and is denoted $A[y, D, \varphi]$.

REMARK. In his pioneering papers Ore [31,33a,b] studied $R = A[y, D, \varphi]$ for a sfield A. Then A is a left Ore domain, which is a right Ore domain iff φ is an automorphism. Furthermore, given polynomials $f, g \in R$ there exist $q, r \in R$ with $\deg r < \deg g$ and such that $f = qg + r$ (= division algorithm). Thus, R is a left Euclidean domain, hence a left principal ideal domain. See Ore [33a], p.483ff., esp. 487 where the left quotient field Q of R is constructed. Cf. also 7.20.1 below.

Let D be any ordinary derivation of A. Thus, $\varphi = 1$ and A is said to be an **ordinary differential ring**. Then (1) implies

$$ya = D(a) + ay \tag{2}$$

and

$$y^n a = \sum_{k=0}^{n} \binom{n}{k} D^k(a) y^{n-k}, \tag{3}$$

where $D^2 = D \circ D, \dots, D^n = D^{n-1} \circ D = D \circ D^{n-1}$ and $\binom{n}{k}$ is the binomial coefficient. In this case, the ring is called the **ring of differential polynomials** over A in the derivation D and is denoted by $A[y, D]$, or $A[y,']$, where the prime represents the derivation.

7.15 Hilbert Basis Theorem for Differential Polynomials. *If A is a left Noetherian ring with an ordinary derivation D, then the differential polynomial ring $A[y, D]$ is left Noetherian.*

The proof of the Hilbert Basis Theorem applies here *mutatis mutandi.*

An ordinary differential ring A is a **Ritt Algebra** if A is an algebra over $\mathbb{Q}$. Thus, any simple ordinary differential ring of characteristic 0 is a Ritt Algebra.

7.16 Theorem. *If A is a Ritt algebra with derivation D, then the differential polynomial ring $R = A[y, d]$ is simple iff D is outer and A has no non-trivial D-invariant ideals.*

See Cozzens-Faith [75], p.43ff.

The next theorem is a corollary of 7.16.

7.17 Theorem (Amitsur [56]). *If A is a simple ring with characteristic 0 and an outer derivation D, then the differential polynomial ring $R = A[y, D]$ is simple.*

Cf. Hauger [77] for a generalization to n commuting derivations.

The **D-constants** of a differential ring A is the set $\{a \in A \mid D(a) = 0\}$.

7.18 Theorem. *If A is a commutative ordinary differential domain of characteristic $p \neq 0$, then $A[y, D]$ is simple iff A is a field of infinite dimension over the subfield k of D-constants.*

The Weyl Algebra

The case where A is a field, and $D(y) = 1$ is called the **Weyl Algebra** A_1. See Rinehart [62] and Webber [60] for some important examples and counterexamples. Also, see Björk [72], Roos [72] and Goodearl [74]. (Cf. 14.15, esp. (8)–(10).)

7.19 Proposition. *Let S be a simple ring of characteristic 0, and let $A = S[x]$ be the ordinary polynomial ring. Let $R = A[y,']$ be the ring of differential polynomials such that $\phi = 1$. Then R is a simple ring.*

See the author's Algebra I for a proof, Proposition 7.30, p.354.

7.20 Remarks. (1) If A is a field, and if D is a derivation, then $R = A[y, D]$ has a left and right division algorithm; for example, if f and g are elements of R, then there exist elements q and r in A such that

$$f = gq + r,$$

$\deg r < \deg g$. This implies that R is a principal left and right ideal domain;

2. If A is not a field, then $A[y, D]$ is not necessarily a principal ideal ring (for example, if $A = k[x]$ is the polynomial ring over a field k, and if $Dx = 1$. Cf. 7.19);

3. Every right and left ideal of $R = k[x][y, D]$, where $Dx = 1$, is projective if the field k has characteristic 0. In this case, every maximal right (left) ideal is principal and R is hereditary. (If k has characteristic $\neq 0$, then R has *gl. dim* $= 2$. Cf. §14);

4. (a) If D is a derivation of A, then the inner derivation D_y of $A[y, D]$ induces D.

(b) The center of $A[y, D]$ is isomorphic to the polynomial ring $T[y]$, where T is the subring of center A consisting of all a such that $D(a) = ay - ya = 0$.

(c) Amitsur [56] shows that the units of $A[y, D] \subseteq A$. Conclude that $x^{-1}Ax \subseteq A$ for all units x of $A[y, D]$, but that A is not contained in the center of $A[y, D]$ when $D \neq 0$. Cf. 2.16A.

When Modules Are Direct Sums of a Projective and a Noetherian Module

A right R-module M is a **Chatters module** if M is a direct sum of a projective module P and a Noetherian module N, with P or N possibly zero.

7.21 THEOREM (CHATTERS [82]). *A ring R is right Noetherian iff each cyclic right R-module is a Chatters module.*

The necessity is of course trivial.

When Modules Are Direct Sums of an Injective and a Noetherian Module

7.22 THEOREM (HUYNH-SMITH [90]). *A ring R is right Noetherian iff every right R-module is a direct sum of an injective module and a (locally) Noetherian module.*

7.23 REMARKS.

1. Locally Noetherian means every $f \cdot g$ submodule is Noetherian;
2. Unlike Chatters' Theorem 7.21, the theorem fails if the condition of 7.22 is applied to just cyclic right R-modules (*ibid*): Let F be a field, $A = F\langle x \rangle$ the power series ring, and the counter-example R is the ring of all matrices $\begin{pmatrix} a & b \\ o & c \end{pmatrix}$, with $c \in A$, and $a, b \in Q(A)$, the quotient field of A.

A right R-module M is said to be a **special** if M is an injective but not Noetherian uniserial module such that any nonmaximal proper submodule S is Noetherian and M/S is injective.

7.24 LEMMA (HUYNH-SMITH [90]). *If R is a ring, then R is never a special right (or left) R-module.*

PROOF. For suppose R is right special, and let J be the unique maximal right ideal. Then J is not $f \cdot g$. If J *were*, then $J \supset J^2$ by Nakayama's Lemma, and $|J/J^2| < \infty$ by the fact that J is $f \cdot g$, so J whence R would be Noetherian since J^2 is.

So J is not $f \cdot g$, hence if $0 \neq r \in J$; then $J \supset rR$, so $r \approx R/a^{\perp}$ is Noetherian. Furthermore, since R is not Noetherian, then $a^{\perp}$ is not Noetherian so $a^{\perp} \supseteq M$.

Then rR is simple $\forall r \in R$, hence M is semisimple hence simple. But then $|R| \leq 2$, a contradiction. □

7.25 LEMMA (*Op.cit.*). *If R is a ring, and every cyclic right R-module is the direct sum of an injective module and a Noetherian module, then every cyclic indecomposable injective right R-module is either Noetherian or special.*

7.26 THEOREM (HUYNH-SMITH [90]). *The following statements are equivalent for a ring R.*

1. *Every cyclic right R-module is the direct sum of an injective module and a Noetherian module.*
2. *Every finitely generated right R-module is the direct sum of an injective module and a Noetherian module.*
3. *R is a finite direct sum of special (injective) right ideals and Noetherian right ideals.*
4. *There exist a positive integer n and right ideals $A_i (1 \leq i \leq n)$ such that $R = A_1 + \cdots + A_n$ and every homomorphic image of A_i is injective or Noetherian for each $1 \leq i \leq n$.*

Dual Generalizations of Artinian and Noetherian Modules

Following Shock [74], an R-module M is a **max module** if every nonzero submodule has a maximal submodule. We say that M is a **Shock module** if every factor module $M/K \neq 0$ is a max module. Thus, a module M is a Shock module if for every submodule K, every submodule $S \supset K$ has a maximal submodule containing K.

7.27 THEOREM (SHOCK [74]). *For a right R-module M the following are equivalent:*

1. *M is Noetherian*
2. *M is a q.f.d. Shock module*
3. *M is a Shock module and all factor modules have $f \cdot g$ socles (possible $= 0$).*
4. *Every factor module of M has $f \cdot g$ radical and $f \cdot g$ socle.*

PROOF. Shock [74], Theorems 3.7 and 3.8.

Dual to max module is the concept of a **min module**, i.e., every nonzero factor module contains a minimal submodule.

7.28 THEOREM (*Ibid.*). *The following are equivalent conditions on a right R-module*

1. *M is artinian*
2. *M is a q.f.d. min module*
3. *M is a min module and every factor module $\neq 0$ has $f \cdot g$ socle*

PROOF. See Theorem 3.1 *ibid.*

7.29 THEOREM (*Ibid.*). *(1) An Artinian module M is Noetherian iff M is a max module; (2) Every Artinian submodule of a max module is Noetherian.*

7.30 REMARKS. (1) Camillo's Theorem 5.20B implies Shock's theorem and its proof uses the latter. The proof of the next theorem requires both; (2) 3 of Theorem 7.28 is equivalent to the condition of the Beachy-Vámos theorem 3.61, that is, every factor module is finitely embedded.

7.31 THEOREM (FAITH [98]). *A right R-module M is Noetherian iff M is q.f.d. and satisfies the acc on subdirectly irreducible submodules.*

See 16.50.

Completely Σ-Injective Modules

A right R-module is said to be *completely injective* provided that every factor module is injective. The theorem of Cartan-Eilenberg 3.22A states that a ring R is right hereditary iff every injective right R-module is completely injective. For a right Noetherian ring R, we show that this criterion for a right hereditary ring can be reduced to the requirement of completely injectivity of a single module, namely injective hull $\hat{R} = E(R)$ of R_R.

7.32 PROPOSITION. *In any ring R, any direct sum of injective right R-modules is an epic image of a direct sum of copies of the injective hull $\hat{R}$ of R_R.*

PROOF. Since a direct sum of a direct sum is a direct sum, and any direct sum of epics is an epic, it suffices to show this for a single injective module $M \neq 0$. Then, $M \approx \mathrm{Hom}_R(R, M) \neq 0$, and any morphism $f : R \to M$ extends by injectivity of M to a morphism $\hat{R} \to M$. Therefore, the so-called trace T of $\hat{R}$ in M

$$T = \sum_{f:\hat{R}\to M} f(\hat{R})$$

obviously equals M, and hence there is an epic $\hat{R}^{(I)} \to M$, where $I = \mathrm{Hom}_R(\hat{R}, M)$. □

7.33 COROLLARY. *Assume R is a ring such that $\hat{R}$ is Σ-completely injective. Then so is any injective R-module M, and R is then right Noetherian and right hereditary.*

PROOF. By 7.32, every direct sum of copies of M is an epic image of copies of $\hat{R}$. The rest follows from 3.4B and 3.22A.

7.34A THEOREM. *Let R be a right Noetherian ring R. If $\{E_i\}_{i\in I}$ is a family of completely injective right R-modules then the direct sum of the family is completely injective.*

PROOF. Let $E = \Sigma_{i\in I} \oplus E_i$ be the direct sum, and K a submodule $\neq E$, and for each submodule S of E, let $\overline{S} = (S+K)/K$. Choose a family $\{F_i\}_{i\in J}$ of submodules of E such that $\{\overline{F}_i\}_{i\in J}$ is a maximally independent family of submodules of $\overline{E} = E/K$, and let E' be the sum $\Sigma_{i\in J} \oplus \overline{F}_i$. The F_i are completely injective so $\overline{F}_i$ is injective for all $i \in J$. Since R is right Noetherian, then E' is injective by 3.4B, hence

$$E' \text{ is a summand of } \overline{E}.$$

Furthermore: if $E' \neq E$, then there is some $i \in I$ such that $\overline{E}_i$ is not contained in E'. But, E', being injective, is a direct summand of the submodule H of $\overline{E}$ generated by E' and $\overline{E}_i$, say $H = E' \oplus F$. But, then

$$F \approx H/E' \approx (E' + \overline{E}_i)/E' \approx \overline{E}_i/(E' \cap \overline{E}_i) \neq 0$$

is an epic image of $\overline{E}_i$, hence is injective, violating the maximality of $\{\overline{F}_i\}_{i\in J}$. Thus, $\overline{E} = E'$ is injective as required. □

Remark The same proof suffices to prove the following:

7.34B THEOREM. *Any direct sum of completely Σ-injective modules is completely Σ-injective.*

7.34C COROLLARY. *A right Noetherian ring R is right hereditary iff $\hat{R}$ is completely injective, and also iff every indecomposable injective right module is completely injective.*

PROOF. By 7.34A, $\hat{R}$ completely injective implies that $\hat{R}$ is Σ-completely injective, so R is right hereditary by 7.33. Also, if every indecomposable injective is completely injective, then every injective module is by 3.4B and 7.34 so 3.22 applies. □

Ore Rings Revisited

A ring R is **right Ore** if R has a classical right quotient ring $Q^r_{c\ell}(R)$, denoted $Q(R)$ for short.

7.35 THEOREM. *Let R be any ring with right quotient ring $Q = Q(R)$.*
(1) *Then Q is a flat left R-module.*
(2) *Every injective right Q-module is an injective right R-module.*

PROOF. (1) In effect, if I is a right ideal, and if $y = \Sigma_{i=1}^n x_i \otimes q_i$ lies in the kernel of the canonical map $I \otimes_R Q \to IQ$ (sending $y \mapsto \Sigma_{k=1}^n x_i q_i$), there exists a regular element $t \in R$ with $q_i t \in R, i = 1, \ldots, n$. Thus,

$$\sum_{i=1}^n x_i q_i = 0 \Rightarrow y = \sum_{i=1}^n x_i q_i t \otimes 1 = 0$$

and so $y = 0$. This proves (1) since this implies that $I \otimes_R Q \approx IQ$ canonically; (2) follows from the fact that the hypothesis implies that the inclusion functor mod-$Q \hookrightarrow$ mod-R has exact left adjoint $\otimes_R Q$. See Proposition 4.C. □

7.36 THEOREM. *If R is right Goldie semiprime, then $Q(R)$ coincides with $\hat{R}$.*

PROOF. Since $Q = Q(R)$ is semisimple by 3.13, hence right self-injective, then Q is an injective R-module by 7.35(2) hence $Q = \hat{R}$ as asserted.

REMARK. If R is right nonsingular, then $\hat{R}$ is a right self-injective VNR ring. (See §12.) In this case, if R is right Goldie, then $Q(R) = \hat{R}$ is semisimple Artinian.

7.37 THEOREM. *Let R be a prime right Goldie ring with simple Artinian right quotient ring $Q = D_n$, and let F be a minimal right ideal of Q. Then every injective right R-module is an epic image of a direct sum of copies of F.*

PROOF. Apply 7.32, 7.36, and the fact $Q \approx F^n$. □

On Hereditary Rings and Boyle's Conjecture

A right R-module is quasi-injective iff M is fully invariant in its injective hull $\hat{M} = E(M)$ (see **sup.** 3.9A).

A ring R is **right** QI if every quasi-injective module is injective. Any right QI ring R is right Noetherian (see 3.9A), and since each simple module is quasi-injective, R is also a right V-ring (see 3.19A).

Boyle's Theorem 3.9B states that a ring R is right and left QI iff R is right and left hereditary Noetherian V-ring. Boyle conjectured:

(BC) *Every right QI ring is right hereditary.*

Theorem 3.20 reduces the question to the case of simple rings.

A theorem of Webber 7.1, as generalized by Chatters 7.2 states that any (2-sided) Noetherian hereditary ring satisfies the restricted right minimum conditions:

(RRM) *R satisfies the d.c.c. on essential right ideals.*

Hence, for right and left QI rings, RRM, and its left-right symmetry, RLM, are necessary conditions for the truth of BC.

A structure theorem of Michler and Villamayor [73] classifies right V-rings of right Krull dimension (K-dim)≤ 1 (see §14) as finite products of matrix rings over right Noetherian, right hereditary simple domains, each of which are restricted semisimple, i.e., R/I is semisimple for each essential right ideal I. These rings are thus right hereditary, verifying Boyle's conjecture for these rings and in particular for RRM rings. We apply our criterion for right hereditary rings and verify Boyle's conjecture for a class of rings satisfying the **restricted right socle condition**:

(RRS) *If I is an essential right ideal $\neq R$, then R/I has socle $\neq 0$.*

This condition is easily seen to be weaker than K-dim $R \leq 1$; however, it turns out to be equivalent for right QI rings insasmuch as $RRS \Rightarrow RRM$ in Noetherian rings. (This follows because then R/I is semiArtinian and q.f.d. Cf. 13.62(2).)

We begin with a lemma and theorem of general interest.

7.38 LEMMA. *Let I be a right ideal in a ring R which is maximal in the set of all right ideals K such that $\widehat{R/K}$ is not semisimple. Then, I is an irreducible right ideal. Moreover, $E = \widehat{R/I}$ is restricted semisimple provided that E has no nontrivial fully invariant submodules* (**= NFI**).

PROOF. A brief argument shows that I is irreducible, that is, R/I is uniform, hence $E = \widehat{R/I}$ is indecomposable. Next assumming E is NFI, then $E = DS$, where $S = R/I$, and $D = \text{End}E_R$. Therefore, for any submodule $K \neq 0$, E/K is generated by the submodules $\{[dS+K]\}_{d\in D}$, where $[dS+K]=(dS+K)/K$ is the submodule of E/K generated by the cosets $\{[dS+K]\}_{s\in S}$.

Now $(dS+K)/K \approx dS/dS\cap K$ is an epic image of dS, and dS is an epic image of S. Moreover, $dS\cap K\neq 0$ since E is uniform, so that $(dS+K)/K$ is a proper epic image of $S=R/I$, which, by the choice of I implies that $(dS+K)/K$ is semisimple. Therefore, E/K, a sum of semisimple modules, is semisimple. □.

7.39 THEOREM. *Let R be any right QI ring, which is not semisimple. Then, R has an indecomposable completely injective right module E which is restricted semisimple, but not semisimple.*

PROOF. If every cyclic module R/I had semisimple injective hull $\widehat{R/I}$ then every cyclic module, hence R, would be semisimple, contrary to hypothesis. Since R is right QI, then every indecomposable injective module is NFI. Inasmuch as every semisimple module in a QI ring is injective, then Lemma 7.38 supplies the module E we want. □

7.40 THEOREM. *Any right QI ring R with restricted right socle condition is right hereditary.*

PROOF. By 3.20, we may assume R is simple and right Noetherian, hence by Goldie's Theorem 3.13, R has a simple Artinian right quotient ring

$$Q=Q(R)\approx D_n\approx F^n$$

where F is a minimal right ideal of Q. As outlined earlier, it suffices to prove that any injective right R-module M is completely injective. Thus, by Lemma 15, M is an epic image of a direct sum of copies of F. Hence, in order to prove that M is completely injective, it suffices to show that F is Σ-completely injective, and by Proposition 7.34A we need only show that F is completely injective.

Now if R is semisimple, there is nothing to prove. Otherwise by Theorem 7.39, there is an indecomposable completely injective right module E which is restricted semisimple but not semisimple. Now $E=\widehat{R/I}$ for an irreducible right ideal I (= R/I is a uniform module), and I cannot be an essential right ideal, since if so, then by the RRS assumption, there is a right ideal $J\subset I$ such that $V=J/I$ is simple. But R right $QI \Rightarrow V$ is injective, and then $\hat{V}=E=V$ is simple. But E is not semisimple. Therefore I is not an essential right ideal, hence I has a nonzero relative complement right ideal K. Now $K\hookrightarrow R/I$ canonically, so

$$E=\widehat{R/I}=\hat{K}\subseteq\hat{R}.$$

But $\hat{R}=Q$ by Theorem 7.36, and hence E is a direct summand of $Q=F^n$, so by the Krull-Schmidt Theorem (for direct sums of indecomposable modules with local endomorphism rings), it follows that $E\approx F$, so that F is completely injective too, which is what we wanted. □

As stated, Cozzens [70] constructed right QI rings R in which every indecomposable injective E either embeds in Q, or is simple. These rings were principal right ideal domains, hence right hereditary. The next corollary establishes a converse.

7.41 COROLLARY. *Any right QI ring R such that every indecomposable injective right module either embeds in $Q(R)$, or is simple, is necessarily right hereditary.*

PROOF. The point of the proof of Theorem 7.40 was to show that the module E given by Theorem 7.39 embeds in Q. As in the proof of Theorem 7.40, we may assume that R is not semisimple, and that E is not simple. Thus, E must embed in Q. □

7.42 REMARK. The next proposition shows that the conditions of Corollary 7.41 actually imply that R is right QI when R is right Noetherian semiprime, that is, one does not have to assume it.

A consequence of the Matlis-Papp Theorem 3.4C and the Johnson-Wong theorem **sup.** 3.8s is that a ring R is right QI iff every R is right Noetherian and every indecomposable injective right R-module is NFI. For a right Noetherian semiprime ring this can be weakened to:

7.43 PROPOSITION. *Let R be a right Noetherian semiprime ring that every indecomposable injective right R-module either embeds in Q, or else if NFI. Then R is right QI.*

PROOF. A result of Faith [72b], Prop. 35A, p.178, asserts that an indecomposable injective module E over a semiprime right Goldie ring R is NFI, provided that E embeds in Q. Thus, every indecomposable injective right module is NFI, so R is right QI. □

Finally, we make explicit a corollary of the proof of Theorem 7.43.

7.44 COROLLARY. *If R is a right Noetherian V ring, and if every indecomposable injective right R-module is restricted semisimple, then R is right hereditary.*

PROOF. For then every indecomposable injective right module is completely injective inasmuch as in a right Noetherian right V ring R, every semisimple module is injective. Then, Theorem 7.34C applies. □

REMARK. Huynh and Rizvi [97], p.272, Prop.4, elaborate on Theorem 7.40.

Δ-Injective Modules

The notation $\mathcal{A}_r(M,R)$ for a right R-module M: each $I \in \mathcal{A}_r(M,R)$ is the right annihilator $r_R(X)$ of a subset X of M. If $M = R$, then $r_R(X) = X^{\perp}$. An injective right R-module E is **Δ-injective** provided $\mathcal{A}_r(E,R)$ satisfies the dcc. (Cf. 3.10A.)

Below, we let $\operatorname{soc}_r R$ denote the right socle of R.

7.45 THEOREM. *Let E be a right Δ-injective over R, and let $Q = \operatorname{Biend} E_R$. If Q is right Kasch, then Q is right Artinian. Necessary and sufficient conditions on $J = rad\ Q$ for Q to be right Kasch right Artinian are the following:*

(1) $J = rad\ Q \in \mathcal{A}_r(E,Q)$.

(2) $J \in \mathcal{A}_r(R,R)$.

Moreover, the following four conditions are sufficient for Q to be right Kasch right Artinian.

(3) $r_Q(soc_r Q) = J$.

(4) $J = Z_\ell(Q)$, *the left singular ideal of* Q.

(5) $J = r_Q(soc_\ell R)$.

(6) $soc_r Q$ *is an essential left ideal.*

(7) $soc_r Q \supseteq soc_\ell Q$.

Furthermore, the following conditions imply that Q is right Artinian:

(8) $J^2 \in \mathcal{A}_r(Q, Q)$.

(9) $J^2 \in \mathcal{A}_r(E, Q)$.

(10) J/J^2 *is finitely generated* (Examples: $|J/J^2| < \infty$, or Q/J^2 is right Goldie).

(11) R *is commutative.*

(12) Q *is primary-decomposable.*

PROOF. See the author's Lectures [82b], Part I, p.39, Theorem 9.5.

7.46 REMARKS. (1) By Hansen's Theorem 3.11A, Q is a semiprimary ring; (2) E is Δ-injective over R iff E is Δ-injective over Q; (3) A **primary-decomposable** ring R is a finite product of primary rings, that is, a finite product of full $n \times n$ matrix rings over completely primary rings, that is, local rings with nilpotent radical. By a theorem of Asano [49] (Cf. 5.1A'), this happens for a semiprimary ring R iff the prime ideals of R commute; (4) Any completely primary ring R is right and left Kasch.

To see (4), first note that if n is the index of nilpotency of J, then $J^{n-1} \neq 0$ and

$$J^{n-1}J = JJ^{n-1} = 0$$

so $xJ = Jx = 0$ for any $0 \neq x \in J$, hence

$$R/J \approx x^\perp \subseteq R$$

so R/J embeds in R (either side).

Secondly, Kasch is a Morita invariant property: the $n \times n$ matrix ring over a Kasch ring is Kasch; and remark (4) follows easily from this fact.

A ring R is a **right Σ-ring** (resp. **Δ-ring**) provided that $E(R_R)$ is Σ-injective (resp. Δ-injective). Thus, any right Δ-ring is a right Σ-ring, by the Teply-Miller Theorem 3.10.

7.47 THEOREM. *Let R be a ring with maximal right quotient ring $Q = Q^r_{\max}(R)$. Then:*

(1) *R is a right Σ-ring (resp. Δ-ring) iff Q is a right Σ-(resp. Δ)-ring.*

(2) *R is a right Δ-ring iff R is a right Σ-ring and Q is semiprimary.*

(3) *If R is a right Δ-ring, then Q is right Artinian under any of the following conditions:*

  (A) *Q is primary decomposable*

  (B) *The prime ideals of Q are commutative*

  (C) *R is commutative*

(4) *Conversely, if Q is right Artinian then R is a right Δ-ring.*

PROOF. This follows easily from 7.45. Cf. the author's [82b], p.45, Theorem 10.18, for additional results and proofs. □

7.48 REMARKS. (1) Any right uniform right Δ-ring R has local right Artinian Q (*ibid.*, p.42, Corollary 9.22).

(2) If R is right nonsingular, then the f.a.e.c.'s: (1) R is a right Σ-ring; (2) R is a right Δ-ring; (3) Q is semisimple (*ibid.*, p.58, Theorem 11.4.9).

Co-Noetherian Rings

A ring R is **right co-Noetherian** provided that every *f.e.* right R-module is Artinian.

7.49 THEOREM (JANS [69]). *If R is right Noetherian and right co-Noetherian, then $\bigcap_{n=1}^{\infty} J^n = 0$, where $J = rad\ R$.*

Cf. Mueller [69].

7.50 THEOREM (VÁMOS [68]. *A commutative ring R is co-Noetherian iff R_m is Noetherian for all maximal ideals m.*

Cf. 16.33-4.

REMARK. Vámos (*ibid.*) proved a companion piece: Every f.e. R-module is $f \cdot g$ iff R_m is Artinian for all maximal ideals m.

CHAPTER 8

Completely Decomposable Modules and the Krull-Schmidt-Azumaya Theorem

If M is a direct sum of indecomposable modules, then M is said to be *completely decomposable.* (Thus: an indecomposable module is an example!)

Any semisimple module is completely decomposable, and any Jordan-Hölder module is another example; in fact, any Artinian or Noetherian module is completely decomposable. If

$$M = \oplus_{i \in I} M_i \tag{1}$$

the sum of indecomposable R-modules M_i, and also if

$$M = \oplus_{j \in J} N_j \tag{2}$$

is another decomposition of M into a direct sum of indecomposable modules, then the decomposition (1) is said to be *unique* if for every decomposition (2) there is a bijection $\pi : I \to J$ and a set of isomorphisms $\varphi_i : M_i \to N_{\pi(i)}$ of R-modules. If it holds for finite sets I, then R is said to **satisfy the Krull-Schmidt theorem**. If this holds for all sets I, then R is said to **satisfy the Azumaya-Krull-Schmidt theorem**. This holds whenever $\text{End}(M_i)_R$ is a local ring $\forall i \in I$ (Krull [28]-Schmidt [25,26]-Azumaya [48,50]). In this case we say that M has an **Azumaya diagram (=AD)**; when I is finite, then we say M has a **finite AD**.

8.A THEOREM. *A right R-module M has a finite AD iff A = End M_R is semiperfect, equivalently, A is semilocal and idempotents modulo the Jacobson radical of A lift.*

In part, this theorem dates back to Krull-Schmidt, and Jordan-Hölder. Cf. Faith [76], p.45, 18.26, which is stated for an object M in an Abelian category (Cf. Swan [68]). Also see Osofsky [70] for a bit of pathology regarding completely decomposable rings.

8B. THEOREM (ZELINSKY [53]-SANDOMIERSKI [72]). *Any right linearly compact ring R is semiperfect.*

Herbera-Shamsuddin and Camps-Dicks Theorems

The next theorem is a grand generalization of Schur's lemma.

8.C THEOREM (HERBERA-SHAMSUDDIN [95]). *If M has finite Goldie and finite dual Goldie dimension, e.g., if M is a linearly compact (=l.c.) right module over a ring, then M has semilocal endomorphism ring.*

The proof uses results of Camps and Dicks [93], especially their characterization of semilocal rings.

8.D THEOREM (CAMPS-DICKS [93]). *An Artinian module M over any ring R has semilocal endomorphism ring. Furthermore, M cancels from direct sums: $M \oplus A \approx M \oplus B \Rightarrow A \approx B$, for R-modules A and B.*

For the last part of 8.D, see Evans' theorems 6.3F–G. Cf. 8.8ff.

8.E THEOREM. *Any l.c. R-module M over a commutative ring has semiperfect endomorphism ring, hence M has a finite AD (Cf. 8A).*

The proof of 8.E given in Herbera-Shamsuddin (following Corollary 7 on p.3597) is due to 8.C and a theorem that states that pure-injective (e.g. l.c.) modules have F-semiperfect endomorphism rings (Jensen-Lenzing [89], p.180, Corollary 8.27). See 6.52.

Swan's Theorem

Unless otherwise noted, the Krull-Schmidt theorem is stated for finite direct sums of $f \cdot g$ indecomposable modules. The next theorem is a corollary of Theorem 8E.

8.F THEOREM (SWAN [62]). *Any Noetherian complete local commutative ring satisfies the Krull-Schmidt theorem, in fact, every $f \cdot g$ module has an AD.*

Cf. Wiegand [98].

Evans' Theorem

A commutative local ring R with residue field $\overline{R}$ is **Henselian** provided that for any monic polynomial $f(x)$, any factorization of $\bar{f}(x)$ in $\overline{R}[x]$ into a product of relatively prime monics lifts to one in $R[x]$.

Swan gave an example of a non-Henselian local ring over which Krull-Schmidt fails. In this sense Evans' theorem is definitive:

8.G THEOREM (EVANS [73]). *A commutative local ring R is Henselian iff every local ring A that is a module-finite R-algebra, i.e., $A = Ra_1 + \cdots + Ra_n$ for $a_1, \ldots, a_n \in A$, satisfies Krull- Schmidt.*

Evans' theorem is used in the proof of:

8.H THEOREM (WARFIELD [78]). *If R is a non-Henselian discrete valuation ring whose completion $\hat{R}$ is not algebraic over R, then the Krull-Schmidt theorem fails for torsion-free R-modules of finite rank.*

Matlis' Problem

(MP): If module E is a direct sum of indecomposable injectives, is every direct summand M of E completely decomposable?

8.1 MATLIS-PAPP THEOREM. *If R is right Noetherian, then MP has an affirmative answer.*

Since M is itself injective, this follows from the Matlis-Papp theorem 3.4C: If R is right Noetherian, every injective right R-module M is completely decomposable.

8.2A THEOREM (FAITH-WALKER [67]). *If E is a completely decomposable injective (or quasi-injective) module over a ring R, then MP has an affirmative answer.*

8.2B THEOREM (FAITH-WALKER [67] AND WARFIELD [69B]). *If E is a completely decomposable module, $E = \oplus_{i \in I} E_i$, where E_i is countably generated indecomposable with local endomorphism ring (e.g. if E_i is injective), then each direct summand of E has the same property, so MP has an affirmative answer.*

REMARK. This follows from the Crawley-Jónsson theorem (see 8.4) as Warfield pointed out (*loc.cit.*).

The Exchange Property and Direct Sums of Indecomposable Injective Modules

A module M has the *exchange property* (Crawley-Jónsson [64]) if for any module A and any two direct sum decompositions

$$A = M' \oplus N = \sum_{i \in I} \oplus A_i$$

with $M' \approx M$, there exist submodules $A_i' \subseteq A$, such that

$$A = M' \oplus \sum_{i \in I} \oplus A_i'.$$

The module M has the *finite* exchange property if this holds whenever the index set I is finite. Examples of modules which have the exchange property are quasi-injective modules and modules whose endomorphism rings are semi-perfect (see Examples 8.4F).

DEFINITION. A direct summand M of a module A has the *exchange property* in A if for any direct sum decomposition $A = \sum_{i \in I} \oplus A_i$, there exist submodules $A_i' \subseteq A_i$, such that $A = M \oplus \sum_{i \in I} A_i'$.

8.3 THEOREM. *Matlis' problem (MP) has an affirmative answer for any direct summand M' with the exchange property in a direct sum of indecomposable injectives.*

As Warfield pointed out, this follows trivially inasmuch as each A_i' is necessarily indecomposable injective, hence $A_i' = A_i$ or $A_i' = 0$, so then $M' = \oplus_{j \in J} A_j$, where $J = \{i \in I \mid A_i' \neq 0\}$.

Crawley-Jónsson Theorem

Two direct sum decompositions

$$A = \oplus_{i \in I} A_i = \oplus_{j \in J} B_j \tag{3}$$

of a right R-module A are said to have *isomorphic refinements provided* that there exist submodules $\{C_{ij} \mid i \in I, j \in J\}$

$$\text{so that} \quad A = \oplus_{i \in I, j \in J} C_{ij},$$

$$A_i = \oplus_{j \in J} C_{ij},$$

$$\text{and} \quad B_j = \oplus_{i \in I} C_{ij}.$$

Obviously two direct sum decompositions of A into indecomposable modules $\{A_i\}_{i\in I}$ and $\{B_j\}_{j\in J}$ have isomorphic refinements iff the decomposition is unique.

8.4 THEOREM (CRAWLEY-JÓNSSON [64]). *Any two direct sum decompositions of a right R-module M have isomorphic refinements if $M = \oplus_{i\in I} M_i$ is a direct sum of any number of countably generated modules $\{M_i\}_{i\in I}$ with the exchange property.*

Warfield, Nicholson and Monk Theorems

A (finite) *exchange module* is one with the (finite) exchange property. An *exchange ring* R is one such that R_R is an exchange module. This is right-left symmetric by Warfield [72b], who proved

8.4A WARFIELD'S THEOREM [72B]. *A right R-module M is a finite exchange module iff End M_R is an exchange ring; in particular finite exchange rings are exchange rings.*

This is also a corollary of a theorem of Nicholson [77] (see 8.4B).

A ring R is a *ring suitable for building idempotents* (= SBI) provided that all idempotents of $R/\text{rad}R$ lift, e.g. any ring with nil Jacobson radical (Jacobson [56,64]), or any semiperfect ring.

A ring R is **suitable** (Nicholson [77]) if idempotents of R/I lift for any left ideal I. (Thus, if $x^2 - x \in I$ for $x \in R$, there exists an idemipotent $e \in R$ so that $x - e \in I$.) A ring R is suitable iff $R/\text{rad}R$ is suitable and R is SBI (*loc. cit.*).

The endomorphism ring of a quasi-injective R-module is suitable by Theorem 4.2A. The next theorem shows suitable is right-left symmetric:

8.4B THEOREM (NICHOLSON [77]). *The following are equivalent conditions on a right R-module M:*

(1) *End M_R is right suitable,*
(2) *End M_R is left suitable,*
(3) *M is a finite exchange module.*
(4) *End M_R is an exchange ring* (Cf. 8.4A).

Combined with another theorem of Nicholson [75], Theorem 4.3, one has:

8.4C COROLLARY. *A ring R is semiperfect iff R is an exchange ring with no infinite set of orthogonal idempotents.*

8.4D COROLLARY (WARFIELD [69C]). *An indecomposable right R-module M is an exchange module iff End M_R is a local ring.*

Monk's theorem is also obtained by Nicholson as a corollary of 8.4B.

8.4E THEOREM (MONK [72]). *A right R-module M is an exchange module iff given $\alpha \in A = \text{End}M_R$ there exists γ and σ in A such that*

$$\gamma\alpha\gamma = \gamma \qquad \textit{and} \qquad \sigma(1-\alpha)(1-\gamma\alpha) = 1 - \gamma\alpha.$$

π-Regular Rings

A ring R is **π-regular** if for each $a \in R$ there exists $n = n(a)$ and $x \in R$ so that $a^n x a^n = a^n$. **Strongly π-regular** means R satisfies the dcc on the set $\{a^n R\}$ for any $a \in R$. Any strongly π-regular ring is π-regular (Azumaya [54]). A ring R is **semi-π-regular** if $R/\text{rad}R$ is π-regular and R is SBI.

8.4F THEOREM (HARANO AND PARK [93]). *If R is a left self-injective strongly π-regular ring, then $J(R)$ is nil and $R/J(R)$ is a finite direct product of full matrix rings over self-injective strongly regular rings.*

REMARKS. (1) A left self-injective algebraic algebra R over a field is strongly π-regular, and the structure of $R/J(R)$ was more fully described by Menal [81], Proposition 2.3; (2) Theorem 8.4F essentially follows from the Kaplansky-Armendariz-Steinberg Theorem 4.5.

8.4G EXAMPLE 1. (NICHOLSON [77]–STOCK [86]). Any semi-π-regular ring is an exchange ring.

EXAMPLE 2. (YU [97]). An exchange ring R with Artinian prime factor rings is strongly π-regular. (Yu [94] studied the countable exchange property.)

EXAMPLE 3. Any semiperfect module is a finite exchange module (Nicholson, *loc.cit.*).

EXAMPLE 4. A Dedekind finite exchange ring is semiperfect by 8.4C and 4.6B.

EXAMPLE 5. By Theorems 4.2A, 8.4A and 8.4B, any quasi-injective module is an exchange module (Fuchs [69], Theorem 3, p.545, Kahlon [71], and Warfield [69a], who showed (1) that any SBI ring R that is VNR modulo Jacobson radical is suitable, and (2) then any projective R-module is a direct sum of one-sided ideals, generated by indempotents; (2) generalized theorems of Kaplansky for VNR rings and local rings.

8.4H THEOREM (GOODEARL-WARFIELD [76]). *Let R be a locally module-finite algebra over a commutative zero-dimensional ring S, and let A be any finitely generated left R-module. Then A has the finite exchange property and any two direct sum decompositions of A have isomorphic refinements. More generally, if M is a left R-module which is a direct sum of finitely generated submodules, then any direct summand of M is again a direct sum of finitely generated modules, and any two direct sum decompositions of M have isormophic refinements. In particular, any projective left R-module is a direct sum of left ideals generated by idempotents.*

8.4I REMARKS. (1) Examples due to Evans [73] show that in Theorem 8.4H one cannot replace the requirement that S is zero-dimensional with the requirement that $S/J(S)$ is von Neumann regular, or even that $R = S$ and S is a (Noetherian) local ring;

(2) The proof uses a characterization of exchange rings due to Goodearl (unpublished): E is an exchange ring if and only if for every pair of left ideals I and J such that $I + J = E$, there is an idempotent $e \in I$ such that $1 - e \in J$.[1]

---

[1]This is essentially Prop. 1.1 of Nicholson [77]. See Theorem 8.4B above.

Yamagata's Theorem

We recall that for a ring R a right ideal I is *(meet) irreducible* provided $I \neq R$ and $I = I_1 \cap I_2$ implies $I = I_1$ or $I = I_2$ for all right ideals I_1 and I_2 of R, equivalently, R/I is uniform. Let $A \subseteq' B$ denote that A is an essential submodule of B.

8.5A THEOREM (YAMAGATA [74]). *The following conditions are equivalent:*

(*i*) *A ring R satisfies the ascending chain condition for irreducible right ideals.*
(*ii*) *Any direct sum M of indecomposable injective modules has the exchange property.*
(*iii*) *Any direct sum M of indecomposable injective modules has the finite exchange property.*
(*iv*) *Any direct summand of the module M which is a direct sum of indecomposable injective modules has the exchange property in M.*
(*v*) *For any direct sum M of indecomposable injective modules, the Jacobson radical of the endomorphism ring $End_R(M)$ is $\{f \in End_R(M) | Ker\, f \subseteq' M\}$.*

Moreover, (ii), (iii) and (v) are equivalent for any such M.

The proof depends on work of Harada and Sai [70] and Harada [71].

8.5B COROLLARY. *If R has the acc on irreducible right ideals, then any direct summand M of a direct sum of indecomposable injectives is completely decomposable, i.e., MP holds.*

REMARK. Various other conditions, e.g. when M is a nonsingular module, implies that MP holds true. (See, e.g. Kahlon [71], Harada [72], and Yamagata [73].)

Cf. Acc on irreducible ideals, §16, esp. 16.39.

Decompositions Complementing Direct Summands

A decomposition

$$A = \oplus_{i \in I} A_i \tag{1}$$

into a direct sum of a right R-modules A_i is said to **complement direct summands** (= cds) provided that each $A_i \neq 0$, and for any direct summand M there exists a subset $J \subseteq I$ so that

$$A = M \oplus (\oplus_{j \in J} A_j).$$

In this case each A_i is necessarily indecomposable. The decomposition (1) *complements maximal direct summands* (= cmds) if for each direct sum decomposition

$$A = M \oplus N \tag{2}$$

with $N \neq 0$ indecomposable, then there exists $i_0 \in I$ so that $A = M \oplus A_{i_0}$.

8.6 THEOREM (ANDERSON-FULLER [72]). *If the decomposition (1) complements direct summands (resp. cmds), then all decompositions of A into a direct sum of indecomposable modules cds (resp. cmds), and so does any direct summand of A. Moreover, if (1) cmds then the decomposition is unique.*

Cf. *loc.cit.* Lemma 1, Theorem 2 and Corollary 3.

8.7 COROLLARY. *A right R-module A has a decomposition (1) that cds iff every direct summand of A has the exchange property in A.*

Cf. *loc.cit.*, p.152.

8.8 THEOREM (WARFIELD [69C] AND ANDERSON-FULLER [72]). *If an injective module A is completely decomposable, then that decomposition cds.*

REMARK. By the Matlis-Papp theorem 3.4C, every injective right R-module A has a decomposition (1) that cds iff R is right Noetherian.

Fitting's Lemma and the Krull-Schmidt Theorem

Fitting's Lemma states that any indecomposable R-module M of finite length has local, in fact, completely primary, endomorphism ring, hence any direct sum of indecomposable modules of finite lengths satisfies the Krull-Schmidt-Azumaya theorem, e.g. any $f \cdot g$ module over a right Artinian ring.

The corresponding question for an Artinian module was raised by Krull [32], and disproved by Facchini, Herbera, Levy and Vámos [95] (denoted FHLV below), employing results of Camps and Dicks [93], who also proved that *any* Artinian R-module over any ring R has semilocal endomorphism ring. Facchini [96] showed that Krull-Schmidt also fails for serial modules. Camps and Menal [91] proved that any commutative semilocal Noetherian domain can be represented as the endomorphism ring of a necessarily indecomposable Artinian module. This was generalized in FHLV: *if A is a module finite algebra over a semilocal Noetherian commutative ring, then $A \approx \mathrm{End} M_R$ where M_R is Artinian, for some ring R.* This is the key to disproving Krull-Schmidt for Artinian modules. Cedó's theorem in Camps-Dicks [93] states that a ring R is semilocal if R satisfies the dcc $\perp$ and acc $\perp$ on point annihilators, and every right or left regular element is a unit. Cedó's theorem generalized Stafford's theorem [82] for Noetherian quotient rings. A commutative acc $\perp$ ring R has semilocal Kasch quotient ring $Q(R)$ (Faith [91b]).

A Very General Schur Lemma

As stated (8.C), Herbera and Shamsuddin [95] in a grand generalization of Schur's lemma proved that any module M that is linearly compact in the discrete topology has semilocal endomorphism ring. While this generalized the Camps-Dicks theorem, the proof makes essential use of the Camps-Dicks characterizations of semilocal rings, and also the concept of dual Goldie dimension.

Faith and Herbera [97] study the question of when over a commutative linearly compact ring every linearly compact R-module has R-linearly compact endomorphism ring; and prove it for Noetherian and valuation rings. Moreover, the question is equivalent to the condition that $M \otimes_R N$ is linearly compact for any two linearly compact R-modules M and N.

Rings of Finite and Bounded Module Type

A good deal of module theory is aimed at the description of the indecomposable finitely generated modules (at least over right Noetherian rings when every finitely generated module decomposes into a direct sum of indecomposable modules!) Let M be an indecomposable module over a right Noetherian ring R, assume that M is finitely generated, and let $g(M)$ be the least cardinal of any set of generators of M. In general, there exist indecomposable modules M with ever larger $g(M)$. Indeed, by Higman's theorem [54], this happens whenever R is the group algebra in characteristic p with noncyclic p-Sylow subgroup G of finite order n; in particular, finite rings can have this property. (However, in the case of cyclic p-Sylow subgroup, n is a bound on the "number" of indecomposable modules (Kasch-Kneser-Kupisch [57]).)

Next assume a bound on the $\{g(M)\}$. This is a reasonable finiteness condition which one frequently encounters in classical algebra, for example, as we have seen, it holds over FGC rings. Such a ring is said to be **right FBG**, or bounded module type. A commutative local FBG ring R has linearly ordered ideals (Warfield [70]) (cf. 6.5), that is, R is a chain ring.

Another kind of finiteness condition that frequently occurs in the theory of finite dimensional algebras and Artinian rings: does right FBG imply finiteness of the isomorphism classes of indecomposable finitely generated right modules? A ring with the latter property is said to be **right FFM**, or finite module type. (Serial rings are right and left FFM rings.) In this notation the question just stated can be stated as the validity of the implication $FBG \Rightarrow FFM$. For algebras of finite dimension over a field this was called the Brauer-Thrall conjecture, and was proved by Roiter [68]. For Artinian rings, Auslander [74] proved the conjecture utilizing notably different methods. Any FFM algebra, in fact, FFM Artinian ring, has finite lattice of ideals (Jans [56], Tachikawa [60] and Colby [66]). By Yamagata [75], a semiperfect ring R has FFM (= finite representation type) iff every direct sum of indecomposable modules enjoys the exchange property.

Auslander ([74,Cor.4.8]) and Ringel and Tachikawa in Tachikawa [73, p.129, Cor.9.5] prove: Let R be a right Artinian right FFM ring. Then, every indecomposable right R-module is finitely generated, and every right module is a direct sum of indecomposable modules. Then, by Zimmermann-Huisgen [79] (cf. Theorem 6.56 in the text) R is right pure-semisimple (cf. 6.57).

Herzog [96b] studies whether pure-semisimple implies FFM, and verifies this for PI-rings and Morita rings (see §13 for Morita rings).

Moreover, Tachikawa [73] also shows over a right FFM ring that all modules have decompositions which complement direct summands (cds), see §8, esp. 8.6ff. Fuller-Reiten [75] (Cf. Fuller [76]) prove a converse for rings over which right and left modules have decompositions which cds. Auslander [74] showed that Artin algebras are FFM provided only that every indecomposable left module is finitely generated. Moreover, the theorem of Faith-Walker [67] (cf. 3.5B) states that if every injective left module is a direct sum of finitely generated modules, then R must be left Artinian. This property characterizes commutative Artinian rings by the Morita Theorems [58]: every commutative Artinian ring R has a Morita duality and every indecomposable injective R-module is $f \cdot g$ (Cf. 13.4B). As Rosenberg and Zelinsky [59] showed, in general the injective hull of a $f \cdot g$ module (e.g. any indecomposable injective) need not be $f \cdot g$. For when injective hulls of cyclics are cyclic, see Faith [66b], Caldwell [67], and Osofsky [68b]; (they include Köthe's uniserial rings–see Faith [66b]).

CHAPTER 9

Polynomial Rings over Vamosian and Kerr Rings, Valuation Rings and Prüfer Rings

A commutative ring R is *Vamosian* if each finitely embedded R-module M (= has finite essential socle) is linearly compact. (See Vámos [75] where the rings are called "classical".) Any linearly compact ring R is Vamosian, and so is any locally Noetherian ring (e.g. any VNR ring R). It can be shown, for any locally Noetherian ring R, that the polynomial ring $R[x]$ is locally Noetherian (e.g. Faith [86], where Vamosian rings were nominated, or Huckaba [88], p. 73, Lemma 13.1). Vámos [75] showed that classical (= Vamosian rings) are SISI (= subdirectly irreducible factors are self-injective). (Cf. FSI rings, 5.9ff.) Faith [86] showed that SISI rings are not preserved under polynomial extension, and asked the same question regarding Vamosian rings. Xue [96] has given a negative answer.

9.1 THEOREM (XUE [96]). *For a commutative linearly compact ring R the following are equivalent:*

(1) *R is Noetherian*
(2) *R is locally Noetherian*
(3) *$R[X]$ is Vamosian*
(4) *$R[X]$ is co-Noetherian.*

By Zelinsky's theorem [53] (cf. Sandomierski [72]) R is a finite direct product of local rings, hence locally Noetherian implies Noetherian (cf. Xue [92]). Xue's theorem shows that any non-Noetherian linearly compact ring R has non-Vamosian $R[X]$.

Kerr Rings and the Camillo-Guralnick Theorem

A ring R is a (right) **Kerr** ring provided that the polynomial ring $R[X]$ satisfies the acc on annihilators, namely acc $\perp$ (in which case R is an acc $\perp$ ring). Any subring of a Noetherian ring is Kerr.

9.2 THEOREM (KERR [90]). *Not every commutative Goldie ring R has Goldie polynomial ring; in fact, not every Goldie ring R is a Kerr ring.*

The second statement followed from the first since a polynomial ring $R[X]$ has acc$\oplus$ (= finite Goldie dimension) iff R does (Shock [72]).

9.3 THEOREM (CAMILLO-GURALNICK [86] AND ROITMAN [89,I]). *If R is an acc$\perp$ algebra over an uncountable field, then R is a Kerr ring and so is any polynomial ring over R in any number of variables.*

The trick in the proof is that any countably generated subring of $R[X]$ embeds in R. (It is necessary and sufficient for acc $\perp$ for any countably generated subring to have acc $\perp$.)

Notes: (1) The theorem is stated in both papers in greater generality: all that is needed is the existence of a set S in the center of R consisting of uncountably many elements, such that $s - t$ is not a zero divisor for all $s \neq t \in S$. (2) A domain A is **Mori** if A satisfies the acc on integral divisorial ideals (= intersections of cyclic A-submodules of $K = Q(A)$), or equivalently, $A/(a)$ has acc$\perp$ for any nonzero $a \in A$ (Roitman *loc.cit.*, p. 248, Theorem 2.2). Roitman applies 9.3 to obtain that for any Mori domain A over an uncountable field, a polynomial ring in any set of variables is also Mori.

9.4 THEOREM (FAITH [93,94,96]). *A commutative acc $\perp$ ring R is Kerr when R has Goldie dimension 1, or when R is Goldie and its classical quotient ring Q has nil Jacobson radical, or when Q is finitely embedded (= f.e.). Then, in all cases, Q is Artinian.*

Actually a f.e. acc $\perp$ ring **is** Artinian (See the first paper cited above.)

9.5 COROLLARY (LOC. CIT.). *If R is a Goldie commutative ring such that Q is an algebraic algebra over a field k or has dimension over k strictly less than the cardinality of k, then Q is Artinian hence R is a Kerr ring.*

9.6 THEOREM (FAITH [96]). *Any locally Noetherian acc $\perp$ ring R is Kerr, and then so is $R[X]$.*

The proof requires Beck's theorem 3.16C.

9.7 THEOREM (CEDÓ AND HERBERA [95]). *There exist Kerr rings R, such that for any integer $n \geq 1$, the polynomial ring $R[X_1, \dots, X_n]$ in n variables is Kerr, but that in $n+1$ variables is not Kerr.*

9.8 THEOREM (ROITMAN [94, II]). *There exists a Kerr ring R such that $R\langle x\rangle$ does not satisfy acc$\perp$.*

This was proved independently by Cedó [95].

Rings with Few Zero Divisors Are Those with Semilocal Quotient Rings

A commutative ring is said to have **few zero divisors** iff the set $z(R)$ of zero divisors is a finite union of prime ideals. The concept originated in 1962 in a paper of E. Davis [65]. (Cf. Huckaba [88], p.49.)

9.9 THEOREM (DAVIS [64]). *A commutative ring R has few zero divisors iff its classical quotient ring $Q(R)$ is semilocal.*

PROOF. See *loc.cit.*, p. 204, Remark 1. □

Note: Davis' theorem corrects the author's paper [96c], i.e., $Q(R)$ is not necessarily Kasch. (I have T. Y. Lam [98] to thank for calling my attention to this.)

REMARKS. (1) Any ring R with the acc on annihilator ideals has semilocal Kasch $Q(R)$, hence few zero divisors (Faith [91b], Cor. 3.7), and also $R[X]$ has few zero divisors, even though $R[X]$ need not have acc on annihilators. (The latter result is a theorem of Kerr [90], see 9.2.); (2) R has semilocal Kasch $Q(R)$ iff the same is true of the polynomial ring over R. (See Faith [91b]). Cf. 16.28-16.32, and 16.42, 16.42f.

Manis Valuation Rings

Griffin [70],p.56, defines a ring to have few zero divisors if it has just "finitely many maximal prime ideals of 0" and remarks that the "principal property" of such a ring R is that for any element $z \in Q(R)$, and for each regular element $a \in R$ there exists $u \in R$ such that $z + au$ is regular, ("Choose u in all maximal 0-primes not containing z and in no 0-primes containing z").

A **Manis valuation** v is an onto map $v : K \to \Gamma \cup \{\infty\}$ of a ring K and a totally ordered group Γ, and symbol ∞, such that

(v1) $$v(xy) = v(x) + v(y)$$

(v2) $$v(x + y) \geqq \min\{v(x), v(y)\}.$$

A subring A is a **Manis valuation ring** if there is a valuation v of K such that $A = \{x \in K \mid v(x) \geq 0\}$. In this case $P = \{x \in K \mid v(x) > 0\}$ a prime ideal, and A is a maximal subring B of K containing A that has a prime ideal P' that contracts to P. Then (A, P) is a called a **max pair** in K, and P is the **valuation prime ideal**. The converse also holds: any max pair (A, P) defines a Manis valuation ring. For domains this is due to Krull [32]. (Cf. 9.10ff. Also see Manis [67] and Griffith [70]. See 9.11–12 and 19.18ff. below.)

Any subring A of K that is maximal with respect to excluding some unit x of K is a Manis valuation ring with valuation prime ideal $P = \sqrt{x^{-1}A}$. Then A is called a **conch subring** and is said to **conch** x in K (Faith [84b,c]).

Conversely, if (A, P) is a max pair in K and if $P = \sqrt{x^{-1}A}$ for a unit x of K, then A conches x and K (Faith [84c]).

Davis [64] defines a subring B of K to be a **quasi-valuation ring** provided that for each regular element of K either x or x^{-1} belongs to B. Furthermore, if A has few zero divisors, then A is a quasi-valuation ring iff A is a Manis valuation (Griffin [70], Lemma 2). By Davis' Theorem 9.9, we have then the:

9.10 COROLLARY. *If A has semilocal quotient ring K, then A is a Manis valuation ring iff A is a quasi-valuation ring.*

In this case, any two ideals I and J of A, one of which contains a regular element, are comparable. In the classical case where A is a domain then any two ideals are comparable, i.e.., A is a valuation domain in the classical sense (cf. Krull [32], where the subject may be said to have originated. Also see 9.11 and 9.19 (1)).

Integrally Closed Rings

A ring A is **integrally closed** in $Q = Q(A)$ if every $q \in Q$ that satisfies a monic polynomial in $A[x]$ belongs to A.

9.11 Theorem (Krull [32]). *Any integrally closed domain A is the intersection of valuation overrings.*

A ring A is a **paravaluation** ring if it is a valuation ring of $Q = Q(A)$ for a Manis valuation v of Q in which $v : Q \to \Gamma \cup \{\infty\}$ need not be onto (as required in the definition).

9.12 Theorem (M. Griffin; see Huckaba [88], p.54, Corollary 9.2). *A ring R is integrally closed in $Q = Q(R)$ iff R is the intersection of paravaluation rings of Q.*

Remark. Cf. the author's paper [84b], and Review [90b] of Huckaba's book [88].

9.13 Theorem (Faith [79a], Faith-Pillay [90]). *Any commutative FPF ring R is integral closed.*

Kaplansky's Question

(K) If A is a chain ring, is $A \approx R/I$ where I is an ideal, and R is a chain domain?

Fuchs and Salce [85] gave a counter-example. (Cf. Osofsky's counter-example [91].)

Hungerford [68] and MacLean [73] gave affirmative answers for A with Noetherian quotient ring $Q(A)$. Also see Fuchs and Shelah [89].

Local Manis Valuation Rings

A Manis valuation ring A of $K = Q(A)$ need not be a local ring, hence need not be a chain ring. A **sandwich** subring of K is a subring that contains the Jacobson radical $J = \text{rad} K$ of K.

9.14 Theorem (Faith [84b]). *If a Manis valuation ring A of $K = Q(A)$ is a local ring then A is a sandwich ring, and conversely if K is a local ring.*

See *op.cit.*, pp.20–21, for this and the next theorem.

9.15A Local Conch Ring Theorem. *Let A conch x in $K = Q(A)$. Then A is a local ring with maximal ideal $\sqrt{x^{-1}A}$ iff K is a local ring and A is a sandwich ring.*

9.15B Corollary. *For a Manis valuation ring A of a chain ring $K = Q(A)$, the following are equivalent:*

1. *A is a local ring,*
2. *A is a sandwich subring of K,*
3. *A is a chain ring (= classical valuation ring).*

See the author's *op.cit.*, p.25; and Froeschl's theorem:

9.15C Theorem (Froeschl [79]). *If $K = Q(A)$ is a chain ring, and A is a valuation ring of K, then A is chained iff $A \supseteq$ the set $z(K)$ of zero divisors of K. Otherwise A has exactly two maximal ideals, one of which is $z(A)$.*

REMARKS. (1) Any chain ring A is a Manis valuation ring for $K = Q(A)$ (see the author's *op.cit.*, p. 22,§14).

(2) If A conches x in K, then A is a maximal subring of K iff $K = A[x]$. If K is a field, this happens iff A is a rank 1 classical valuation ring. If A is Noetherian, then by the Principal Ideal Theorem 2.22, A is conch in K iff A has dimension 1.

(3) Froeschl's Theorem yields (2) $\Leftrightarrow$ (3) of 9.16B.

9.16 THEOREM. *If A is a local FPF ring, then A is a Manis valuation ring for $Q(A)$, and conversely if $Q(A)$ is self-injective.*

See the author's *op.cit.*, pp.20 and 24.

9.17 EXAMPLE (VALENTE [87], COMMUNICATED TO THE AUTHOR BY VITULLI [86]). In general, a Manis valuation v is defined on a ring K not necessarily quotient ring (Manis [67]), and furthermore v is not necessarily extendable to $Q(K)$. First consider the $\mathbb{Z}$-adic valuation $w : \mathbb{Q} \to \mathbb{Z}_\infty = \mathbb{Z} \cup \{\infty\}$ (e.g. $w(a)$ is the max n so that $2^n \mid a$ for all $0 \neq a \in \mathbb{Z}$, and $w(a/b) = w(a) - w(b)$.) Then a Manis valuation $v : \mathbb{Q}[x] \to \mathbb{Z}_\infty$ assigns $v(f)$ the value $w(f(0))$, and the **core** (or infinite prime ideal) is:

$$H = \{f \mid v(f) = \infty\} = x\mathbb{Q}[x].$$

Then, v extends to $\mathbb{Q}[x]_H$ but no further. In general, a Manis valuation v of a ring K extends to $\mathbb{Q}(K)$ iff the core H consists entirely of zero divisors. (This corrects a misstatement in the author's paper [86c].)

Domination of Local Rings

9.18 DEFINITION. If (R, m) is a local ring, then a local ring (S, n) **dominates** (R, m) in case $S \supseteq R$, and $n \cap R = m$, equivalently $m \subseteq n$.

9.19 THEOREMS. *(1) (Krull [32]) Every local domain is dominated by a valuation domain.*

(2) If (R, m) is a local domain and F is a subfield of $Q(R)$, then $R \cap F$ is a local ring dominated by R.

(3) Every Noetherian local ring is dominated by a complete local ring.

(4) (Chevalley [54]) Every Noetherian local domain is dominated by a rank-1 discrete valuation domain.

(5) (Gilmer-Heinzer [97]) Every Noetherian local ring R is dominated by a 1-dimensional Noetherian local ring S so that regular elements of R are regular in S, and each associated prime of S contracts to an associate prime of R.

See Gilmer-Heinzer for references and proofs.

REMARKS. (1) is the key result in the proof of Krull's Theorem 9.11. Also see Nagata [62], (11.9); (2) See Nagata, (17.6) for the proof of (3); (3) See Cahen-Houston-Lucas [96] for a generalization of (4).

Marot Rings

A **Marot** ring R is a commutative ring such that every regular ideal (= one containing a regular element) is generated by regular elements. (Named by Huckaba [88], p.31. We also refer to Huckaba, sections 7–9, for historical background and proofs of this and the following section on Krull rings).

9.20 EXAMPLE. Any ring R with few zero divisors, equivalently by 9.9, any ring R with semilocal quotient ring $Q(R)$, is a Marot ring, e.g., if R has acc $\perp$. (See Remark following 9.9; also see Huckaba, *op.cit.*, p.32, Theorem 7.2.)

Krull Rings

A commutative ring R is a **Krull ring** if R is Marot and either $R = Q_{c\ell}(R)$ or else there exists a family $\{v_i\}_{i\in I}$ of discrete rank 1 valuations of R such that:

1. $R = \bigcap_{i\in I} V_i$ of the corresponding valuation rings $\{V_i\}_{i\in I}$.
2. For each regular element $x \in Q(R), v_i(x) = 0$ for just finitely many i, equivalently x is a unit in all but finitely many of the V_i.

Cf. Ratliff [98] (also Fossum [73]) for other possible generalizations of Krull domains.

9.21 THEOREM. *If R is Marot, then R is integrally closed iff R is the intersection of the valuation overrings of R.*

See Huckaba [88], p.54, Theorem 9.2. Cf. Griffith Theorem 9.12.

9.22 THEOREM (HUCKABA [76]). *The integral closure of a Noetherian ring R is a Krull ring.*

9.23 THEOREM. *The polynomial ring $R[X]$ over a commutative Marot ring is a Krull ring iff R is a finite product of Krull domains.*

See Fossum [73]. Also see 9.26Aff.

9.24 RELATED RESULTS. (1) (**Nagata** [62], Theorems 33.2 and 33.12) *If D is a Noetherian domain of dimension 2, then the integral closure D' of D is Noetherian;*

(2) (**Huckaba** [88],§11,esp. 11.6) *(1) fails if D is an arbitrary Noetherian ring of dimension ≤ 2, but in this case every regular ideal of D' is $f \cdot g$;*

(3) (**Akiba** [80]], Theorem 13.11 in Huckaba, *op.cit.*) *If R is an integrally closed reduced McCoy ring, then so is the polynomial ring $R[X]$;*

(4) (**Endo** [61]) *If $Q(R)$ is a VNR ring, then R is integrally closed iff R_m is an integral closed domain for all $m \in maxR$;*

(5) (**Quentel** [72]) *If R has compact min spec, then $R[X]$ is integrally closed iff $Q(R)$ is VNR.*

9.25 THEOREM. *If R is a Noetherian commutative ring, then $R[X]_W$ is Noetherian for any set X of variables, where $R[X]_W$ is $R[X]$ localized at the set W of monic polynomials.*

REMARK. (1) Thus, for an infinite set X, $R[X]$ is a non-Goldie subring of the Noetherian ring $R[X]_W$. Furthermore, $R[X]$ is an acc$\perp$ ring, in fact, Kerr (cf. 9.2ff); (2) 9.25 is a remark of D. D. Anderson in Camillo [90],p.75.

Rings with Krull Domain Centers: Bergman and Cohn Theorems

9.26A THEOREM (BERGMAN [71]). *Any right Noetherian right hereditary ring R is a finite product of indecomposable rings each having a Krull domain as center.*

9.26B REMARK. Small and Wadsworth [81] show that the centers in Bergman's theorem need not be Noetherian even for a PI-ring R.

9.26C THEOREM (BERGMAN AND COHN [71]). *Any Krull domain R is isomorphic to the center of some right and left principal ideal domain.*

9.26D REMARK. Herbera and Menal [89] pointed out that this provided new examples showing that the center of an FPF ring need not be FPF.

Annie Page's Theorem

9.26E THEOREM (ANNIE PAGE [84]). *Any commutative Krull domain A arises as the center of a right hereditary right Noetherian PI-ring.*

9.26F REMARK. The example has the form

$$\begin{pmatrix} B & k(X,Y) \\ 0 & k(Y) \end{pmatrix}$$

where B is a PID with quotient field $k(X)$, the field of rational functions in the variable X over $k = Q(A)$, the quotient field of A, such that $B \cap k = A$.

The Maximal Quotient Ring of a Commutative Ring

Let R be a commutative ring, with quotient ring $Q(R)$. An ideal I of R is **dense** if I is faithful, i.e., $I^{\perp} = 0$.

Let J_1 and J_2 be (finitely generated) dense ideals of R and let $f_1 \in \operatorname{Hom}_R(J_1, R)$ and $f_2 \in \operatorname{Hom}_R(J_2, R)$. Then $J_1 J_2$ is a (finitely generated) dense ideal so that we may define both $f_1 + f_2$ and $f_1 f_2$ as homomorphisms on $J_2 J_2$. Define f_1 and f_2 to be *equivalent* if they agree on a dense ideal of J of R. From Lambek [66] (13, Lemma 1, p.38) one sees that f_1 and f_2 agree on a dense ideal J if and only if they agree on $J_1 \cap J_2$ and hence if and only if they agree on $J_1 J_2$. The elements of the **maximal quotient ring** $Q_{\max}(R)$ are defined to be equivalence classes of the homomorphisms $\operatorname{Hom}_R(J, R)$ for dense ideals J. (Cf. §12 for the maximal quotient rings of noncommutative R, which may be constructed as equivalence classes of homomorphisms defined via "dense" right ideals J.)

Since $J = xR$ is dense for each regular element $x \in R$, then one sees that canonically

$$R \subseteq Q(R) \subseteq Q_{\max}(R).$$

Furthermore, if $E = E(R)$ is the injective hull of R, and $S = \operatorname{End} E_R$, then $Q_2 = Q_{\max}(R)$ can be identified with

$$Q_2 = Q_{\max}(R) \approx \operatorname{Biend} E_R = \operatorname{End}_S E$$

(Lambek [63]), and also

$$Q_2 \approx \operatorname{ann}_E \operatorname{ann}_S(R)$$

(Findlay-Lambek [58]. See Lambek [66], and also §12, **sup.** 12.A).

9.27 REMARKS. (1) If R is Kasch, then R is the only dense ideal, hence $R = Q_{\max}(R)$ in this case; (2) As an exercise in Noetherian ideal theory, one shows that $Q(R)$ is Kasch for a Noetherian (more generally any acc$\perp$) commutative ring R hence $Q(R) = Q_{\max}(R)$ in this case (see 16.31).

The Ring of Finite Fractions

If we restrict the above construction for $Q_{\max}(R)$ to homomorphisms $f \in \mathrm{Hom}_R(J, R)$ for $f \cdot g$ dense ideals J, then we obtain the **ring $Q_0(R)$ of finite fractions** of R, and

$$R \subseteq Q(R) \subseteq Q_0(R) \subseteq Q_{\max}(R)$$

canonically.

The name "ring of finite fractions" comes from the fact that each element of $Q_0(R)$ can be identified with an element of $Q(R[X])$, the classical quotient ring of the polynomial ring $R[X]$. For $f \in Q_0(r)$ let $J = (b_0, \dots, b_n)$ be a dense ideal of R such that $f \in \mathrm{Hom}(J, R)$. Let $b(X) = b_n X^n + \cdots + b_n$ and $a(X) = a_n X^n + \cdots + a_0$, where $f(b_i) = a_i$ for each i. Then since $b_j f(b_i) = f(b_j b_i) = b_i f(b_j)$, $[a(X)/b(X)]b_i = a_i$ for each i. Whence f can be identified with multiplication by $a(X)/b(X)$. Since J is dense, $b(X)$ is a regular element of $R[X]$. Even though $b(X)$ is a regular element of $R[X]$, J need not contain a regular element of R. When J does contain a regular element r of R, then f 'reduces' to the element $s/r \in Q(R)$ where $fr = s$. For integral domains and Noetherian rings such a reduction always occurs since in both cases an ideal either has a nonzero annihilator or contains a regular element. However, there are many examples where $Q_0(R)$ properly contains $Q(R)$ (see, for example, Lucas [93]).

An ideal I of R is said to be *regular* if it contains a regular element and *semiregular* if it contains a finitely generated dense ideal. If the only semi-regular ideals of R are the regular ones, then R is McCoy, in which case $b(X) = b_n X^n + \cdots + b_0 \in R[X]$ is regular if and only if $c(b) = (b_0, \dots, b_n)$ is a regular ideal of R. This establishes the following:

9.28 THEOREM (LUCAS [93]). *If R is a commutative McCoy ring, then the ring $Q_0(R)$ of finite fractions coincides with the quotient ring $Q(R)$.*

The converse is false (Lucas [93]).

Prüfer Rings and the Davis, Griffin and Eggert Theorems

An ideal I of a commutative ring R is **invertible** (in $Q(R)$) if $IK = R$ for an R-submodule K of $Q(R)$. This implies, e.g., that I is $f \cdot g$ projective. An integral domain R is Prüfer iff R is semihereditary. (N.B.) Cf. 9.33.

A ring R is **Prüfer** (Griffin [69]) if every regular ideal is invertible. An **overring** of R is a subring of $Q(R)$ that contains R. A Dedekind ring R is a Noetherian Prüfer ring.

A **discrete valuation domain** (= **DVD**) is a Noetherian valuation domain. Any Noetherian valuation ring is a Dedekind ring, and is a principal ideal ring. (Cf. Hinohara's theorem stated 6.3Af.)

9.29 THEOREM (DAVIS [64]). *An integral domain R is Prüfer iff every overring of R is integrally closed.*

9.29B THEOREM (RICHMAN [65]). *An integral domain R is Prüfer iff every overring is flat (cf. 12.3).*

9.30 THEOREM (GRIFFIN [69]). *A commutative ring R is Prüfer iff every overring of R is integrally closed.*

A commutative ring R is an **I-ring** if every subring of $Q_{\max}(R)$ containing R is integrally closed.

9.31 THEOREM (EGGERT [76]). *A reduced ring R is an I-ring iff R is Prüfer and $Q(R) = Q_{\max}(R)$.*

9.32 COROLLARY. *A reduced ring R is an I-ring iff R is semihereditary.*

9.33 THEOREM (GRIFFIN [70]-[74]). *A commutative ring R is semihereditary iff R is Prüfer and $Q(R)$ is VNR. Every integrally closed subring R in a VNR quotient ring $Q(R)$ is the intersection of semihereditary Manis valuation rings.*

9.34 THEOREM (FAITH [84B]). *A reduced ring R is FPF iff R is semihereditary and $Q(R)$ is self-injective. Any Manis valuation ring R with VNR ring $Q(R)$ is FPF.*

REMARK. For connections between FPF and Prüfer noncommutative rings, see e.g. the author's [77b,82b], and [84] (with Page).

Strong Prüfer Rings

A ring R is a **strong Prüfer ring** if every $f \cdot g$ semiregular ideal is locally principal. A $f \cdot g$ regular ideal is invertible iff it is locally principal, hence any McCoy Prüfer ring R is strong.

9.35 THEOREM (LUCAS [93]). *A commutative ring R is a strong Prüfer ring iff the ring $Q_0(R)$ of finite fractions of R is McCoy and every ring between R and $Q_0(R)$ is integrally closed in $Q_0(R)$.*

Discrete Prüfer Domains

A Prüfer domain R is **discrete** if every primary ideal of R is a power of its radical (Gilmer [72], p.295.) Moreover, a domain R is discrete iff R_M is a generalized discrete valuation domain for each $m \in$ mspec R (*op.cit.*), that is $H_1/H_2 \approx \mathbb{Z}$ for any two consecutive groups H_1 and H_2 of the value group of R_m $\forall M \in$ mspec R (*op.cit.*, p.205).

Strongly Discrete Domains

A Prüfer domain is **strongly discrete** (=SD) if $P \neq P^2$ for every prime ideal $P \neq 0$ (Popescu [84]). Moreover (*op.cit.*) a Prüfer domain R is SD iff R is Discrete and satisfies acc$P$(= **acc on prime ideals**). Furthermore this, too, is a local property by Fontana-Popescu [95].

Generalized Dedekind Rings

An integral domain R is **generalized Dedekind** (= GD) iff R is SD and every prime ideal P is the radical of a finitely generated ideal (Popescu [84]), Theorem 2.5); equivalently, R is (i) locally a generalized discrete valuation domain; (ii) R has acc$P$; and (iii) min spec R/I is finite for every ideal I.

Facchini's Theorems on Piecewise Noetherian Rings

A commutative ring R is **piecewise Noetherian** (= PWN) iff R satisfies: (i) R has acc$P$; (ii) R has acc on P-primary ideals for each prime ideal P; and (iii) minspecR/I is finite for every ideal I.

9.36 THEOREM (FACCHINI [94]). *A Prüfer domain R is GD iff R is PWN. Furthermore, any Prüfer domain with Krull dimension is GD.*

See *op.cit.* 3.1 and 3.2.

9.37 THEOREM (FACCHINI [94], p.163, **sup.** Prop.2.3). *For a Prüfer domain R, the canonical correspondence* $P \to E(R/P)$ *is 1-1 between prime ideals and isomorphism classes of indecomposable injectives iff R is strongly discrete.*

Cf. 16.36ff.

CHAPTER 10

Isomorphic Polynomial Rings and Matrix Rings

Let R and S denote rings. Following Brewer and Rutter [72], one says that R is *n-invariant* if an isomorphism

$$\sigma : R[X_1, \dots, X_n] \approx S[Y_1, \dots, Y_n]$$

of the polynomial rings in n-variables implies an isomorphism of R and S. If $\sigma(R) = S$, for any isomorphism σ, then R is *strongly n-invariant. If R is (strongly) n-invariant for all n, then R is (strongly) invariant* (e.g. $\mathbb{Z}$ is strongly invariant).

Hochster's Example of a Non-unique Coefficient Ring

10.1 THEOREM (HOCHSTER-MURTHY [72]). *Not every commutative integral domain R is 1-invariant.*

The example is given by Hochster [72] who states that Murthy discovered a "similar example (unpublished)."

Brewer-Rutter Theorems

10.2 THEOREM (BREWER-RUTTER [72]). *If the center A of R has Krull dimension* 0 *(equivalently A/N(A) is a VNR ring) then R is invariant.*

10.3 COROLLARY (JACOBSEN AND BREWER-RUTTER [72]). *If R is a VNR ring, then R is invariant.*

The proof of the theorem uses a lemma that states that if R is commutative, and $\sigma(R) \subseteq S$ then $\sigma(R) = S$ (Lemma 2, *loc. cit.*). In general, if $A/N(A)$ is strongly n-invariant, where A is the center of R, then R is n-invariant (*loc. cit.*) (Theorem 1, *loc.cit.*). Thus, the proof of the theorem reduces to the case where R is a commutative VNR.

10.4 THEOREM (BREWER-RUTTER [72]). *If the center of R is a finite product of local rings, then R is invariant.*

The Theorems of Abhyankar, Heinzer and Eakin

10.5 THEOREM (ABHYANKAR, HEINZER, EAKIN [71]). *If A is a domain of transcendence degree* 1 *over a field k, then A is invariant, and strongly invariant unless A is the polynomial ring k[X].*

If $A = k[X]$, and $B = k[Y]$, then $A[Y] = B[X]$ for a variable Y, but the identity isomorphism does not carry A onto B, so A is not strongly invariant.

10.6 EXAMPLE (EAKIN AND HEINZER [73]). *There exists an integral domain R such that R is strongly invariant; R[X] is invariant but not strongly invariant and R[X, Y] is not invariant.*

10.7 THEOREM (EAKIN AND HEINZER [73]). *If A is a not strongly invariant Dedekind domain, then there exists $s \neq 0$ in A so that $A[1/s] \approx k[X]$ for some field k and variable X. Furthermore if A contains $\mathbb{Q}$, then $A \approx k_0[Y]$ for some field k_0 and variable Y.*

10.8 REMARK. Hochster's examples in 10.1 are locally polynomial rings (see Eakin and Silver [72]; Cf. Eakin and Heinzer [73]).

Isomorphic Matrix Rings

Let n be an integer ≥ 2. A ring R is said to have n-matrix cancellation provided that whenever $R_n \approx S_n$ for a ring S, then $\approx S$. Then R is said to be n-**cancellable**. If this holds for all $n \geq 2$, then R has **matrix cancellation**. According to Lam [98c], §8B, Cohn[66] gave the first example of a ring R that is not 2-cancellable. The construction is given by Lam, *ibid.*, of a domain R with $R^2 \approx R^4$. Thus, R^2 fails to have a unique basis number. Moreover

$$R_2 \approx \text{End}_R R^2 \approx \text{End}_R R^4 \approx R_4 \approx (R_2)_2$$

yet $R \not\approx R_2$, since R is a domain (see Cohn, *ibid.*).

Lam's Survey

Lam [95] cites a number of classes of matrix cancellable rings:

(1) R commutative
(2) R right Artinian
(3) R is a right and left self-injective ring (Gentile [67]).

Moreover, Lam [98c] cites:
(1) R is a right self-injective VNR (Goodearl and Boyle)
(2) R is semilocal
(3) R is VNR of bounded index

REMARK. The proofs of (5) and (6) depend on "(weak) cancellation" of $f \cdot g$ projective modules (cancellation of $f \cdot g$ projective modules implies the latter, *ibid.*).

CHAPTER 11

Group Rings and Maschke's Theorem Revisited

In this chapter we revisit Maschke's theorem, generalized in 11.10. The next theorem generalized the classical theorem that a group algebra kG of a finite group G over a field k is QF (see 11.4).

Connell's Theorems on Self-injective Group Rings

11.1 THEOREM (CONNELL [63]). *If G is finite, then the group ring AG over a ring A is right self-injective (= SI) iff A is right SI.*

Moreover:

11.2 THEOREM (CONNELL [63], FARKAS [73], RENAULT [70], JAIN). *If a group ring AG is right SI, then G is finite.*

11.3 THEOREM (CONNELL [63]). *A group ring AG is right Artinian iff A is right Artinian and G is finite.*

11.4 COROLLARY. *A group ring AG is QF iff A is QF and G is finite. Thus, any group algebra kG of a finite group over a field k is QF.*

11.5 THEOREM. *A group ring AG is right PF iff G is finite and A is right PF.*

This theorem follows from 11.1, 11.2 and a theorem of Louden [76].

Perfect and Semilocal Group Rings

11.6 THEOREM (RENAULT [73] AND S. M. WOODS [71]). *A group ring AG is right perfect iff G is finite and A is right perfect.*

11.7A THEOREM (LAWRENCE [75]). *If a group algebra RG is semilocal, and k properly contains the algebraic closure of $\mathbb{Z}_p$, then G is a finite extension of a p-group.*

11.7B THEOREM (LAWRENCE-WOODS [76]). *If k is a field of characteristic 0, then kG is semilocal only if G is finite.*

11.8 THEOREM (ZIMMERMANN [82]). *A group ring RG is right pure-injective iff R is right pure-injective and G is finite.*

REMARK. If R is left linearly compact (= l.l.c.) and G is finite, then RG is l.l.c. as an R-module, hence a l.l.c. ring. Conversely, if RG is l.l.c., then RG is right pure-injective (r.p.i.) by Zimmermann's theorem 6.D', so G is finite and R is r.p.i. by 11.8.

von Neumann Regular Group Rings

The *orbit of a group* is the set of orders of its elements.

11.9 THEOREM (AUSLANDER [57] - MCLAUGHLIN [58] - CONNELL [63]). *A group ring AG is VNR iff A is VNR, G is locally finite and every n in the orbit of G is a unit of A.*

11.9′ DOMANOV'S THEOREM [77,78]. *There exists a prime VNR group algebra kG over an arbitrary field k which is not primitive.*

The next corollary characterizes semisimple group rings:

11.10 COROLLARY (CONNELL [63]). *A group ring AG is semisimple iff G is a finite group of unit order in A, and A is a semisimple ring.*

For when AG is a V-ring, see e.g. Hartley [77].

11.10B THEOREM (CONNELL [63]). *A group ring RG is prime iff R is prime and G has no finite normal subgroups $\neq 1$.*

Jacobson's Problem on Group Algebras

If $R = kG$ is a group algebra over a field k of characteristic 0, is the Jacobson radical $\text{rad}R = 0$? Amitsur's theorem was reported on in Jacobson's book [64], also see my book, vol. II, p. 260–265: the answer is "yes" if k is not algebraic over $\mathbb{Q}$, e.g. if k is non-denumerable. A large number of theorems state a positive result under various assumptions on G, e.g. for any ordered group, hence any Abelian group (Theorem of B. H. Neumann [49]. Cf. Ribenboim [69], Theorems 23 and 24, p. 153ff).

Isomorphism of Group Algebras: The Perlis-Walker Theorem

If G and H are Abelian groups of the same finite order n over an algebraically closed field k of characteristic 0, then by Maschke's theorem,

$$kG \approx kH \approx k^n$$

are isomorphic even though G and H need not be. Thus, the group algebra kG of a finite Abelian group G over the field k does not determine the group. The beauty of the following Perlis-Walker theorem is that the rational group algebra $\mathbb{Q}G$ does determine G.

11.11 THEOREM (PERLIS-WALKER [50]). *If G and H are finite Abelian groups such that $\mathbb{Q}G \approx \mathbb{Q}H$, then $G \approx H$.*

See e.g. Passman [77], Theorem 1.2, p. 645.

Dade's Examples

Dade [71] showed that there exist two non-isomorphic metabelian groups G and H such that $kG \approx kH$ for every field k. As Dade (*loc.cit.*) remarks, the Whitcomb-Jackson theorem (see below) implies $\mathbb{Z}G \not\approx \mathbb{Z}H$ for the groups in his example (see Passmann [77], p. 661. Also see Passman's theorem on p. 658).

Higman's Problem

A problem posed by Graham Higman in 1940 and later by Richard Brauer in 1963 (see Roggenkamp and Scott [87]): Does the integral group ring $\mathbb{Z}G$ determine the group G? That is, if G and H are groups, does the following hold:

$$(HC) \qquad \mathbb{Z}G \approx \mathbb{Z}H \Rightarrow G \approx H.$$

A. Whitcomb in 1967 (U. of Chicago Ph.D. thesis) and D. A. Jackson in 1969 independently verified HC assuming that one of the two groups is metabelian (= has Abelian derived group). See Passman [77], p.669. Roggenkamp and Scott [87] verified HC for: (1) finite p-groups G over $\mathbb{Z}$ or $\hat{\mathbb{Z}}_{(p)}$ (the ring of p-adic integers); (2) for nilpotent groups G.

Theorems of Higman, Kasch-Kupisch-Kneser on Groups Rings of Finite Module Type

A group ring kG of a finite group G over a field k is semisimple iff the characteristic p of k does not divide $n = |G|$ (theorem of Maschke–see 11.10). When $p \mid n$, then a theorem of Higman [56] states that kG has finite representation type (= FFM) iff the p-subgroups of G are cyclic. Since the p-Sylow subgroups are conjugate, this holds iff G has a cyclic p-Sylow subgroup P. In this case, Higman found a bound $b(n)$ on the number of isomorphism classes of indecomposable finitely generated modules independent of G, and Kasch-Kneser-Kupisch [57] sharpened this by showing that $b(n) = n$ is an upper bound, and that $b(n) = n$ iff $G \rhd P$ and G/P is Abelian of exponent m, and k contains all of the mth roots of unity. Janusz [69] constructed all of the indecomposable modules, and also showed that whenever kG is FFM, then every indecomposable module has squarefree socle.

Janusz and Srinivasan Theorems

Prime examples of FFM rings are serial rings. Janusz [69] characterized when kG is serial assuming that k is a splitting field for G. This happens iff the p-Sylow subgroup of G is cyclic, and every simple kG module F is the tensor $k \otimes F^\star$, for some R-module $F^\star$, where R is a complete local domain of characteristic 0 with residue field $\approx k$. (Cf. Srinivasan [60] who showed that every indecomposable module embeds in kG when G is p-solvable with cyclic p-Sylow subgroup. Also see Janusz [70,72a].)

Morita's Theorem

Morita [56] determined necessary and sufficient conditions for the radical J of a left Artinian ring R to be a principal left, and a principal right, ideal $J = Ra = bR$. To wit: R is serial and quasi-primary-decomposable in the sense that R is a finite product of rings whose principal indecomposable left ideals have the same "multiplicity". Let G be a finite group with p-Sylow subgroup P, and let H be the largest normal subgroup with order prime to p. Then the group algebra kG over an algebraically closed field k of characteristic p has the stated property iff HP is a normal subgroup of G and P is cyclic.

Roseblade's Theorems on Polycyclic-by-Finite Group Rings

One of the most famous problems on group rings is the **zero divisor question**: if G is torsionfree, is KG an integral domain? Yes, if G is supersolvable (Formanek), a result generalized by D. Farkas and R. Snider to G polycyclic-by-finite. (There is a vast literature on this subject–see Passman [77].)

A **capital** of a ring R is a factor ring R/A where A is a maximal ideal. (If R is commutative then R/A is a field.) A field K is **absolute** if K has prime characteristic and K is absolutely algebraic.

11.12 THEOREM (ROSEBLADE [73]). *If K is any absolute field and G is polycyclic-by-finite, then any simple KG module is finite dimensional over K.*

A Weak Nullstellensatz

11.13 REMARK. The "weak Nullstellensatz" states that every simple R-module of a polynomial ring $R = K[x_1, \dots, x_n]$ over a field K is finite dimensional over K, so as Roseblade suggests (*loc.cit.*, p.309) 11.12 is a weak Nullstellensatz for KG.

This generalized a theorem of P. Hall for finitely generated nilpotent G. Moreover, by the work of Hall cited by Roseblade (*loc. cit.*), this implies that the simple modules of the integral group ring $\mathbb{Z}G$ are all finite (Cf. Jategaonkar [74b] who proves residual nilpotence of the augmentation ideal.) Furthermore:

Hilbert Group Rings

11.14 COROLLARY (ROSEBLADE). *If R is any commutative Noetherian Hilbert ring all of whose capitals are absolute and if G is polycyclic-by-finite, then any simple RG module is finite dimensional over a capital of R.*

Roseblade also proved, for any ring R, that RG is a Noetherian Hilbert ring iff R is, generalizing various forms of the Hilbert Nullstellensatz (see Roseblade, *loc. cit.*, p. 309).

CHAPTER 12

Maximal Quotient Rings

A ring R is *right nonsingular* (= n.s.) if every essential right ideal I has zero left annihilator.[1] Nonsingular rings (but not the terminology) were introduced by R. E. Johnson in the early 1950's (see below). The most important examples are (1) simple rings (with identity); (2) semihereditary rings (e.g. Von Neumann regular ring); (3) commutative rings without nilpotent elements $\neq 0$; (4) integral domains (commutative or not, e.g. free algebras over a field), and (5) Goldie semiprime rings. The latter rings, by the theorems of A. W. Goldie and L. Lesieur-R. Croisot include: (5a) any right order in a semisimple Artinian ring, e.g. (5b) any semiprime (or prime) right Noetherian ring (see 3.13).

A remarkable fact about nonsingular rings is that these rings are large subrings, in a technical sense presently defined, of self-injective Von Neumann regular rings.

The right singular ideal *sing* R of a ring R consists of all elements $x \in R$ such that the right annihilator $x^{\perp}$ is large (= essential) in R in the sense that $x^{\perp} \cap I \neq 0$ for all right ideals $I \neq 0$. This is a 2-sided ideal, and the left singular ideal may not equal the right singular ideal. The ring R is right nonsingular (n.s.) if sing $R = 0$. (These are Johnson's $J_r = 0$, or J-rings.) A commutative ring R is n.s. if and only if R has no nilpotent elements except 0 ($= R$ is *reduced*). When R satisfies the *acc* on right annihilators (= acc$^{\perp}$), or more generally, the *acc* on the set $\{x^{\perp} | x \in \text{sing} R\}$, then sing R is nil, hence is contained in the Jacobson radical rad R. (From this one easily deduces that semiprime right Goldie rings are right non-singular.) Furthermore, if R is right self-injective (= the canonical right R-module R is injective), then Y. Utumi [56] proved that sing R coincides with the Jacobson radical and that $R/$sing R is Von Neumann regular, hence n.s. In 1964 Osofsky [Canad. J. Math. **20** (1968), 405–413] constructed the first semi-primitive ring R which is not n.s.; thus, sing R is not always contained in rad R. In 1974 J. Lawrence constructed a primitive ring $R =$sing R and indicated that Osofsky's original example was also primitive.

The complexity, depth, and extent of the structure theory of R. E. Johnson's n.s. rings rivals, but naturally is unimaginable without, N. Jacobson's structure theory of primitive and semiprimitive rings published in the mid-forties, and in N. Jacobson's colloquium volume [56, 64]. The main point of similarity is between Johnson's n.s. prime rings R with uniform right ideal U, and Jacobson's primitive rings with linear transformations (l.t.'s) of finite rank. To wit, a theorem of Utumi (*loc.cit.*) states that the maximal right quotient ring $\hat{R}$ of R is the full ring of l.t.'s of $E(U)$ as a vector space over $\Delta = \text{End}_R E(U)$; and more to the point is Johnson's transitivity theorem which generalizes the Chevalley-Jacobson

[1] Many topics treated in §12 have been taken up in earlier chapters. Please consult the index.

density theorem (roughly) as follows: if $x_1, \dots, x_n \in U$ are linearly independent over $D = \text{End}_R U$, and if $y_1, \dots, y_n \in U$, then there exists $k \in D$, $r \in R$ such that $kx_i = y_i r$, $i = 1, \dots, n$. In 1951 Johnson proved this under the added assumption of a uniform left ideal $V \neq 0$, and for $D = VU \subset U \cap V$. This theorem has now been put into a completely module-theoretic setting by J. Zelmanowitz [76–77] in a theorem which makes the analogy explicit, since it implies both Chevalley-Jacobson's and Johnson's theorems. An important ingredient of the proof of the transitivity-density and Zelmanowitz theorems (the latter **not** incidentally yield a proof of Goldie's theorem and the Faith-Utumi theorem simultaneously) is the Jacobson and Johnson-Wong double annihilator condition (d.a.c.) for quasi-injective M; namely, $\text{ann}_M \text{ann}_A S = RS = \Sigma_{i=1}^n Rx_i$ for any finite subset $S = \{x_1, \dots, x_n\}$ of M (see Prop. 3.8). This is the basis for the Johnson transitivity theorem, and the Johnson-Wong transitivity theorems which imply the Chevalley-Jacobson density theorem. Cf. also "weak primitivity" in Zelmanowitz [81] and [84].

The most significant and imaginative departure from the Jacobson structure theory is Johnson's concept of the maximal quotient ring of an n.s. ring. In modern terminology, the injective hull $E = E(R)$ of a right nonsingular R is a right self-injective Von Neumann regular ring $\hat{R}$ containing R as a subring, called the maximal right quotient ring of R. (In Johnson's early papers, some of these concepts were not available.) That $\hat{R}$ is right self-injective was proved by Johnson and Wong only in 1959, or several years after Utumi [*op.cit.*, 1956] generalized Johnson's concept of the maximal quotient ring of an n.s. ring to construct the maximal right quotient ring $Q^r_{\text{max}}(R)$ of any ring R (see below), containing the classical right quotient ring $Q^r_{\text{cl}}(R)$ in the event R is a right Ore ring. As stated, the nice aspect of the Johnson theory is that any right n.s. R can be embedded as a large submodule into a Von Neumann regular (and right self-injective) ring. In honor of this fact, these rings were awarded various appellations such as "nice" or "neat", but nonsingular seems to have won out.

The Maximal Quotient Ring

In modern terminology, Utumi's maximal right quotient ring Q^r_{max} of a ring R is the maximal rational extension $\overline{R}$ of R. Here, the terminology and concept of a **rational extension** M of a submodule N is that defined by G. D. Findlay and J. Lambek [58], namely, that $\text{Hom}_R(S/N, M) = 0$ for all between modules S of $M \supseteq N$. Equivalently for all $x, y \in M$, and $y \neq 0$ there corresponds $r \in R$ with $xr \in M$ and $yr \neq 0$. Thus, any rational extension is essential, that is, N is large in M. In fact, in Utumi's theorem, one can take $Q_{\text{max}} = \text{ann}_E \text{ann}_S R$, where $S = \text{End}_R E$, and $E = E(R)$ (and similarly for the maximal rational extension of any module). Furthermore, Lambek [63] showed that Q_{max} is canonically the biendomorphism ring of E, that is $Q_{\text{max}} = \text{End}_S E$. In the special case that R is right n.s., we have that $Q_{\text{max}} = \hat{R}$; and in the special case that R is commutative and Noetherian or, even weaker has acc⊥; (see Faith [91b]), then Q_{max} coincides with the full ring Q_{cl} of quotients of R, namely $Q^r_{\text{cl}} = \{a/b \mid a, b \in R, b \in R^\star\}$, where $R^\star$ is the set of non zero-divisors. For the ring $R = T_n(K)$, the lower triangular matrices over a field K, R is right n.s. and $Q_{\text{max}} = \overline{R} = K_n$, the full $n \times n$ matrix ring. (For $n = 2$, the ideal $S = \left\{\begin{pmatrix} 0 & 0 \\ x & 0 \end{pmatrix}\right\}$ is the left singular ideal, where the right singular ideal is zero. Hint: The *left* annihilator of any nilpotent ideal is an essential right ideal.) Note, $T_n(K)$ is Artinian, so is $Q^r_{\rm cl} = T_n(K)$, and so is $K_n = Q^r_{\max} \neq Q^r_{\rm cl}$. When R is semiprime right Goldie, then the three quotient rings coincide: $Q^r_{\rm cl}(R) = Q_{\max}(R) = \hat{R}$.

$Q_{\max}(R) = R$ for any right Kasch ring since $\mathrm{Hom}(R/I, R) \neq 0$ for any maximal right ideal I, hence the same is true for any right ideal $I \neq R$. Any acc$\perp$ commutative ring R has semilocal Kasch classical quotient ring (see theorem 16.31), hence $Q_{\max}(R) = Q_{\rm cl}(R)$, for any such ring, e.g. any commutative subring R of a Noetherian ring. Also see Armendariz and MacDonald [72] on rings with Kasch $Q_{\max}$. (Kasch rings were called S (for skalar) rings by Kasch).

The fact that $Q_{\max}$ can be defined for any ring R, and includes as a special case the classical quotient ring $Q_{\rm cl}$, means that the theory is applicable to arbitrary commutative rings *inter alia.* (Cf. **sup** 9.27.) Moreover, to a large extent, following Gabriel and Lambek, the emphasis after the mid-sixties has been on $Q_{\max}$, and more general quotient rings, rather than on the restrictive class of n.s. rings. One reason for this is that after Goldie's important break-through contructing $Q_{\rm cl}$ for right Noetherian semiprime rings, there naturally has been an impetus to extend the theory to rings more general than n.s. Goldie rings (i.e., drop the n.s.). This program has been carried out by Goldie himself, Lambek and Michler, Jategaonkar, Robson, Small and others, localizing e.g. in Noetherian or PI-rings at prime or semiprime ideals. B. Müller has given a summary of some of this in a McMaster Univ. Lecture Notes article. Also see F. L. Sandomierski [77], J. A. Beachy [76] and Jategaonkar's book [86].

When $Q^r_{\max}(R) = Q^\ell_{\max}(R)$: Utumi's Theorem

DEFINITION. A ring R is right **cononsingular** provided that every right ideal I with ${}^\perp I = 0$ is essential. (Right singular is, of course, the converse implication: ${}^\perp I = 0$ if I is essential.)

12.0A THEOREM (UTUMI [63B,C]). *If R is right nonsingular, then R is right cononsingular iff every annihilator right ideal is a complement right ideal. Moreover, if R is also left nonsingular, then*

$$Q = Q^r_{\max}(R) = Q^\ell_{\max}(R)$$

iff every right or left annihilator is a complement, equivalently, R is right and left cononsingular.

REMARK. In any right nonsingular ring R, any complement right ideal I is an annihilator: since IQ is a direct summand of $Q = Q^r_{\max}(R)$, then IQ is an annihilator right ideal of Q, hence $IQ \cap R = I$ is a right annihilator of R.

COROLLARY. *A right and left nonsingular cononsingular ring R is Dedekind Finite.*

PROOF. This follows from Utumi's Theorem 4.7B. Since Q is right and left selfinjective, Q hence R is DF. □

COROLLARY. *Any Abelian VNR ring R is right and left cononsingular, and Q is also Abelian.*

PROOF. Utumi, *ibid.* Cf. Goodearl [79], p.30, Theorem 3.8. □

Courter's Theorem on When All Modules Are Rationally Complete

A right R-module M is **rationally complete** provided that M is its maximal rational extension. (See the paragraph "Maximal Quotient Ring" above.)

12.0B THEOREM (COURTER [69]). *The following conditions on a ring R are equivalent.*

1. *Every right and left R-module is rationally complete.*
2. *R is a product of finitely many full matrix rings over right and left perfect local rings.*
3. *R is right perfect and every right R-module is rationally complete.*
4. *R is right and left perfect and every right and left R-module is co-rationally complete (in a sense dual to rationally complete).*

Snider's Theorem on Group Algebras of Characteristic 0

The following theorems are related to Amitsur's Lemma 2.39.

12.0C THEOREM (SNIDER [76]). *If F is a field of characteristic zero, then any group algebra $R = FG$ is right and left nonsingular, hence $Q_r = Q^r_{\max}(R)$ and $Q_\ell = Q^\ell_{\max}(R)$ are VNR.*

PROOF. The proof follows from the following three lemmas of Snider's. □

$Z(R)$ denotes the right singular ideal. Regarding the two lemmas below, compare the proof of Amitsur's Theorem 2.40.

1. LEMMA. *Let R be a ring with involution. If $xx^\star = 0$ implies $x = 0$, then the singular ideal $Z(R) = 0$.*

PROOF. If $0 \neq x$ in $Z(R)$, then pick $0 \neq x^\star r$ in $x^\perp \cap x^\star R$. We have $r^\star x x^\star r = (r^\star x)(r^\star x)^\star = 0$ which implies $r^\star x$ and $x^\star r = 0$, a contradiction. □

REMARK. Cf. the proof of Amitsur's Lemma 2.39.

2. LEMMA. *If F is a field of characteristic 0 closed under complex conjugation, then $z(FG) = 0$.*

PROOF. FG has an involution defined by $(\sum_i \alpha_i g_i)^\star = \sum_i \overline{\alpha}_i g_i^{-1}$. Also the coefficient of 1 in $(\sum_i \alpha_i g_i)(\sum_i \alpha_i g_i)^\star$ is $\sum \alpha_i \overline{\alpha}_i \geq 0$. Therefore the above lemma is satisfied. □

3. LEMMA. *If F and K are fields with $F \subseteq K$, then $Z(F[G]) \subseteq Z(K[G])$.*

PROOF. Let $[k_i]_{i \in I}$ be a basis for K over F. If I is an essential right ideal of $F[G]$, then $I' = \sum_i k_i I$ is an essential right ideal of $K[G]$. □

The next result suffices for Snider's Theorem.

4. THEOREM. *If F is a field of characteristic* 0 *and G is any group, then* $Z(FG) = 0$.

PROOF. Let K be the algebraic closure of F, and let P be the real closed field inside of K. Then $K = P(i)$, so lemmas 2 and 3 apply.

REMARK. Snider, *ibid.* points out that if G is locally finite then $Z(FG) = \text{rad}(FG)$.

This remark establishes Jacobson's conjecture for locally finite G:

12.0D THEOREM. *If G is locally finite, and F is a field of characteristic* 0, *then FG is semiprimitive.*

REMARK. This was known to Jacobson.

Galois Subrings of Quotient Rings

12.A THEOREM (HARČENKO [74]-COHEN [75]). *If R is a semiprime ring with a finite group G of automorphisms without $|G|$-torsion (= no elements of additive order dividing $|G|$), then R is right Goldie iff the Galois subring R^G is right Goldie.*

Cf. Tominaga [73] and Kitamura [76,77]; and Fisher-Osterburg [78], who reprove 12.A.

The next theorem generalized the author's theorem [72c] for Ore domains (see 6.30).

12.B THEOREM (HARČENKO [74,75]). *If G is a finite group of automorphisms of a ring R without nilpotent elements $\neq 0$, then R is right Goldie iff R^G is right Goldie.*

12.C THEOREM (FARKAS AND SNIDER [77]). *If G is a finite group of automorphisms of a semiprime ring R without $|G|$-torsion, and if R^G is right Noetherian, then so is R; in fact, R is a $f \cdot g$ right R^G-module.*

12.D$_1$ THEOREM (FISHER-OSTERBURG [78]). *Let R be a semiprime ring, and G a finite group of automorphisms of R. Then:*

(1) If R is without $|G|$-torsion, then R is right Goldie iff R^G is right Goldie. Moreover,

$$Q^r_{c\ell}(R^G) = Q^r_{c\ell}(R)^{G^{ex}},$$

where G^{ex} is the canonical extension of G to $Q^r_{c\ell}(R)$.

(2) If R is right Noetherian, and $|G|^{-1} \in R$, then R^G is right Noetherian. (See Fisher-Osterburg, op.cit., p.488).

REMARKS. (1) R semiprime is necessary for $12.D_1$ even assuming $|G|$ is a unit; (2) Similarly R^G right Noetherian does not imply R right Noetherian even assuming $|G|^{-1}$ exists (*ibid.*, p.492, Example 2).

12.D$_2$ THEOREM (FISHER-OSTERBURG, *ibid.*). *If R^G has right Krull dimension* (§14), *then:*

1. *R satisfies acc on semiprime ideals,*
2. *Every semiprime factor ring of R is right Goldie,*

3. *Nil subrings of R are nilpotent,*
4. *Each ideal of R has just finitely minimal primes,*
5. *Every ideal of R contains a product of prime ideals.*

REMARK. R need not have right Krull dimension even if R^G does (same example for Remark (2) above suffices).

12.E THEOREM (LEVITZKI [35]). *If G is a finite group of automorphisms of a right Artinian ring R, and if $|G|^{-1} \in R$, then R^G is right Artinian; if R is semisimple then so is R^G.*

The next theorem is a partial converse of the second part of 12.E.

12.F THEOREM (COHEN-MONTGOMERY [75]). *If G is a finite group of automorphisms of a semisimple ring R without $|G|$-torsion, then R is semisimple Artinian if R^G is.*

A number of theorems relating chain conditions of a ring R and a subring S are given by Björk [73]. In particular the next theorem answers a question raised in that paper.

12.G THEOREM (FORMANEK AND JATEGAONKAR [74]). *If S is a subring of R such that $R = \Sigma_{i=1}^{n} u_i S$ for finitely many elements u_i such that $u_i S = S u_i$ $\forall i$, then a right R-module M is Noetherian (resp. has finite length as an R-module) iff the same is true as a right S-module. Moreover, every simple right R-module is semisimple of finite length over S.*

This generalized theorems of P. M. Eakin (1968) and D. Eisenbud (1970).

Localizing Categories and Torsion Theories

In another direction, by 1962 the work of A. Grothendieck [57], Gabriel, J.-P. Serre and others on a general localization theory for Abelian categories with generators and exact direct limits (carried out in Gabriel's thesis [62]) had given impetus to study localizations of the category mod-R of all right modules with respect to more general localizing subcategories.

In this direction, Gabriel showed that there is a one-to-one correspondence $\mathcal{G} \Leftrightarrow \mathcal{F}_{\mathcal{G}}$ between the localizing subcategories $\mathcal{G}$ and the idempotent, topologizing filters $\mathcal{F}_{\mathcal{G}}$ of right ideals of R (l.c.). Moreover, there is a canonical localizing functor T: $(\text{mod-}R)/\mathcal{G} \rightsquigarrow \text{mod-}R_{\mathcal{G}}$, where $R_{\mathcal{G}}$ is called the (usually less than maximal) quotient ring of the localization and in classical situations (e.g. when $\mathcal{F}_S$ contains a cofinal set of finitely generated right ideals), T is an equivalence induced by $\otimes_R R_S$.

This emphasized the torsion theoretic aspect of localization foreshadowing the work in the subject by S. E. Dickson [66] and others. Thus there is a one-to-one correspondence between certain torsion theories T, called hereditary, and localizing subcategories $\mathcal{G}_T$, and $\mathcal{G}_T$ is the localizing subcategory corresponding to the obvious filter $\mathcal{F}$ defined by T. (Thus, $I \in \mathcal{F}$ if and only if $T(R/I) = R/I$.)

In 1971 Lambek gave an elegant, elementary presentation of torsion theories and quotient rings, showing that in "nice" cases $R_{\mathcal{G}}$ can be realized, as before, as the biendomorphism ring of an injective module. Moreover, $Q_{\max}$ appears as the $R_{\mathcal{G}}$ with respect to the largest torsion theory with respect to which R is torsion free. Also see Walker and Walker [72] in this connection.

Ring Epimorphisms and Localizations

12.1 THEOREM (SILVER [67]). *A ring homomorphism $f : A \to B$ is an epimorphism in the category RINGS of rings iff the canonical map $B \otimes_A B \to B$ is an isomorphism.*

This is also characterized in Stenström [75], p.226, via fullness of the canonical functor mod-$B \rightsquigarrow$ mod-A. Cf. Storrer [73]. The next theorem appears as Corollary 1.3, p.226 of the cited book.

12.2 THEOREM (STENSTRÖM [75]). *If $f : A \to B$ is an epimorphism in RINGS, then any injective right A-module M is canonically an injective right B-module.*

Cf. Dade's Theorem 3.17B where this fails for classical localizations. Cf. also Silver [67], pp. 48–49ff on localization theory.

A ring homomorphism $f : A \to B$ is a *right localization* if f is a ring epimorphism, that is, $B \otimes_A B \approx B$ canonically, and ${}_AB$ is flat.

12.3 THEOREM (RICHMAN [65]). *If A is a commutative domain, the localizations of A are the subrings B of $K = Q(A)$ containing A such that B_A is flat (= B is a flat overring of A).* (Cf. 9.29B.)

Cf. Silver [67], p.47, where he also remarks that by a theorem of O. Goldman, every flat overring B of a Dedekind domain A is a partial quotient ring AS^{-1} for a m.c. subset of A iff the ideal class group of A is torsion. Cf. Storrer [68].

12.3A THEOREM (KOBAYASHI [85]). *A ring R is right non-singular and right FPF iff R is right bounded, $R \hookrightarrow Q^r_{\max}(R)$ is a right localization* (Cf. 12.1, **sup.** 12.2), *and $R = Tr_R(I) \oplus I^\perp$ for any $f \cdot g$ right ideal I, where $Tr_R(I) = \sum_{f \in I^\star} f(I)$ and $I^\star = Hom_R(I, R)$.*

See Yoshimura [94,98] for a discussion of these results, and additional results and generalizations. See Remark 12.14.

Continuous Regular Rings

A lattice L is **upper continuous** if L is complete and

$$a \wedge (\vee_{i\in I} b_i) = \bigvee_{i\in I} (a \wedge b_i)$$

for all $a \in L$ and all chains $\{b_i\}_{i\in I}$ of subsets. Dually for **lower continuous**, and L is **continuous** if both upper and lower continuous. The term (upper, lower) $\aleph_0$-**continuous** applies if these conditions hold with the restriction that $|I| \leq \aleph_0$.

A VNR ring R is **right continuous** if the lattice $L(R_R)$ of principal right ideals is upper continuous. R is **continuous** if R is both right and left continuous, equivalent, if $L(R_R)$ is continuous. (Then $L({}_RR)$ is continuous.) Similarly for (right) $\aleph_0$-continuous.

REMARK. In a VNR ring, every $f \cdot g$ right ideal is generated by an idempotent, hence is principal (**sup.** 4.A).

Complemented and Modular Lattices

A lattice L is **distributive** provided that $\forall a, b, c \in L$

$$a \cap (b \cup c) = (a \cap b) \cup (a \cap c)$$

L is **modular** when the distributive law holds whenever $a \geq b$:

$$a \cap (b \cup c) = b \cup (a \cap c).$$

EXAMPLES. (1) The lattice of subsets of a set $S \neq \emptyset$ is distributive, as is any linearly ordered set; (2) The lattice $L(M)$ of submodules of an R-module is modular. See Jacobson [51], pp.193–4.

A lattice L is **complemented** provided that L has a greatest element 1 and at least element 0, and every $a \in L$ has a **complement** a', that is, an element a' with $a \cup a' = 1$ and $a \cap a' = 0$.

If L is a complemented modular lattice, and if $a \in L$, then

$$L_a = \{b \in L \mid b \leq a\}$$

is also a complemented modular lattice under the induced order. Then any complement of $b \in L_a$ is called a **complement of b relative to** a (see Jacobson [51], p.205, for these concepts).

A lattice L is **self-dual** in case there is a 1-1 order reversing mapping $f : L \to L$, also called an **involution** of L, and we write $f(a) = a^\star \ \forall a \in L$. In this case, L is **orthocomplemented** provided that $a^\star$ is a complement of a $\forall a \in L$.

Von Neumann's Coordinatization Theorem

A **continuous geometry** (= **CG**) is a complemented modular lattice which is both upper and lower continuous. A CG is **irreducible** if 0 and 1 are the elements with unique relative complements.

12.4 THEOREM (VON NEUMANN [36A,B]). *Every irreducible continuous geometry L with a homogeneous basis of order $n \geq 4$ is isomorphic to the lattice of right ideals of a (unique up to isomorphism) VNR ring $\mathcal{R}(L)$.*

12.4′ THEOREM (KAPLANSKY [55]). *Every orthocomplemented complete module lattice is a continuous geometry.*

12.4″ THEOREM (UTUMI [65]). *$\mathcal{R}(L)$ is right self-injective.*

12.4‴ THEOREM (GOODEARL [74]). *$\mathcal{R}(L)$ is also left self-injective, and, moreover, a simple ring. Conversely, any simple right and left self-injective VNR ring $R \approx \mathcal{R}(L)$ for an irreducible continuous geometry L.*

von Neumann's Dimension Function

Goodearl, *ibid.*, p.84, points out that if L is a continuous Geometry, then by von Neumann [60], p.58, Corollary 1, p.60 and Theorem 7.3, there is a unique normalized dimension function D on L, and the range of D is either $\{0, 1/n, 2/n, \dots , 1\}$ for some positive integer n, or else the entire unit interval $[0, 1]$. The first case is referred to as **Case** n, the second as **Case** ∞. As shown, *ibid.*, p.83, L falls into Case ∞ if and only if the DCC fails in L, which yields a way to test for Case ∞ without reference to the dimension function. We note that von Neumann used the term "continuous geometry" only to refer to an irreducible continuous geometry in Case ∞.

REMARKS. (1) It can be shown, e.g. in von Neumann, *ibid.*, that any complemented modular lattice (or projective geometry) L with a "homogeneous basis of order ≥ 4" is isomorphic to the lattice of principal right ideals of a (unique up to isomorphism) VNR ring $\mathcal{R}(L)$. Cf. the theorems of Stephenson and Camillo, 3.51′ and 3.51′f; (2) von Neumann [60], Theorem 18.1, p.237, characterized those VNR rings R for which the lattice of principal right ideals is an irreducible continuous geometry: this holds if and only if R is indecomposable as a ring *and* is complete with respect to a certain rank-metric. In case ∞, theorem 12.4‴ shows that metric completeness may be replaced by self-injectivity, which is a type of algebraic completeness; (3) Let L be an irreducible continuous geometry in Case ∞. (Examples are constructed by von Neumann [36b].) Inasmuch as L has order n for any $n > 0$ ([60],p.93), von Neumann's coordinatization theorem 12.4 says that there exists a regular ring R whose lattice of principal right ideals is isomorphic to L. By Theorem 12.4‴, R is a simple, right and left self-injective ring which is not Artinian.

L is **irreducible** iff $L \not\approx L_1 \times L_2$ for two nonzero modular lattices L_1 and L_2. Furthermore, $\mathcal{R}(L)$ is indecomposable *qua* ring iff L is irreducible.

Utumi's Characterization of Continuous VNR Rings

12.4A THEOREM (UTUMI [60,61]). *The following are equivalent conditions for a VNR ring R.*

1. *R is right continuous.*
2. *R contains all idempotents of its maximal right quotient ring $Q^r_{\max}(R)$.*
3. *Every right ideal of R is essential in a principal right ideal eR generated by an idempotent e.*

12.4B THEOREM (UTUMI [60,61]). *If R is a right continuous VNR ring, then R is right self-injective under any of the following assumptions:*

1. *R has no nonzero Abelian idempotents.*
2. *$M_n(R)$ is right continuous for any $n > 1$ (Utumi [60]).*
3. *R is directly indecomposable qua ring (e.g. R prime).*
4. *Every primitive factor ring is Artinian.*

Also see Goodearl [79].

Semi-continuous Rings and Modules

A ring R is right **semi-continuous** if R satisfies the condition:

(C 1) Every right ideal essentially generates an idempotent, i.e., 12.4A (3) holds.

Equivalently, every complement right ideal is a direct summand.

Moreover, R is right continuous if R is right semi-continuous and satisfies:

(C 2) If a right ideal B is isomorphic to a right ideal eR, where e is an idempotent, then B is generated by an idempotent.

Note that (C 2) holds in any VNR ring R, so any right semi-continuous regular ring is right continuous.

REMARK. Continuous and semi-continuous rings are generalized to modules in a number of papers, where (C1), the condition that every submodule of M is essential in a direct summand, equivalently, every complement submodule is a direct summand, is also called an **extending** or **CS** module; (C2) is the condition that every submodule isomorphic to a direct summand is actually itself a direct summand; and a third condition (C3) is that if two submodules A and B are direct summands such that $A \cap B = 0$, then $A \oplus B$ is a direct summand.

M is **continuous** if M satisfies (C1) and (C2), and **quasi-continuous** if M satisfies (C1) and (C3).

Mohammed and Mueller [90] show that continuous modules have the exchange property. (Cf. Oshiro and Rizvi [96].)

DEFINITION. Let $\mathcal{P}$ be a property of modules. A right R-module M is **quotient-$\mathcal{P}$** if every factor module M/K has property $\mathcal{P}$.

12.4C THEOREM (OSOFSKY-SMITH [91]). *A quotient-CS cyclic right R-module is quotient finite dimensional (= q.f.d.).*

COROLLARY. *If R is a ring, and if R_R is quotient-CS, then R is right q.f.d.*

The next theorem generalizes Osofsky's Theorem 3.18A. **Right duo** means right ideals are ideals.

12.4C′ THEOREM (OSOFSKY-SMITH [91]). *Every cyclic right R-module is continuous iff R is product of a semisimple S and a product of a finite number n of zero dimensional right duo right chain rings (where possibly $S = 0$ or $n = 0$).*

PROOF. *Ibid.* Corollary 9. □

REMARK. A local ring R is zero dimensional iff $J = \text{rad } R$ is nil. Furthermore, by a remark of Osofsky-Smith, *ibid.*, a right chain ring with nil radical is a right duo ring. They also show that the property in 12.4C′ is not right-left symmetric.

COROLLARY (AHSAN-KOEHLER). *If every cyclic right R-module is quasi-injective then R is a finite product as in Theorem 12.4C′, where the n chain rings are right linearly compact.*

PROOF. See Osofsky-Smith, *(loc.cit.)*.

12.4D THEOREM (HUYNH-RIZVI-YOUSIF [96]). *Consider the following conditions on a ring R: (1) Every $f \cdot g$ right R-module is CS; (2) Every 2-generated right R-module is CS; (3) R is right Noetherian and $(\text{rad } R)^2 = 0$. Then (1) $\Rightarrow$ (2) $\Rightarrow$ (3). Furthermore, if $R/\text{rad } R$ is VNR, then R is right Artinian.*

Cf. Dung-Smith [95], and Huynh-Dung-Wisbauer [91].

12.4E THEOREM (HUYNH-JAIN-LOPEZ-PERMOUTH [96]). *If every cyclic singular right R-module is CS, and R is simple, then R is right Noetherian.*

12.4F THEOREM (SANTA CLARA-SMITH [96]). *The following are equivalent for a ring R:*

1. *R has finite right Goldie dimension and every direct sum of an injective right module and a semisimple right module is CS.*
2. *R has finite right Goldie dimension and every direct sum of a continuous right module and a semisimple right module is CS.*
3. *R is right Noetherian and $R/(soc\ R_R)$ is a right V-ring.*

REMARK. If every singular right R-module is injective (= R is right SI), then every direct sum of a CS-module and a semisimple module is CS (*ibid.*).

Furthermore:

12.4G THEOREM (*ibid.*). *If R is a right nonsingular ring, then every direct sum of an injective right module and a semisimple right module is CS iff $R/(soc\ R_R)$ is a right Noetherian right V-ring.*

12.4H THEOREM (P. F. SMITH [97]). *If R is semiprime right Goldie, and M is a nonsingular right CS-module, then M is a direct sum of an injective module and a finite direct sum of uniform modules. If R is commutative, and $\overline{R} = R/N(R)$ is Goldie, then a nonsingular CS-module M is a direct sum of an injective $\overline{R}$-module and finitely many uniform modules, where $N(R) =$ prime rad R.*

These theorems generalized some of the theorems of Kamal-Mueller [88a,b,c]. (*Vide* also Mohammed-Mueller [90].)

A ring R is **top-regular** if $R/\text{rad}R$ is VNR. A right self-injective ring is top-regular by Utumi's Theorem 4.2. So is any algebraically compact ring (Zimmerman-Huisgen and Zimmerman [78]).

12.5 THEOREM (UTUMI [65]). *If R is right continuous, then R is top-regular. Any right self-injective ring is right continuous.*

REMARK. 1) A right continuous VNR ring R splits (as in 4.3A) into a product $R_1 \times R_2$ where R_1 is Abelian and R_2 is right self-injective with no Abelian idempotents; 2) (Zimmerman-Huisgen and Zimmerman [78]) Any algebraically compact ring is top-regular; 3) The rings R that satisfy (C1), called semi-continuous rings in the author's paper [85], have the characterizing property that every complement right ideal is generated by an idempotent. These rings are called **CS-rings** in Chatters and Hajarnavis [77].

12.6 THEOREM (FAITH [81B]. *Every commutative FPF ring is CS.*

12.7 THEOREM (RIZVI [88]). *If R is commutative and quotient finite dimensional, then every continuous module is quasi-injective iff $Q(R/I)$ is self-injective for every irreducible ideal.*

12.8 THEOREM (RIZVI AND YOUSIF [90]). *The following conditions are equivalent: (1) singular right R-modules are continuous; (2) R/I is semisimple for each essential right ideal I.*

Cf. 4.1G.

12.8A THEOREM (GÓMEZ PARDO AND GUIL ASENSIO [97C]). *Let R be a right CS-ring. Then:*

(1) If R is right Kasch, then R_R is finitely embedded;
(2) If R is a right cogenerator, then R is right PF.

12.8B THEOREM (HUYNH, JAIN AND LÓPEZ-PERMOUTH [98]). *Let R be a prime ring. Then:*

(1) If R is right Goldie of Goldie dimension ≥ 2, and if R is right CS, then R is left Goldie.
(2) R is right Goldie left CS iff R is left Goldie right CS.

CS Projective Modules

12.9 THEOREM (JØNDRUP [76], LAZARD [74], SAHAEV [77], VASCONCELOS [69], ZÖSCHINGER [81]). *Let R be an arbitrary ring with Jacobson radical J. The following are equivalent conditions, and are satisfied whenever R is commutative:*

1. *If M is a $f \cdot g$ flat right R-module such that M/MJ is projective, then M is projective.*
2. *If M is a projective right R-module, such that M/MJ is $f \cdot g$, then M is $f \cdot g$.*
3. *If M is a projective right R-module, then every $f \cdot g$ submodule $S \neq M$ is contained in a maximal submodule.*
4. *Every $f \cdot g$ projective right R-module M is a CS-module.*

(n') *The condition (n) above for left R-modules, $n = 1, 2, 3, 4$.*

PROOF. For commutative R, (1) is due to Vasconcelos [69]; (2) is due to Lazard [74]; for arbitrary R, (1) $\Leftrightarrow$ (2) $\Rightarrow$ (3) is due to Jøndrup [76]; and Zöschinger [81] proved that (1) $\Leftrightarrow$ (2) $\Leftrightarrow$ (3) $\Leftrightarrow$ (4) and that (4) is right-left symmetric. See Mohammed-Sandomierski [89] for a discussion of all this, and the next theorem.

12.10 THEOREM (MOHAMMED-SANDOMIERSKI [89]). *A projective right R-module M is a CS-module iff $A = EndM_R$ is a right CS-ring. Moreover, if M is free of infinite rank, this is equivalent to the condition that every $f \cdot g$ projective right A-module is a CS-module.*

Strongly Prime Rings

A ring R is right **strongly prime (= SP)** if to each nonzero $a \in R$, there is a finite subset F of R so that $(aF)^{\perp} = 0$. Thus all SP rings are prime. Domains, right Goldie prime rings, and simple rings are examples of SP Rings provided in the papers of Handelman and Lawrence [75] (referred to as [HL] below, Rubin [73] and Viola-Prioli [75] (Cf. his Rutgers U. Ph.D., 1973). These rings are called **Absolute Torsionfree** (= AFT) by Rubin [R] and Viola-Prioli [VP].

12.11 THEOREM (RUBIN [73]). *If R is module finite over center, or right nonsingular and has finite right Goldie dimension, then R is right AFT iff R is prime. Any polynomial ring $R[x]$ over a right AFT ring is right AFT.*

12.12 THEOREM (VIOLA-PRIOLI [73,75]). *A ring is right AFT iff for each $f \cdot g$ projective module P, and nonzero submodule S, there is an embedding $P \hookrightarrow S^n$ for some integer $n > 0$.*

12.13 THEOREM (HANDELMAN AND LAWRENCE [75]). *(1) Any VNR SP ring is simple; (2) If R is right SP, then R is right nonsingular, $Q^r_{\max}(R)$ is simple, and there exists a group G so that the group ring RG is primitive. Furthermore, (3) if RG is right SP, then G has no nontrivial locally finite normal subgroups. Partial converse: (4) if R is right SP, and G is torsion-free abelian, then RG is right SP.*

Some of these results and those of Rubin [73] were obtained independently by Viola-Prioli (*op.cit.*).

EXAMPLE (CEDÓ [97]). There exists a ring R so that $R[[x]]$ is right and left SP, but R is neither right nor left SP.

12.14 REMARK ON FLAT EPIMORPHISMS. *Interalia* Yoshimura [98] characterized a commutative ring R with the property that over overring R between R and $Q_{\max}(R)$ is a flat epimorphism of R by the property that every overmodule of R in $Q_{\max}(R)$ is projective, equivalently, every overring is integrally closed in $Q_{\max}(R)$. Cf. Griffin's theorem 9.30 and Eggert's theorem 9.31.

CHAPTER 13

Morita Duality and Dual Rings

See e.g. Morita [58], Müller [69], Zelmanowitz and Jansen [88], Xue [92], or Faith [76], for the background and definitions of Morita duality, defined by an injective cogenerator E_R of mod-R such that ${}_SE$ is an injective cogenerator for S-mod, the category of left S-modules, where $S = \text{End}E_R$. A module M_R is *E-reflexive* if the canonical map $M \to \text{Hom}_S(\text{Hom}_R(M, E))$ is an isomorphism, and symmetrically for a left S-module N.

This functor defines a duality between the full subcategories consisting of the E-reflexive modules of mod-R and S-mod, and then R (resp.S) is said to be a *right (left) Morita ring.*

If the functor $\text{Hom}_R(-, E)$ is, or defines, a duality and $\text{Hom}_S(\ \ , E)$ is the inverse, then we say E_R *defines* or *induces* a (Morita) duality, and $\text{Hom}_R(\ \ , E)$ is the duality functor. If $S \approx R$, then E_R defines a **Morita self-duality**. A ring R with such a Morita duality is said to be a right **Morita ring**. See Arhangelskii, Goodearl, and Huisgen-Zimmermann [97] for connections with Pontryagin duality.

REMARKS. (1) If E induces a right Morita duality for R, and if I is an ideal of R, then $F = \text{ann}_E I$ induces a right Morita duality for R/I; (2) If $S = \text{End}E_R$ in (1), then center $S \approx$ center R. (See the author's *Algebra II*, Exercises 23.17(c) for (1), and 23.35 for (2).

13.1 THEOREM (MÜLLER [70]). *A right R-module E defines a (Morita) duality iff R_R is linearly compact and E_R is a linearly compact and finitely embedded injective cogenerator. Moreover, then the E-reflexive modules are precisely the linearly compact modules.*

See Xue [92], pp.33–34.

13.2 THEOREM (OSOFSKY [66]). *Any ring R with Morita duality is semiperfect and, in fact, so is the endomorphism ring of any finitely embedded injective cogenerator.*

See, e.g. Xue, p.11, Proposition 1.19.

13.3 THEOREM (ZELINSKY [53]-SANDOMIERSKI [72]). *Any right linearly compact ring is semiperfect.*

See, e.g. Xue, p.29, Corollary 3.14

13.4A. THEOREM (MORITA [58]). *Any Artinian commutative ring R has a self-duality.*

In this case, we may suppose R is a finite product $R = \Pi_{i=1}^n R_i$ of Artinian local rings, hence suppose R is local with maximal ideal m. Then $E = E(R/m)$ induces the Morita duality, and the E-reflexive modules are just the $f \cdot g$ R-modules.

13.4B REMARK. (1) By theorems of Azumaya [59] and Morita [58], this also holds true for any non-commutative Artinian Morita ring: all $f \cdot g$ modules are E-reflexive and conversely. See Xue, p. 94, Theorem 1.11. (2) The duality for a finite dimensional algebra A over a field k, that sends any $f \cdot g$ right A-module M onto its k-dual $M^\star = \mathrm{Hom}_R(M, R)$, is a Morita duality induced by $A^\star$. Thus, $A^\star$ is an injective cogenerator in both mod-A and A-mod, and

$$A \approx \ \mathrm{End}_A A^\star \approx \mathrm{End} A^\star_A.$$

Cf. my *Algebra II*, p.198,23,32-33.

13.4C THEOREM (MATLIS [58]). *Let R be a commutative Noetherian ring, and P a prime ideal, and $\hat{R}_P$ be the completion of R_P in the PR_P-adic topology. Then*

$$\hat{R}_P \approx End_R E(R/P)$$

has a Morita self-duality induced by $E(R/P)$. Moreover, $E(R/P)$ is an Artinian R_P-module with simple socle $\approx R_P/PR_P \approx Q(R/P)$.

REMARK. See 5.4B: $\hat{R}_P$ is linearly compact. Matlis proved this without recourse to Morita's Theorem [58] (which of course implies 13.4C). Also 13.10 below implies 13.11A and B each of which implies 13.4C.

13.4D THEOREM. *If R is a commutative ring, and if P is a prime ideal such that $E(R/P)$ is Σ-injective, then R_P is a Noetherian local ring, $E(R/P)$ is canonically an injective cogenerator for R_P which induces a Morita duality for $\hat{R}_P = End_R E(R/P)$.*

PROOF. Apply Corollary 3.15B and Theorem 13.4C in the case $R = R_p$.

The concept of Δ-injectives used below is defined in 3.10A.

13.4E. THEOREM (FAITH [82B]). *If a Δ-injective right R-module E induces a right Morita duality for R, then (1) R is right Artinian, and (2) $R = Q^r_{\max}(R)$ is its own maximal quotient ring.*

PROOF. *op.cit.*, p.47, Corollaries 10.14 and 10.15. ((1) follows essentially from Osofsky's Theorem 4.22.)

For a study of Δ-injective modules see the author's Lectures [82a], Part I. Also see Nastasescu [80], on Artinian objects in Grothendieck categories; also references in my Lectures [82a] and a discussion of them on pp.64–65. Cf. also Albu [80] on a related "dual" theorem .

13.5 THEOREM (MÜLLER [70]). *If R is a commutative ring, and if R has a duality, then R is linear compact and has a self-duality.*

13.6 THEOREM (ÁNH [90]). *A commutative ring R has a duality iff R is linearly compact.*

This theorem answered a question dating back to Zelinsky [53] and Müller [69] on the structure of rings with duality. Cf. Orsatti and Roselli [81] and Dikranjan and Orsatti [84].

13.7 THEOREM (NAKAYAMA [39–41], AZUMAYA [59,66] AND MORITA [58]). *Any PF (= right and left PF) ring R has a self-duality, in particular any QF ring has a self-duality, induced by* $Hom_R(\,, R)$.

This follows from Theorem 4.20: R is a f.e. injective cogenerator (both sides). Cf. 13.13-14.

13.7A PROPOSITION (NAKAYAMA [40], GOURSAUD [70], FAITH [72A]). *If A is a ring which is completely right PF in the sense that every factor ring is right PF, then A is a primary-decomposable Artinian serial ring.*

PROOF. See the author's *Algebra II* [76], p.238, Theorem 25.4.6A. □

13.7B PROPOSITION (KÖTHE [35], ASANO [39,49], NAKAYAMA [40], FAITH [66B]). *A ring is a primary decomposable serial ring if and only if the following equivalent conditions hold*

(a) *R is right and left Artinian principal ideal ring.*
(b) *R is a serial Artinian right principal ideal ring.*
(c) *R is a right principal ideal QF ring.*
(d) *R is a principal right ideal ring, and every left prindec of R is serial of finite length.*
(e) *Every factor ring of R is QF.*
(f) *R is left or right artinian, and the injective hull of each cyclic right R-module is cyclic.*
(g) *R is right Noetherian, and the injective hulls of cyclics are cyclic in mod-R.*

PROOF. See the author's *Algebra II*, p.238, Theorem 2.5.4.6B □

13.8 THEOREM (VÁMOS [77]). *Let S be a ring that is a linear compact module over a ring R contained in its center (e.g. if* S_R *is* $f \cdot g$ *and R is l.c.). If E induces a self-duality for R, then* $Hom_R(S, E)$ *induces a self-duality for S.*

See, e.g. Xue [92], p.57, Theorem, 7.6 and Chapter 2 (*loc.cit.*) for other duality theorems for ring extensions.

13.9 THEOREM (AMDAL AND RINGDAL [68] AND WASCHBUSCH [86]). *Any serial ring has a self-duality.*

Cf. Oshiro [87] for additional results and comments.

13.10 THEOREM (FAITH AND HERBERA [97]). *If R is commutative, and M a finitely embedded R-module, then* $A = End_R M$ *is a linearly compact R-module and has a self-duality.*

13.11A COROLLARY (BALLET [81]). *If M is an Artinian module over a commutative ring R, then* $A = End_R M$ *is Noetherian and has a self-duality.*

Thus, in both 13.10 and 13.11, A is a semiperfect ring by 13.2.

Another corollary for 13.10:

13.11B THEOREM (FACCHINI [81]). *If R is commutative, and M is Artinian with simple socle, then* $A = End_R M$ *is a complete local commutative ring with a Morita duality induced by M.*

Dual Rings

A ring R is a *right dual* (= right D) ring if every right ideal I is an annihilator right ideal, equivalently, R/I is a torsionless right R-module for all I (see 1.5). A dual ring (= D-ring) is one that is both left and right dual.

13.12 EXAMPLE. If R is a cogenerator ring, e.g. a right PF ring (see 4.20), then every right R-module is torsionless, hence R is a right D-ring. If R is a right and left cogenerator ring, then R is a D-ring and in fact PF (see 13.14).

13.13 THEOREM (KATO [68A]). *A ring R is right PF iff $E(R_R)$ is torsionless and R is right and left Kasch.*

This implies the next theorem since any right cogenerating ring is right Kasch.

13.14A THEOREM (KATO [68A,B] AND ONODERA [68]). *If R is a right and left cogenerator, then R is PF, and conversely.*

By the Morita equivalence $X \mapsto X^n$ between mod-R and mod-R_n which carries an R-module X generated by n elements onto a cyclic R_n-module X^n, one has the corollary of 4.23 (3):

13.14B COROLLARY. *The following are equivalent conditions:*

1. *R is PF,*
2. *R_n is a D-ring for all n,*
3. *$f \cdot g$ right or left R-modules are torsionless.*

13.15A THEOREM (NAKAYAMA [41B]). *Let R be a ring for which the lattice of right ideals is anti-isomorphic (i.e., dual) to the lattice of left ideals by a mapping d. If R is right or left Artinian or Noetherian, then R is QF and $d(I)$ is the left (right) annihilator for each right (left) ideal.*

Cf. Baer [43].

13.15B COROLLARY (*loc.cit.*). *If R is an algebra of finite dimension over a field, and every right ideal is an annihilator, then R is QF.* (Cf. 13.19)

Skornyakov's Theorem on Self-dual Lattices of Submodules

13.15C THEOREM (SKORNYAKOV [60]). *If M is an R-module, and if the lattice $L(M)$ of submodules of M is complemented and self-dual then M has finite Goldie and finite dual Goldie dimension.*

REMARK. It follows that M in 13.15C is quotient finite dimensional. Cf. Theorem 16.50. It also follows that M has semilocal endomorphism ring. See Theorem 8.C. This implies that a D-ring is semilocal: see the next theorem.

Hajarnavis-Norton Theorem

13.16 THEOREM (HAJARNAVIS AND NORTON [85]). *A D-ring R is semilocal, right and left $f \cdot g$ injective, every $f \cdot g$ R-module has finite Goldie dimension and R/J^ω is Noetherian, where J is the Jacobson radical and $J^\omega = \bigcap_{n\in\omega} J^n$.*

REMARK. Thus, any D-ring is quotient-finite dimensional, and quotient-finite-dual Goldie dimensional. Cf. 5.20B and 7.21ff.

A right R-module M satisfies $AB5^\star$ if

$$\bigcap_{i\in I}(N+M_i) = N + \bigcap_{i\in I} M_i$$

for all submodules N and inverse sets of submodules $\{M_i\}_{i\in I}$. (Lemonnier [79] simplified the proof of 13.16 using AB5*.) Any l.c. R-module M satisfies $AB5^\star$ by a theorem of Leptin [14], Satz 1 (Cf. Xue [92], Cor.3.9).

REMARK. (1) If M_R satisfies $AB5^\star$, and the set $\mathcal{S}(M)$ of non-isomorphic simple images of M is finite, then End M_R is semilocal. See Herbera and Shamsuddin [95], Corollary 7, for the proof. (2) If a ring R satisfies right $AB5^\star$ (e.g. if R is right l.c.) then

1. R/J^ω is right Noetherian,
2. $J^\omega = 0$ if R is right and left Noetherian,
3. Every factor ring of R is a D ring iff R is uniserial (Hajarnavis and Norton (*loc.cit.*),
4. R is a dual ring if and only if R satisfies $AB5^\star$ and is an essential extension of its socle on both sides and R-dual takes simple to simple,
5. A one-sided Noetherian ring R is a quasi-Frobenius ring if and only if it is an essential extension of its socle and satisfies $AB5^\star$ on both sides and R-dual takes simple into simple.

Note:

1. (1) is due to Mueller [70], and
2. (2) to Menini [86]. See Herbera and Shamsuddin (*loc.cit.*)
3. (4) and (5) owe to Ánh-Herbera-Menini [97].

13.17 THEOREM (HAJARNAVIS AND NORTON [85]). *A D-ring R is QF iff J is transfinitely nilpotent.*

13.18 EXAMPLE (FAITH-MENAL [92]). There exists a right Noetherian right D-ring (= *right Johns ring*) that is not Artinian: Let D be a countable EC sfield with center k (Cf. 6.24), and let $A = D \otimes k(x)$, where $k(x)$ is the field of rational functions over k. Then D is an A-bimodule that is simple as a right A-module, and the split-null extension $R = A \ltimes D$ is a non-Artinian right Johns ring. As stated in our paper, the proof depends on the cited work of Cohn on EC fields, and also Resco [87] who proved that A is a non-Artinian V-domain, thereby solving a problem posed by Cozzens and Faith [75].

Faith-Menal Theorem

13.19 THEOREM (FAITH-MENAL [92]). *A right Noetherian right D-ring R is right Artinian provided that either (1) R is semilocal, or (2) R has finite left Goldie dimension, or (3) if $x \in R$ then x is a unit iff $x^\perp = 0$.*

REMARK. As we pointed out (*loc.cit.*), (1) is immediate from the work of Johns [77] who made an error in his paper that, as pointed out by S. M. Ginn, resulted from using a false lemma of R. P. Kurshan (see *loc.cit.*); (2) can be weakened to "R has no infinite direct sum of minimal left annihilators" (*ibid.*).

13.20 THEOREM (FAITH-MENAL [92]). *If R is a right Noetherian ring in which principal right ideals are annihilators, and*

$$(I_1 \cap I_2)^{\perp} = I_1^{\perp} + I_2^{\perp}$$

for every pair of left ideals I_1 and I_2, then R is QF.

This was stated in Johns [77], but the theorem was in doubt in view of the remark above and the counter-example to one of Johns' theorems (see 3.18).

Commutative Rings with QF Quotient Rings

In this paragraph R will denote a commutative ring.

13.21 THEOREM (BASS [63]). *A commutative Noetherian ring R has QF quotient ring Q iff the zero ideal is unmixed and all of its primary components are irreducible.*

DEFINITION. A commutative ring R has the **commutative endomorphism property** (= CEP) if every ideal of R has commutative endomorphism ring.

13.22 REMARKS. (1) Any faithful ideal I of a commutative ring R has CEP (see, e.g. Faith [82a,b], p.76).

(2) If R is reduced (= R has no nilpotents except 0), then R has CEP (Cox [73]).

(3) If R is self-injective, then R has CEP.

(4) (Alamelu [73], Remark 2, p.30) If S is a m.c. set of regular elements, then $f \cdot g$ ideals of R has CEP iff the same is true of RS^{-1}.

(5) (Vasconcelos [73]) If M is any $f \cdot g$ faithful module over a reduced ring R, and if $\text{End}_R M$ is a commutative reduced ring, then $M \hookrightarrow R$.

On a Vasconcelos Conjecture

13.23 THEOREM (COX [73] AND ALAMELU [73]). *Let R be a commutative ring with Noetherian quotient ring Q. Then every ideal of R has commutative endomorphism ring iff Q is QF.*

This theorem verified a conjecture of Vasconcelos [70c].

13.24A THEOREM (ALAMELU [76]). *If R is a coherent commutative p-injective ring, then every ideal of R has commutative endomorphism ring.*

13.24B PROPOSITION (loc.cit., Prop. 4). *If $R = Q(R)$ is coherent, and R induces $\text{End}_R I$ for every (2-generated) ideal I, then I is self-injective* (resp. p-injective).

13.25 COROLLARY. *If R has IF quotient ring Q, then every $f \cdot g$ ideal of R, and every ideal of Q has commutative endomorphism ring.*

This follows from 13.22.4 and the fact that an IF ring is FP-injective, hence p-injective, and coherent (see Theorem 6.9).

13.26 THEOREM (FAITH [96B]). *A commutative ring R has local QF quotient ring $Q(R)$ iff R is a uniform acc$\perp$ ring.*

REMARK. 13.26 generalizes Shizhong's Theorem 7.11.

Kasch-Mueller Quasi-Frobenius Extensions

The next concept generalizes group rings $R = SG$ for a finite group G.

13.27 DEFINITION AND THEOREM (KASCH [54,61] AND MUELLER [64,65]). *A ring R is a* **left quasi-Frobenius extension** *of a subring S provided that ${}_SR$ is $f \cdot g$ projective and the S-dual module $R^\star = Hom_S(R, S)$ generates R-mod in such a way that R is an (R, S)-direct summand of copies of $R^\star$. Then R is a QF ring iff S is a QF ring.*

13.28 REMARK. Kasch (*op.cit.*) proved this for Frobenius extensions and rings (not discussed here, but see the author's *Algebra II*, pp.220–221ff).

Balanced Rings and a Problem of Thrall

Thrall [48] proposed the classification of finite dimensional algebras, called $QF-1$ algebras over which all $f \cdot g$ faithful right R-modules are balanced. These properly contain the class of QF-algebras.

Camillo [70] generalized the property by asking when every right R-module is balanced, and called these **balanced rings**.

13.29 CAMILLO'S THEOREM [70]. *Let R be balanced. Then:*
1. *R is semiperfect with nil Jacobson radical.*
2. *If R is Noetherian (resp. commutative), then R is Artinian (resp. QF).*

13.30 CAMILLO-FULLER THEOREM [72]. *Let R be an algebra of finite dimension over a field.*
1. *If R is local and $QF-1$, then R is QF.*
2. *If R is balanced, then R is uniserial.*

REMARK. Dlab and Ringel independently obtained (2) of the theorem, according to the author's note on p.376 of *op.cit.* See 13.30D below.

A ring R is **finitely right balanced** if every $f \cdot g$ right R-module is balanced (see "A general Wedderburn Theorem", 3.52 ff., and also **sup.** 13.29).

13.30A THEOREM (NESBITT AND THRALL [46]). *Every uniserial ring is balanced.*

13.30B THEOREM (DLAB AND RINGEL [72]). *The following are equivalent conditions on a ring R: (1) R is right balanced; (2) R is finitely right balanced; (3) R is a direct sum of uniserial rings and full matrix rings over "exceptional" rings* (**vide** *op.cit.*).

13.30C COROLLARY. *Any finitely right balanced ring is left balanced, and Artinian.*

13.30D THEOREM (CAMILLO-FULLER [72], AND DLAB-RINGEL [72]). *A balanced ring R that is module-finite over center is uniserial.*

Cf. 13.30.

13.30E THEOREM (CAMILLO [70]-RINGEL [74]). *A commutative Noetherian $QF-1$ ring is QF.*

Cf. 13.29.

13.30F THEOREM (MORITA [76], ONODERA [76]). *For a right linearly compact ring, the following conditions are equivalent:*

(1) Every (faithful) $f \cdot g$ projective right module is balanced.

(2) Every (faithful) finitely embedded injective right R-module is balanced.

(3) Every (faithful) injective right R-module with essential socle is balanced.

(4) Every (faithful) projective left R-module with superfluous (= the dual of essential) radical is balanced.

When Finitely Generated Modules Embed in Free Modules

R is **right FGF** (resp. **CF**) if all $f \cdot g$ (resp. cyclic) right R-modules embed in free modules. Any right CF ring R is right Kasch, hence $R = Q^r_{\max}(R)$. By Theorem 3.5C, any right and left CF is QF, in which case, by Theorem 3.5B, every R-module embeds in a free module.

A conjecture of the author's is that $FGF \Rightarrow QF$. This has been verified in a number of special cases, and we propose to discuss some of these.

13.31 THEOREM. *Let R be right FGF ring. The following are equivalent:*

(FGF 1) *R is QF,*
(FGF 2) *R is a subring of a right Noetherian ring,*
(FGF 3) *R is semilocal with essential right socle,*
(FGF 4) *R has finite essential right socle,*
(FGF 5) *R is right self-injective,*
(FGF 6) *R is left CF,*
(FGF 7) *R is left Kasch,*
(FGF 8) *R satisfies acc⊥,*
(FGF 9) *R satisfies dcc⊥.*
(FGF 10) *Every countably generated right R-module has a maximal submodule.*

REMARKS. (1) $(FGF1) \Leftrightarrow (FGF5)$ is due to Tol'skaya [70] and Björk [72]; and (FGF 6) implies (FGF 5), assuming R is right FGF in both cases; (2) By Lemma 20 of Rutter [74], and also by the Colby-Würzel Theorem 6.8, any right FGF ring is right IF, as reported in Jain [73]. Moreover, Jain (*op.cit.*) deduced from a result of Simon [72] that conversely a right IF ring that is either a subring of a right Noetherian ring, or a right perfect ring, is right FPF; (3) See [82e] of the author for additional details, e.g. $(FG2) \Rightarrow$ acc $\perp$; and dcc$\perp \Rightarrow$ right Artinian $\Rightarrow$ right Noetherian and (FGF7) is *ibid.*, Theorem 3.7; (4) See Pardo-Asensio [98] for (FGF 10). Also see 13.35.

13.32 THEOREM (MENAL [82]). *If R has projective injective hull $E(R)$, and if R is right CF, then R is QF.*

REMARK. (1) Menal's proof makes heavy use of Osofsky [66]; (2) Menal *op.cit.* characterized when $H = \text{End } R_R^{(\omega)}$ is left coherent while Lenzing [70] characterized when H is right coherent. Moreover, Menal shows that if R is a right cogenerator of mod-R, then H is left FP-injective.

13.33 THEOREM (MENAL [82], FAITH [66] AND THE WALKERS [66]). *The following are equivalent conditions on $H = \text{End } R_R^{(\omega)}$.*

1. *H is left IF,*
2. *H is right FP-injective,*
3. *R is QF,*
4. *H is right self-injective.*

REMARK. Menal proves the equivalence of (1), (2), and (3), and (3) $\Leftrightarrow$ (4) is due to Faith [66] and the Walkers [66].

A ring R is **right FGEP** if every $f \cdot g$ right R-module is an essential submodule of a projective. Cf. Jain and López-Permouth [90] who considered right **CES** rings, where "cyclic" replaces "$f \cdot g$" above. See 13.34A.

13.34 THEOREM (GÓMEZ PARDO AND ASENSIO [97]). *The following are equivalent:*

1. *R is QF,*
2. *R is right $FGEP$,*
3. *R is left $FGEP$.*

REMARK. The idea of the proof uses (FG4) of 13.31.

13.34A THEOREM (JAIN AND LÓPEZ-PERMOUTH [89], AND GÓMEZ-PARDO [97A]). *The following conditions on a ring R are equivalent:*

(1) *R is right CES, i.e., every cyclic right R-module embeds as an essential right ideal in a direct summand of R_R.*

(2) *R is of one of the following types:*

(a) R is uniserial as right R-module.

(b) R is an $n \times n$ matrix ring over a right self-injective ring of type (a), or (c) R is a direct sum of rings of types (a) or (b).

PROOF. This was proved by the first two authors *loc.cit.* assuming R is semiperfect, and the second two authors proved that assumption was redundant. $\square$

A Theorem of Pardo-Asensio and A Conjecture of Menal

In [83], Question 1, Menal asked if there existed a cardinal α such that every α-generated right R-module embeds in a free R-module, is then R right QF? This is partially answered affirmatively:

13.35 THEOREM (GÓMEZ PARDO-GUIL ASENSIO [98]). *If $\alpha \geq |R|$, and every α-generated right R-module embeds in a free right R module, then R is QF.*

REMARK. This paper also generalized Menal's Theorem 13.32 in several ways.

Johns' Rings Revisited

As stated, R is **right Johns** if R is a right Noetherian right D-ring (= every right ideal is an annihilator); R is **strongly right Johns** if every full $n \times n$ matrix ring R_n is right Johns. A right FA (= **finitely annihilated**) ring R is a right D ring for which every right ideal $I = X^{\perp}$ for a finite subset X of R. Cf. 1.5.

13.36 THEOREM (RUTTER [74]). *For a right Artinian ring, the following are equivalent:*

1. *R is QF,*
2. *R_n is a right D-ring for all n,*
3. *R is strong right Johns.*

13.37A FAITH-MENAL THEOREM [94]. *The following are equivalent:*

1. *R is strongly right Johns,*
2. *R is left FP-injective and right Noetherian,*
3. *Every finitely generated right R-module is Noetherian torsionless.*

In this case R is right FPF.

13.37B COROLLARY (*op.cit.*). *Let R be strongly right Johns. The following are eqivalent:*

1. *R is QF,*
2. *R is semilocal,*
3. *R has finite left Goldie dimension,*
4. *R is left Noetherian,*
5. *R_n is right FA for all n,*
6. *$J = radR = X^{\perp}$ for a finite subset X of R.*

REMARK. The semilocal case is due to Johns [77].

13.38 THEOREM (FAITH-MENAL [95]). *If R is a right Johns ring, then R/J is a right V-ring, where J is the Jacobson radical.*

Two Theorems of Gentile and Levy on When Torsionfree Modules Embed in Free Modules

An element x of a right R-module is **torsion** if $xa = 0$ for a regular element a (a is not a right or left zero divisor). M is **torsionfree** if 0 is the only torsion element of M.

13.39 FIRST THEOREM (GENTILE [60], LEVY[63B]). *If the set $t(M)$ of torsion elements is a submodule for each right R-module, then R is a right Ore ring.*

REMARK. Gentile's theorem is for a domain R. A ring R is **right FG(T)F** if each $f \cdot g$ (torsionfree) right R-module embeds in a free module.

13.40 SECOND THEOREM (*ibid.*). *If R is a right Ore ring, and if R is right FGTF, then R is right FGF. Furthermore, if R is semiprime, then* $Q = Q^r_{c\ell}(R) = Q^\ell_{c\ell}(R)$ *is semisimple, that is, R is right and left Goldie.*

13.41 REMARK. Gentile's theorem is for a domain R.

When an Ore Ring Has Quasi-Frobenius Quotient Ring

13.42 THEOREM (FAITH [82E]). *Under the same assumptions as (the Second Gentile-Levy) theorem, then Q is QF under any of the conditions of Theorem 3.31, e.g. assuming R is also left Ore and left FGTF.*

PROOF. For then R is right FGF, and so Theorem 13.31 applies. If R is also left Ore and left FGTF, then R is also left FGF, so (FGF 6) of 13.31 applies. (This is an application of the Faith-Walker theorem 3.5D.)

13.43 COROLLARY. *A commutative ring R is FGTF iff $Q(R)$ is QF.*

13.44 REMARK. For when $f \cdot g$ torsionless modules embed in free modules, see the author's paper [90].

Levy's Theorem

13.45 THEOREM (*ibid.*). *If R is right Ore, then torsion-free divisible modules are injective iff* $Q = Q^r_{c\ell}(R)$ *is semisimple; (2) Moreover, if R is also left Ore, then every divisible right R-module is injective iff Q is semisimple and R is (right) hereditary; (3) Any right Goldie right hereditary ring is right Noetherian.*

REMARK. (1) and (2) of 13.45 follow from Theorems 13.39 and 13.40, and the theorem of Cartan-Eilenberg 3.22A.

CHAPTER 14

Krull and Global Dimensions

The (classical) Krull dimension of a commutative Noetherian ring R is defined by:

$$\dim \mathrm{R} = \sup\{\text{rank } P \mid P \in \text{Spec } R\}$$

where Spec R is the set of prime ideals of R. (Cf. 2.21f.) Thus dim $R = 0$ iff every prime ideal is maximal; and dim $R = 1$ iff dim $R/P = 0$ for every prime ideal P, etc. A Noetherian ring R may have finite or infinite dimension, but, e.g. by Bass' Theorem 2.24, one sees that R has at most countable dimension.

If $M \in \text{Spec} R_P$, then $M = N_P$ for a unique $N \in \text{Spec} R$, hence

$$\dim R_P = \text{rank } P.$$

The inequality

$$\text{rank } I + \dim R/I \leq \dim R$$

holds for all proper ideals I (Exercise).

The **radical** of an ideal I of R is the ideal:

$$\sqrt{I} = \{r \in R \mid \exists n,\ r^n \in I\}.$$

Thus $\sqrt{I}/I$ is the maximal nil ideal or **nil radical** of R/I, hence contained in the Jacobson radical of R/I.

14.1 THEOREM. *For any commutative ring R, and ideal I,*

$$\sqrt{I} = \bigcap_{\substack{P \in \text{Spec} R \\ P \supseteq I}} P.$$

PROOF. See Prop. 2.35A. □

14.2 THEOREM. *If R is a Noetherian local ring with maximal ideal P, the following are equivalent:*

1. *dim $R = n$,*
2. *rank $P = n$,*
3. *n is the least number of elements $\{x_1, \ldots, x_n\}$ so that,*

$$\sqrt{(x_1, x_2, \ldots, x_n)} = P,$$

4. *n is the least number of elements $\{x_1, \ldots, x_n\}$ of P so that $R/(x_1, \ldots, x_n)$ is Artinian.*

This follows easily from 2.23 and 2.25. See Bruns and Herzog [93], p.367.

Homological Dimension of Rings and Modules

An injective resolution (inj. res.) of a module M is an infinite exact sequence

$$\text{(inj. res.)} \quad 0 \to M \to M_0 \to M_1 \to \cdots \to M_n \to \cdots$$

such that M_n is injective, $n \geq 0$. If $M_k \neq 0$, and if $M_n = 0$ $\forall n > k$, then the inj. res. is said to have length k. Otherewise, the length is defined to be ∞.

The **injective dimension** of M, abbreviated inj. dim M, is the l.u.b. of the lengths of all possible injective resolutions. For example, M is injective and $\neq 0$ if and only if inj.dim $M = 0$. By agreement, inj.dim $0 = -1$.

Projective-resolution and projective dimension (proj. dim) are defined by dualism. The **projective dimension** of a module M over a ring R is also called *homological dimension*, abbreviated $\dim_R M$, or simply dim M.

The **right global dimension** of a ring R is defined to be

$$\text{r.gl. dim } R = \sup\{\dim M \mid M \in \text{ mod-}R\}.$$

For example, R is semisimple if and only if r.gl.dim $R = 0$. The left global dimension is denoted l.gl. dim. One can prove that

$$\text{r.gl. dim } R = \sup\{\text{inj. dim } M \mid M \in \text{mod-}R\}$$

(see Cartan-Eilenberg [56], Northcott [60], or MacLane [63]).

14.3 PROPOSITION. *Every module M has an injective resolution and a projective resolution.*

14.4 PROPOSITION. *Projective, injective, and global dimension are Morita invariants.*

14.5 INEQUALITY THEOREM. *Let A be a right R-module, and B a submodule. Then:*

1. *If dim $A >$dim B, then dim $A =$dim A/B.*
2. *If dim $A <$dim B, then dim $A/B = 1+$dim B.*
3. *If dim $A =$dim B, then dim $A/B \leq 1+$dim A.*

For proof, Kaplansky [69b], p.169.

14.6 COROLLARY. *If $B \subseteq A$, then dim $B \leq$max {dim A, dim A/B}. Moreover, dim $B =$dim A/B implies dim $B =$dim A.*

14.7 COROLLARY. *Let T be a subring of a ring R, and assume that T is a summand of R as a (T, T)-bimodule. Then:*

$$\textit{r.gl. dim } T \leq \textit{r.gl. dim } R + \dim_T R.$$

For proof, see Kaplansky [69b].

A ring R of r.gl. dim≤ 1 is **right hereditary**. By 8.12, the semisimple rings are the rings of global dimension 0. The next result has the corollary (8.16) that

$$\text{r.gl. dim } R[x] = \text{r.gl. dim } R + 1,$$

where $R[x]$ denotes the polynomial ring. Thus, $R[x]$ is right hereditary whenever R is semisimple.

14.8A CHANGE OF RINGS THEOREM. *Let R be a ring, and x a central element which is not a zero divisor in R. Let R/x denote the factor ring $R/(x)$. If A is a nonzero R/x-module, and if $dim_{R/x}A = n$ is finite, then $dim_R A = n+1$. Hence:*

$$r.gl.\ dim\ R \geq 1 + r.gl.\ dim\ R/x.$$

For proof, see Kaplansky [69a], p.172.

14.8B SECOND CHANGE OF RINGS THEOREM. *Under the same assumptions on R and x, for any R-module A such that $ann_A x = 0$, we have*

$$dim\ _{R/x}(A/xA) \leq dim\ _R A.$$

14.8C THIRD CHANGE OF RINGS THEOREM. *Under the same assumptions on R, x and A, as in 14.8B, if R is left Noetherian, and if x belongs to the radical of R, then*

$$\dim_{R/x} A/xA = \dim_R A.$$

The next theorem for the case of a field R is a theorem of Hilbert, and for general rings, a theorem of Eilenberg-Rosenberg-Zelinsky [57].

The Hilbert Syzygy Theorem

14.9 THE HILBERT SYZYGY THEOREM. [1] *For any ring $R \neq 0$, the global dimension of the polynomial ring $R[x_1, \dots, x_n]$ in n indeterminates is:*

$$r.gl.\ dim\ R[x_1, \dots, x_n] = n + r.gl.\ dim\ R.$$

PROOF. By 14.8A, r.gl. dim $R[x] \geq 1+$ r.gl. dim $R[x]/x$. But, $R[x]/x \approx R$, so r.gl. dim $R[x] \geq 1+$ r.gl. dim R. The lemma below reverses this inequality, and proves the theorem for the case $n = 1$. The general result then follows by induction on n. □

14.10 LEMMA. *If M is any module over a polynomial ring $R[x]$ over any ring R, then*

$$\dim_{R[x]} M \leq \dim_R M + 1.$$

14.11 PROPOSITION (AUSLANDER [55]). *Let A be a right B-module, let $\mathcal{J}$ be a nonempty well ordered set, and let $\{B_i \mid i \in \mathcal{J})$ be a family of submodules of A such that if $i, j \in \mathcal{J}$ and $i \leq j$, then $B_i \subseteq B_j$. If $A = \cup_{i\in\mathcal{J}} B_i$ and $\dim_R(B_i/\cup_{j<i} B_j) \leq n$ for all $i \in \mathcal{J}$, then $\dim_R(A) \leq n$.*

For proofs see the author's *Algebra I*, p.374ff. Also see Osofsky [68], [74], and esp. [78] for a number of generalizations.

14.12 GLOBAL DIMENSION THEOREM (AUSLANDER [55]). *If R is any ring, then*

$$r.gl.\ dim\ R = \sup_{I \subseteq R}\{dim\ R/I\},$$

the supremum of the projective dimensions of cyclic right R-modules.

---

[1] From the *American Heritage Dictionary*: Syzygy derives from the Greek *suzugia*, union, and *suzugos*, yoked: either of two points in the orbit of a celestial body in opposition or conjunction with the sun. (Music) combining of two metrical feet into a single metrical unit in prosody.

PROOF. Let $d =$ r.gl. dim$R$, and let n denote the right side of the equality. If A is an R-module, well-order the elements of A, and let B_i (resp. B_i') be the submodule of A generated by all x_j such that $j \leq i$ (resp. $j < i$). Then the factor module B_i/B_i' is cyclic $\forall i$ and, hence, has dim $\leq n$. Therefore, Auslander's proposition implies $d \leq n$. Thus $d = n$. □

14.13 COROLLARY. *For any ring R,*

$$r.gl.\ dim\ R = \sup\{proj.\ dim\ M \mid M\ \ f \cdot g\ right\ R\text{-}module\}.$$

14.14 COROLLARY. *For any ring R, if R is not semisimple, then*

$$r.gl.\ dim\ R = 1 + \sup_{I \subseteq R}\{dim\ I\}.$$

REMARK. In particular, any non-semisimple right hereditary ring has global dimension 1, and conversely.

14.15 REMARKS AND RELATED RESULTS.

(1) (Auslander [55]) If R is right and left Noetherian, then the right and left global dimensions are equal.

(2) (Kaplansky [58b]) The right and left global dimensions are unequal for a general ring R. Indeed, a right hereditary ring may not be left hereditary.

(3) (a) For any prime p, and integer $n > 1$, gl.dim $\mathbb{Z}/p^n\mathbb{Z} = \infty$; (b) A ring of infinite right global dimension may have an arbitrary integer $n > 0$ as left global dimension.

(4) (Hochschild [58]) For any ring R, and set S, the (semi)group algebra $R[S]$ of the free (semi)group $[S]$ on S has

$$\text{r.gl. dim}R[S] = 1 + \text{r.gl. dim}R.$$

In particular, the **free ring** on S, defined to be the semigroup ring $\mathbb{Z}[S]$, has gl.dim= 2. *Thus, any ring T is isomorphic to the factor ring A/I of a ring A of gl.dim*= 2.

(5) The free algebra over a field is right and left hereditary, in fact, a fir. (See Cohn [71b].)

(6) The ring $T_n(R)$ of lower triangular matrices over a ring R satisfies l.gl.dim $T_n(R)$ = 1+l.gl.dim R. Thus, $T_n(R)$ is left hereditary if and only if R is semisimple.

(7) (Osofsky [71b,74]) If R is right Noetherian, then

$$\text{r.gl. dim } R = \sup\{\text{inj. dim } R/I \mid I \subseteq R\}.$$

(8) (Rinehart [62]) (a) Let R be the algebra (ring) of differential polynomials over a field k. Then

$$t = \text{gl.dim}\left(\overset{n}{\otimes}_k R\right) \leq \text{gl.dim } R + 2(n-1),$$

and also $t \leq 2n$. If k has characteristic $\neq 0$, then $t = 2n$;

(b) The projective dimension of R as a module over the "enveloping" algebra $R \otimes_k R^{op}$ is 2.

A *differential ideal* P is an ideal consisting of constants, i.e., $D(P) = 0$.

(9) (Goodearl [74]) Let A be a commutative Noetherian Ritt algebra with gl.dim $A = n < \infty$. If $k = \sup\{pd_A(R/P) \mid P$ is a prime differential ideal of $A\}$, then r.gl. dim $A[y, D] = \max\{n, k+1\}$.

Recall from 7.18f that the **Weyl algebra** $A_1(A)$ over a commutative ring is the differential polynomial ring $A[x][y, D]$ where $D(x) = 1$.

(10) Let A be a commutative Noetherian ring with gl.dim $A = n < \infty$. If A is an algebra over $\mathbb{Q}$, then r.gl.dim $A_1(A) = n+1$, while if A has characteristic $p > 0$, then r.gl.dim $A_1(A) = n+2$. This generalizes (8).

With the obvious modification, this also holds for Weyl algebras $A_t(A)$ of arbitrary degree t. See 14.46. Also see Goodearl (*loc.cit.*) for, *i.a.* references to Björk [72], N.S. Gopalakrishnan and R. Sridharan [66], Rinehart and Rosenberg [76], and Roos [72].

(11) (Goodearl [74], Rinehart and Rosenberg [76]) Let S be the differential ring in n commuting derivations over a Noetherian ring R. If M is an S-module that is $f \cdot g$ as an R-module, then

$$pd_S(M) = n + pd_R(M)$$

where $pd =$ projective dimension.

(12) See Rosenberg and Stafford [76] on the global dimension of a differential polynomial ring $R[x, D]$ over a right and left Noetherian ring R.

Regular Local Rings

A local ring R with maximal ideal m, and residue field $k = R/m$, is said to be **regular** if R is Noetherian commutative and

$$\dim R = \dim_R(m/m^2)$$

where dim R is the Krull dimension of R and $\dim_k(m/m^2)$ is the dimension of the vector space m/m^2 over k. Moreover, a ring R is **regular** if R is Noetherian commutative and R_m is regular for all maximal ideals m of R.

14.16 THEOREMS (AUSLANDER-BUCHSBAUM [57–59]). *Let R be commutative.*

AB 1 *A local Noetherian ring R is regular iff gl.dim$R < \infty$. In this case, R is an integral domain and R_P is regular for each prime ideal P.*

AB 2 *Any regular local ring R is a unique factorization domain (= UFD).*

AB 3 *A Noetherian commutative ring R of finite global dimension is regular, and gl.dim$R = \dim R$.*

AB 4 *A Noetherian commutative ring of finite Krull dimension dim R is regular iff gl.dim $R < \infty$.*

AB 5 *If R is regular of dimension n, then the polynomial ring $R[X]$ is regular of dimension $n+1$.*

REMARK. All the results of 14.6 appear in [57] except AB 2 which appeared in [59]. (It is called the Auslander-Buchsbaum-Serre Theorem.) A number of generalizations of AB 2 appear in Kaplansky [70], e.g. p.185, Theorem 184, and the following.

14.17 THEOREM (KAPLANSKY, *l.c.*, THEOREM 185). *If R is a regular domain in which every invertible ideal is principal, then R is a UFD.*

REMARK. By AB 1, any Noetherian commutative local ring with zero divisors has infinite global dimension. Osofsky [69] showed there exists non-Noetherian local rings of finite global dimension that are not domains. However, in the sequel [69b], she proved that any chain ring of finite global dimension must be a domain.

14.18 THEOREM. *If R is a Noetherian local ring with maximal ideal m, then R is regular iff its completion $\hat{R}_m$ is regular.*

This follows from the natural isomorphisms:

$$R/m \approx \hat{R}/m\hat{R} \quad \text{and} \quad m/m^2 \approx m\hat{R}/m^2\hat{R}.$$

14.19 COROLLARY. *The power series ring $R\langle x_1, \dots, x_n\rangle$ in n variables over a regular local ring R is a regular local ring.*

REMARK. 1) When k is a field, then $k\langle x_1, \dots, x_n\rangle$ is a regular local ring of dimension n; 2) If R is a Noetherian local ring, of dimension 0, then R is regular iff R is a field; and when R has dimension 1, then R is regular iff R is a discrete valuation domain.

Here is a curious result of my colleague W. Vasconcelos, and a student of his, B. Greenberg.

14.20 THEOREM. *If R is a commutative coherent ring of global dimension two, then the polynomial ring $R[x_1, \dots, x_n]$ is coherent.*

This appears in his book [76], p.88. The $n = 1$ case appears in his paper [73a].

So what is curious about this? In general, one does not have a characterization of R for which $R[x]$ is coherent. Isn't that curious? Also, according to my esteemed colleague it is not known if $R[x]$ coherent implies $R[x, x_2, \dots, x_n]$ coherent.

Another curious thing about 14.20 is that nothing is said about global dimension 1. What happens in this case is:

14.21 THEOREM. *If R is a commutative semihereditary ring, then the polynomial ring $R[x_1, \dots, x_n]$ is coherent.*

This theorem, I am told, goes back to Nagata (in his solution of Hilbert's 14th problem). In particular, a result of Raynaud and Gruson [71], p.25, establishes it for any valuation domain (= VD), and a semihereditary ring is locally a VD.

Noncommutative Rings of Finite Global Dimension

If R is a left Artinian ring of finite left global dimension, then Green [80] showed that eRe, e a primitive idempotent, may not have finite global dimension. However:

14.22 THEOREM (ZACHARIAS [88]). *If R is left Artinian and eRe has finite left global dimension for every idempotent e, then R is a quotient ring of a hereditary Artinian ring.*

Stafford showed there exist a left Noetherian local ring R of left global dimension 2 that is not a domain, and not right Noetherian (See Chatters-Hajarnavis [80], p. 138.)

14.23 THEOREM (RAMRAS [74], SNIDER [88]). *If R is a right and left Noetherian local ring of global dimension $n \leq 3$, then R is a domain.*

REMARK. Additional positive results are obtained by Brown, Hajarnavis and MacEacharn [82] for rings of finite global dimension that satisfy the Artin-Rees (=AR) condition.

Classical Krull Dimension

Classical Krull dimension for a commutative Noetherian ring R is defined by counting the lengths of chains. For example, a prime ideal P is a chain of length 0, and prime ideals $P \supset Q$ form a chain of length 1. The classical Krull dimension of R is the supremum of the lengths of all chains of prime ideals in R (see **sup** 2.22). Infinite classical Krull dimension has been defined similarly by transfinite induction by Krause [70]:

14.24 DEFINITION. (Krause [70], Definition 11) For convenience, consider -1 as the least ordinal (instead of 0), and for a ring R, define sets X_α of prime ideals, for each ordinal α, such that $X_{-1} = \emptyset$. For each ordinal $\alpha \geq 0$, if X_β has been defined for each $\beta < \alpha$, define X_α to be the set of prime ideals P such that all prime ideals $Q \supset P$ are contained in $\bigcup X_\beta$. (e.g. X_0 is the set of maximal ideals of R). If some X_γ contains all prime ideals of R, then the classical Krull dimension $C\ell.K.\text{dim}(R)$ is defined to be the smallest such γ.

Examples of Krull dimension 0: any simple ring, and any right Artinian ring; Krull dimension 1: any Dedekind domain not a field.

14.25 THEOREM. *If R is a ring, then $C\ell.K.dim(R)$ exists iff R satisfies the acc on prime ideals.*

In this case $C\ell.K.\text{dim}R$ is the first γ such that $X_\gamma = X_{\gamma+1}$. Cf. Goodearl and Warfield [89], p.210, Prop. 12.1, for the proof of ($\Leftarrow$). The sufficiency ($\Rightarrow$) is proved in a number of exercises illustrating $C\ell.K.\text{dim}$ (Exercise 12A). Cf. Albu [74].

14.25A COROLLARY. *The infinite polynomial ring R over any ring with unit fails to have classical Krull dimension.*

Cf. *op.cit.* Ex. 12B.

Krull Dimension of a Module and Ring

The concept of Krull dimension for Noetherian rings has been generalized by Gabriel [62], p. 382, for Abelian categories, Rentscher and Gabriel [67] for finite ordinals, Krause [70] as discussed above, Gordon and Robson [74], and Lemonnier [72] for arbitrary posets.

14.26 DEFINITION. **The Krull dimension of a module M**, denoted $K.\dim(M)$ is defined by transfinite induction:

1. $K.\dim(M) = -1$ iff $M = 0$
2. If α is an ordinal ≥ 0, and if the set of modules with $K.\dim = \beta$ has been defined for all ordinals $\beta < \alpha$, then $K.\dim M = \alpha$ iff $K.\dim(M) = \beta$ has not been defined for some $\beta < \alpha$, and for every countable chain (S), $M_0 \supseteq M_1 \supseteq \cdots \supseteq M_n \supseteq \cdots$ of submodules, there holds
$$K.\dim(M_i/M_{i+1}) < \alpha$$
for all but finitely many indices i.

Thus, $K.\dim(M)$, is the smallest ordinal α such that $K.\dim(M_i/M_{i+1}) < \alpha$ $\forall i$, for any chain (S).

It follows that $K.\dim(M) = 0$ iff $M \neq 0$ and M is Artinian. Also, $K.\dim(M) = \alpha \geq 0$ iff $K.\dim(M) \geq \alpha$ but no descending chain (S) of submodules satisfies $K.\dim(M_i/M_{i+1}) \geq \alpha$ for infinitely many i. (See Goodearl and Warfield [89], p.223ff.)

14.27 THEOREM (GABRIEL [62]). *Any Noetherian module M has Krull dimension, denoted $K.dim$ M.*

PROOF. If not, then by induction we may assume that all proper homomorphic images of M have Krull dimension, and let $\alpha = \sup\{K.\dim M/N\}$ where N ranges over all submodules of M. Let M_i be a submodule of M in the descending chain (S) defined above, and we show that $K.\dim\,(M_i/M_{i+1}) \leq \alpha$ for all i. If $M_n = 0$, then $K.\dim\, M_i/M_{i+1} = -1$ for all i, so the assertion is true in this case. Otherwise, all $M_i \neq 0$, in which case
$$K.\dim\,(M_i/M_{i+1}) \leq K.\dim\,(M/M_{i+1}) \leq \alpha \quad \forall i,$$
hence $K.\dim\, M \leq \alpha + 1$, contrary to the assumption that M does not have Krull dimension.

□

The **right Krull dimension** of a ring R is defined to be $K.\dim\,(R_R)$ and denoted $r.K.\dim\,(R)$.

14.28 THEOREM. *If R is right Noetherian, then*
$$K.dim\,(M) \leq r.K.dim\,(R)$$
for any $f \cdot g$ right R-module M.

14.28A THEOREM (RENTSCHER-GABRIEL [67], GORDON-ROBSON [73]). *If R is right Noetherian, and if M is a nonzero $f \cdot g$ R-module, then*
$$K.dim\,(M[x]) = 1 + K.dim\,(M)$$
and hence
$$r.K.dim\,(R[x]) = 1 + r.K.dim\,(R).$$

PROOF. See proof of Gordon-Robson in Goodearl-Warfield [89], theorem 13.17, p.237, and historical note on p.240.

14.28B THEOREM (NOUAZÉ-GABRIEL, AND RENTSCHLER-GABRIEL [67]). *Let k be a field of characteristic 0, and $A_n(k)$ the Weyl algebra. Then*

$$K.dim\ (A_n(k)) = n.$$

PROOF. See Goodearl-Warfield, p.238, Theorem 3.18 and historical note on p.240.

REMARKS. (1) If R is a Noetherian ring of finite global dimension, is

$$K.\dim(R) \leq r.gl.\dim R?$$

Answer is affirmative for commutative R in which case equality holds. Refer to Matsumura [80], 18.G, or Northcott [62], p.208, Theorem 24. Other positive results are discussed by Goodearl and Warfield, *ibid.*, p.284ff;

(2) If R is Noetherian, does R satisfy dcc on prime ideals?

Again, answer is affirmative for commutative R, by Krull's Theorem 2.23 and Corollary 2.23A, but false for one-sided Noetherian R by Jategaonkar [69], Theorem 4.6. However, (2) *obviously is true whenever:* $K.\dim R < \infty$ *or* $Cl.K.\dim R < \infty$.

14.29A THEOREM (LEMONNIER [70], GORDON-ROBSON [73]). *If M is a nonzero module with Krull dimension, then M has finite Goldie dimension* (**sup.** *3.13), hence M is quotient finite dimensional (= q.f.d.).*

See (*loc.cit.*) p.7, 1.3. Cf. 5.20B and 7.21ff.

REMARK. If a q.f.d. module M has acc on subdirectly irreducible submodules (= accsi), then M is Noetherian (Faith [98], Cf. 16.50). In particular, any accsi module M with Krull dimension, or any accsi linear compact module M, is Noetherian.

14.29B THEOREM (GORDON-ROBSON [73]). *The polynomial ring $R[x]$ over a ring R has right Krull dimension iff R is right Noetherian.*

See (*loc.cit.*) p.60, 9.1.

14.29C COROLLARY. *The infinite polynomial ring $R[x_1, \dots, x_n, \dots]$ does not have Krull dimension.*

14.29D THEOREM (LENAGAN [73]). *If R has right Krull dimension, then nil subrings of R are nilpotent.*

14.30 THEOREM (GORDON, LENAGAN AND ROBSON [73], AND GORDON AND ROBSON [73]). *If R has right Krull dimension, then the prime radical of R is nilpotent.*

14.31 THEOREM (GORDON-ROBSON [73], COROLLARIES 3.4 AND 5.7). *Suppose R has right Krull dimension. (1) If R is semiprime then R is right Goldie, hence $Q^r_{c\ell}(R)$ exists and is semisimple. (See 3.13); (2) $Q^r_{c\ell}(R)$ exists and is right Artinian iff R satisfies the regularity condition* (**sup.** *3.55).*

REMARK. (1) of 14.31 is also proved by Goldie-Small [73]. Moreover:

14.31A THEOREM (GOLDIE-SMALL [73]). *If R has right Krull dimension, then*

$$r.K.\dim(R) \geq c\ell.K.\dim(R)$$

14.31B THEOREM. *If R is any right FBN ring, then*

$$r.K.dim(R) = c\ell.K.dim(R).$$

PROOF. See Gordon-Robson, *ibid.*, or McConnell-Robson [87], 6.4.8. □

14.31C THEOREM (KRAUSE, MICHLER). *If a right R-module has Krull dimension, then:*

$$K.dim(M) \leq sup\{1 + K.dim(M/N) \mid N \underset{\text{ess}}{\subset} M\}.$$

14.32A THEOREM. *If M has $K.dim \leq \alpha$, then $K.dim(N) \leq \alpha$ and $K.dim(M/N) \leq \alpha$ for all submodules N. Furthermore:*

$$K.dim(M) = max\{K.dim(N),\ K.dim(M/N)\}.$$

14.32B THEOREM (HUYNH-DUNG-SMITH [90]). *Let M be a module and α be an ordinal. Then M has Krull dimension at most α, if and only if every factor module of M contains an essential submodule with Krull dimension at most α.*

14.32C REMARK. The $\alpha = 0$ case implies Theorem 7.9.

Critical Submodules

A right R-module M is α **critical** if K.dim $(M) = \alpha \geq 0$ and K.dim $(M/N) < \alpha$ for all submodules $N \neq 0$ (in this case N is α-critical). Any α-critical module is uniform, and every non-zero module with Krull dimension contains a critical submodule (Goodearl-Warfield, *op.cit.*, pp. 227–228.) The concept originated with R. Hart [71] and the terminology with Goldie [72]. See Notes, (*op.cit.*) p.239 for these references, and for the following:

A **critical composition series** for M is a chain

$$M_0 = 0 < M_1 < \cdots < M_n = M$$

such that M_i/M_{i-1} is critical, $i = 1, \ldots, n$, and

$$K.\text{dim}(M_{i+1}/M_i) \geq K.\text{dim}(M_i/M_{i-1})$$

for $i = 1, \ldots, n-1$. Then the series has **length** n, and M_i/M_{i-1} are the **critical quotients**, $i = 1, \ldots, n-1$.

14.33 THEOREM (JATEGAONKAR [74B]-GORDON [74]). *If M is a nonzero Noetherian R-module, then M has a critical series. Two critical series have the same length, and their critical quotients can be paired so they have isomorphic nonzero submodules.*

Acc on Radical Ideals (Noetherian Spec)

As stated for a commutative ring R, and ideal I, the radical of I is denoted $\sqrt{I}$, and $\sqrt{I}/I = \text{nil rad} R/I$. An ideal I is a **radical** ideal if $I = \sqrt{I}$, equivalently, I is a semiprime ideal (see 2.25ff and 2.37). We let $\sqrt{\text{acc}}$ denote the acc on radical ideals.

14.34 THEOREM (KAPLANSKY [74], THEOREMS 87 AND 88). *If R is a commutative ring with $\sqrt{acc}$, then every radical ideal is a finite intersection of prime ideals, and for each ideal I there are just finitely many prime ideals minimal over I.*

14.35 REMARKS. (1) This generalizes results for Noetherian R. (See 2.27 and 2.28.); (2) The condition $\sqrt{\text{acc}}$ is referred to as **Noetherian Spectrum**, e.g. in the paper cited below:

14.36 THEOREM (PENDLETON-OHM [68]). *A commutative ring R has $\sqrt{acc}$ iff every radical ideal $\sqrt{I} = \sqrt{I_1}$ for a $f \cdot g$ ideal $I_1 \subseteq I$. In this case the polynomial ring $R[X]$ has $\sqrt{acc}$.*

Cf. The result of W. W. Smith 16.8B.

14.37 THEOREM (GORDON-ROBSON [73]). *Any commutative ring with Krull dimension has $\sqrt{acc}$.*

14.38 THEOREM (PUSAT-YILMAZ, AND SMITH [96]). *A commutative ring R has $\sqrt{acc}$ iff R has acc on prime ideals and every radical ideal is the intersection of finitely many prime ideals.*

Goodearl-Zimmermann-Huisgen Upper Bounds on Krull Dimension

This section extends the results of Bass [71], who linked Krull dimension of commutative Noetherian rings and ordinal lengths of well-ordered descending chains of ideals, and those of Gulliksen and Krause [73], who refined and extended Bass' results to Noetherian modules over arbitrary rings. Cf. Goodearl-Zimmermann-Huisgen [86] for an account of this and 14.39–14.45 below. Occasionally we abbreviate this reference as [GZH], or (GZH) [86].

For a module M, let $L(M)$ be the lattice of submodules of M, and $L^\star$ the *dual* of a lattice L, that is, the lattice having the same elements as L but carrying the reverse ordering. For any module M, we define $\kappa(M)$ to be the least ordinal κ such that $[0, \kappa)^\star$ (that is, the dual of the interval $[0, \kappa) = \{\text{ordinals } \alpha \mid 0 \leqq \alpha < \kappa\}$) cannot be embedded in $L(M)$. In the case of a noncommutative ring R considered as a right module over itself, we write $\kappa_r(R)$ for $\kappa(R_R)$.

14.39 REMARK. The possibility $\kappa(M) = \alpha + 1$ for a limit ordinal α is ruled out: for $\kappa(M) > \alpha$ means that $[0, \alpha)^\star$ is isomorphic to a chain $L \subseteq L(M)$; since α is a limit ordinal, all the submodules in L are nonzero, whence $[0, \alpha+1)^\star \simeq L \cup \{0\}$ and so $\kappa(M) > \alpha + 1$. **In summary, if $\kappa(M) > \alpha$ for a limit ordinal α, then $\kappa(M) \geqq \alpha + 2$.**

14.40 REMARK. The existence of a well-ordered descending chain of submodules of maximum length in M is equivalent to $\kappa(M)$ being a successor ordinal. This makes $\kappa(M)$ a more precise measure for the complexity of the submodule lattice of M than the supremum $o(M)$ of all lengths of well-ordered descending chains, as considered in 14.26. $o(M)$ may equal a limit ordinal α in either of two situations: there exists a chain of length α, or there exist chains of length β for all $\beta < \alpha$ but none of length α. In the first case, $\kappa(M) = \alpha + 2$, while in the second case, $\kappa(M) = \alpha$.

14.41 THEOREM (GOODEARL-ZIMMERMANN-HUISGEN [86]). *For a module M, the following statements are equivalent:*

(a) *The Krull dimension of M exists and is countable.*

(b) *$\kappa(M)$ is countable.*

If conditions (a) and (b) hold, and $K.dim(M) = \alpha \geqq 0$, then

$$\omega^{\beta} < \kappa(M) \leqq \omega^{\alpha+1}$$

for all ordinals $\beta < \alpha$. In case α is finite, the lower bound for $\kappa(M)$ can be improved to $\kappa(M) > \omega^{\alpha}$.

In the Noetherian case, the stated improvement in 14.41 extends to arbitrary countable Krull dimensions. For if M is a Noetherian module with countable Krull dimension $\alpha \geqq 0$, then M has an α-critical factor M', and results of Gulliksen and Krause [73] Corollary 2.5 and Proposition 2.12; [73], Proposition 4.2, show that $\kappa(M') > \omega^{\alpha}$, when $\kappa(M) > \omega^{\alpha}$.

The following examples show that the bounds obtained for $\kappa(M)$ in the theorem are best possible in general. (However, see [GZH] for sketches of the corresponding submodule lattices.) In each of these examples, the lattice of submodules is a chain. To construct modules with suitable submodule lattices, we use the result that every complete compactly generated chain appears as the submodule lattice of some module (Pierce [71], Theorem 2.1). (See Goodearl [80], Corollary 1.5, for an easier construction in the case of a countable chain.) Recall that an element x in a lattice is *compact* provided that whenever x equals the supremum of a subset X of the lattice, then x is the supremum of some finite subset of X. A lattice is *compactly generated* if each of its elements is the supremum of a set of compact elements.

14.42 REMARKS. $K.\text{dim}(\mathbb{Z}_{p^\infty}) = 0$ while $\kappa(\mathbb{Z}_{p^\infty}) = \omega$. This illustrates 14.41 in the case of $\alpha = 0$.

EXAMPLE A (*op.cit.*). There exists a module A such that $K.\text{dim}(A) = \omega$ and $\kappa(A) = \omega^{\omega}$.

PROOF (*ibid.*). Choose a module A for which $L(A)$ is a chain containing submodules $A_1 = 0 < A_2 < \dots$ such that $\cup A_n = A$ and $L(A_{n+1}/A_n) \simeq [0, \omega^n]^\star$ for all n. By Lemma 10.2, each A_{n+1}/A_n has Krull dimension n. Thus all proper submodules of A have finite Krull dimension, and $K.\text{dim}(A) = \omega$.

Observe that

$$\omega^n < \kappa(A_{n+1}) = \omega^n + \omega^{n-1} + \cdots + \omega + 1 < \omega^{\omega}$$

for all n. In particular, $[0,\omega^\omega)^\star$ cannot be embedded in $L(A_{n+1})$ for any n. Since any proper submodule of A is contained in some A_n, and since $[1,\omega^\omega)^\star$ is isomorphic to $[0,\omega^\omega)^\star$, we see that $[0,\omega)^\omega_\star$ cannot be embedded in $L(A)$. Therefore $\kappa(A)=\omega^\omega$.

In the case of a module M with $K.\dim(M)$ a countable successor ordinal α, the invariant $\kappa(M)$ can even lie strictly below ω^α, with $\omega^{\alpha-1}+2$ being the lower value allowed by the theorem.

EXAMPLE B (*op.cit.*). there exists a critical module B such that $K.\dim(B)=\omega+1$ and $\kappa(B)=\omega^\omega+2$.

One chooses for B a module for which $L(B)$ is a chain containing submodules $B_1=B>B_2>\dots$ such that $\cap B_n=0$ and $L(B_n/B_{n+1})\simeq L(A)$ for all n, where A is as in Example A. Obviously $K.\dim B>\omega$, in fact $K.\dim B=\omega+1$ as stated. See *op.cit.* for the proof of this and the rest.

EXAMPLE C (*op.cit.*). Given any ordinal $\alpha\geqq 0$, there exists a module C such that $K>\dim(C)=\alpha$ and $\kappa(C)=\omega^{\alpha+1}$.

PROOF (*ibid.*). Choose a module C for which $L(C)$ is a chain containing submodules $C_1=0<C_2<\dots$ such that $\cup C_n=C$ and $L(C_{n+1}/C_n)\simeq[0,\omega^\alpha n]^\star$ for all n. By Gordon-Robson [73] Lemma 10.2, each C_{n+1}/C_n has Krull dimension α. Thus all proper submodules of C have Krull dimension at most α, and $K>\dim(C)=\alpha$.

Since $\kappa(C_{n+1}/C_n)>\omega^\alpha n$ for all n, we have $\kappa(C)\geqq\omega^{\alpha+1}$. On the other hand, as

$$\omega^\alpha n+\omega^\alpha(n-1)+\cdots+\omega^\alpha+1<\omega^{\alpha+1}$$

for all n, we see that $\kappa(C_n)\leqq\omega^{\alpha+1}$. But any proper submodule of C being contained in some C_n, we conclude that $\kappa(C)=\omega^{\alpha+1}$.

□

Finally, there are examples for which κ lies strictly between the bounds given in the theorem.

EXAMPLE D ([GZH) [86]). Given any ordinal $\alpha\geqq 0$, there exists a module D such that $K.\dim(D)=\alpha$ and $\kappa(D)=\omega^\alpha+2$.

PROOF. Choose a module D such that $L(D)\simeq[0,\omega^\alpha]^\star$. □

EXAMPLE E. There exists an $(\omega+1)$-critical module E such that $\kappa(E)=\omega^{\omega+1}$. See (*op.cit.*).

EXAMPLE F (GORDON-ROBSON [73]). For each countable ordinal α there exists a commutative Noetherian ring with Krull dimension $=\alpha$.

14.43 THEOREM ((GZH) [86]). *(1) If R is a ring whose right Krull dimension is a countable limit ordinal α, then $\kappa_r(R)>\omega^\alpha$; (2) If R is a right fully bounded ring with countable right Krull dimension α, then $\kappa_r(R)>\omega^\alpha$.*

14.44 COROLLARY. *If R is a commutative integral domain with countable Krull dimension α, then $\kappa(R)=\omega^\alpha+2$.*

REMARK. The proofs make heavy use of Gordon-Robson [73], and the following

14.45 LEMMA. *If M is an α-critical module, for some ordinal α, then $\kappa(M)\leq\omega^\alpha+2$. If α is countable, then equality holds.*

McConnell's Theorem on the n-th Weyl Algebra

Assume that $\dim R$ denotes either right global dimension or right Krull dimension, and assume below that $\dim R < \infty$. Let $A_n(R)$ denote the n-th Weyl Algebra

$$A_n(R) = R[x_1, y_1, \ldots, x_n, y_n]$$

where $x_i y_j = y_j x_i \ \forall i \neq j$, and $x_i y_i - y_i x_i = 1 \ \forall i$.

(1) If R is right Noetherian, then

$$\dim(R) + n \leq \dim A_n(R) \leq \dim R + 2n.$$

(2) If $mR = 0$ for some integer $m > 1$, then

$$\dim A_n(R) = n + \dim R.$$

(3) If k is a field of characteristic 0, then dim $A_n(k) = n$ (Cf. 14.15 (10) and 14.28B) whereas if $Q = Q(A_n(k))$ is the quotient sfield of $A_n(k)$, then dim $Q = 2n$.

See McConnell [84] for pertinent references to (1)–(3).

14.46 THEOREM (MCCONNELL [84]). *Let k be a field of characteristic* 0.

*(a) If R is the enveloping algebra of a finite dimensional Lie Algebra over k; or (b) the group algebra $R = kG$ of a polycyclic by finite group G; or (c) an affine Noetherian PI-algebra over k (**affine** means $f \cdot g$ **qua** algebra over k), then*

$$\dim A_n(R) = n + \dim(R).$$

14.47 REMARK. For the primitive (maximal) ideals I of the universal enveloping algebra U of a finite dimensional complex nilpotent Lie algebra g, consult Dixmier [68]. (The quotients U/I are the Weyl algebras A_n, for the same n for "almost all I.") See J. C. McConnell [74,75] and Borho-Gabriel-Rentschler [73] for more general Lie algebras, and generalizations.

Historical Note

Eilenberg-Nagao-Nakayama [56] proved a hereditary semiprimary ring has the property that every factor ring has finite global dimension. Jans-Nakayama [56] and Chase [60] characterized semiprimary rings with the property that every factor ring has finite global dimension: they are triangular in the sense that there is a set $e_1, \ldots, e_n$ of orthogonal indecomposable idempotents with sum $= 1$ such that $e_i(\text{rad } R)e_j = 0$ for $i \geq j$. This happens iff $R/(\text{rad } R)^2$ has finite global dimension, and then gl.dim R is strictly less than the number r of simple rings in the Wedderburn-Artin decomposition of $R/\text{rad } R$. (Actually, gl.dim $(R) \leq$ gl.dim $(R/(\text{rad } R)^2) < r$. See Chase [60, p.22].) Moreover, when $R/\text{rad } R$ is separable, then any semiprimary ring with $R/(\text{rad } R)^2$ of finite global dimension is a factor ring of a unique hereditary semiprimary ring (Jans-Nakayama [56], cited by Chase (*loc.cit.*)). This theory is generalized further by Harada [64], who completely determines the structure of a hereditary semiprimary ring R as a generalized triangular matrix ring over a semisimple ring. Moreover, for any ideal I,

$$\text{gl.dim } (R/I) \leq r - s + 1$$

where s is the number of simple ideals in a direct sum decomposition of I modulo rad R. Also, if n =index of nilpotency of rad R, then

$$\text{gl.dim } (R/I) \leq n - 1.$$

Harada applies these results to give another proof of the main structure theorem for a hereditary order R over a rank 1 discrete valuation ring (Harada [63]). (See Reiner [75, p.358] for this and other results on the structure of hereditary orders, including the theorem of Jacobinski [71] stating that hereditary orders are "extremal.")

CHAPTER 15

Polynomial Identities and PI-Rings

Let k be a field and $F = k\langle x_1, \dots, x_d\rangle$ the free k-algebra on $x_1, \dots, x_d$. The elements of F are called *polynomial* and a k-algebra A is said to satisfy the polynomial identity

$$p(x_1, \dots, x_d) = 0, \tag{1}$$

if p is an element of F which vanishes for all values of the x's in A. If A satisfies a polynomial identity where p is not the zero polynomial, then A is called a **PI-algebra**. Many of the results proved for commutative or Noetherian rings, and finite dimensional algebras over fields, have their counterparts for PI-algebras. It will simplify matters to assume a field of coefficients, but it is possible to consider more general coefficient rings, e.g. every ring R is an algebra over any subring of its center C, and also, e.g. over $\mathbb{Z}$.

15.1 Remarks and Some Basic Results. Jacobson attributes the concept of a PI-algebra to Dehn [22].

1. Every commutative ring satisfies the identity $[x, y] = xy - yx = 0$ and so is a PI-algebra over $\mathbb{Z}$.
2. Every Boolean ring satisfies the identity $x^2 - x = 0$.
3. An algebra A is called **almost nil** if it has the form $k \cdot 1 + N$ where N is a nil ideal and it is almost nil of **bounded index** if every element of N is of bounded index, that is, there exists an integer $n > 0$ such that $x^n = 0$ for all $x \in N$. If A is almost nil, the commutator $[x, y] \in N$ for all $x, y \in A$. Hence if A is almost nil of bounded index it satisfies an identity $[x_1, x_2]^n$ and N satisfies x^n, for some n.
4. An interesting identity for algebras is Wagner's identity for k_2. Note that if

$$a = \begin{pmatrix} p & q \\ r & -p \end{pmatrix}$$

then tr $(a) = 0$, and

$$a^2 = \begin{pmatrix} p^2 + qr & 0 \\ 0 & p^2 + qr \end{pmatrix}$$

commutes with every matrix. Since tr $(a) = 0$ for all a, b we have $[[ab]^2, c] = 0$ for all $a, b, c \in k_2$. Hence

Eq.(1) $$(x_1x_2 - x_2x_1)^2 x_3 - x_3(x_1x_2 - x_2x_1)^2$$

is an identity for k_2. This is **Wagner's identity** published in a paper appearing in 1937. Wagner also gave identities for k_n.

5. Every finite-dimensional algebra A over a field k satisfies an identity. If $r = \dim A < n$, then A satisfies the identity

Eq.(2)
$$S_n(x_1, \ldots, x_n) = \sum_\sigma \operatorname{sgn}(\sigma) x_{1\sigma} x_{2\sigma} \ldots x_{n\sigma} = 0,$$

where σ runs over all permutations of $1, \ldots, n$ and $\operatorname{sgn}(\sigma)$ is 1 or -1 according as σ is even or odd. S_n is called the **standard polynomial** and (2) standard identity of degree n. For any $a_1, \ldots, a_n \in A$, $S_n(a_1, \ldots, a_n) = 0$ if two of the a_i's coincide. Now take a k-basis $u_1, \ldots, u_r$ of A; then for any $a_1, \ldots, a_n \in A$, $S_n(a_1, \ldots, a_n)$ can by linearity be written as a linear combination of terms $S_n(u_{i_1}, \ldots, u_{i_n})$. Since $r < n$, at least two of the u_i's must coincide, so all terms vanish. This shows that A satisfies Eq.(2).

6. If A is a commutative k-algebra, then A_n satisfies the standard identity of degree $n^2 + 1$, since there is an A-basis of n^2 elements, so the result follows (as in #5).

7. If every element of A is algebraic over a field k, of degree at most n, each element of A satisfies an equation

Eq.(3a)
$$x^n + a_1 x^{n-1} + \cdots + a_n = 0, \qquad \text{where } a_i \in k.$$

Then A is algebraic over k of bounded degree. If the equation for some element of A has degree less than n, we can bring it to the form (3a) by multiplying by a power of x. Writing $[x, y] = xy - yx$, we obtain from (3a),
$$[x^n, y] + a_1[x^{n-1}, y] + \cdots + a_{n-1}[x, y] = 0.$$
Thus the commutators in this expression are linearly dependent and so A satisfies the identity

Eq.(3b)
$$S_n([x, y], [x^2, y], \ldots,]x^n, y]) = 0.$$

Eq.(3b) also holds for an algebra A over a commutative ring k if all of the elements $a \in A$ satisfy Eq.(3a), that is, a monic polynomial of degree $\leq n$.

8. Any subalgebra or homomorphic image of a PI-algebra is again a PI-algebra.

9. A integral domain R contains a copy of either $\mathbb{Z}$ or $\mathbb{Z}_p$ for a prime p in its center C, according as R has characteristic 0 or p.

   Every integral domain PI-algebra over $\mathbb{Z}$ or $\mathbb{Z}_p$ is a left and right Ore domain (see 15.7A).

10. A polynomial p and the corresponding identity $p = 0$ is said to be multilinear if it is homogeneous of degree 1 in each variable. In a PI-algebra A, there exist multilinear identities of the same degree as a given identity $p = 0$. (See Cohn [91], p.376. Corollary 5.3.)

11. To illustrate 10, suppose A satisfies the identity

Eq.(4)
$$x^2 = 0.$$

Here we do not restrict our algebra to have a unit element (for otherwise it would have to be trivial, by (4)). Then A also satisfies $(x+y)^2 - x^2 - y^2 = 0$, i.e.,

Eq.(5)
$$xy + yx = 0,$$

and this is multilinear. In characteristic not equal to 2 the identities Eq.(4) and Eq.(5) are equivalent, for (5) $\Rightarrow$ (4) by putting $y = x$, which gives $2x^2 = 0$.

12. Let R be an algebra over a field k with center C. If R contains a subalgebra A which is a PI-algebra such that $R = AC$, then R is a PI-algebra.

    By 10, A satisfies a multilinear identity $p(x_1, \ldots, x_n) = 0$. Let $\{u_\lambda\}$ be a k-basis for R and put $a_i = \Sigma\alpha_{i\lambda}u_\lambda$, where $\alpha_{i\lambda} \in C$. Then

$$p(a_1, \ldots, a_n) = \Sigma\alpha_{1\lambda_1} \cdots \alpha_{n\lambda_n} p(u_{\lambda_1}, \ldots, u_{\lambda_n}) = 0,$$

    by multilinearity, thus p vanishes on R.

13. Let A be a finite-dimensional algebra over an infinite field k Then any polynomial identity $p(x_1, \ldots, x_d) = 0$ for A also holds in $A_E = A \otimes_k E$, for any extension field E of k.
14. Let K be a commutative $\mathbb{Z}$-algebra and $A \in K_n$. If tr $(A^r) = 0$ for $r = 1, \ldots, n$, then $A^n = 0$. Conversely. (This involves just elementary matrix theory.)
15. Any PI-algebra satisfies an identity in 2 variables:

$$p(x_1, \ldots, x_n) = 0$$

    implies $p(xy, xy^2, \ldots, xy^n) = 0$; and the trick is to show the elements $\{xy^i\}$ are free, $i = 1, \ldots, n$.

Amitsur-Levitzki Theorem

15.2 Amitsur-Levitzki Theorem. *The standard polynomial S_{2n} is an identity for the $n \times n$ matrix ring k_n over any commutative ring k. Thus, any faithful algebra A over a commutative ring k having a finite free k-basis of n elements satisfies S_{2n}.*

15.3 Staircase Lemma (Kaplansky [48,95]). *The $n \times n$ matrix ring k_n over a commutative ring k satisfies no identity of degree $< 2n$.*

See Herstein [68], p.157, Lemma 6.31, and Cohn [91], p.379, Lemma 5.9.

Kaplansky-Amitsur Theorem

A polynomial f is called a **proper identity** if f is an identity for A and no coefficient of f annihilates A. If K is a field this is equivalent to: f is an identity and $f \neq 0$. If f has a coefficient 1, it is a proper identity for every algebra for which it is an identity. All the examples of 15.1 are of this type. In this section we state a fundamental theorem on polynomial identities:

15.4 Theorem (Kaplansky [46,95]-Amitsur). *If A is a primitive algebra which has a proper identity of degree d, then the center C of A is a field, A is simple and $[A : C] \leq [\frac{d}{2}]^2$.*

Proof. See Herstein [68], p.157, Theorem 6.31.

15.5 COROLLARY. *An algebra A satisfying a proper identity, and having no nil ideals $\neq 0$ embeds qua ring into a matrix ring B_m over a commutative ring B.*

Moreover, B can be chosen to be a product of fields. (See Herstein [68], p.159, Theorem 6.3.2. There is a misprint in the statement: replace direct sum by direct product.)

Posner's Theorem

DEFINITION. A subalgebra A of an algebra Q is called a **left (right) order** in Q if every regular element of A has an inverse in Q and every element of Q has the form $a^{-1}b, a, b \in A$ $(ba^{-1}, a, b \in A)$. Then $Q = Q(A)$, the classical left (right) quotient ring of A.

15.6 POSNER'S THEOREM [60]. *Let A be a prime algebra satisfying a proper identity. Then:*

1. *The algebra A_0 of central quotients of A is finite dimensional simple over its center and the center is the quotient field F of the center C of A.*
2. *A is a left and right order in A_0.*
3. *A and A_0 satisfy the same identities.*

REMARKS. 1) By "central quotients" one means that $A_0 = \{ac^{-1} \mid a \in A, c \in C\}$. Posner [60] proved that $Q(A)$ is simple of finite dimension over its center B, and the proper and central PI parts are due to Rowen and Formanek. (See Jacobson [75], p.57, Theorem 2.); 2) If D is a sfield of finite dimension n over center k then $D \otimes_R D^0 \approx k_n$ by the Brauer Group Theorem that states that $[D^0] = [D]^{-1}$ in $Br(k)$ (2.6Bf. and 4.16As).

15.7A COROLLARY. *If A is a prime algebra satisfying a proper identity, then A is right and left Goldie.*

15.7B COROLLARY. *Any integral domain R satisfying a proper identity is a right and left Ore domain.*

PROOF. This follows from 15.6. Cf. 6.29.

15.8 THEOREM (ARTIN [69]-PROCESI [72]). *A ring R is a rank n^2 Azumaya algebra over a commutative ring k iff R satisfies all the identities of $n \times n$ matrices over $\mathbb{Z}$ but no nonzero factor algebra satisfies the $(n-1) \times (n-1)$ identities over $\mathbb{Z}$.*

Nil PI-Algebras Are Locally Nilpotent

An identity $P = 0$ will be called **strongly regular** if the non-zero coefficients of P are units (invertible elements) of k. If A satisfies a strongly regular identity, then so does every sub-algebra and every homomorphic image. Strongly regular identities are proper. If A satisfies a strongly regular identity, then A satisfies a multilinear one of no higher degree (Jacobson [75], p.17, Lemma 2).

15.9 THEOREM (KAPLANSKY [48]). *A nil algebra satisfying a strongly regular identity is locally nilpotent (= every $f \cdot g$ subalgebra is nilpotent).*

15.10 THEOREM. *Let A be an algebra satisfying a strongly regular identity of degree d. Then any nil subalgebra B of A satisfies $B^{[d/2]} \subset N(0)$ the sum of the nilpotent ideals of A.*

See Jacobson [75], p.34, Theorems 1 and 2.

Note: 15.9 and 15.10 follow from the next two theorems if k is a field.

The next theorem gives an affirmative answer to the Kurosh problem for PI algebras.

15.11 THEOREM (KAPLANSKY [46] AND LEVITZKI [46]). *If A is an algebraic algebra over a field k satisfying a polynomial identity then A is locally finite* ($= f \cdot g$ subalgebras are finite dimensional).

See Herstein [68], pp.167–168 for the proof.

15.12 THEOREM. *If A is an algebraic algebra of bounded degree over a field k, then it is locally finite.*

PROOF. By 15.1(7), A satisfies a polynomial identity, Eq.(3b), hence the result follows from the Theorem. □

15.13 THEOREM (REGEV [72]). *The tensor product of two PI-algebras over a field k is again a PI-algebra.*

See Cohn [81], p.383, Theorem 6.4. There is an informative review by J. L. Fisher in Small [81], p.230, #11.01.018.

Rowen PI-Algebras

Let $k\langle X\rangle$ denote the **free algebra**[1] in the non-commuting variables $\{X\}$ over a commutative ring k. Below are three equivalent definitions of PI-algebras over k.

DEFINITION 1. (ROWEN). An algebra A over a commutative ring k is called a **PI-algebra** (or algebra with polynomial identity) if there exists a polynomial $f(x_1, \dots, x_m) \in k\langle X\rangle$ which is a proper identity for every non-zero homomorphic image of A.

DEFINITION 1′. An algebra over k is called a PI-algebra if there exists an identity f for A such that $S_f A = A$ for S_f the set of coefficients of f.

The condition $S_f A = A$ is equivalent to: there exist $a_i \in A$ such that

$$\alpha_1 a_1 + \dots + \alpha_r a_r = 1$$

where $S_f = \{\alpha_1, \dots, \alpha_r\}$. Let $\prod A_j$ be the direct product of copies A_j of A and let $\bar{a}_j$ be the element of $\prod A_j$ that has a_j in every place. Then $\Sigma \alpha_j \bar{a}_j = 1$. Hence $\prod A_j$ is a PI-algebra.

It is clear from the definitions that a homomorphic image of a PI-algebra is PI. However, despite Regev's Theorem 15.13, it is not clear that subalgebras are PI or that tensor products with commutative algebras are PI.

[1]While we sometimes have used $k\langle X\rangle$ to denote power series elsewhere in the text, there ought to be no confusion to revert to this use for the rest of this chapter.

15.14 THEOREM (AMITSUR). *If A is a PI-algebra, A satisfies an identity S_{2n}^m for some n and m.*

DEFINITION 1″. An algebra A is called a PI-algebra if it satisfies an identity S_{2n}^m for suitable n and m.

Definition 1″ shows that every subalgebra of a PI-algebra is PI. One can use Amitsur's result and the linearization method to obtain a multilinear identity whose non-zero coefficients are ± 1. It follows from this that if A is a PI-algebra, then so is $A^C = C \otimes_k A$ for any commutative algebra k.

See Jacobson [75], pp. 58–60.

Generic Matrix Rings Are Ore Domains

The generic matrix ring of degree n in $x_1, \dots, x_d$,

$$F_{(n)} = k\langle x_1, \dots, x_d \rangle_{(n)}$$

is the algebra over k on $x_1, \dots, x_d$ which is universal for homomorphisms into $n \times n$ matrix rings over commutative rings. The elements of $F_{(n)}$ may themselves be thought of as matrices, so that $F_{(n)}$ is a ring of $n \times n$ matrices, generated by $x_1, \dots, x_d$ and every mapping $x_i \mapsto a_i \in A_n$ where A is commutative, can be extended to a unique k-algebra homomorphism.

15.15 PROPOSITION. *Let $F = k\langle t_1, \dots, t_d \rangle$ be the free algebra over k, $F_{(n)} = k\langle x_1, \dots, x_d \rangle_{(n)}$ the generic matrix ring of degree n and $v : F \to F_{(n)}$ the k-algebra homomorphism in which $t_i \mapsto x_i$. Then $p \in F$ vanishes identically on every $n \times n$ matrix ring over a commutative k-algebra if and only if $pv = 0$.*

15.16 THEOREM. *The generic matrix ring $k\langle x_1, \dots, x_d \rangle_{(n)}$ is a left and right Ore domain.*

Generic Division Algebras Are Not Crossed Products

From this result it follows that $F_{(n)}$ has a skew field Q of fractions, called the **generic division algebra of degree n over k**. Also $[Q : C] = n^2$, where C is its center. These concepts have been used by Amitsur [72] to prove that not every division algebra is a crossed product: if the generic division algebra of degree n were a crossed product, with Galois group G, then every division algebra of degree n is a crossed product with group G; Amitsur then constructs two division algebras of degree n (for certain n, and in characteristic 0) which cannot be expressed as crossed products with the same Galois group; it follows that the generic division algebra is not a crossed product. Cf. Cohn [81], p. 386–389, for this, and 15.15–16.

When Fully Bounded Noetherian Algebras Are PI-Algebras

A ring R is **(right) fully bounded** if every prime factor ring R of R is right bounded (i.e., every essential right ideal contains an ideal).

15.17 THEOREM (AMITSUR-SMALL [96]). *If R is a (right) fully bounded Noetherian algebra over an algebraically closed field k of cardinal $> dim_k R$, then A is a PI-algebra.*

Cf. 3.43.

Notes on Prime Ideals

Going beyond the classical correspondence between prime ideals in polynomial rings over algebraically closed fields and algebraic varieties, the importance of prime ideals in general commutative rings no doubt stems from the information about the ring obtained "locally" via the local ring R_P at a prime ideal P. However, this cannot guarantee the importance of prime ideals in noncommutative rings inasmuch as the local ring may not exist. For example, a prime ring R may not have a right or left (classical) quotient ring $Q(R)$. Although the point of the Goldie-Lesieur-Croisot theorem is that $Q(R)$ exists if R is, say, right Noetherian, nevertheless, there may not exist a "local" ring R_P at a prime ideal P (see, however, the local "envelope" at P as defined in Goldie [67], and developed in Lambek-Michler [73,74] and Jategaonkar [74b]: Jategaonkar characterizes when this exists by a very simple condition on the injective hull of R/P. See his Theorem [4.5]).

This difficulty is partially surmounted if we restrict our attention to polynomial identity (PI) rings, since, by a theorem of Posner [60] and Small [72], one may construct R_P (in the canonical way) provided that R and R/P satisfy exactly the same polynomial identities.

Jategaonkar [74c] extended the rank theorem for prime ideals for Noetherian commutative rings to Noetherian PI algebras. Kaplansky has given a historical and lucid account of Krull's Principal Ideal Theorem (in Kaplansky [68]); and Jategaonkar [75] has generalized this to PI-rings.

Historical Notes

(1) See Kaplansky [85] for historical notes on PI's, esp. central PI's "for matrices of any size independently constructed by Formanek [72] and Razmyslov [73]."

(2) If K is a field, then KG satisfies a polynomial identity iff G has an abelian subgroup of finite index. (Theorem of S. A. Amitsur, I. M. Isaacs, D. S. Passman, and M. Smith.) This is related to group algebras KG all irreducible representations of which are of bounded degree. (See Kaplansky [49b] and Amitsur [61]; also Snider [74] for a generalization.)

CHAPTER 16

Unions of Primes, Prime Avoidance, Associated Prime Ideals, ACC on Irreducible Ideals and Annihilator Ideals in Commutative Rings

If R is a ring, and S is a subset $\neq \emptyset$ closed under the operations of addition and multiplication, we say that S is a subring -1 (= **subring minus 1**).

McCoy's Theorem

16.1 THEOREM (MCCOY [57A]). *Let R be a commutative ring, and $I_1, \dots, I_n$, $n \geq 2$, finitely many ideals such that at least $n-2$ of them are prime. If S is a subring-1 (e.g. an ideal) of R contained in $I_1 \cup \cdots \cup I_n$, then $S \subseteq I_k$ for some k.*

16.2 REMARK. If $n = 2$, then I_1 and I_2 can be arbitrary ideals. Theorem 16.1 is "Theorem 81" of Kaplansky [74], and the next theorem, a reformulation of McCoy's theorem, is "Theorem 82." Also see Eisenbud [96], p.90.

16.3 THEOREM. *Let R be a commutative ring, S a subring -1, and I an ideal of S. If $P_1, \dots, P_n$ are finitely many prime ideals of R, and*

$$S \backslash I \subseteq P_1 \cup \cdots \cup P_n$$

then $S \subseteq P_k$ for some k.

16.4 REMARK. In the theorem, $S \subseteq I \cup P_1 \cup \cdots \cup P_n$, so McCoy's theorem applies.

16.5 PRIME AVOIDANCE THEOREM. *If $P_1, \dots, P_n$ are finitely many prime ideals of a commutative ring, and if S is a subring -1 such that $S \not\subseteq P_i$, $\forall i = 1, \dots, n$, then $S \not\subseteq \bigcup_{i=1}^{n} P_i$, hence some element $s \in S$ lies outside, that, is, some element of S avoids every $P_i, i = 1, \dots, n$.*

The Baire Category Theorem and the Prime Avoidance Theorem

If V is a vector space over an infinite sfield, then V is not the union of a finite number of proper vector spaces. (Note, however, that $\mathbb{Z}_2 \oplus \mathbb{Z}_2$ is a union of three vector subspaces generated by its nonzero elements.) A similar result holds for finite groups (see McCoy [57a]).

A subset S of a metric space V is **nowhere dense** if the closure of S contains no interior point of V.

16.6 BAIRE'S CATEGORY THEOREM. *If V is a metric space, then V is not the union of a countable family of nowhere dense subsets.*

PROOF. See Capson [68], pp.50–51.

This is used in the proof of the:

16.7A SHARP-VÁMOS THEOREM [85]. *Let (R, m) be a commutative Noetherian complete local ring, and let $\{P_i\}_{i=1}^{\infty}$ be a countable family of prime ideals. Then, if I is an ideal, and $x \in R$ is such that $x + I \subseteq \bigcup_{i=1}^{\infty} P_i$, then $x + I \subseteq P_j$ for some $j \geq 1$.*

PROOF. (*Ibid.*) The m-adic topology on R is induced by a metric one. For $x \neq y \in R$, let $\|x - y\| = 1/2^t$ where t is the greatest integer so that $x - y \in m^t$. Since each ideal I of R is closed in the m-adic topology, and R is a topological ring with respect to this topology, it follows that each coset $x + I$ of an ideal I is a complete metric space. Since

$$x + I = \bigcup_{i=1}^{\infty}((x + I) \cap P_i),$$

then by Baire's Category Theorem there exists $j \geq 1$ such that $(x + I) \cap P_j$ is *not* nowhere dense in $x + I$. Let c be an element of the interior of $(x + a) \cap P_j$. Thus, there exist $d \in R$ and $t > 0$ such that

$$c \in (x + I) \cap (d + m^t) \subseteq (x + I) \cap P_j.$$

From this we deduce that $I \cap m^t \subseteq P_j$. Since P_j is prime, then $I \subseteq P_j$ or $m^t \subseteq P_j$. In the latter case $m \subseteq P_j$ whence $m = P_j$, and there is nothing to prove. In the former case, $c \in P_j$, hence $x - c \in I \subseteq P_j$, so $x \in P_j$, hence $x + I \subseteq P_j$ as asserted. □

16.7B PRIME AVOIDANCE THEOREM FOR COMPLETE LOCAL RINGS (*ibid.*). *Let $\{P_i\}_{i=1}^{\infty}$ be a countable family of prime ideals of a commutative Noetherian complete local ring (R, m), let I be an ideal and $x \in R$ be such that $(x)+I \not\subseteq \bigcup_{i=1}^{\infty} P_i$. Then there exists $y \in I$ such that $x+y \notin \bigcup_{i=1}^{\infty} P_i$. On the other hand, if $I \subseteq \bigcup_{i=1}^{\infty} P_i$, then $I \subseteq P_j$ for some $j \geq 1$.*

REMARKS. (1) As the authors state, this contains a lemma of Burch [72]; (2) the theorem fails for a non-complete Noetherian local ring (R, m). (See Sharp-Vámos, *ibid.*)

W. W. Smith's Prime Avoidance Theorem and Gilmer's Dual

16.8A THEOREM (W. W. SMITH [71]). *Let R be a commutative ring, I an ideal, and $\mathcal{P}$ any nonempty set of prime ideals. The following conditions are equivalent*

(1) $I \subseteq \cup\mathcal{P}$ iff $I \subseteq P$ for some $P \in \mathcal{P}$.

(2) A prime ideal $I \subseteq \cup\mathcal{P}$ iff $I \subseteq P$ for some $P \in \mathcal{P}$.

(3) For each $P \in \mathcal{P}$, $P = \sqrt{x}$ for some $x \in R$.

REMARK. This theorem generalized that of Reis and Viswanathan for Noetherian R.

16.8B COROLLARY (W. W. SMITH [71]–GILMER [97]). *(1) of 16.8A holds for an ideal I and all $\mathcal{P}$ iff $\sqrt{I} = \sqrt{x}$ for some $x \in R$.*

PROOF. See Gilmer *loc.cit.*, Prop. 4, p. 330.

REMARK. By the Pendleton-Ohm Theorem 14.6, 16.8B implies that (1) of 16.8A holds for all ideals I only if R has acc on radical ideals.

16.8C THEOREM (GILMER [97]). *The condition dual to (2) holds (namely for a given prime ideal I, it is true that $I \supseteq \cap \mathcal{P}$ iff $I \supseteq$ some $P \in \mathcal{P}$) iff R is semilocal and zero-dimensional.*

Irreducible Modules Revisited

For convenience, we repeat an earlier definition.

16.9A DEFINITION. A right R-module M is **irreducible** or **uniform**, if $M \neq 0$ and satisfies the equivalent conditions.

1. if A and B are submodules such that $A \cap B = 0$, then $A = 0$ or $B = 0$.
2. every nonzero submodule of M is an essential submodule.
3. the injective hull $E(M)$ of M is indecomposable.

When this is so then every nonzero submodule of M is uniform, and the annihilator $I = \text{ann}_R x$ of any nonzero $x \in M$ is an irreducible right ideal of R. Moreover, $A = \text{End } E(M)_R$ is a local ring.

PROOF. Straightforward module theory. See e.g. Matlis [58], Sharp and Vámos [71], p.51 ff., or the author's *Algebra I*, pp.205 and 343. Cf. 3.14A. □

Cf. Subdirectly irreducible modules 2.17C.

16.9B PROPOSITION AND DEFINITION. *A right R module M is said to have* **finite Goldie dimension** *if M satisfies the equivalent conditions:*

1. *every independent set of submodules of M has finite cardinal a.*
2. *there exists a finite independent set $\{U_i\}_{i=1}^n$ of submodules of M such that $\Sigma_{i=1}^n U_i$ is an essential submodule of M.*
3. *the injective hull $E(M)$ of M is a direct sum of finitely many indecomposable modules $\{E_i\}_{i=1}^m$.*

In this case, $a = m = n$, and $E(M) = \bigoplus_{i=1}^{n} E(U_i)$. Furthermore, M has finite Goldie dimension iff End $E(M)_R$ is semiperfect.

PROOF. See e.g. the author's Algebra I, *ibid.*, and *Algebra II*, 18.24-25, p.44ff.

(Subdirectly) Irreducible Submodules

A submodule S of M is said to be **irreducible** if $S \neq M$, and M/S is an irreducible module. Thus M is an irreducible module iff 0 is an irreducible submodule. The same terminology applies *mutatis mutandis* for subdirectly irreducible submodules.

Similarly, for a right ideal I of a ring R, I is irreducible (resp. subdirectly irreducible) if R/I is an irreducible (resp. subdirectly irreducible) module. Cf. 2.25s, **sup.** 8.5A and 2.17Cs.

If E is right R-module with endomorphism ring A, then any left A-submodule of E is called a **counter submodule**.

16.9C PROPOSITION AND DEFINITION. *An ideal I is an irreducible ideal of R iff there exists an indecomposable injective right R-module E and an element x of E whose annihilator is I. In this case we say that I is a* **point** *annihilator of E.*

For any injective module E, we let $\mathcal{A}_p(E)$ denote the set of all point annihilators. Then $\mathcal{A}_p(E)$ satisfies the acc iff the set cyclic countermodules Ax of E satisfies the dcc, for any injective right R-module E.

PROOF. The proof follows from (1)–(4) below:

1. A $f \cdot g$ counter-submodule M of E satisfies the double annihilator condition (= *dac*) by the theorem of Jacobson [64] and Johnson-Wong [61], namely:

$$\text{ann}_E\ \text{ann}_R M = M.$$

(This holds for any quasi-injective module E. See, e.g. the author's *Algebra II*, p.66, 19.10.)

2. An injective module E is indecomposable iff every submodule $S \neq 0$ is **uniform** in the sense that $xR \cap yR \neq 0$ for any $x \neq 0$ and $y \neq 0$ in S.
3. Moreover, an injective module E is indecomposable iff E is the injective hull of a uniform module $\neq 0$. (Then every submodule $\neq 0$ is a uniform module.) If xR is a non-zero cyclic submodule of an indecomposable injective module E, then the fact that the $xR \approx R/\text{ann}_R x$ is uniform shows that $\text{ann}_R x$ is an irreducible right ideal of R.
4. Conversely, if I is a right ideal, then $R/I \hookrightarrow E$ iff $I = \text{ann}_R x$ for some $x \in E$.

Two irreducible ideals I and K are **related** (Notation: $I \sim K$), provided that $R/I \hookrightarrow E(R/K)$. In this case, $E(R/K) = E(R/I)$, hence $R/K \hookrightarrow E(R/I)$, and one concludes that "$\sim$" is an equivalence relation on the poset of irreducible ideals. We let $[I]$ denote the equivalence class determined by I, and let acc$[I]$ denote the acc on the set $[I]$).

A prime ideal P is said to be **Noetherian** if the local ring R_P is a Noetherian ring.

16.9D COROLLARY. *If P is a Noetherian prime ideal, then R satisfies acc$[P]$.*

PROOF. By Prop. 16.9C (and its proof) the ideals $I \in [P]$ are in 1-1 correspondence with the cyclic counter modules of $E(R/P)$, and since these coincide with the cyclic counter modules of $E(R/P)$ *qua* R_P-module, we conclude that $[P]$ satisfies acc.

The next theorem is well known. As before, R is commutative, $E(M)$ denotes the injective hull of an R-module M, and $\mathcal{A}(E)$ is the set of ideals of R that are the annihilators of subsets of E.

16.10 LEMMA. *Let M be an R-module. If P is a maximal element of $\mathcal{A}(M)$, then P is a maximal element of $\mathcal{A}_p(M)$, and conversely, a maximal element of $\mathcal{A}_p(M)$ is maximal in $\mathcal{A}(M)$. In this case, $R/P \hookrightarrow M$ canonically.*

PROOF. Suppose P is as stated, say $P = \text{ann}_R X$, where $X \subseteq M$, and if $0 \neq x \in X$, then $\text{ann}_R x \supseteq P$, and $\text{ann}_R \neq R$, hence, by maximality of P, $\text{ann}_R x = P$, so $P \in \mathcal{A}_P(M)$. Then $R/P \hookrightarrow M$ via the map $r + P \mapsto xr\ \ \forall r \in R$.

Conversely, let $P = \text{ann}_R x$ be maximal in $\mathcal{A}_p(M)$, and let $P \subseteq P'$ where $P' \in \mathcal{A}(M)$. Then, for each $0 \neq y \in\ \text{ann}_M P'$, $\text{ann}_M P'$, $\text{ann}_R y \supseteq P$, so $\text{ann}_R y = P$ by maximality of P in $\mathcal{A}_p(M)$. This proves that $P = P'$, hence P is maximal in $\mathcal{A}(M)$. □

Associated Prime Ideals

16.11 DEFINITIONS. (1) An element a of R is a **zero divisor** of an R-module M provided $ax = 0$ for some $0 \neq x \in M$. We let $zd(M)$ be the set of zero divisors of M. Trivially $zd(M) = zd(E(M))$.

(2) If M is an R-module over a commutative ring R, any prime ideal P of R that is the annihilator of an element of M is called an **associated prime ideal** of M. The set of associated prime ideals of M is denoted $\text{Ass}(M)$.

(3) For any prime ideal P of R, P is the unique associated prime ideal of $E(R/P)$, and consists of $zd(E(R/P))$.

Note: $\text{Ass}(M) = \text{Ass}(E(M))$, and is possibly empty. The set of maximal associated primes is denoted $\text{Ass}^\star(M)$,and by 16.12 below consists of all P maximal in $\mathcal{A}(M)$. Note: $\text{Ass}^\star(M) = \text{Ass}^\star(E(M))$. Also $\text{Ass}(M^n) = \text{Ass}(M)$ and $A^\star(M^n) = \text{Ass}(M)$ for any integer $n \geq 1$.

EXAMPLE. $M = \mathbb{Z} \oplus \mathbb{Z}_2$ has two associated primes; 0 and $2\mathbb{Z}$, hence $\text{Ass}M \neq \text{Ass}^\star M$.

16.11A PROPOSITION. *In a commutative reduced ring R, every prime annihilator ideal $P \neq 0, R$, is a maximal associated prime, hence $Ass(R) = Ass^\star(R)$.*

PROOF. Let $0 \neq x \in P^\perp$. Then $x^\perp \supseteq P$, and if $x^\perp \neq P$, then primeness of P implies $x \in P$, so $x^2 = 0$, contradicting the assumption that R is reduced. Thus, $x^\perp = P$, so $P \in \text{Ass}^\star(R)$. □

REMARK. The author is indebted to T. Y. Lam [98b] for raising the question of when $\text{Ass}(R) = \text{Ass}^\star(R)$, and related questions. See Lam [98a].

Cf. 16.42f. Also 16.53.

16.11B THEOREM. *Let M be a $f \cdot g$ nonzero module over a Noetherian commutative ring R. Then:*

(1) *$Ass(M)$ is finite, nonempty, and contains every prime minimal over $ann_R M$.*

(2) *The union of $Ass(M)$ consists of all zero-divisors on M and 0.*

(3) *There is a chain*

$$0 = M_0 \subset M_1 \subset \cdots \subset M_n = M$$

of submodules and $\{P\}_{i=0}^{n-1}$ of primes of M so that $M_{i+1}/M_i \approx R/P_i \; \forall i$, and each associated prime of M is one of the P_i.

(4) *For each m.c. set S of R,*

$$Ass_{RS^{-1}}(MS^{-1}) = \{PRS^{-1} | P \in Ass(M), \text{ and } P \cap S = \emptyset\}$$

PROOF. See Eisenbud [96], p. 89, Theorem 3.1 and p. 93, Prop. 3.7. □

16.11C EXAMPLE. Let R be a commutative Noetherian ring, with classical quotient ring Q having non-nilpotent Jacobson radical J. Then Q is semilocal Kasch and J is the intersection of all the maximal associated prime ideals, whereas, by Prop. 2.36, the maximal nilpotent ideal N is the intersection of all the minimal prime ideals of Q. By (1) of Theorem 16.11B, there exists an associated prime, a minimal prime, that is not a maximal ideal, that is, $\text{Ass}(Q) \neq \text{Ass}^\star(Q)$. It follows that $\text{Ass}(R) \neq \text{Ass}^\star(R)$. For an explicit example see 16.53.

16.12 THEOREM. *If R is a commutative ring, then (1) any maximal element P of $\mathcal{A}(E)$ is a maximal associated prime ideal for any R-module E; (2) if E is indecomposable injective, then any maximal element P of $\mathcal{A}(E)$ is unique, and $E = E(R/P)$; (3) for any prime ideal $P, E(R/P)$ is canonically an injective R_P-module and cogenerates R_P.*

PROOF. For (1), see 16.10 and consult McAdam [70] for a more general result. (Also see Kaplansky [74], Theorem 6, and Exercise 29 on p.66.) For (2), see 16.11(3); also see Sharp and Vámos [71], p.52, 2.31. For (3) see the author's Lectures [82a], p.50, 11.3.1.

16.13 REMARKS. (1) Any maximal element P of $\mathcal{A}(E)$ is a point annihilator, hence a maximal element of $\mathcal{A}_p(E)$, and by 16.10, conversely. (2) Classically it is known for any ring A that $Q(A)$ is a local ring iff the set of zero divisors of A is a prime ideal P. (See 16.6.) In this case $Q(A) = A_P$. This is indeed the case whenever A is a **uniform ring** (= 0 is an irreducible ideal). Furthermore, by a theorem of the author [96a], then $Q(A)$ is Artinian iff A satisfies acc$\perp$, the acc on annihilator ideals (and then $Q(A)$ is QF); (4) For any R-module M, $E(M)$ and M have the same associated primes.

16.14A PROPOSITION. *If M is an uniform R-module over a commutative ring R, then: (1) the set $zd(M)$ of zero divisors of M is a prime ideal P, and (2) $E = E(M)$ is canonically an R_P-module. If E is Σ-injective then: (3) there exists $x \neq 0$ in E so that $xP = 0$, equivalently, $R/P \hookrightarrow E$; in this case (4) P is the unique associated prime ideal of M, E is the minimal injective cogenerator for R_P; and (5) R_P is a Noetherian ring.*

PROOF. (1) is an exercise, and (2) follows from the fact that if $s \in R\backslash P$ then injectivity of E implies that s induces a unit $\bar{s}^{-1}$ in $A = \text{End}_R E$; (3) is a corollary of Lemma 16.10 and the well-known theorem of the author [66a] (also see 3.7A and 3.14) to the effect that Σ-injectivity of E is equivalent to acc on the set $\mathcal{A}(E)$; (4) follows from (2) and (3) of Theorem 16.12 and (5) is a corollary, since any (right) ideal I of R belongs to $\mathcal{A}(E)$ for any cogenerator E for R. See Theorem 3.15C. □

REMARK. If $E(R)$ is $\sum$-injective, then $|\text{Ass } R| < \infty$. Cf. Cailleau's Theorem 3.14, and Beck's Theorems 3.15A, B and C. Also see 16.17.

16.14B THEOREM. *Assume the notation of (1) of Proposition 16.14A, and assume that $\mathcal{A}p(E)$ satisfies the acc. Then $P = zd(M)$ is the unique associated prime ideal of M.*

PROOF. For any $0 \neq x \in M$, let P_x be a maximal annihilator containing $\text{ann}_R x$. Then $P = zd(M)$ is the union of the P_x's. However, if $0 \neq y \in M$, by uniformity of M, there exists $0 \neq z \in xR \cap yR$. Moreover, $\text{ann}_R z \supseteq \text{ann}_R x$ so $\text{ann}_R z = \text{ann}_R x$ by maximality of $\text{ann}_R x$. Since $\text{ann}_R x \supseteq \text{ann}_R y\ \forall y$, then $P_x = \text{ann}_R x = P$. Now apply 16.10 and 16.11(3). □

16.15 PROPOSITION (FAITH [91B], 3.8). *If $S = \{L_\alpha\}_{\alpha\in\Lambda}$ is a nonempty set of maximal associated primes of a module M over a commutative ring R such that $\{W_\alpha\}_{\alpha\in\Lambda}$, where $W_\alpha = ann_M L_\alpha\ \forall\alpha$, is a maximal independent set of annihilator submodules of M, then S is the set of all maximal associated primes of M.*

PROOF. Let $L \in \text{Ass}^\star(M)$, and let $W = \text{ann}_M L = {}^\perp L$. Then $L = \text{ann}_R W = W^\perp$, in fact $L = w^\perp$ for any $0 \neq w \in W$. By maximality of $\{W_\alpha\}_{\alpha\in\Lambda}$,

$$W \cap \sum_{\alpha\in\Lambda} W_\alpha \neq 0$$

hence, there exists $0 \neq w \in W$ and finitely many elements $w_{\alpha_i} \in W_{\alpha_i}$, $i = 1, \dots, n$, so that

$$w = w_{\alpha_i} + \cdots + w_{\alpha_n}.$$

But this implies that

$$\cap_{i=1}^{n} w_{\alpha_i}^{\perp} = \cap_{i=1}^{n} L_{\alpha_i} \subset L = w^{\perp}.$$

Since L is prime, then by 16.1 there exists at least one $L_{\alpha_{i_0}} \subset L$, and then $L_{\alpha_{i_0}} = L$ by maximality.

16.16 DEFINITION. If M is a right R-module, and if $E(M) = \oplus_{\alpha\in\Lambda} E_\alpha$ is a direct sum of indedecomposable injective modules $\{E_\alpha\}_{\alpha\in\Lambda}$, then we define the **uniform or Goldie dimension of** M as $d(M) = |\Lambda|$, which is unique by the Krull-Schmidt-Azumaya Theorem (8.As). This definition agrees with Definition 16.9B (for finite Goldie dimension) in case $|\Lambda| < \infty$. Furthermore, $E(M)$ coincides with $H = \oplus_{\alpha\in\Lambda} E(U_\alpha)$, where $U_\alpha = E_\alpha \cap M$ is a uniform submodule of M.

16.17 THEOREM. *If M is a module over a commutative ring R, then*

$$d(M) \geq |Ass^{\star}(M)|.$$

If $E(M)$ is directly decomposable, then $d(M) \geq |Ass(M)|$.

PROOF. The first statement follows from Theorem 16.15. In the second statement, we may assume that $E(M)$ has the form H in the above remark. Thus, if $P \in \text{Ass}(M)$, then by the Krull-Schmidt-Azumaya theorem cited above, $E(R/P) \approx E(U_\alpha)$ for some $\alpha \in \Lambda$. This shows that $|\Lambda| = d(M) \geq |\text{Ass}(M)|$. □

16.18 COROLLARY. *If M has finite Goldie dimension $d(M)$, then $d(M) \geq |Ass(M)|$.*

REMARKS. (1) Let $A = k\langle X\rangle$ be the ring of power series in an infinite set X of variables over a field k, and let R denote A modulo the ideal generated by all xy, $x, y \in X$. Then R has infinite Goldie dimension but $|\text{Ass}(R)| = 1$, that is, Goldie dimension is not in general a good upper bound for $|\text{Ass}(R)|$; (2) If M is an R-module with Krull dimension, then M is (quotient) finite dimensional hence $|\text{Ass}M| < \infty$; (3) If $E(M)$ is Σ-injective, then $E(M)$ is completely composable by Cailleau's Theorem 3.14. This happens by Theorem 3.7A if R satisfies the acc on annihilators of subsets of $E(M)$, e.g. when R is Noetherian.

Chain Conditions on Annihilators

16.19 THEOREM (FAITH [66A]). *If M is a right R-module, then the set $\mathcal{A}_r(M, R)$ of right ideals of R that are annihilators of subsets of M satisfies the acc iff for every right ideal I there is a $f \cdot g$ right ideal $I' \subseteq I$ such that*

$$ann_M I = ann_M I'.$$

PROOF. See also my *Algebra II*, p. 112, 20.2A.

The **countermodule** of a right R-module M is the left A-module M, where $A = \text{End}M_R$.

16.20 DUAL THEOREM. *A right R-module M satisfies the dcc on $\mathcal{A}_r(M,R)$ iff M satisfies acc on the set $\mathcal{A}_\ell(M,R)$ of counter submodules that are the annihilators of subsets of R, and iff for every $S \in \mathcal{A}_\ell(M,R)$, there is a $f \cdot g$ counter submodule $S' \subseteq S$ such that*

$$ann_R S = ann_R S'.$$

PROOF. Same proof as the Theorem, *mutatis mutandis.*

16.21 COROLLARY. *If M_R is quasi-injective right R-module, then $\mathcal{A}_r(M,R)$ satisfies the dcc iff the countermodule M is Noetherian.*

PROOF. By (1) in the proof of 16.9C, every $f{\cdot}g$ counter submodule of M belongs to $\mathcal{A}_\ell(M,R)$, hence the acc in $\mathcal{A}_\ell(M,R)$ implies that M is left Noetherian over $A = \mathrm{End} M_R$. The converse is obvious. □

16.22 DEFINITION. An injective right R-module E is called **Δ-injective** provided that $\mathcal{A}_r(E,R)$ satisfies the dcc.

16.23 REMARK. The Teply-Miller Theorem 3.10 implies that every Δ-injective module is Σ-injective. (See my Dekker Lectures [82a] for another proof.) By the proof of 16.21, this also implies the implication

$$\text{dcc in } \mathcal{A}_r(M,R) \Rightarrow \text{the countermodule } M \text{ has finite length}$$

for quasi-injective M_R.

16.24 THEOREM (FAITH [66A]). *A ring R has acc⊥ iff to each right ideal I there is a $f \cdot g$ ideal $I' \subseteq I$ such that ${}^\perp I = {}^\perp(I')$.*

PROOF. Immediate corollary of Theorem 16.19. □

16.25 THEOREM. *If M is an R-module over a commutative ring R, and if R satisfies the dcc on anihilators of finite subsets of M, then $|Ass^\star(M)| < \infty$.*

PROOF. If $P_i \in \mathrm{Ass}^\star(M)$, then $P_i = \mathrm{ann}_R x_i$ for some $x_i \in M$. Since

$$\mathrm{ann}_R\{x_1, \dots, x_n\} = P_1 \cap P_2 \cap \dots \cap P_n = L_n$$

by the dcc on finite annihilators in R, there exists n such that L_n is minimal in the set S of all such finite intersections. But if $P \in \mathrm{Ass}^\star(M)$, then $P \cap L_n \in S$, hence $P \cap L_n = L_n$, that is, $P \supseteq L_n$. Since P is prime, and P contains the product $P_1 P_2 \dots P_n$, then P contains one of the P_i, say $P \supseteq P_{i_0}$. By definition of maximal associated primes, then $P = P_{i_0}$, that is, $P_1, \dots, P_n$ are the only maximal associated primes of M. □

REMARK. Under the same assumptions the proof of 16.25 shows that there is a finite subset $P_1, \dots, P_n$ of $\mathrm{Ass}(M)$ such that every associated prime P contains some P_{i_0}, so there are just finitely many minimal associated primes (see 16.54).

An annihilator right ideal I is a **finite** (resp. **point**) **annihilator** provided that $I = X^\perp$ for some finite subset (resp. for some element) X of R.

16.26 COROLLARY. *If R is a commutative ring, and if R is either (Goldie) finite dimensional, or has dcc on finite annihilators, then $|Ass^\star R| < \infty$.*

PROOF. Corollary 16.17 and Theorem 16.25. □

Semilocal Kasch Quotient Rings

As defined in Chapter 6, **sup.**6.32A, a commutative ring R has the *zero intersection property for annihilators* (= zip) provided that to every faithful ideal I there corresponds a $f\cdot g$ faithful ideal $I_1 \subseteq I$. Any commutative dcc$\perp$ ring R is zip, by Zelmanowitz [76b], who initiated the concept of zip rings (but not the terminology).

As defined in 6.38 (5)ff, a ring R is McCoy provided that every $f\cdot g$ faithful ideal contains a regular element. Any polynomial ring $R = A[X]$ over any ring A is McCoy (see, e.g. Huckaba [88] where McCoy rings are called *rings with Property A*).

A commutative ring R has $\mathcal{E}_{\max}$ (= enough maximal annihilators) provided that every annihilator ideal I is contained in a maximal associated prime (= maximal annihilator) ideal. Clearly, $\mathcal{E}_{\max}$ is equivalent to the dual concept $\mathcal{E}_{\min}$. We let **accpa** denote the acc on point annihilators.

We note a corollary of 16.12.

16.27 PROPOSITION. *A commutative accpa ring R has $\mathcal{E}_{\max}$, hence $\mathcal{E}_{\min}$.*

16.28A THEOREM. *A commutative ring R has Kasch quotient ring $Q(R)$ iff R is zip McCoy. Furthermore, over any zip ring R, the polynomial ring $R[X]$ in any set of variables X has Kasch quotient ring.*

PROOF. See the author's paper [91b] theorems 1.2 and 1.4.

16.28B COROLLARY. *A commutative ring R is zip iff the polynomial ring $R[x]$ in a variable x has Kasch quotient ring.*

An ingredient of the proof of the next theorem is the same as in the Noetherian case, namely McCoy's Theorem 16.1: an ideal I that is contained in a finite union of prime ideals must be contained in one of them.

16.29A THEOREM. *The f.a.e.c.'s on a commutative ring R with $Q = Q(R)$.*
1. *Q is semilocal Kasch.*
2. *R has $\mathcal{E}_{\max}$, and $Ass^\star R$ is finite.*
3. *R is zip and $Ass^\star R$ is finite.*
4. *Q is semilocal and rad Q is a finite annihilator.*

PROOF. See Theorem 3.6 of the author's paper [91b].

16.29B REMARK. 16.29A corrects Theorem 3.6, *loc.cit.*, where the Ass(R) is erroneously defined as Ass$^\star(R)$.

We abbreviate "annihilator right ideal" by right **annulet**, and "maximal (minimal) right annulet" by **maxulet** (**minulet**). See 2.37Es.

16.30 THEOREM. (*Ibid.*). *Any right zip ring R has right $\mathcal{E}_{\min}$, hence left $\mathcal{E}_{\max}$.*

PROOF. Let A be a right annulet $\neq 0$. If A does not contain a right minulet, then there exists an infinite chain

$$A \supset A_1 \supset \cdots \supset A_n \supset \cdots$$

of right annulets $A_n \neq 0$. Moreover, $B = \bigcap_{n=1}^{\infty} A_n$ is a right annulet. Since R is right zip, then $B \neq 0$. For suppose that $B = 0$, and that $L = \cup_{n=1}^{\infty} {}^{\perp}A_n$. Then $L^{\perp} = 0$, so $L_1^{\perp} = 0$ for a finitely generated ideal $L_1 \subseteq L$. Then $\exists n$ with $L_1 \subseteq {}^{\perp}A_n$, and consequently we have that

$$0 = L_1^{\perp} \supseteq ({}^{\perp}A_n)^{\perp} = A_n$$

so $A_n = 0$, a contradiction. Thus B is an annulet *not* $= 0$. By transfinite induction we may construct chains of nonzero right annulets $\{A_\alpha\}_{\alpha\in\Lambda}$ of every larger cardinality with nonzero intersections, involving a contradiction. □

Acc⊥ Rings Have Semilocal Kasch Quotient Rings

16.31 THEOREM (FAITH [91B]). *If R is a commutative acc⊥ ring, then R has just finitely many maximal associated primes $P_1, \dots, P_n$, and $Q = Q(R)$ is a semilocal Kasch acc⊥ ring with just the maximal ideals $P_1Q, \dots, P_nQ$.*

PROOF. By 16.26, $|\text{Ass}^\star R| < \infty$, and by 16.30, R has $\mathcal{E}_{\max}$, hence Q is semilocal Kasch by 16.29. Since Q is semilocal Kasch, then the set $M_1, \dots, M_n$ of maximal ideals is finite, and are associate primes. By the 1-1 correspondence (via contraction) between annulets of Q and those of R, then $P_i = M_i \cap R, i = 1, \dots, n$, are the maximal associated primes of R. □

REMARKS. (1) In an exercise in Lambek's book [68],p.113 (attributed to L. Small), it is proved that in an acc⊥ ring R every dense ideal contains a regular element, i.e., $Q(R)$ is Kasch; (2) Cf. 2.37G.

16.32 THEOREM. *If a commutative ring R has finite Goldie dimension, the following conditions are equivalent:*

1. *$Q = Q(R)$ is Kasch.*
2. *R has $\mathcal{E}_{\max}$.*
3. *R is zip.*
4. *R has $\mathcal{E}_{\min}$.*

In this case Q is semilocal and R is McCoy.

PROOF. By Theorem 16.26, $|\text{Ass}^\star R| < \infty$, and then (1) – (4) are equivalent by theorems 16.27 and 16.29.

REMARK. By 16.10, the acc on point annihilators implies $\mathcal{E}_{\max}$.

16.32A COROLLARY. *A finitely embedded commutative ring R has semilocal Kasch $Q(R)$.*

PROOF. R is zip, and evidently finite dimensional.

16.32B COROLLARY. *If R is a commutative finite dimensional zip ring, then $Q(R[X])$ is semilocal Kasch.*

PROOF. *Ibid.*, 4.5, p.1883.

Beck's Theorem

Beck defined **Noetherian depth, n.d.$_R R$**, of a ring R via the length d of a minimal injective resolution for which all the injectives are Σ-injective and similarly for n.d.$_R M$ of an R-module M.

16.33 THEOREM (BECK [72]). *The following conditions on a commutative ring R are equivalent:*

(1) *$n.d._R R \geq 0$.*
(2) *R has a flat embedding in a Noetherian ring.*

(3) $|Ass\ R| < \infty$, *every* $P \in Ass\ R$ *is Noetherian and*

$$\cup(P \in Ass\ R)$$

is the set of zero divisors of R.
Similarly, $n.d._R M \geq 0$ *implies* $|Ass_R M| < \infty$.

16.34 REMARKS. (1) By (2) of 16.33, any ring with $n.d._R R \geq 0$ must have acc$\perp$, hence $Q(R)$ is semilocal Kasch by 16.31; (2) The effect of the last assertion in (3) of 16.33 is that R has $\mathcal{E}_{\max}$. Thus, by 16.29, this could be replaced by: *R is zip.*

ACC on Irreducible Right Ideals

We now consider a condition denoted **right acci** (resp. accsi), namely the acc (resp. dcc) on irreducible right ideals, encountered previously in 8.5A ff. Right **accsi** denotes the acc on subdirectly irreducible (= SDI) right ideals.

16.35 COROLLARY. *If R satisfies right acci (resp. dcci), then every indecomposable injective right R-module E satisfies dcc (resp. acc) on cyclic counter submodules. Furthermore, then any maximal element P of $Ap(E)$ is an associated prime, is unique, and $E = E(R/P)$.*

PROOF. See 16.9C, 16.11(3), 16.12, and Theorem 16.14B. □

16.36 THEOREM. *If R is a commutative acci ring, then*

$$P \mapsto E(R/P)$$

is a 1-1 correspondence between prime ideals P and indecomposable injectives.

PROOF. Same. Also see 16.43. □

Cf. Facchini's Theorem 9.37.

16.37 COROLLARY. *An acci Prüfer domain is strongly discrete.*

PROOF. This follows from Facchini's theorem 9.37 and Theorem 16.36.

Nil Singular Ideals

The **right singular ideal of a ring** R is the set

$$Z_r(R) = \{x \in R \mid x^\perp \subseteq' R\}$$

where $A \subseteq' B$ denotes A is an essential submodule of B. $Z_r(R)$ is an ideal, and generally *not* $= Z_\ell(R)$.

The next proposition, derived from Goldie's work, is explicit in the author's Springer Lectures [67].

16.38 PROPOSITION. *If R satisfies the acc on point annihilators $x^\perp$ $\forall x \in Z_r(R)$, then $Z_r(R)$ is a nil ideal.*

PROOF. Let $x \in Z_r(R)$ and choose n so that $(x^n)^\perp = (x^{n+1})^\perp$. If $x^n \neq 0$, then $Y = (x^n)^\perp \cap (x^n) \neq 0$ hence let $0 \neq y = x^n a \in Y$. Then $x^n y = x^{2n} a = 0$, hence $a \in (x^{2n})^\perp = (x^n)^\perp$, that is, $y = x^n a = 0$, a contradiction. Thus, $x^n = 0$, so $Z_r(R)$ is nil. □

Primary Ideals

16.39 PROPOSITION. *If I is an irreducible ideal, and P/I is the set of zero divisors of R/I, then I is primary iff $P = \sqrt{I}$, equivalently, $Q(R/I) = R/I_{P/I}$ has nil radical.*

PROOF. Immediate from 2.25(3).

16.40 THEOREM. *An irreducible ideal I of a commutative ring R is primary under any one of the conditions:*

(1) *R/I satisfies the acc on point annihilators.*

(2) *R/I is an acci ring.*

PROOF. We may assume $I = 0$. Since R is uniform, $Z_r(R) = Z(R)$ is the set P of zero divisors which is a prime ideal. Thus (1) implies P is nil so R (that is, 0) is primary by Proposition 16.39; (2) Since R is uniform, $R/x^{\perp} \approx xR$ is uniform for all $x \in R$, hence $x^{\perp}$ is irreducible. Thus, acci implies the conditions (1), so (2) follows from (1). □

16.41 COROLLARY. *A subdirectly irreducible ideal I of a commutative accsi ring R is primary.*

PROOF. Same proof as theorem 16.40 (1) and (2). □

If a ring R is Kasch then every maximal ideal M is an annihilator, equivalently an associated prime ideal (equivalently $R/M \hookrightarrow R$).

16.42 PROPOSITION. *A local ring R_P at an associated prime ideal P is Kasch.*

PROOF. Suppose P is an associated prime ideal, and say $P = x^{\perp}$ for $x \in R$. Then $PR_P = (x/1)^{\perp}$ in R_P. This follows since $(x/1)^{\perp} \supseteq PR_P$, and $\neq R_P$ since this would imply the existence of $s \in R/P$ so that $sx = 0$. But this is impossible since $x^{\perp} = P$. Thus, R_P is Kasch. □

REMARKS. (1) Lam [98b] pointed out that the converse of Theorem 16.42 is false, but does hold assuming that P is $f \cdot g$, e.g. it holds for Noetherian R. (This is an exercise in Lam's new book [98a].)

(2) A good counterexample (also suggested by Lam) to the converse of 16.42 is the direct product $R = \mathbb{Q}^{\omega}$, and a maximal ideal P containing the direct sum $\mathbb{Q}^{(\omega)}$. Thus, $P^{\perp} = 0$, while R_P is a field, hence Kasch. (Cf. Kaplansky's Theorem 3.19B.)

16.43 THEOREM. *If R is an acci (resp. accsi) ring, and I is an irreducible (resp. SDI) ideal, then the set P/I of zero divisors of R/I is the unique (associated) prime ideal of R/I and*

$$Q(R/I) = (R/I)_{P/I}$$

is Kasch.

PROOF. Immediate from 16.40, 16.41 and 16.42. Also see 2.32. □

16.44 REMARK. The proof shows that the conclusion of Theorem 16.43 holds assuming that R/I satisfies the *acc* on point annihilators.

16.45 THEOREM. *If R_P is Noetherian (or an acc$\perp$ ring) for all prime ideals P, then $Q(R/I)$ is QF and I is primary for all irreducible ideals I.*

PROOF. $Q(R/I) = R_P/I_P$ is Noetherian (resp. an acc$\perp$ ring) and irreducible, hence a QF ring, so the result follows by the proof of 16.44. $\square$

A right R-module M is a **chain module** if the set of submodules is linearly ordered by inclusion, equivalently, every submodule $S \neq M$ is irreducible. It follows then that a chain module M has the acc (resp. dcc) on irreducible submodules iff M is Noetherian (resp. Artinian). However, the same is true for a chain module M that satisfies the acc (resp. dcc) on subdirectly irreducible (= SDI) submodules, a result that is a corollary of the next proposition.

16.46 PROPOSITION. *If M is a chain R-module, then between two different submodules there lies a subdirectly irreducible submodule.*

PROOF. If $M_1 \subset M_2$ are submodules and if either M_1 or M_2 is SDI, there is nothing to prove, hence assume M_1 and M_2 are not SDI. Now, by Birkhoff's theorem, 2.17C, every submodule $\neq M$ is intersection of SDI submodules, hence suppose

$$M_1 = \cap_{\alpha \in A} S_\alpha$$

where S_α is SDI. Since M is a chain module, either $S_\alpha \supseteq M_2$ or $S_\alpha \subseteq M_2$, hence there is a cofinal subset of $\{S_\alpha\}_{\alpha \in A}$ lying inside M_2. This proves the proposition. $\square$

16.47 COROLLARY. *A chain module M has acc (dcc) on SDI submodules iff M is Noetherian (Artinian). Hence a valuation ring R has accsi (resp. dccsi) iff R is Noetherian (Artinian).*

16.48 EXAMPLE OF A NON-NOETHERIAN LOCAL ACCSI RING. Let $R = k[\overline{x}_1, \dots, \overline{x}_n, \dots]$ be the epimorphic image of the infinite polynomial ring

$$k[x_1, \dots, x_n, \dots]$$

modulo the ideal generated by all $\{x_i x_j\}_{i,j=1}^{\infty}$. As Vámos observed, R is a local SISI ring over which every SDI ideal I has colength ≤ 2, that is $|R/I| \leq 2$, hence R has accsi, but R is not Noetherian. This also holds for any non-Noetherian local ring (R, m) with $m^2 = 0$.

16.49 REMARKS. (1) The example in 1.10 also is an acci ring; (2) a submodule S of a module M is a **waist** if every submodule N, either $N \supseteq S$ or $S \supseteq N$. The substance of Proposition 16.46 is that every waist $S \neq 0$ of a module M contains a SDI submodule.

Characterization of Noetherian Modules

16.50 THEOREM (FAITH [98B]). *A right R-module M is Noetherian iff M is quotient finite dimensional (= q.f.d.) and satisfies the acc on subdirectly irreducible submodules (= accsi).*

PROOF. The necessity is trivial. Conversely, if M is q.f.d., by Camillo's theorem 5.20B, for any submodule A there is a $f \cdot g$ submodule S so that A/S has no maximal submodules, equivalently, $\text{rad}(A/S) = A/S$. Thus, if M is not Noetherian we may suppose $A \neq S$, an assumption that will lead us to a contradiction.

Choose $x \in A\backslash S$, and let B be a submodule $\supseteq S$ maximal w.r.t. $x \notin B$. Then, as in Birkhoff's Theorem 2.17C, B is a subdirectly irreducible (= SDI) submodule, that is, $\overline{M} = M/B$ has essential simple socle $\overline{x}R$, where $\overline{x}$ is the image of x in $\overline{M}$. Since $x \notin B$ then $A + B \neq B$. Note that $\overline{A} = (A + B)/B$ is not $f \cdot g$, in fact has no maximal submodule, otherwise A would have a maximal submodule. This means we can repeat this procedure in $\overline{M}$ and choose a $f \cdot g$ submodule $\overline{S}_1$ so that $\overline{A}/\overline{S}_1$ has no maximal submodule, e.g. take $\overline{S}_1 = \overline{x}R$, that is, $S_1 = B + xR$. Then, we choose an element $\overline{x}_1 \in \overline{A}\backslash\overline{S}_1$ that is, $x_1 \in (A+B)\backslash S_1$, where $A+B \supset S_1 = B+xR$. Next, choose a SDI submodule $B_1 \supset B$ maximal w.r.t. $x_1 \notin B_1$.

An evident induction establishes an infinite ascending chain of SDI submodules, contrary to assumption. This contradiction establishes the theorem. □

Camillo's Theorem

The proof of Theorem 16.50 is patterned after the proof of:

16.51 CAMILLO'S THEOREM [75]. *If R is a commutative ring, then R is Noetherian iff R/I is Goldie for every ideal I.*

REMARK. Theorem 16.50 actually implies 16.51 since R is q.f.d., and furthermore, R is an accsi, in fact acci, ring by the proof of Theorem 16.40 above.

16.52 Historical Note: Theorem 16.50 originally was submitted in November 1997 to the "Shorter Notes Section" of the Proceedings of the A.M.S. When I inquired of Professor Ken Goodearl in January 1998 as to its status, in an e-mail Ken told me it was "too good to be true" (actual quote). However, the referee's report implied that the converse statement was the case: *it was too true to be good*; in other words, it did not fit the PAMS criteria for acceptance! Ken resolved the contretemps by accepting it under his aegis as Communicating Editor of *Communications in Algebra.* I accepted to show my respect for *Communications in Algebra* as a premier mathematics journal.[1] As Professor Jacobson said in regard to my paper [61d] (my first paper at the Institute—see "Snapshots"), it is a sad day for mathematics when a paper is rejected because the proof is too short (and, I might add, even more so if it is "too good to be true!") Besides, since most of the ideas in the proof of 16.50 I owe to Shock and Camillo, and to some of their deep theorems, the fact is that the proof isn't really all that short!

16.53 NOETHER'S EXAMPLE: WHEN $\text{Ass}^{\star}R \neq \text{Ass } R$. *The primary decomposition for* (0) *in the ring* $R = \mathbb{C}[x, y]/(xy, y^2)$ *has an embedded prime, hence a minimal (associated) prime ideal not a maximal associated prime.* (I am indebted to T. Y. Lam [98a] for suggesting this illuminating example.)

16.54 Open Questions: (1) Does acc$\perp$ in R imply $|\text{Ass } R| < \infty$? Since acc $\perp \Leftrightarrow$ dcc $\perp$ in commutative R, then acc$\perp$ implies that $|\text{Ass}^{\star}R| < \infty$ by 16.26 (Cf. 16.11A);

(2) Same question as (1) assuming R is Goldie.

(3) Is $d(M) \geq |\text{Ass}M|$? See Theorem 16.17.

---

[1] Another aspect of the "criteria" was that the paper belonged in a Journal specializing in algebra! I do believe that Ken Goodearl would have accepted the paper but that he would not overrule the referee.

CHAPTER 17

Dedekind's Theorem on the Independence of Automorphisms Revisited

The Galois Theory of a commutative field K contains Dedekind's theorem[1] on the (linear) independence over K of automorphisms $g_1, \dots, g_n$ as functions $K \to K$, i.e., it is impossible for elements $k_i \in K, i = 1, \dots, n$ to exist that satisfy the identity

$$(R\,K) \qquad \sum_{i=1}^{n} k_i g_i = 0 \quad \text{on} \quad K \text{ with } k_i \text{ not all zero,}$$

or, equivalently, impossible that

$$(R'K) \qquad \sum_{i=1}^{n} k_i g_i(x) = 0, \quad \forall x \in K \quad \text{with some} \quad k_i \neq 0.$$

This implies one inequality, $[K : K^G] \geq n$, needed for the important dimension relation $[K : K^G] = (G : 1) = |G|$. (See Artin [55]), henceforth referred to as Artin.) Redoing the proof of Artin for an arbitrary commutative ring yields the:

17.0 THEOREM (FAITH [82C]). *(Dependence Theorem for a Commutative Ring K) Every dependence relation $(R\,K)$ implies*

$$(DR\,K) \qquad k(g-1) = 0 \quad \text{on } K$$

for some $0 \neq k \in K$ and $1 \neq g$ belonging to the group G of automorphisms of K. Expressed otherwise, g induces the identity in the factor ring $K/k^{\perp}$, where the exponent denotes annihilation. Moreover, if G is a torsion group, then G is dependent iff for some $1 \neq g \in G$ either the fixring K^g, or the group Y of elements with zero (g)-trace, contains a non zero ideal of R (Cf. 17.8B). *Furthermore, if R is reduced, then K^g must contain a nonzero ideal* (Cf. 17.8).

COROLLARY. *Automorphisms belonging to a group G are independent over a local ring K if $g = 1$ is the only $g \in G$ that induces the identity (id.) in the residue field $\overline{K}$.*

Conventions

Linear independence of automorphisms is assumed in the Galois Theory for commutative rings of Auslander-Goldman [60] and is also the starting point for that in Artin. Usually, we drop the modified "linear" and speak of independence corresponding to a group G of automorphisms. If $G = (g)$, set $K^g = K^G$. The fact

[1](For Nathan Jacobson and Sam Perlis) The first part of this chapter was dedicated to "Jake" in [82c] on his 70th birthday, and the last part to "Sam" in [87].

that each automorphism g of a commutative ring K has a unique extension g^{ex} to the quotient ring $Q = Q_{c\ell}(K)$, shows that not only does Dedekind's theorem hold for integral domains, but that a set G of automorphisms of K is independent over K iff the corresponding set G^{ex} is independent over Q.

Résumé of Results

For any commutative ring K when G is a torsion group, then $(DR\ K)$ implies that K is n-radical over $K^{g,k} = K^g + k^{\perp}$ in the sense that $x^n \in K^{g,k}$, where $n = |(g)|$; and, dually, K is n-torsion over $K^{g,k}$. Thus, denying the conclusion, (e.g. by requiring that G be a torsion group with $|(g)|$ a unit) yields independence theorems for automorphisms over K.

Furthermore, by a theorem of Kaplansky [51] on division rings (see 2.9B), if K is a local ring which is not extended from its radical J by some Galois subring K^g with $1 \neq g \in G$, then $(DR\ K)$ implies that $\overline{K} = K/J$ has prime characteristic p, that $|(g)| = p^e$, and that $\overline{K}$ is purely inseparable over $\overline{K}^g$ of exponent e. (Finiteness of $\overline{K}$, another possibility offered by Kaplansky's theorem is ruled out by the fact that $\overline{K}$ is p^e-radical over $\overline{K}^g$.)

Similar theorems hold for the structure of the classical quotient ring Q of K, e.g. if $|(g)|$ is a non zero-divisor of K, then $(DR\ K)$ implies $Q = Q^{g^{ex},k}$, where "ex" denotes the "extension of." If Q is a local ring, meaning that the zero divisors of K form an ideal $zd\ K$, and if no $g \neq 1$ induces id. on $K/(zd\ K)$, then the automorphisms are independent. Furthermore, if $(zd\ K)^{\perp} \neq 0$, e.g. when Q is a local Kasch ring not a field, then the converse holds, as the Dependence Theorem and $(DR\ K)$ readily shows, since the radical of Q is annihilated by some $0 \neq k \in K$.

Another type of dependence of an automorphism group occurs when some K^g for $g \neq 1$ contains a nonzero ideal I of K. But, then, if $0 \neq k \in I$, we have

$$\text{(FI)} \qquad kg(x) = g(kx) = kx \qquad \forall x \in K$$

so $(DR\ K)$ holds. Conversely, for a reduced ring K, $(DR\ K)$ implies that K^g contains an ideal $\neq 0$ (Proposition 17.8); in fact, K^g contains $N_g(K)$, the norm of k with respect to (g). Thus, a reduced ring K has dependent automorphism group G iff some K^g with $1 \neq g \in G$ contains a nonzero ideal of K. Furthermore, if a reduced ring K is n-power radical over $K^{g,k}$, where K has prime characteristic n, then $(DR\ K)$ holds when $g \neq 1$, and $k \neq 0$ (Corollary 17.4D).

A third kind of dependence of a group G occurs when G is a torsion group and there exists $1 \neq g \in C$ such that the K^g-submodule $Y(g)$ consisting of all elements with zero (g)-trace contains a nonzero ideal. In fact, by Theorem 17.8B G is dependent iff K^g or $Y(g)$ contains a nonzero ideal, for some $1 \neq g \in G$.

Dependent Automorphisms of Polynomial and Power Series Rings

We can easily classify dependent automorphism groups generated by translations or rotations of a polynomial ring $K = F[x]$ in a single variable, e.g. if g is the translation sending $x \mapsto x + a$ where $0 \neq a \in F$, then (g) is dependent iff some nonzero multiple ma is a zero divisor. (Similarly for a rotation $x \mapsto ax$.) Significantly, the switching automorphisms of $K = F[x_1, \dots, x_n]$ generates independent automorphism groups.

Similar results hold for automorphisms of a power series ring $K = R[[x]] = R\langle x\rangle$. It is known that an automorphism g that leaves the elements of R fixed maps x onto $\sum_{1=0}^{\infty} a_i x^i$, where a_1 is a unit, and a_0 belongs to the ideal $I_c(R)$ consisting of all $a \in R$ for which there exists a homomorphism $R[[x]] \to R$ sending x onto a (see Eakin and Sathaye [80]). The structure of $\mathrm{Aut}_R R[[x]]$ is unknown in general, even when R is a domain (see e.g. Atkins and Brewer [80]). However, when R contains the rational number field, and G is a torsion subgroup of $\mathrm{Aut}_R R[[x]]$, then by Eakin and Sathaye [80], G is conjugate to the subgroup of the **circle group**, where the circle group consists of all "rotations" of finite order, that is, automorphisms sending $x \mapsto cx$ for an n-th root of unity c. Thus, a necessary and sufficient condition for G to be independent is that some primitive n–th root of unity c, for some $n > 1$, be such that $c^m - 1$ is a zero divisor in R, for $1 < m < n$.

Normal Basis

In the original version of this paper (see the footnote, p.243) I "thought" I proved a normal basis theorem for a local ring K with finite independent automorphism group G.[2] Since then S. Endo showed me a "true" theorem: A semilocal ring K has a normal basis $g_1(u), \dots, g_n(u)$ over K^G iff K is free of rank n over K^G, $G = (g_1, \dots, g_n)$ is independent, and the trace map $t_G : K \to K^G$ is onto (Letter of July 1980).

However, under the hypothesis of the Independence Theorem (i.e., no $g \neq 1$ induces id. in $\overline{K}$), K is actually Galois[3] over K^G, hence has a normal basis over K^G (Chase et al [65]; Cf. DeMeyer-Ingraham [71]).

The Dependence Theorem

We begin this section with the Dependence Theorem stated in the introduction.

17.1 Dependence Theorem. *When $(R\ K)$ holds, then there is a "shortest" relation of the form*

$$(SRK) \qquad k(g_i - g_j) = 0 \quad \textit{for some } i \neq j \textit{ and } 0 \neq k \in K$$

and, in fact, there is a relation of the form
(DRK)

$$(DR\,K) = (DR\,K)(k, g) \quad k(g-1) = 0 \textit{ for some } 1 \neq g \in G, \textit{ and } 0 \neq k \in K,$$

that is,

$$(DR'K) = (DR'K)(k, g) \qquad kg(x) = kx \quad \forall x \in K.$$

[2]In a letter of September 1998, S. Endo constructed a counterexample for a regular local ring K of characteristic $p > 0$ dividing n.

[3]This was pointed out to me by F. DeMeyer (Letter of February, 1981).

PROOF. We note that $(SR\ K)$ is obtained from $(R\ K)$ by the familiar technique of replacing x in $(R'K)$ by $x = uy$ for a fixed $u \in K$. First allowing y to range over K, we obtain another $(R\ K)$ of the form

$$(R\ K_u) \qquad \sum_{i=1}^{n} k_i g_i(u) g_i = 0 \qquad \text{on } K.$$

Next, multiply $(R\ K)$ by $g_{i_0}(u)$, where i_0 is any i such that $k_i \neq 0$, and assuming that some $k_i g_{i_0}(u) \neq 0$, subtract the result from $(R\ K_u)$ to obtain a "shorter" relation inasmuch as the coefficient of g_{i_0} is now $= 0$. Thus, in this case, all coefficients must $= 0$, that is, for all i,

$$(DR'K_u) \qquad k_i g_i(u) = k_i g_{i_0}(u)$$

holds. If this is true for all u, then

$$k_i(g_i - g_{i_0}) = 0 \quad \text{on } K,$$

so $(SR\ K)$ holds as asserted. Then $(DR\ K)$ follows by letting $k = g_{i_0}^{-1}(k_i)$ and $g = g_{i_0}^{-1} g_i$. Since the only assumption made on u was that some $k_i g_{i_0}(u) \neq 0$, for some i_0 for which $k_i \neq 0$, then denying the assumption yields $k_i g_j(u) = 0$ for all i, j, which shows that $(DR'K_u)$ holds without restriction completing the proof. □

The Skew Group Ring

The *trivial crossed product* or *skew group ring* $K \star G$ of K and G is the ring consisting of all finite linear combinations $\sum_{i=1}^{n} k_i g_i$, with $k_i \in K$, and $g_i \in G$, with multiplication defined by the formula

$$(kg)(ph) = kg(p)gh \qquad \forall k, p \in K \quad \text{and} \quad g, h \in G$$

and its implications. there is a canonical homomorphism

$$h : K \star G \to \text{End}_F K, \qquad \text{where } F = K^G$$

$$\sigma = \sum_{i=1}^{n} k_i g_i \to \sigma', \quad \text{where } \sigma'(x) = \sum_{i=i}^{n} k_i g_i(x) \quad \forall x \in K.$$

The next result gives equivalent conditions for h to be a monomorphism. As stated, a group of automorphisms of K will be called *dependent (independent) accordingly as its elements are linearly dependent (independent) over* K.

17.1A COROLLARY. *For a group G of automorphisms of K, the following conditions are equivalent:*

1. *K is faithful as a canonical left $K \star G$ module.*
2. *$K \star G \hookrightarrow$ End K_F, canonically, where $F = K^G$.*
3. *G is independent over K.*
4. *Every cyclic subgroup of G is independent over K.*
5. *Every pair $\{1, g\}$ is linearly independent over $K, \forall 1 \neq g \in G$.*

When this is so, then K is a faithful module over the group ring $R = K^G G$ of the ring K^G and the group G.

PROOF. (1) $\Leftrightarrow$ (2) by the preceding remarks, (2) $\Leftrightarrow$ (3) is obvious, and both (3) $\Leftrightarrow$ (4) and (3) $\Leftrightarrow$ (5) by the Dependence Theorem. The last statement follows, since $R \hookrightarrow K \star G$ canonically. □

We record the following curiosity.

17.1B PROPOSITION. *If G is a torsion group, H a finite normal subgroup, and if G induces a dependent group of automorphisms of $L = K^H$, then G is dependent over K.*

PROOF. Let $\overline{G}$ denote the group of automorphisms of L induced by G, that is, $\overline{G} \approx G/H$ canonically), and suppose $(DR\ L)(\overline{g}, k)$ holds, for some $0 \neq k \in L$, so

$$k\overline{g}(y) = ky \qquad \forall y \in L.$$

Then

$$kgT_{(g)}(x) = kT_{(g)}(x) \qquad \forall x \in K$$

defines a dependence relation $(R\ K)$ over K where

$$T_{(g)}(x) = x + g(x) + \cdots + g^{n-1}(x)$$

is the (g)-trace of $x \in K$. □

17.1C REMARK. Regarding Theorem 1: $(DR\ K)$ is equivalent to the statement that the ideal $((1-g)K)$ generated by the set $(1-g)K = \{g(x) - x \mid x \in K\}$ has nonzero annihilator. (Note: in general $(1-g)K$ is not an ideal!). Thus, G is independent iff

$$((1-g)K)^{\perp} = 0 \qquad \forall 1 \neq g \in G. \tag{1}$$

This happens, e.g., if

$$((1-g)K) = K \qquad \forall 1 \neq g \in G. \tag{2}$$

The Induction Theorem

Another formulation of the Dependence Theorem is the:

17.2 INDUCTION THEOREM. *G is independent over K iff no $1 \neq g$ in G induces the identity in $K/k^{\perp}$, for any $0 \neq k \in K$.*

17.3 PROPOSITION. *Assume that $Q = Q_{c\ell}(K)$ is a Kasch local ring. Then an automorphism group G is dependent iff some $1 \neq g \in G$ induces the identity in $K/zd(K)$.*

PROOF. If $g \neq 1$ induces the identity in $K/zd(K)$, then $g(x) - x \in zd(K)\ \forall x \in K$, so $(DR'K)$ holds for any k such that $k^{\perp} = zd(K)$. Conversely, if C is dependent, then $(DR\ K)$ holds for $0 \neq k \in K$ and $1 \neq g \in G$, so $g(x) - x \in zd(K)$ for all $x \in K$, and hence g induces the identity in $K/zd(K)$. □

REMARK. Any commutative acc$\perp$ ring K has semilocal Kasch Q (Theorem 16.31).

Radical Extensions

When $g \neq 1$ and $k \neq 0$ we set $K^{g,k} = K^g + k^\perp$. Another consequence of the Dependence Theorem is:

$$K = K^{g,k} \Rightarrow (DR\ K)$$

$$(\text{For } K = K^{g,k} \Rightarrow g(x) - x \in k^\perp, \text{ so } (DR'K) \text{ holds.})$$

The next theorem shows that $(DR\ K)$ implies that a "close" relationship exists between K and $K^{g,k}$.

17.4 THEOREM ON RADICAL-TORSION EXTENSIONS. *If a dependence relation $(DR\ K)$ holds on K, and if g has finite order n, then K is n-radical and n-torsion over $K^{g,k}$.*

PROOF. A simple calculation shows that $(DR\ K) = kg^i(x) = kx$ for any i, and hence, $k\beta_x = kx^n$, where $\beta_x = N_g(x)$ is the norm with respect to (g), and $n = |g|$. thus, $\beta_x \in K^g$, hence $x^n \in K^g + k^\perp = K^{g,k}$. Similarly, $nx \in K^{g,k}$.

17.4A COROLLARY. *If G is a dependent torsion group over K, such that $|g|^{-1} \in K \ \forall g \in G$, then $K = K^{g,k}$ for some $g \in G, k \in K$.*

17.4B COROLLARY. *If G is a dependent torsion group over K such that $|g| \notin zd(K) \ \forall g \in G$, then $Q = Q^{g^{ex},k}$ for some $g \in G, k \in K$.*

PROOF. For then G^{ex} is a dependent group over Q, and 4.1 applies.

17.4C COROLLARY. *Assume K is not radical over any proper subring. Then a finite group G of automorphisms of K is dependent iff $K = K^{g,k}$ for some $1 \neq g \in G$ and $0 \neq k \in K$.*

Partial Converse to Theorem 17.4

17.4D THEOREM. *If K is a power of p radical over a subring $K^{g,k}$, where K has prime characteristic p, and if K is reduced, then $(DR\ K)$ holds: $k(g-1) = 0$ on K.*

PROOF. If $x \in K$, then for some $e \geq 1$, $x^{p^e} \in K^g + k^\perp$ so $g(x^{p^e}) - x^{p^e} \in K^\perp$, that is, $k(g(x^{p^e}) - x^{p^e}) = 0$. Then $K(g(x) - x)^{p^e} = 0$, hence $(k(g(x) - x))^{p^e} = 0$. Since K is reduced, then $k(g(x) - x) = 0$, so $(DR\ K)$ holds.

17.4E EXAMPLE. Let $K = P[x,y]/(x^2, xy, y^2)$ be the ring of rational functions in two variables x and y over the prime subfield P, modulo the ideal (x^2, xy, y^2). Let $\overline{u}$ denote the image of $u \in P[x,y]$ under the canonical map $P[x,y] \to K$, and let g be the automorphism induced by switching $\overline{x}$ and $\overline{y}$. The fixring F of (g) is the vector subspace over P spanned by $\overline{1}$ and $\overline{x} + \overline{y}$, and the radical is the vector subspace $J = P\overline{x} + P\overline{y}$. The element $\alpha = \overline{x} + \overline{y} \in F$ annihilates J since $\alpha\overline{x} = 0$ and $\alpha\overline{y} = 0$. Since $0 \neq \alpha K = P(\overline{x} + \overline{y}) \subset F$, then

$$\alpha g(k) = g(\alpha k) = \alpha k \quad \forall k \in K,$$

so (g) is dependent.

Kaplansky's Theorem Revisited

We next investigate the situation for a local ring K with residue field $\overline{K} = K/J$ (with radical $K = J$) when $(DR\ K)(g, k)$ holds for g of finite order $n \neq 1$. For then by the Theorem on Radical-Torsion Extensions, K is radical over $K^g + k^\perp$, hence $\overline{K}$ is radical over $\overline{K}^g$. In the event that $\overline{K} \neq \overline{K}^g$, there is a decisive theorem of Kaplansky 2.9B on the structure of the radical extension $\overline{K}$ over $\overline{K}^g$, namely, there are just two possibilities

(KAP 1) $\overline{K}$ is algebraic over a finite field.

(KAP 2) $\overline{K}$ is purely inseparable over $\overline{K}^g$.

Now in our situation, **(KAP 1)** is ruled out by the following lemma.

17.5 LEMMA ON $|g|$. *If $\overline{K} \neq \overline{K}^h$, for all automorphisms $h \neq 1$, then $(DR\ K)$ (k, g) implies that $|g| = p^e$, for $e \geq 1$.*

Before proving this lemma, we pause to state two theorems.

17.6 PURELY INSEPARABLE RESIDUE FIELD THEOREM. *If $\overline{K} \neq \overline{K}^g$ for a local ring K satisfying $(DR\ K)$, then $\overline{K}$ is purely inseparable over $\overline{K}^g$.*

17.7 PERFECT RESIDUE FIELD THEOREM. *If K is a local ring, and if $\overline{K} \neq \overline{K}^g$ for any $g \neq 1$ in a finite automorphism group G, and if $\overline{K}^g$ is a perfect field $\forall g \in G$, then G is independent over K.*

PROOFS. The proof of the Lemma on $|g|$ follows from Kaplansky's theorem and Theorem 4, since the former implies that $\overline{K}$ has characteristic p, and the latter implies that $n\overline{x} = 0\ \forall \overline{x} \in \overline{K}$, so $p \mid n$. Write $n = p^e n_0$, with $(n_0, p) = 1$, and set $h = g^{p^e}$. Now $(DR'K)(k, g) = kg(x) = kx$ and thus $kh(x) = kx \quad \forall x \in K$, that is, $(DR'K)(k, h)$ holds. By hypothesis, $K \neq K^h + J$ when $h \neq 1$, hence the fact that p does not divide $n_0 = h$ implies by Kaplansky's theorem that $n_0 = 1$, so $g^{p^e} = 1$. Since the Radical Extension Theorem implies that $\overline{x}^{|g|} \in \overline{K}^g$, when $\overline{K}^g$ is perfect (i.e., finite as in (KAP 1)), then $\overline{x} \in \overline{K}^g\ \forall \overline{x} \in \overline{K}$, contrary to assumption. This proves the Perfect Residue Field Theorem; and the Purely Inseparable Residue Field Theorem is a restatement employing Kaplansky's Theorem. □

17.6′ PURELY INSEPARABLE RESIDUE THEOREM FOR $Q = Q_{c\ell}(K)$. *If $Q = Q_{c\ell}(K)$ is a local ring which is not extended from its radical $J(Q)$ by a Galois proper subring, that is, if $Q \neq J(Q) + Q^H$ for some automorphism group H, and if K has a dependent automorphism torsion group G, then $\overline{Q} = Q/J(Q)$ is purely inseparable over $\overline{Q}^{g^{ex}}$ for some $1 \neq g \in G$. Furthermore, if $x \in K$, then $x^{p^e} \in K^g + k^\perp \subseteq K^g + zd(K)$, for some $0 \neq k \in K$, where $|g| = p^e$.*

PROOF. This follows Theorem 17.6 (in view of the fact that G is dependent iff G^{ex} is dependent). The last statement is a trivial consequence. □

Reduced Rings

The next proposition is a partial converse of one part of Corollary 17.A(3).

17.8 PROPOSITION. *If K is a reduced ring, and if g has finite order and satisfies the dependence relation $(DR\ K)$, then K^g contains a nonzero ideal of K.*

PROOF. Suppose $(DR\ K)$ holds, so that $k(g-1)=0$ on K, and by 17.4 $k(\beta - x^n)=0$, $\forall x \in K$, where $\beta_x = N_g(x) = \prod_{i=1}^{n-1} g^i(x) \in K^g$. Take $x=k$, $\beta = \beta_k$, so that $k(\beta - k^n)=0$. If K is reduced, then $k^{n+1} = \beta k \neq 0$, hence $\beta = N_g(K) \neq 0 \in K^g$. Now $(DR'K)$ implies that

$$h(k)hg(x) = h(k)h(x) \qquad \forall h \in (g).$$

Since any $y \in K$ has form $y = h(x)$, then $hg = gh$ implies that

$$h(k)g(y) = h(k)y \qquad \forall y \in K.$$

Hence

$$N_g(k)g(y) = N_g(k)y$$

or

$$g(\beta y) = \beta g(y) = \beta y \qquad \forall y \in K$$

with $\beta \in K^g$. Since $g(\beta y) = \beta y\ \forall y \in K$, then βK is a nonzero ideal of $K \subseteq K^g$.

□

17.8A PROPOSITION. *If $(DR\ K)$ holds on K, then either K^g contains a nonzero ideal of K, or else $nk^2 = 0$.*

PROOF. By dualizing the proof of Proposition 17.8 we have

$$\beta g(y) = \beta y \qquad \forall y \in K,$$

where $\beta = T_g(k) = \sum_{i=0}^{n-1} g^i(k) \in K^g$. Then, as in the proof of Proposition 17.8, either $\beta = 0$, or else βK is a nonzero ideal of K contained in K^g.

Now suppose $\beta = T_{(g)}(k) = 0$. By the proof of Theorem 17.4, $k(T_{(g)}(x) - nx) = 0\ \forall x \in K$; in particular, for $x = k$, we have $k(\beta - nk) = 0$, hence $nk^2 = 0$. □

The Role of Ideals in Dependency

17.8B THEOREM. *Let G be a torsion automorphism group. Then G is dependent over K iff for some $1 \neq g \in G$ either the fixring K^g, or the K^g-submodule Y consisting of all $x \in K$ with $T_{(g)}(x) = 0$, contains a nonzero ideal of K.*

PROOF. If G is dependent, then $(DR\ K)$ holds, and by the proof of 17.8A, either K^g contains the nonzero ideal βK, or else $\beta = 0$. Since $kag(x) = kax\ \forall a \in K$, then if K^g does not contain a nonzero ideal, we must have $ka \in Y$ for all $a \in K$, that is $kK \subseteq Y$.

Conversely, if K^g contains an ideal $\neq 0$, then G is dependent by Corollary 1.3. Moreover, if Y contains the ideal kK for some $k \neq 0$, then $T_g(kx) = 0$ for all $x \in K$, and this implies a relation

$$\sum_{i=1}^{n} g^i(k) g^i = 0 \quad \text{on} \quad K \qquad (n = |g|)$$

with nonzero coefficients, so G is dependent. □

Galois Subrings of Independent Automorphism Groups of Commutative Rings Are Quorite

Let R be a commutative ring, G a finite group of automorphisms, let $A = R^G$ be the Galois subring, let G^{ex} denote the canonical extension of G to the quotient ring $Q = Q_{c\ell}(R)$, and $F = Q^{G^{ex}}$. It is easy to see that F is the partial quotient ring of A with respect to the multiplicatively closed subset S of A consisting of all $a \in A$ that are regular in R, that is, $S = A \cap R^\star$, and thus that G is *quorite* in the sense that $Q_{c\ell}(A) = F$ iff R is *torsion free* over A in the sense that $A^\star \subseteq R^\star$. (See the author's paper [76].) Sufficient ring-theoretical conditions for this are: (1) R is reduced (= semiprime, or nonsingular); (2) R is flat over A. For example, (1) happens if R is semihereditary, and (2) when G is a Galois group.

17.9 THEOREM. *If G is an independent finite automorphism group of R, then G is quorite.*

PROOF. Assume the above notation. Let $a \in A^\star$, and I the annihilator ideal of a in R. Let $T_G(x)$ denote the trace of any x in R under G. Since $I \cap A = 0$, evidently $T_G(x) = 0$ for any x in I, and moreover, if r is any element of R, we have then that $xr \in I$, whence $T_G(xr) = 0$, that is,

$$\sum_{g \in G} g(x)g(r) = 0$$

and therefore

$$\sum_{g \in G} g(x)g = 0 \quad \text{on } R.$$

Since G is independent, it follows that $x = 0$, whence $I = 0$, which proves the theorem. □

17.10 EXAMPLE. The converse of the theorem fails. Let R be the direct product of three fields $F_1 \times F_2 \times F_3$, with $F_1 \approx F_2$, and let g denote the extension of this isomorphism to an automorphism of R with Galois subring $R^g \supseteq F_3$. Since R^g contains an ideal, then by 17.8B the group (g) is dependent, but quorite since R is reduced.

17.11 QUESTION. Let G be a group of automorphisms of a non-commutative ring. If G is finite and independent, is G quorite?

If R is an integral domain, then the answer is yes by Theorem 6.30 without assumption of independence, a result generalized by Theorem 12.A to any ring R such that R^G is semiprime without $|G|$-torsion. See 12.B-F for related results.

The assumption that G is independent is the basis of a number of theorems for what are called **strictly Galois extensions** in Onodera and Tominaga [61] and e.g. Nagahara and Tominaga [70].

Let $Q = Q^r_{\max}(R)$ denote the maximal right quotient ring of R, and G^{ex} now denote the group of automorphisms of Q which extends G to Q. The theorem of Kitamura [76] states

$$Q^{G^{ex}} = Q^r_{\max}(R^G)$$

that is, G is *maximally quorite*, if the trace function $R \to R^G$ is *non-degenerate*, that is, does not vanish on any nonzero right ideal.

17.12 COROLLARY. *If R is commutative, and G independent, then G is maximally quorite.*

PROOF. By Theorem 17.8B, if G is independent, then the trace function is non-degenerate. In fact, Theorem 17.8B states that a torsion group G is dependent over R iff for some $g \neq 1$ in G either the fixring R^g contains a nonzero ideal of R, or else the g-trace function is degenerate. □

Automorphisms Induced in Residue Rings (For Sam Perlis on His 85th Birthday)

In this section we report on extensions of results of the previous sections on independent automorphisms.

Let G be a group of automorphisms of a commutative ring K, and let K^g denote the Galois subring consisting of all elements left fixed by every g in G. An ideal M is *G-stable*, or *G-invariant*, provided that $g(x)$ lies in M for every x in M, that is, $g(M) \subseteq M$, for every g in G. Then, every g in G induces an automorphism $\overline{g}$ in the residue ring $\overline{K} = K/M$, and if $\overline{G}$ is the group consisting of all $\overline{g}$, trivially

$$\overline{K}^{\overline{G}} \supseteq \overline{K^G}. \tag{1}$$

When the inclusion (1) is strict, then G is said to be *cleft* at M, or by M, and otherwise G is *uncleft* at (by) M. When G is cleft at all ideals except 0, then G is cleft, and uncleft otherwise.

The main results on uncleft groups are for G locally finite in the sense that orbit number $n(x) = |Gx| < \infty$ for every x in K. Below let $L(G, M)$ be the inverse image of $\overline{K}^{\overline{G}}$ under the canonical map $K \to \overline{K} = K/M$.

17.13A THEOREM (FAITH [87]). *If G is a locally finite automorphism group of K, and if M is a G-invariant ideal, then $\overline{K}^{\overline{G}}$ is radical-torsion over $\overline{K^G}$; that is, if $\overline{x} \in \overline{K}^{\overline{G}}$, and if $n = |Gx|$, then*

$$\overline{x}^n \in \overline{K^G} \quad \textit{and} \quad n\overline{x} \in \overline{K^G}.$$

17.13B COROLLARY. *If G is a locally finite automorphism group of K with unit orbit lengths, or if K is generated by*

$$\{x^{|Gx|} \mid x \in K\}$$

then G is uncleft. Moreover, if M is a maximal ideal such that $\overline{K} = K/M$ has characteristic not dividing $|Gx|$ $\forall x \in L(G, M)$, then G is uncleft at M.

Employing Kaplansky's theorem on the structure of radical extensions of fields in the same way as in 17.6–7, we obtain:

17.14A THEOREM. *If G is locally finite and cleft at a maximal ideal M, then the residue field $\overline{K} = K/M$ has prime characteristic p, and p divides $|Gx|$ for all x in $L(G, M)$ not in $K^G + M$. Moreover,*

$$x^{p^{e(x)}} \in K^G + M,$$

where $e(x)$ is the exponent of p in $|Gx|$.

A subfield B of a field A is *relatively* perfect if A contains no purely inseparable extension of B (other than B).

17.14B COROLLARY. *Let G be locally finite on K, and M a G-invariant maximal ideal. Then:*

1. *If $\overline{K}^g$ is a relatively perfect subfield of $\overline{K} = K/M$, then G is uncleft at M.*
2. *If G has unit orbit lengths (resp. if $|Gx| \notin M$ for all $x \in K$), then G is uncleft (resp. uncleft at M).*

A number of these results are implicit in the author's [82c] (see 17.0–17.7 above), but under various restrictions such $G = (g)$ cyclic, M a point annihilator ideal, and the requirement throughout that G is a linearly independent group of automorphisms, all of which obscure the generality and beauty of the theorems.

Regarding local properties of uncleftness, Example 6 of the author's [87] shows that uncleftness at prime ideals does not imply uncleftness; and Theorem 7, *ibid.*, shows that uncleftness at a G-stable ideal P implies uncleftness of the extended group at the maximal ideal of the local ring at P. As an application, we proved that any Galois group G (in the sense of Auslander-Goldman [60], and Chase et al [65]) is uncleft.

Results also yield other specific information on the nature of Galois groups. It is known that a finite group G of automorphisms of K is a Galois group provided that for every $1 \neq g \in G$ and every maximal ideal M of K there is an element $x \in K$ so that $g(x) - x \in M$ (*ibid.*). Thus, if G is not a Galois group, and if

$$g(x) - x \in M \quad \text{for all} \quad x \in K \qquad [\text{that is, } \overline{g}(M) = 1]$$

then either (1)

$$K = K^g + M \tag{1}$$

or else

(2)
(2a) $\overline{K} = K/M$ has prime characteristic $p \mid n$ and
(2b) $\overline{K}$ is purely inseparable over $\overline{K}^g$ of exponent equal to that of p in n.

This shows that non-Galois groups for commutative rings bear a close resemblance to those for fields in that, excepting for the case (1) where g acts trivially modulo M, inseparability of field extensions is necessitated. (This also shows that a non-Galois group G must have (g)-stable maximal ideal M for some $1 \neq g \in G$.)

Notes on Independence of Automorphisms

By the Cartan-Jacobson Galois Theory for sfields, the dimension of the vector space over a sfield A generated by a group G of automorphisms is equal to

$$[A : A^G] = (G : \mathcal{I}(G))[T(G) : C]$$

where $\mathcal{I}(G)$ is the group of inner automorphisms, and $T(G)$ is the algebra over the center C of A generated by all x such that $I_x \in G$. (See **sup.** 2.7, and Theorem 2.7.)

In particular, distinct coset representatives of G modulo $\mathcal{I}(G)$ are linearly independent over A. Thus, any group of outer automorphisms is independent over A, a result that holds for prime rings A. (See Mihovski [94], Corollary 10.)

Montgomery [79] proved that the skew group ring $K \star G$ of a finite group G over an integral domain K is prime iff G is independent over K. (Cf. 17.1A.) Furthermore, $K \star G$ is semiprime iff $tr_G(R) \neq 0$. (*Ibid.* See Montgomery [80] for a fuller account of fixrings of automorphism groups. Cf. the informative review by Passman in Small [86].)

Consider the concepts of the symmetric $Q_s(R)$ and Martindale quotient rings Q of a (semi)prime ring R, where $Q = Q_\ell(R)$ is defined with respect to the filter of essential left ideals, and $Q_s(R)$ is the set of all $q \in Q$ such that $qB \subseteq R$ for some ideal B of R.

The main theorem (Theorem 8) of Mihovski [94] states that a set M of automorphisms of a prime ring R is linearly independent over R iff there exist

$$\sigma_0, \sigma_1, \dots, \sigma_n \in M$$

and units $q_1, \dots, q_n \in Q_s(R)$ such that $1 + q_1 + \cdots + q_n = 0$ and $x^{\sigma_i} = q_i^{-1} x^{\sigma_0} q_i$, for all $x \in R$, and $i = 1, \dots, n$. When R is a simple ring, then $Q_s(R) = R$, hence the q_i are units of R.

Letters from Victor Camillo (Excerpts)

1. On Theorem 3.51′, regarding infinite matrices and Morita Equivalence, my best recollection is that the referee's report came back with the statement that the theorem as stated by me had been conjectured by Stephenson in his thesis. I inserted this comment. It seemed fair to me. Other experts have since informed me that by their lights this is not true. I was in fact inspired by Eilenberg's remark as printed in Anderson-Fuller [73] that if R and S are Morita Equivalent then the corresponding infinite matrix rings must be isomorphic. I simply wondered if the converse might be true. On the other hand, Stephenson did prove in his thesis that R and S are Morita Equivalent if and only if the corresponding matrix rings where only a finite number of entries are allowed, are isomorphic. This was a whale of an insight, so that the credit for first looking at infinite matrix rings as characterizing Morita Equivalence certainly belongs to Stephenson.
2. I don't think that my theorem on q.f.d. modules (stated in 5.20B) actually uses Shock's Theorem, though it certainly was inspired by it.
3. Regarding my theorem with Fuller on QF-1 rings (stated in Theorem 13.30), we had to work hard to get this published after its initial rejection. My best recollection is that we received a referee's report that said that our result was good, but had been superseded by an unpublished manuscript of Dlab and Ringel.[4]
4. The reference to Camillo, Fuller, and Voss [79] was superceded by my paper cited in Theorem, 3.51′, and never appeared.[5]

[4]The author wrote a supporting letter to the editor (D. S. Rim) saying that it would be the first time in mathematical history that "the original paper was rejected in favor of the sequel!"

[5]The letters (Camillo [98]) have been edited.

Part II

Snapshots of Some Mathematical Friends and Places

Memory is more indelible than ink
Edward Dahlberg

CHAPTER 18

Snapshots of Some Mathematical Friends and Places

It seems appropriate for me to acknowledge here a number of benefactors and friends. It is germane to this work that I began my graduate studies in midcentury at the University of Kentucky in Lexington, Spring 1950, and continued as a Teaching Assistant at Purdue University in West Lafayette, Indiana, 1951–1955. Previously I had been a radio technician ("RT") in the U.S. Navy (1945–1946) before attending UK as an undergraduate on the GI bill (1946–1950).

Some Profs at Kentucky and Purdue

Incredibly, despite the extensive mathematics required in the Navy's RT program (which lasted an entire year), at UK I was still required to take College Algebra (and from an insipid text). Fortunately Dr. Theodore Adkins enlivened the course with a historical development, tracing algebra back to the Arabs (see Van der Waerden [85], *History of Algebra*, about this). His enormous enthusiasm for the subject, usually thought of merely as a tool of science and higher mathematics, and his countless anecdotes, made an indelible impression on me, and no doubt influenced me to continue the study of algebra.

I also salute Professor James ("Jim") Ward for introducing me to abstract algebra at UK, first using the book by Marie Weiss, and then the Birkhoff and MacLane classic "*A Survey of Modern Algebra.*" These were considered revolutionary if not incendiary books back then! (my sophomore year at UK). (I am grateful to UK's Professor Wimberly Royster for refreshing my memory in a letter in October 1997.)

In addition, I have Professor A. W. Goodman to thank for, e.g. Gauss' quadratic reciprocity law *inter alia*, using Uspensky's book. I later found out why Professor Goodman eschewed using his first name. He was a member of the Institute in 1955-1956, and I found his first name there, Adolph. (*Circa* post WWII, this was not a popular name.)

Professor D. E. South at UK taught me statistics and probability with such delightful gusto that I switched my major to actuarial mathematics under his spell. I still remember the Euler-Maclaurin sum formula, but I switched back to pure mathematics (= mathematics that has yet to be applied?) at Purdue.

Professor Arthur Rosenthal (a former Dean at Heidelberg U.) made the well-ordering of the cardinal numbers in a course in Set Theory at Purdue in Spring 1952 a spellbinding spectacle that lives in my mind's eye these many years after.

When Professor Rosenthal used up the front blackboards, he asked the class rhetorically "Und [sic] now class, vere do ve go from here?' A wag convulsed the class by saying "To the next blackboard!" The professor raised his eyebrows and exclaimed "Ach, so!" and continued the scale of ordinal numbers on the side board.

And *bless* Professor Sidney H. Gould for persuading me to "take the plunge" into Set Theory.

Mama and Sis

How often have I said "I owe everything to Mama." She has a name: Vila Belle Foster; she encouraged my intellectual development during The Great Depression (1929–1939) when neighbor's children were being asked to work. ("To help make ends meet" is the cliché but to put food on the table was the real reason.) She often said to me, *"Carl, get as much education as you can, because it's something that can never be taken away from you"*. (This insecurity was a legacy of the depression), in which banks closed and people were thrown out of work.) My late sister, Louise, volunteered to help make ends meet, and, although a straight A student unfortunately did not finish high school. Sis was devoted to family happiness and bought Mama the appliances which eased her life of ceaseless labor. She also made the most mouth-watering chocolate and lemon meringue pies–I'm sure Sis landed in 'pie heaven' after her death in 1984, exactly 20 years after Mama's. Wherever it is, I want to go there when I die. (Not long after this was written, I prophetically discovered in "Moe's Books" in Berkeley a delightful book by John D. Barrow with the punning title "A Pi in the Sky".)[1]

Perlis' Pearls

Mathematically I owe "everything" to Professor Sam Perlis, who infected me with his love for Galois theory in a memorable course he taught to uncomprehending students. If ever pearls were cast before blind eyes, these beautiful theorems were.

Sam's appeal was so lofty-intellectual that I often wondered if he knew how **little** at times we students comprehended. At the same time, he inspired me to *really* understand this beautiful theory, and in doing so, I was weaned away from the reliance on the rote learning that practically everybody practiced. Later when Paul Mostert, one of Purdue's best students, asked me if finite fields F were Galois over subfields, I looked at him pityingly, scarcely concealing my joy at having learned for myself that, as splitting fields of the separable polynomial $x^n - 1$ (where $n+1$ is the number of elements of F) **they had to be Galois** over any subfield.

With Sam's guidance in 1954–55, I wrote my Ph.D. thesis on Galois theory "Normal bases and completely basic fields" in 1954–55 which was published in the *Transactions of the American Mathematical Society*, two years later in 1957. (Thirty years later, in 1986, my theorems were rediscovered and published by *two* mathematicians (Blassenohl and Johnsen) in the *Journal of Algebra*, and five years

[1] "Moe" Moskowitz was one of Berkeley's fixtures when I spent a year there as a visiting scholar, 1965-1966. Eventually he built his own building opposite Cafe Mediterranean on Telegraph Avenue. Then you could read poetry while gazing at the facade of his building. Moe was a lovable man, always with a stogey in his mouth, and his death in Spring 1997, and his life, have been movingly memoralized in a booklet *On the Finest Shore* by his many literary friends, colleagues, and family.

later in yet another paper in *Archiv der Mathematik* in which I was duly acknowledged).

The Ring's the Thing

Although my thesis was about Galois theory, my thesis problem, which Sam dreamed up, was about a normal basis $g_1(u), g_2(u), \ldots, g_n(u)$ of a Galois field extension K/F with Galois group $G = \{g_1, \ldots, g_n\}$. Another way of saying this is that K and the group algebra FG are naturally isomorphic as FG-modules, a fact that struck me profoundly. Furthermore, ring theoretical properties of FG were of interest, and depended on whether the characteristic p of F divided n or not. (See, e.g. my papers [57] and [58b] in References.)

My "Affair" with Ulla

Originally a stage actress, Ulla Jacobsson attracted world-wide attention by swimming nude in a Swedish lake in her second film "One Summer of Happiness" (1951). Sometime in 1952 Sam and I saw this film in an "Art Theatre" in West Lafayette, Indiana, where Purdue is located. "One Summer of Happiness" was a love story, and we fell in love with Ulla, along with millions of others! The sad, tragic ending in the film, in a senseless moped accident, intensified our feelings. She was the only movie star whose autograph I solicited. Was I ever surprised when she sent me an autographed photograph—but, alas, completely clad!

The first thing I did, when I visited Sweden for the International Congress of Mathematicians in the summer 1962, was swim in a Swedish lake near Linköping at the home of Jan Erik Odnoff, whom I met at the Institute for Advanced Study in the Fall of 1961. As you might guess, the frigid water doused my ardor—for Swedish lakes—but not for Ulla, who died in 1982.

How I Taught Fred to Drive

Without "little Fred" (6'4" tall), I never would have known the meaning of the word "brother". He was loyal (I started to write "faithful"), generous, and fun. At my retirement party at Rutgers on April 30th of this year, he regaled the department with how I taught him to drive. "Carl was the first member of the family to have a car" (a Jeep loaner from Collier Encyclopedia where I had a summer job)–"He tossed me the keys, and said,"remember two things: (1) Don't run into anything; and (2) Don't run over anybody! When I returned in an hour, I had learned to drive!" The Department gave him an ovation.

At this juncture somebody said, "That's the way he teaches calculus", then somebody else said, "He doesn't teach calculus, the students say he teaches Art, Philosophy, Literature, Love of Life, and *everything except calculus.*"

I never before realized how much fun it is to be roasted, or to retire.

"The Old Dog Laughed To See Such Fun"

Antoni Kosinski, Chair, Rick Falk, Acting Chair, my student and colleague, Barbara Osofsky, and colleague Earl Taft[2] enlivened the happy occasion with numerous anecdotes and many kind reminiscences of my thirty-five years at Rutgers. Joop Kemperman as well as two students, John Cozzens and Holmes Leroy Hutson, pleasantly surprised me by their presence. To top it off, Patty Barr, Barbara Miller, Barbara Mastrian and Mary Ellen Mack sang a wonderful ditty composed by Linda York, "A Song For Faith" (see below), who also lent her voice, to make it a quintet. *Was everybody bowled over—the "Old Dog" (me) laughed to see such fun*! (It's a pity Linda had not written an encore.)

C is for the COPIES that you left us.
A is for how much we ADORE you.
R is for the RINGS and things you wrote of.
L is for your LAUGHTER and your LOVE.

F is for FAREWELL old friend we'll miss you.
A we need ASSURANCE YOU'LL VISIT SOON
I your INSPIRATION turns night to noon
....And without the
TIMES of joy we foresee gloom.
H is for our HOPE that you'll be HAPPY
.... and have a HALE and HEARTY life.
For Carl, we're very glad to have known you
Move on good friend...enjoy the life to come
But never, no never, oh never
Forget where you've come from!

My "Lineage"—Math and Other

According to Karen H. Parshall, my math lineage descends from Hubert Newton (without Ph.D.) at Yale, E. H. Moore (about whom she wrote in her article [84]), L. E. Dickson, A. A. Albert, and S. Perlis (all four at Chicago). As an amazing coincidence, three of the five served as President of the American Mathematical Society: E. H. Moore (1901–1903); L. E. Dickson (1917–1919) and A. A. Albert (1965–1967). Who ever has had so much to live up to?

My father was Herbert Spencer Faith who, as did my mother, grew up on a farm outside Paducah, Kentucky. While he never finished high school, he was an omnivorous reader who captivated you with his blue-green eyes and his serene, mildly abstracted disposition. Women adored him and I idolized him. (I had a lot to live up to in this too!)

Big Brother—"Edgie"

I also owe a great debt to my late brother Eldridge ("Edgie"),[3] who was the "guiding light" of the family for working his way through Indiana University and

[2]Earl makes a wonderful toastmaster—he is distantly related to Henny Youngman, the acknowledged "King of the One-Liner". When he died in February 1998 at the age of 91, Henny's last words might well have been "Take my life, please!"–a take-off on his patented joke "Take my wife, please!"

[3]Elderberry was another pet name we adopted for Eldridge–it proved prophetic: he was an elder in his church in Indianapolis where he lived!

becoming a top-notch research chemist for Eli Lilly, Pitman-Moore, Dow and other pharmaceutical companies. He held over 25 patents for sulfa drugs, dating back to pre-World War II, which were the drugs of choice before the discovery of penicillin, and credited with the low death rates for "sulfa dusted" wounded GI's. Way back in 1945 while I was in high school, Edgie helped me write a paper on the medical uses of radioactive elements. Years later I thought of him when I was given a radioactive "cocktail" to drink for a diagnostic procedure. Edgie died in 1995, age 80.

H.S.F. Jonah and C. T. Hazard

On the undergraduate level, these two men shaped my teaching philosophy as much as Sam. Clifton Hazard was in charge of putting teaching in the title Teaching Assistant or TA. His philosophy was the same as Perlis': *learning meant reaching and achieving, not rote regurgitation of letter-perfect lectures all carefully written on the board.* His motto was "make the student fire you as soon as you can", that is, dispense with his need of you. His method was: (1) Force students to read the text, and under no circumstances read it to them; (2) Force them to do the homework, and under no circumstances do it for them. (Hints were allowed, however. *In other words give them the keys and tell them to remember the two things I told Fred?*)

After 4 years and 4 summers of teaching by these principles, I became so excited by the progress students made that I never ever abandoned them, even though other universities (I taught at Michigan State, Penn State...) did not have the same quality of math students as did Purdue, the largest engineering school in the country. And certainly no one with Hazard's mathematical and intellectual integrity vis-a-vis undergrads.

Harold Jonah (father of David Jonah—Detroit U.) backed me up with his impressive eyebrows and visage that could make your knees shake when he frowned at you. I tried to keep this from ever happening to me! And when I left Purdue, the chairman gave me an unbeatable letter of recommendation. But now I wonder if it was entirely deserved, since he consistently beat me at golf. In any event, I soon lost my "wings" when I found that students elsewhere did not have the preparation and motivation of Purdue's "Boilermakers". I wish everyone could have the same keen and kindly instruction in their teaching as these three gave me.

John Dyer-Bennet and Gordon Walker

In my third year at Purdue I signed up for a reading course in A. A. Albert's "Structure of algebras," expecting Sam, who was Albert's student, and who helped him write the book, to guide me through it. Much to my surprise, Sam persuaded John Dyer-Bennet (Professor Emeritus, Carleton College, Northfield, Minnesota) to take his place. This was an enormous piece of serendipity, because I learned what a talented professor John was, which I otherwise would not have known. He took apart the book line by line, and then rewrote it correctly, that is, according to his tastes, which were impeccable. This eliminated a great deal of obscurity in Albert's exposition, and contributed to a deeper understanding of linear associative algebras than I would have acquired otherwise.

Perhaps I never before expressed my gratitude to John for doing this for me, but now, forty-three years later, I do so. Thanks, John.

In an entirely different way, Gordon L. Walker (the late Executive Director of the A.M.S.) guided me through a course in Algebraic Geometry. This being visual, was decidedly more intuitive, and we did not rewrite the text: R. J. Walker's "Algebraic Curves." (R. J. was no relation to Gordon.)

Gordon collaborated with Sam smack dab in midcentury to prove a beautiful and seminal theorem stating that for two finite Abelian groups the rational group algebras determine the groups (see 11.11). This result does not generalize to other fields or groups (see **sup** 11.11 and 11.11ff).

Henriksen, Gillman, Jerison, McKnight, Kohls and Correl

At Purdue, these three, Mel, Lenny and Jerry, initiated students (including Jim McKnight, Carl Kohls, Ellen Correl and me) into the mysteries of "Rings of Continuous Functions" (and of continuing seminars?) There was a lot of excitement generated by them, culminating in the work of Gilman and Jerison [60] with the quoted title. A McGill University Lecture Notes by Nathan Fine, Gillman and Joachim Lambek explicated the maximal quotient ring of $C(X, \mathbb{R})$, for suitable X, as the ring of all functions continuous on dense open subsets. (For a more recent account of certain aspects of this subject, see Dales and Woodin [96], and a highly readable, informative review by my colleague Gregory Cherlin [98].)

Joop and Vilna, Len and Reba

These warm people greatly enriched the lives of students and faculty at Purdue. Joop Kemperman arrived from Holland in 1951, and his beautiful bride, Vilna, came a year later. By the vagaries of academic life, Joop is now a Rutgers Professor Emeritus, after spending 25 years at the University of Rochester. He retired in 1995 shortly after Vilna died.

Len Gillman (U. of Texas, Austin) spent 10 years of his life as a professional pianist before taking up mathematics. He frequently gave recitals at national meetings of the American Mathematical Society. He and his wife, Reba, were active West Lafayette music lovers, and Reba sang in Gilbert and Sullivan operas.

For years, Purdue University was host to the Metropolitan Opera when it still toured the country, making stops in Bloomington and West Lafayette. The program guide proudly boasted that the auditorium sat six more seats than Radio City Music Hall in New York. *But, alas, without the 50 Rockettes*!

Some Fellow Students at Purdue

They were fellow workers and friends who toiled in the mathematical vineyard at Purdue, and in fact, Bob Gambill is a professor emeritus there, while Jack Hale is Director of Applied Mathematics at Georgia Tech. Our real work there was basketball (the Intramural League) and golf, a game Paul Mostert taught me. I called Bob "Shotgun" for his propensity for shooting the ball from any spot or position on the court. Jack was Little All-American at Berea College (Kentucky), one of our farm teams! After several close games we lost the Intramural League Trophy to the undergraduate team. But it wasn't Jack's fault—he played some great ball—*we just ran out of steam.* (David Riney and Chuck Yeager[4] completed the five.)

Ted Chihara (Purdue U. at Calumet) worked under Professor Rosenthal on orthogonal polynomials. Ellen Correl worked with Nelson Shanks in topology, and is an emeritus professor at the University of Maryland. Paul Mostert introduced to me his trick of moving *his* putter, while *you* putted, in order to break your concentration!

Michigan State University (1955–1957)

I have mixed feelings about MSU in East Lansing. The happy memories of the birth of my two daughters, Heidi in 1955 and Cindy in 1957, sustained an overall difficult two years there. (They were born in Sparrow Hospital in Lansing, and as a result I called them "my little sparrows.")

Sam Perlis was so much in love with MSU and the beautiful campus which the Red Cedar River bisected, that he told me that if I didn't accept their offer, then he wouldn't write any more letters of recommendation for me.

This negative input from Sam hit me like a ton of bricks, since heretofore he had been unfailingly supportive. For one thing, instead of the two courses I taught as a TA at Purdue, I was scheduled to teach 4 courses, including a course in **slide rule**! I was getting a stiff dose of reality after grad school.

Second, after the Kentucky Wildcats in Lexington, and the Purdue Boilermakers in West Lafayette, the rah-rah-rah of sports-dominated universities had paled long ago. After UK I no longer went to any of the games or even listened to them on the radio.

Third, because of my successful defense of my Ph.D. Thesis, I felt I had earned a better chance at research development than the then poor soil of MSU.

Sam Berberian, Bob Blair, Gene Deskins, and the Oehmkes

It turned out that my forebodings about MSU were justified: MSU treated young professionals such as Sterling (Sam) Berberian, Robert Blair, W. E. ("Gene") Deskins (who came to MSU in 1958) as so much cannon fodder for frosh-soph courses in largely engineering calculus. When I left for Penn State,[5] and a two course teaching load, and an increase in salary of 50% (to $6500 per academic year), a large number of others left at the same time, including Sam Berberian (U. of Texas, Austin), Bob Blair, and Bob and Theresa Oehmke ended up in Iowa City.

The attrition at MSU was so great that an investigation was made by a select committee and I was asked to fill out a detailed questionnaire as to why I had left. Although I was no longer interested, I filled out the form in the hopes that those who remained (e.g. Marvin Tomber and John G. Hocking) might benefit.

---

[4]I had the pleasure of teaching Calculus to his son Carl twenty (?) years later at Rutgers. He was a chip off the old block and made a solid "A".

[5]Ironically Penn State was also a football and athletic powerhouse: it joined the Big 10 in 1995 to symbolize (solemnize?) this fact!

Nevertheless, I benefited greatly from my association with the fine mathematicians that I have mentioned, and the seminars that I participated in, including Jacobson's Colloquium volume [55], and Cartan-Eilenberg's Homological Algebra [56]. The influence of these two masterpieces has lasted to this very day—40 years later.

"Cupcake"

Gene Deskins (U. of Pittsburgh) was a frat brother of mine at the University of Kentucky. He graduated in 1949, the year before I did, and became a TA at Wisconsin I believe. Another fact of our relationship: I persuaded him to leave OSU to come to MSU. Why? Was it because **misery loves company**? No, I was very fond of Gene and his wife, Barbara. He was like a big brother to me—Barbara called him "cupcake" for some inexplicable reason, but somehow it fit! I also admire him for his deep mathematical ability and knowledge. (He specialized in group theory.)

Leroy M. Kelly, Fritz Herzog, Ed Silverman and Vern Grove

These four were very friendly and helpful to us at MSU. Fully aware of the difficulties that the young professors were encountering, they did as much as they could to alleviate them. Fritz in particular was quite jocular: "Teaching is a calculus problem—you have to minimize," he told me, apropos of the heavy teaching load. (I wish I could remember some of L. M. Kelly's delicious barbs at the MSU administration.)

Vernon Grove, one of the most lovable people that I have met in academia, told me after a tremendous performance of Offenbach's *Gaiété Parisiennes*, by the Royal Canadian Ballet Company, that ballet is nothing but a "glorified leg-show"! I was so shocked that it took me twenty years (when I reached his age then) to agree with him. (Just kidding—think of *The Nutcracker*!)

Orrin Frink

As chair of the Mathematics Department at Pennsylvania State University in State College, Professor Orrin Frink asked me if I would like to attend the International Congress of Mathematicians in Edinburgh in July 1958. He encouraged me to apply for an NSF travel grant advertised on the bulletin board, and by one of life's miracles, I was awarded the grant.[6] I used to come across Orrin's papers referenced in many papers, and when I told him this, he would ask me the titles; then when I complied, he'd say "Yep," as if he **knew** they had to use (as they did) his work to get theirs. Aileen Frink, his wife, also a mathematician, told me, when I called my condolences, that at age 82, he was still working at mathematics up to the very last. Orrin reminded me a lot of my father–he *was* very fatherly to me and other young mathematicians; and in one respect—the blue haze of cigarette smoke enveloping him—he evoked an indelible image of my Dad.

[6]To anticipate somewhat, see the "Indian Idea of Karma" below.

Gottfried Köthe and Fritz Kasch

At the Edinburgh Congress, I presented the results in Galois theory in my thesis, and by another stroke of luck, Professor Gottfried Köthe and Friedrich (Fritz) Kasch, both of Heidelberg University, attended. Later during a boat tour of nearby islands, Fritz (an expert in Galois theory) invited me to apply for a Fulbright-Nato postdoctoral to study with him. Thus, I spent the year 1959–60 in Heidelberg learning from him not just the Cartan-Jacobson Galois theory of skew fields (§2), but more importantly, as it turned out, Homological Algebra.

Romantische Heidelberg

The title "Romantic Heidelberg" is richly deserved. The newly restored and burnished red sandstone medieval castle broods over the city of narrow *Strassen*, *Die Alte Brücke* (the ancient [Roman] bridge) over the Neckar River, *Der Philosophenweg* (Philosopher's Walk) along a hill across the river. (My classics-scholar wife, Molly Sullivan, informs me that "romantic" derives from "Roman"). And suddenly you remember you are in the home of the "Student Prince" when you drink beer in *Die rote Ochsen* (The Red Oxen) and *Vater Rhein* (Father Rhine) where everybody sings "*Du Liegst Mir Im Herzen*" and "*Ist Das Nicht Ein Schnitzelbank?*" *Jawohl!*

Reinhold Baer

Fritz arranged for me to speak at several universities including Frankfurt (home of the hot dog?), Muenster (home of the cheese), and Munich (home of the beer hall Putsch!). The annual meeting of the German Mathematical Society took place in Muenster where Reinhold Baer, newly-replanted in Germany from his exile in Illinois U. in Urbana, gave a letter perfect address, despite his dubious claim that he left his notes at home. (It kept everybody on the edge of their seats hoping for a slight imperfection in the impeccable Baer!)

After my lecture at Frankfurt where Baer was Professor, he arranged for me to stay with him and his wife at his home in Falkenstein, a high tor overlooking the city. I was deeply honored by this sign of friendship and esteem. Professor Baer confided to me that his reasons for leaving Urbana was that the pension for his wife would have been reduced to just 10% of his if he died even a day before official retirement. Germany was offering a lot more, and a guarantee for his wife. (He spent two years at the Institute for Advanced study in 1935–37 shortly after leaving Germany.)

There could be no doubt of the deep love he felt for his wife—he was courtly and romantic, helping her into the car, seating her at the table before he sat down and holding hands, and gazing deeply into her eyes. *This open expression of love affected me profoundly, as did the excellence of his mathematics.*

Before I left, he again confided to me: he didn't like music or art created after the 17th century. I was shocked to my toes by this quixotic aspect of this paragon. I thought of the lovely Impressionists, the bold Expressionists, Cubism, Dadaism, Abstract Expressionism, 20th Century Realism, Van Gogh, Monet, Manet, Matisse all being thrown out the window. Of course that still left, e.g. daVinci, Botticelli,

Michelangelo, Bellini, Titian et al, the Greeks, the Romans, the Celts, the Norse, Chinese, Hindu and Moslem art galore.

I couldn't bear to think of abandoning all that art, not to mention 18th, 19th and 20th century music! But it did explain his courtly love for his wife, I think—that definitely dates back to the 17th century!

Death in Munich (1960)

While in Munich for my talk at the University, I had a close encounter with death. As I stood at a crowded intersection on the Hauptstrasse waiting for the green light, a van careened close by. Instinctively I stepped back, but, horrifyingly, the man standing next to me got hit by the van's rearview mirror, and he fell with a thud on the sidewalk, bleeding profusely from his face. I could see his nose had been severed, and his brains exposed under the welling blood, yet he was still alive—so much so that he kept trying to blow his nose with his handkerchief even though there was nothing there!

I knew there was little hope for him, and tried to comfort him, holding his head up, and talking to him soothingly.

Ruhren Sie sich! Die Hilfe kommt sovort! "Stay calm, help is coming" was all I could think of to say. As people hurried by, sidestepping him, I yelled *"Rufen um Hilfe"*, "Call for help!"

"Death can be so indiscreet when it happens on the street"

This became the line of a poem that I wrote many years later about a heart-attack victim dying on the sidewalk in New York.

Bob Hope used to joke that he missed being handsome "by a nose." In Munich in 1960, I missed death by a nose. It goes without saying that ever since that time I do not stand close to curbsides.

Marston Morse and the Invitation to the Institute

At Heidelberg, I met H. Seifert, F.K. Schmidt, Albrecht Dold, and, one day, miraculously, Marston Morse of the Institute for Advanced Study, in Heidelberg University for a colloquium talk. (I almost laughed outloud at the incongruity of his highpitched, squeaky voice.) At a party for Morse at Seifert's house that night, Morse and I were the only two Americans, so naturally we spent much of the evening talking (we were both homesick). Before the evening was up, Morse inquired about my plans for next year, and when I said that I would be returning to Penn State, he suggested that I apply for a National Science Foundation Postdoctoral Fellowship at the "Institute", as everybody calls it. I still remember my incredulous response: *Me? Apply to the Institute?* He then suggested that on the application I refer to his invitation. Fritz Kasch agreed to write a letter of recommendation. The two together helped me get the NSF and membership at the Institute, where Marston and I continued our friendship until his death in 1977 at the age of eighty-five.

What Frau Seifert Told Me

After a delightful evening of violin solos and piano-violin duets, Rheinwein, Moselles, and the soft delicate *Eiswein* from The Black Forest, Frau Seifert drew me aside (Was I not a representative of the USA?) and said, "*Lieber Herr Doktor, Sie müssen in die Vereinigten Staaten gehen, und den Amerikanern sagen 'Nicht alle Deutschen sind Tiere!'*" ("You must go back to the USA and tell them that not all Germans are animals!") *I was shocked and deeply embarrassed that anyone would judge an individual by the insanity of his compatriots.* Think how the Seiferts and other Internationalists must have suffered to think they were being adjudged as Nazis or Nazi sympathizers.

One thinks of Goethe's dictum: "So noble in the individual, so base as a (whole) nation."

To this day I can't think what answer or reassurance I offered Frau Seifert. I thought that *my being there* was proof of how I regarded Germany and Germans. With sorrow, forgiveness, and hope for the future, I am happy to say, now, 36 years later, that I feel the same way now as I did then.

"Some Like It Hot" (Manche Mög Es Heiss)

That having been said, I have to record a very upsetting experience in fall 1959 that I had in Heidelberg when I attended the Billy Wilder comedy with the stated title starring Marilyn Monroe, Jack Lemon, and Tony Curtis. Before the film began (or after, I can't recall) there was a short documentary on Germany's invasion of Poland on September 1, 1939! Why in the name of sanity would this tragedy be considered a companion piece to a MM hit? But that's not the worse part: the film depicted Polish Calvarymen with lances attacking the German Panzers (tanks) with the inevitable result—the poor Poles were slaughtered. I was simply devastated by the unexplained and unexplainable laughter of the German audience.

This happened almost 40 years ago, and I wish I could forget it as an isolated demonstration of German cruelty, but, as the saying goes, *if wishes were horses (wings) then beggars would ride (fly).* It made me ill to watch those brave Poles dying, fighting for their country with inadequate weapons, and I will never forgive *those* Germans for thinking it funny.

Marston Morse

Marston was born in Waterville, Maine in 1892. He received his Ph.D. from Harvard in 1917, and one year later, he received a medal ("Croix de Guerre") with a silver star for serving in the American Expeditionary Services in WWI. Other honors include the Meritorious Service U. S. Army Ordinance Award 1944, National Medal of Science 1964, Chevalley, Legion of Honor 1952 and twenty honorary doctorates, including Paris '46, Pisa '48, Vienna '52, Rennes '53, Maryland '55, Notre Dame '56, and Harvard '65. He was a joint winner of the Bocher Prize in 1933, and the next year was the Colloquium Lecturer of the American Mathematical society on "Calculus of Variations in the Large". Marston was President of the American Mathematical Society in 1941–43. Was there ever a more honored American mathematician?

I didn't suppose there could be any mathematician who had not heard of Morse Theory, nor not had an inkling of what Morse Theory is about, yet incredibly, one visitor to the Institute asked Marston what field he worked in. Taken back, Marston blurted out "**Why, my theory!**"[7] Another time Marston asked what Bourbaki was writing about. When told "Foundations," Marston retorted, "Always the foundation, never the Cathedral!"

When I gave Marston a complimentary copy of my algebra, after some time, he remarked at lunch with me, "Carl, I don't know what your book is about, but you go to great lengths in it!" To be honest, Morse was not the only professor at the Institute who did not share my enthusiasm for "Categories, Rings, and Modules" (the subtitle). Another professor once ridiculed category theory, again at lunch, as "abstract general nonsense" (a familiar epithet), yet years later gave a lecture at a Rutgers Colloquium that was devoted to characterizing all categorical epimorphisms in the concrete category for which he was famous. (I have a witness of impeccable character[8] who was there for the epithet and one can verify the conversion to category theory by reading the now famous paper!)

These make great stories, even if they were apochryphal, like Thomas Hardy's reply to the question: What is the secret of your success? "Longevity!" was the retort. (Isn't longevity itself a success of a kind?)

Marston and Louise

Marston and Louise Morse exemplified the democratic principle of the Institute. They were delightful people to talk to, and they opened their home to the members for parties that filled their enormous house and garden on Battle Road, near the site of the "Battle of Princeton". There was always live music performed at Louise's parties, especially on the Steinway.

Louise Morse: Picketing IDA

Louise and I often found ourselves in protest marches against the Vietnam War and nuclear testing, however, I didn't join her in picketing the Institute for Defense Analyses (IDA). I was a veteran of World War II whose job in the Navy as Aviation Electronics Technician Mate (i.e., "RT") and the fact that I had worked at IDA in the summer of 1964, made me privy to the kind of work that Princeton's IDA did: breaking the secret codes of USSR and China, **inter alia**. I tried to convince her that IDA had nothing to do with napalm and flechettes, whose use horrified the entire world, but Louise persisted in picketing IDA, then situated on Princeton University land, as a symbol of Government.

"In that case", I asked her, "why not picket the Post Office?" And she said, "Carl, I'm a grandmother, mother and a wife. My family needs me, and I cannot afford to spend time in jail."

But the picketing and marches did have success: (1) Princeton withdrew its hospitality to IDA (which moved across town to Thanet Road); (2) USA and USSR discontinued nuclear testing above ground.

[7] I have my colleague Gregory Cherlin to thank for this and the following anecdote.

[8] Georgia Višnjić Triantfillou, a member of the Institute from Greece, 1978–1979 (she told me that she felt sorry for the professor!)

Kay and Deane Montgomery

Although Deane Montgomery of the Institute was, of course, not a dean, he was one of the first professors, and by his love of mathematics, he did so much to encourage young mathematicians. His wife Kay hosted legendary parties, not at the Institute but in their home on Rollingmead in Princeton Township, *where we really got to know the other members.*[9]

Deane Montgomery's prolific and monumental mathematical contributions were described by Armand Borel in the Notices of the American Mathematical Society and in a memorial given at the Institute on November 13, 1992. A memorial brochure included addresses by a number of other close friends and associates, notably L. Zippin with whom Deane collaborated on the solution of Hilbert's Fifth Problem.

Armand gives warm tribute to Deane's human qualities:

> "He was always seeking out and encouraging young mathematicians. ...Maybe remembering his own beginnings in an out-of-the-way place [Weaver, Minnesota in 1909], he had a special interest and talent in finding out people with considerable potential among applicants from rather isolated places, about whom not much information was available."

Among his honors were President of the American Mathematical Society (1961–63), the National Academy of Sciences in 1955, and the Steele Prize in 1988. His Presidential speech in Cincinnati in 1961 must have been the shortest in history–about 5 minutes–in which he elaborated on how hard the founders worked to establish and foster the Society, and then just as he got everybody on the edge of their seats straining to hear this wonderful eulogy, he abruptly sat down. Deane may have been "gregarious" as Armand said, but he was a man of few words: *he liked to listen but was not one to talk much. His quick intelligence and wry humor did the talking for him.*

Kay was incredibly funny and death on humbug—she'd break you up with her witty asides. In undermining pomposity, she upheld the principle of democracy: *all people are created equal.*

"Leray Who?"

Once at lunch I was introduced to the French mathematician, Jean Leray, whose name I didn't quite catch. (It was pronounced as one word: *jeanleraý*!) Some idiot snickered, "*You* don't know Leray?" He then reddened when Leray said, "Why should he know me any more than I know him?" And then he said his name so that I could understand it.

Don't you just love people who share their humanity with you? A high school English teacher of mine, Dorothea Stephens, used to say, "Well, I have to show that I'm human don't I?" And I would shake my head sideways because otherwise she couldn't have taught Shakespeare with such passion.

---

[9]Some of the things we learned were not so nice. I remember one alcoholic philandering member who persisted in putting his arm around my wife. When she objected he said, "I meant no offense!" I was amazed at the quantity of alcohol that mathematicians consumed in Princeton.

How Deane Helped Liberate Rutgers

As a member of Rutgers Research Council (1962–64, 66–67), I was asked to recruit Deane for the Rutgers University Science Advisory Council, comprised of a dozen or more top science leaders from New Jersey. Deane was not eager to accept.

Deane was a dedicated professor of the Institute who came to *work*, that is, *think, every morning at 5 o'clock*! (Armand Borel movingly recalls this fact in his warm obituary for Deane in the *Notices of the American Mathematical Society.*)

Nevertheless, Deane acceded to my request when I pointed out to him that Kay would not be allowed to matriculate at Rutgers College which was then all men. As a consequence of his enjoining the struggle for women's rights, Rutgers College became co-educational in 1971, just three years after Deane joined the Council. Subsequently, my daughter Heidi's diploma reads "Rutgers University." Even though she was in Douglass (an all women's) College, she was able to take a sizable share of her credits at Rutgers College seated in the largely men's classes. She is very proud of this application of the Declaration of Independence "All people are created equal." And I am very proud that Rutgers freed itself of its ancient bias against women.

And more than that, we have Deane Montgomery to thank for the decisive role that he, as a Professor of the Institute for Advanced Study, played in freeing women at Rutgers. Another personal consequence of this emancipation: in 1982, my wife-to-be, Molly Sullivan, received her second Bachelor Degree (the first from Texas) from Rutgers.

Hassler Whitney

I used to take my two young daughters—Heidi was five and Cindy was three—to Hass' (everybody called him Hass or Has)[10] house once a month on Sunday afternoons in keeping with a custom of my native state. Later, his charming wife told me that while at first he was chagrined by our uninvited visits, he gradually came to keenly anticipate them, and the playmates they provided for his two daughters.

I owe to the Whitneys two of my most enduring friendships—with Joan and Charles Neider—that began in 1962 at one of *their* legendary parties: when the beautiful wife of one of the local M.D.s chinned herself from the beams in the ceiling, I realized how wonderful life can be if you just let yourself go. Joan and Charles joined in with the mirth and—bingo—we started being friends! Isn't this really true-to-life? *Fun and Friendship, The Inseparable Twins.*[11] One aspect of Hass that attracted me: his slightly abstracted demeanor that reminded me of my father. (My father had died suddenly in February 1952 of a coronary thrombosis and I was still grieving for him.) An indelible picture I have of him: at tea he ate his cookies while perched on the side of a chair.

[10] Hassler has no *z*-sound, whereas "Has" has, so I prefer to write "Hass" instead.

[11] I had previously met the Neider's swimming at the "Y", and I already knew Charles as the editor of Mark Twain's "Short Stories" when I was at Penn State (1957–1959).

John Milnor

When Deane stepped down from the council, we were able to persuade Jack Milnor at the Institute to take his place. He, too, was effective in establishing a coherent science program for Rutgers. (I know because as a member of the Rutgers Research Council I attended many of the meetings when Jack stood up for the Rutgers Mathematics Department and its Center for Excellence program.)

Well do I remember Whitney's speech at the ICM-Stockholm, 1962 in which he explained the work of John Milnor, one of the Fields Medalists: he drew large imaginary diagrams in the air with his hands, and the audience roared their approval.

Jack subsequently was interviewed by a Princeton reporter, and asked if his work had any practical applications. Jack truthfully answered, "No, not to my knowledge."

Paul Fussell

Recently I glowed with pleasure to read on p.254 of Paul Fussel's *Doing Battle: The Making of a Skeptic*, (Little, Brown and Company, Boston 1996):

> Among the largesse showered on Rutgers in the sixties was a dramatic increase in research funds. In 1965 I was granted a research year to work on my book tentatively entitled **Samuel Johnson, Writer...**.

I am quite proud that I was one of the Council, and voted for Fussell's sabbatical.

Subsequently, in the summer of 1971, Fussell (now Professor Emeritus of Pennsylvania University in Philadelphia) took notes in the Imperial War Museum in London of masses of letter and papers from WWI soldiers for his multi-award winning, e.g. the National Book Award, *The Great War and Modern Memory.* I would like to think his spending time in the British Museum in 1965-6 inspired this.

Hetty and Atle Selberg

I found out from Atle Selberg that his given name is Norse for Atila. Was there ever a worse misnomer? He was the antithesis of the raging, rampaging, ravishing Hun, for Atle was quiet, soft-spoken and reflective. The late Hetty (Hedwig) was a Romanian exile when they met in Sweden at the Mittag-Leffler Institute. We often went to their house for parties whose guests formed an International Who's Who, perhaps with a Nordic, that is, Scandinavian flavor (Carleson, Gärding,...). Atle and I would go for long rambles in the Institute Woods where I discovered he knew the names of every tree! And that's not all—he was a learned man of almost omniscient bent. For example, he foresaw the backlash against the liberalism of President Lyndon B. Johnson and Vice-President Hubert Humphrey, that enabled Richard Milhous Nixon to squeak to a victory in 1968 and again in 1972. The SDA and the student uprisings that vented youth's understandable anger at the Vietnam war added fuel to the flames, and Atle said the same thing had happened in Europe, e.g. the collapse of the Weimar Republic in Germany led to Hitler's ascendency. (Not that Nixon could be compared to Hitler!)

Atle was a member of the Institute in 1947–48, became a professor in 1949, and a year later won the Fields Medal at the ICM (1950) in Amsterdam.

Hetty was a computer scientist who thoroughly enjoyed her work so much that the Selbergs often stayed put in Princeton despite Atle's love for the island of the coast of Southern Norway where they had a summer home. She was universally admired for her wit, intelligence, and great warmth.

The Selberg children, Ingrid and Lars, attended Princeton High School with Heidi and Cindy.

Another Invitation to the Institute

Sometime in the Spring of 1961 Hass approached me about my plans for the academic year 1961–1962. When I answered "back to Penn State," he told me that the Institute had a policy of offering an additional year's membership to young mathematicians on the theory that we try to get too much squeezed into a single year and thereby miss the "idea of the Institute."

The Idea of the Institute As an Intellectual Hotel

And so I came to stay on at the Institute for the second year. When I expressed apprehension at losing my job at Penn State, both Hass and Deane reassured me of the worth to me, and the mathematical community: it was the chance of a lifetime—take it!

When I asked Robert Oppenheimer, the Institute's director, for his advice he referred to the Institute as an Intellectual Hotel, where members get away from the distractions of the mundane world.[12] He said, for this reason, for the longest time phones were not permitted in the members' study. So he gave me his blessing too.

The greatest gift that I received from my years (1960–62, 1973–74, 1977–78) and intervening summers was the discovery: **you didn't have to die to get to heaven!**

This echoed what Abraham Flexner, the Founding Director, said of the Institute: it is *paradise on earth.*

Oppie and Kitty

"Oppie and Kitty", as everyone called Oppenheimer and his wife, loved to dance and attended Institute parties. In October 1960, Institute Freshman that I was, I was the first person to show up at the dance in the Common Room of Fuld Hall, and so had the honor of asking Kitty for the first dance. It was divine—*she danced as light as a feather.*

Gaby and Armand Borel

Gaby Borel frequently gave dinner parties for smaller groups of members and visitors, and Armand introduced us to his vast collection of Jazz LP's that lined

[12]This description reminds me of Woody Allen's delicious spoof "The Whore of the Mensa" about an intellectual woman who gives not her body but her mind for hire!

the walls of their home. Gaby was an inspired cook, and Armand entertained us with the sound and lore of the great jazz musicians. However, it was in the *Penguin Book of Jazz* that I learned that the French root of Jazz had a sexual meaning, as in "Don't jazz me", or "Don't give me any of that jazz!!"

Gaby

On their several trips to the Tata Institute in Colaba (a suburb of Bombay), India, Gaby was so overcome by the desperate poverty she saw everywhere she turned, that she helped organize U.S. outlets for the colorful clothes that the Indians made. She devoted herself to helping others help themselves.

Gaby also had an artistic side that one of her daughters, Anne, inherited. At the Princeton Art Association's "Sundays with a Model", Gaby would join Vic and Barbara Camillo and me in "Life Studies". After two of mine were selected in a juried show at McCarter, I flirted briefly with the idea of becoming another Picasso. I think it was Freeman Dyson who (at my invitation) came to the opening of the show and, at my house afterwards, talked me out of a career change by citing Hilbert's advice to a young mathematician: *Write a novel, you do not have the imagination to become a mathematician.* Years later I wonder: do I? When Vic Camillo wrote a congratulatory letter to the Math Department for my retirement claiming that I "could have been either an artist, writer or mathematician", I found it amusing, wondering: *which did I choose?*

Alliluyeva

Armand also tickled the keys himself, as was noted by Svetlana Alliluyeva in her memoir "Only One Year" (Random House, 1969).

Svetlana was Stalin's daughter who left Moscow in December 1966 to deposit her husband's ashes in India, and instead of returning to the USSR, she fled to the United States, aided and abetted by George F. Kennan and others at the Institute. When she arrived in Princeton she exclaimed "Princeton is more of a park than a city", a familiar sensation of newcomers.

George F. Kennan

One of the most influential and admired professors at the Institute was George F. Kennan, a former ambassador to the USSR. He wrote "American Diplomacy, 1900–1950" (U. Chicago Press, 1951), "Soviet-American Relations, 1917–1920, 1 & 2" (Princeton U. Press, 1965), "Soviet Foreign Policy, 1917–1941," "Russia and the West under Lenin and Stalin" (Little, Brown and Company, 1961), and several volumes of Memoirs including an autobiography.

When I met him, in 1960 at lunch at the Institute (where else?), I knew him as the author of the famous "Containment Policy" vis-a-vis the USSR, which originally had appeared in the forties in **Foreign Affairs**, a magazine I subscribed to (this was a most famous article by the author who signed himself "X"). The Containment Policy was a brilliant original concept of how to deal with USSR imperialism. The policy eschewed a purely military containment, and espoused the true strength of the United States, namely its democratic ideals, its constitution, its freedom, liberty and justice as our most potent weapons against Soviet totalitarianism.

To quote Michael Howard in his October 31, 1997 review in the Times Literary Supplement of John Lewis Gaddis' book "We Now Know: Rethinking Cold War History" (Oxford U. Press, 1997):

> Stalin was, as Alan Bullock reminded us, as pathological as Hitler himself. But whereas Hitler was determined to achieve his objectives within his own lifetime, **so oder so**, and knew he could not do so without fighting, Stalin was a Marxist true believer who knew history was on his side, and that he could afford to wait. ...It took the insight of George Kennan to see that Stalin was wrong: *it was the West that could afford to wait. (My emphasis.)*

Liberty, freedom, independence and nationalism proved contagious, and the USSR was dissolved in 1991 under Mikhail Gorbachev leadership and his policy of **Glas'nost**.

After my year in Germany I was deeply conscious of the fear that Europeans had for the USSR. The Berlin Wall[13] which was built August 1961 to keep East Germans and other satellite citizens from escaping from the "empire of evil", an epithet of President Ronald Reagan. Kennan himself in a telegraphic message of February 22, 1946 reprinted in his Pulitzer Prize-winning "Memoirs: 1925–1950" wrote:

> World communism is like a malignant parasite which feeds on diseased tissue.

During my years at the Institute, I took every opportunity I could to join Kennan at his table at lunch to discuss the Cold War. He has emerged as the Saviour of the West, and as our foremost champion of American ideals.

Kennan's Memoirs

Kennan's *Memoirs* are fascinating to read, particularly "A Personal Note" in Chapter 1, on his Milwaukee youth and poverty, and his Princeton University days, 1921–1925

> ...forbidden participation in sports, too poor to share the most comon avocations...I remained, therefore, an oddball on campus, not eccentric, not ridiculed, just imperfectly visible to the naked eye. ...In these circumstances, Princeton was for me not the sort of place reflected in [Fitzgerald's] *This Side of Paradise.*

For those who worry about their children's lack of definite career choices, this might interest you:

> My decision to try for entry [in the Foreign Service] was dictated mainly...*by the feeling that I did not know what else to do. ...Milwaukee held no charms for me* (my emphasis).[14]

Isn't that beautiful?

---

[13]It was to last twenty-seven years—until November 1989! It stretched 870 miles and divided the entire country—not just Berlin—and guarded by 1000 (!) dogs, teams of police, watchtowers and minefields (see "Word Watch" by Anne H. Soukhanov, Holt, N.Y., 1995.)

[14]Youthful doubts about career choices and role models were identified by Erik Erikson in "Identity, Youth and Crisis," W. W. Norton, New York, 1968.

Kurt Gödel

Nobody would guess that I was able to communicate with the world's greatest logician, and I couldn't, at least not in the rarified atmosphere of his "Consistency of the Continuum Hypothesis" and "The Incompleteness Theorem". The latter shattered Hilbert's conjecture that mathematics axioms were complete, or could be completed, by proving that any logical system that was rich enough to axiomatize arithmetic would inevitably lead to questions that could not be answered, **pro or no**, within the system. Gödel's trick lay in the numbering of the axioms and the propositions, each with its own **Gödel number**, and then asking a question about the Gödel number of one of the propositions which patently could not be answered inside the systems.[15] The method is called **Gödel's Diagonal Method**.

Often I chatted with Gödel in front of the Institute's main building, Fuld Hall, that was modeled after Independence Hall in Philadelphia (Mrs. Felix Fuld was the married name of the sister of Louis Bamberger, who funded the Institute in 1930 with a grant of six million dollars). Our conversations were long or short depending on how long he had to wait for the taxi he took to get home! (There was an Institute station wagon that ran at regular intervals but along a fixed route.)

P. J. Cohen

Paul Cohen's Theorem [66] together with Gödel's, established the consistency of the Continuum Hypothesis (= CH). Paul was a neighbor of ours who lived in an apartment on Olden Lane next to the house at 43 Einstein Drive where we lived. He was our babysitter on occasions when the babysitter pool went dry, and amused our children with things like Cantor's and Russell's paradoxes. (If you don't believe it, then you don't know Paul!)

It was at Stanford U. where he developed the "Forcing Principle" establishing that the denial of CH **also** was independent of Zermelo-Frankel (= ZF) set theory. When he came to the Institute to lecture on his discovery (and to receive Gödel's blessing?), he told me that a Stanford U. colleague had provoked him into the proof by ridiculing the Forcing Principle so he set out determined to prove himself right! If only we all had such determination and such ideas!

Kurosch Meets Witt

Alexandr Kurosch gave a lecture in Fuld Hall, in the Spring of 1961 I believe, relating to a theorem of Ernst Witt, who was a member at the time. Since some lectures were given at Fine Hall on Princeton University campus, Witt went there instead, and arrived just as Kurosch was leaving the lecture room Fuld 114. Never at a loss for words, I put in my oar and introduced the two mathematical giants and told Witt that Kurosch had spoken *inter alia* on a theorem of his. Witt smiled and said "I proved that theorem when I was in the USSR". Kurosch did a doubletake

[15]Alfred Tarski [56] proved that *logical systems, as well, are semantically incomplete.* See e.g. Barrow [92], p.125, who also discusses Skolem's contribution regarding *non-isomorphism of arithmetic systems characterized by a finite set of axioms.* In 1949, Tarski proved that the theory of algebraically closed fields is not finitely axiomatizable. See Bell and Slomson [71],p.101, and also their historical note on p.106.

and said, in a friendly way, "Why, I never knew you visited the USSR. When was that?" "When I was in the *Wehrmacht*;" Kurosch turned on his heel and left without a word. Witt could be very chilling, at times.

I might add that Witt has been criticized for having been at age 20(?) a member of Hitler Youth, and even for having been a Nazi or Nazi supporter (Witt was born in 1914). My recollection of his response is that practically all German youths joined or were forced to join, not only Hitler Youth, but also either the *Wehrmacht* or the *Luftwaffe* (literally air weapon, i.e., air force), as Fritz Kasch had been. Could one be a loyal German without espousing the Nazis? And after WWII there were no Nazis (just neo-Nazis).[16]

Hitler's View of the Institute

I once read somewhere, maybe in John Toland's "Adolph Hitler," that Hitler described the Institute as a haven for Negroes and Jews.[17] As an "Intellectual Hotel", the Institute is indeed a haven, for intellectuals regardless of race or religion, and it would be sheer nonsense to suppose that the Institute would invite Witt in 1960–1961, if indeed he had been a Nazi. But, as I implied, after WWII there were no Nazis.

The Interesting Case of Threlfall

The co-author with Seifert of "Lehrbuch der Topologie" was an Englishman, William Threlfall, who fell in love with Germany after WWI, and especially Heidelberg, German mathematics, and, of course, the Seiferts. As WWII progressed, he was ordered to leave Germany by the British Government, or suffer the consequences of being charged with aiding the enemy. You could love Germany and hate the Nazis as Threlfall did.

My Friendship with Witt

When I met Witt at the Institute in Fall '61 we often spoke German. (Although seemingly like all Germans, he spoke English fluently, *I still tended to think in German.*) Nevertheless, many members of the Institute did not warm to Witt either because of the War, or because of Witt's lack of tact. Since I had expressed my openness to Germany and Germans through my Fulbright-Nato Postdoctoral at Heidelberg, and could communicate with him in German, I became his *de facto* friend. P. J. Cohen, who lived in the same apartment building as Witt, also befriended him, so we three often found ourselves in each others' company.

Regarding paranoia over Germany and Germans, one member went so far as to say at tea time before a large group that Germany in 1960 was still the world's biggest threat to peace. However, I polled the group, and he found himself isolated: everybody else thought that the USSR was (or maybe the USA?).

---

[16]Forgetting one's Nazi past is called Waldheimers disease: "selective forgetfulness...40 years later [Waldheimer, the Austrian President, forgot] he was a Nazi." From "Word Watch" by Anne H. Soukhanov, Holt, New York, 1995.

[17]Somewhere (maybe in *Mein Kampf*?) Hitler expressed this view of humanity: *everybody has his price, and you would be amazed at how cheap it is.*

My First Paper at the Institute: Communicated by Nathan Jacobson

Witt showed his genius in his broad reach and deep insights into mathematics. Even my paper [61d], "Derivations and Finite Extensions", submitted to the Bulletin of A.M.S. in May 1961, interested him a great deal. It was my first paper at the Institute. The main theorem states:

> *The dimension function on the vector space of derivations of finitely generated field extensions F over k is monotone.* From this one obtains: *The minimal generator function on F over k is monotone not only when the transcendence degree of F/k is ≤ 1 (e.g. whenever F/k is algebraic) but also whenever Lüroth's theorem holds, e.g. by the theorem of Castelnuovo-Zariski, whenever tr.d. $F/k \leq 2$ and k is algebraically closed of characteristic zero.*

"Proofs Too Short"

Previously this paper had been turned down at *Proceedings of the A.M.S.* because "the proofs were too short", but Nathan Jacobson accepted it for the more prestigious *Bulletin of the A.M.S.*, and wrote me that it was a sad day for mathematics when theorems like these were rejected for having simple proofs. I am very proud that the paper appears *"Communicated by Nathan Jacobson."* (See Theorem 1.32 in Part I.)

Caroline D. Underwood

The Executive Secretary of the Institute was Caroline Underwood (Carol-'line, can't you hear me calling', Carol-'line? is the first line of a lovely song that explicated the Southern pronunciation of her name). A lieutenant in the WAVES, she stayed in the WAVES Active Reserve until her retirement in the 80's. She was the most helpful person imaginable—no task was too small or large for her. Everybody at the Institute raved about her proficiency and helpfulness, but what meant the most to me—and others I am sure—was her great kindness and radiant sunniness. God Bless You Carol-'line, Carol-'line, Carol-'line.

Mort and Karen Brown

They were across-the-street neighbors on Einstein drive in 1960–61, and became dear friends. We occasionally baby-sat for each other, and this brings you closer than math does. Mort (Michigan U., Ann Arbor) was the author of a very short proof of Schoenfliess' Theorem that somewhat upstaged Marston Morse's proof. (I know that Marston was deeply impressed by this evidence of Mort's youthful genius.)

The Brown's were two scintillating people–a terrific couple. Karen was a sociologist who introduced me to the theories of Emil Durkheim. More than that, they were both articulate, intellectually stimulating people you could count on for interesting conversations.

If I fell a little in love with them, I wasn't the only one of the members to do so–they were about the most popular couple there—you found them at almost every party (which were in abundance–at least once a week).

One subject, my favorite, was taboo with Mort. Once in a circle of friends at tea, Mort challenged me by putting a teaspoon in my empty teacup and said, "There, try to make something sexual out of that!" Everybody roared with laughter, and then Mort reddened as he suddenly realized the Freudian overtones of his action.

Leah and Clifford Spector

A brilliant logician, Cliff was another close neighbor of ours on Einstein Drive. His sudden death in the Spring of '61 cast a deep pall over the Institute. On a Thursday, I took Cliff with me to the Princeton YMCA-YWCA pool for a swim. I thought I was hot stuff, and challenged Cliff to a race–my backstroke vs. his crawl–and he beat me by a stroke. I was properly chastened. Then Friday he was put in the hospital for unexplained pains, and suddenly on Saturday he died of unsuspected leukemia.

According to his wife, Leah, he was alert to the very end, and his mind traced out countless explanations for his then mysterious malady.

John Ernest

John Ernest (U. of California, Santa Barbara) was their closest neighbor, and he and his wife, as we all did, rallied around Leah and her children to give them comfort. John spoke with a dignity and solemnity founded in his own faith that touched us all.

"I Like This Motel"

The houses at the Institute were ultra modern ranch-style houses designed by Breuer with enough glass to qualify them as a fish bowl–you could amuse yourself watching your neighbor making coffee, reading the paper, or whatever. Our first night at 43 Einstein Drive, Heidi said, "I like this motel—how long are we staying here?" *Once a polite European mathematician doffed his hat while my wife was standing in the bedroom.* During 1961–1962, our neighbors on the fishbowl side, across the quadrangle, were the Bishops, Errett and Jane, with whom we shared baby-sittings and occasional cookouts.

Institute Cats

When members left the Institute at the end of their year, they frequently left their pet cats behind to fend for themselves in the high grass of the Institute fields, so there was a constant supply of semiwild cats to adopt. Our first we called *Zebra*, and the second, *Midnight*, because of his nocturnal caterwauling, as well as for his ink black fur.

Because everybody fed them, I called them: **The fat cats of the Institute, root-a-toot-toot!**

Yitz

The late Professor Israel Nathan Herstein (of Chicago U.) preferred to be called Yitz. While at Purdue I met Yitz at the frequent mathematical meetings that took place the University of Chicago. Often Yitz would take us to Mama Luigi's for a late dinner, where we talked for hours. He had a love for Italian food and Italy—we always managed to end up Italian: Dimaggio's in San Francisco, Valerio's in Cincinnati, Otello in Rome, Sardis or Manganaro's in New York, the "Annex" in Princeton. (Yitz kept an automobile in storage in Rome for his frequent visits there—but he was equally at home in all of the world's great cities.)

How Yitz came to help me get the Fulbright-NATO Postdoc at Heidelberg is a long story I can't go into here, but the records will show that he wrote in favor of my application.

Like many, many mathematicians, I fell under the spell of his brilliance, and as a matter of fact, I owe him for recommending to me my first Ph.D. student, Barbara Osofsky. When we were in Stockholm for the ICM in summer 1962, he told me that she had been at Cornell but married "Abe", and was now at Rutgers, where I was headed, having completed membership at the Institute 1960–1962. He urged me to invite her to my seminars, which I did in the fall of 1962.

Injective Modules and Quotient Rings

The above is the title of Lecture Notes in Mathematics, vol. 49, finished on Valentine's Day, 1964, but not published until 1967 by Springer-Verlag. It would have been vol. 5 had I given it to the editors Albrecht Dold and Beno Eckmann at the time of its acceptance in 1964, but I delayed making out the index until my daughter Heidi, age 12, got impatient with my procrastination and did it for me. I am grateful to her for its publication and the enthusiastic reception by the mathematical community, which encouraged me to write my two volumes *Algebra* in the Springer Grundlehren series, also edited by Dold and Eckmann.

Fritz, Bruno, Rudy, and Ulrich

I began the *Lectures* at Pennsylvania State University in Spring 1962. Frink had asked me to help recruit Fritz while I was still in Germany, and Fritz had accepted the offer of full professor at Penn State in Fall '61 while I was still at the Institute. He brought with him not only Bruno Mueller, but two gifted students Rudy Rentschler and Ulrich Oberst with him. They proved to be intellectually challenging auditors. When I was vague or handwaved a proof, Rudy announced that "the direction we are going is clear but it's the details on how to get there that are unclear." This was about a course that I gave on "Fields of Algebraic Functions," based on Chevalley's book. (Poor Rudy!)

The High Cost of Living in Germany (1959–1960)

How I recruited Fritz for Penn State makes an interesting story: meat was still scarce and expensive in Germany in 1959–60, but because of my Fulbright I was able to purchase it at the Army PX at American prices. Now Fritz was very fond of steak, and the size of the steaks we fed him in our Pension in *Neuenheim* made his eyes pop out. I used to joke that Penn State bought Fritz for the price of a steak! To illustrate the belt-tightening of the German people even then, years after the war: when occasionally I did buy a pound of steak at the *Fleischerei* (Butchers) for my family, the Fleischer would say, "*Sie meinen en Viertel Pfund, nicht wahr?*" (You mean a fourth of a pound, don't you?) They weren't joking: few Germans in Neuenheim could afford an entire pound.

Steve Chase

During 1961–62, I was fortunate to meet Stephen U. Chase, a visiting professor at Princeton, who came to the Institute several afternoons a week to lectures and to tea. His brilliant grasp of mathematics greatly benefited me and my muddle-headed tendencies. I owe him a great debt which I acknowledged in the introduction to my *Lectures*.

The Institute and Flexner's Idea

My association with the Institute did not end in 1962 but continued with visitorships for the years 1973–74 and 1977–78 and every summer 1960–1979, when I met my wife Molly Sullivan. Moreover, I became an Associate Member of the Institute for Advanced Study (AMIAS) when it was founded in the early 80's to give support to the Institute, and as the happy acronym suggests "Keep in Touch."

The Members and Faculty of the Institute are listed in "A Community of Scholars 1930–1980." According to Pamela Hughes, Institute Development Officer, this is currently being brought up-to-date for the millennium. The 1980 edition with a Foreword by Harry Woolf, the Director, contains the historical development as an idea and a place. (See J. Mitchell [80].) Oswald Veblen was the first professor and Abraham Flexner the first Director. It appears that Bamberger was thinking of founding another Rockerfeller University type institution, but Flexner, who was an M.D. who greatly influenced American medicine, thought differently. He was steeped in the Classics and Philosophy and he thought that America needed a place where the world's greatest thinkers gathered to converse free of the distractions of classes, lectures, and students. Bamberger agreed and to this day the professors there do not offer courses, although they may offer extensive lectures, such as André Weil's "History of Number Theory" which was widely attended, and later published.

Lunch with Dyson, Lee, Yang and Pais

The democratic principle at the Institute encouraged free exchange of ideas in an informal relaxed setting, notably at lunch, tea or dinner. Often I had lunch with three physicists, Freeman Dyson, Chen Ning Yang, Abraham Pais, and others (often Tsung Dao Lee) in the dining room on top of Fuld Hall. The topics careened from Dyson's idea of perpetual motion utilizing twin stars to back to earth crushing of the human rights movement in China, or any improbable thing. I am reminded of how in Alice in Wonderland, the Red Queen thought of "six impossible things before breakfast." When you lunch with four genuises, it's intellectual hardball, and I was completely out of it.

Lee and Yang shared a Nobel for their devising the experiment which disproved parity in laws of physics. There's an (apochryphal?) story about a son of Yang when asked at school what he wanted to do when he grew up, he quipped "Win a Nobel all by myself!"

Helen Dukas

Helen was Einstein's assistant, who lived with him and his wife at 112 Mercer Street in Princeton. After his death, in 1955, she moved into an office in Fuld Hall, and catalogued his papers, writing English abstracts since everything was in German. She helped dozens of Einstein scholars, and physicists to access Einstein's work and letters, and as such, she was herself an Einstein scholar in the deepest sense. I wrote about her accomplishments in letters to the local Princeton papers when in her obituary they referred to her as his secretary. In doing so, I was basing my remarks on 25 years of friendship with Helen, but I wasn't the best person to say so. Was I ever happy when Abraham Pais backed me up in the strongest terms—he described how invaluable Helen had been in aiding his own research for his award-winning book on Einstein and his work, "*Subtle is the Lord.*"

Helen was a living saint—I think the right meaning still holds true: completely and utterly unselfish, without malice, or bitterness, helpful and kind to everybody. She adored children and babysat *gratis* for members of the Institute.

She co-authored with Banesh Hoffman a book on the wit and wisdom of Einstein (*Albert Einstein, The Human Side*, Princeton, 1979). Once Einstein, lecturing in Zurich without a fee, noticed that the superintendent was charging Hörgelt (an admission fee) and asked him, why do you charge admission, when I don't take a fee? The Super replied, "Herr Professor, think about the heat!" Helen told me this story in German, "*Denken Sie mal, Herr Professor, an die Heizung!*"

Another story that Helen told me. Einstein's first paper was rejected by a Swiss Physics Journal as being *too* short. He thought about it, *added one sentence*, and *then* the paper was accepted!

Arthur and Dorothy Guy

This reminiscence is out of sync for reasons that will become clear when you read the following one about Pat Woolf.

Everybody has had the experience of meeting someone from the past without either one being able to remember exactly where.

The first time this happened to me, I later found out that the person seated two seats away from me in a Nuclear Physics graduate course at Purdue in 1952 was the Airplane ID instructor when I was an R.T. at the Navy's Dearborn, Michigan Radio School. At Dearborn, he would come into the back of the room behind us and flick off the lights and begin 10-second slides of aircraft, enemy and ours, for rapid ID. He had a nasalized voice and his accent was definitely Bostonian (like JFK's "Hahvahd") *but he himself I never saw...I only recognized his voice.*

At various times we tried without success to place each other–it was then 6 years after I had been demobilized (demobbed as they call it) from the Navy, and neither guessed the *Secret* until one day after class I literally bumped into him at the A& P (in West Lafayette) and shouted: **It was at Dearborn**!

What an amazing coincidence! We became the best of friends at Purdue, with Arthur Guy and his wife Dorothy, and visited each other over the years. He was a gourmet who had a nose for wine which he twirled in the glass and warmed with his hands. He also taught us to warm up cheese manually. Unfortunately he inherited a fatal genetic condition, from his family in Wooster, Massachusetts, and died in his thirties.[18] Dorothy, a *summa cum laude* from Stevens College in Missouri, visited us after his death at the Institute in 1962. The adage that says that *genes are a poor man's will*, needs revising. In the age of DNA and the *genome* project, we learn that it's everybody's fate—rich or poor.

Patricia Kelsh Woolf

She burst into the Institute life the way she did at Purdue University when she was an undergrad and I was a TA: **she was always on the run, bubbling with energy**.

As Harry Woolf's wife (the Directress?) I did not recognize her at first, and we went through some painful: *were you at's*? This kept up for over a year, but one day I was sitting on the hood of my automobile outside the ECP (= Electron Computer Project), when she jogged down Olden Lane from the Director's House with pig-tails flying, and when she reached me I said, "*You are Pat Kelsh whom I knew at Purdue and who gave me a Boilermaker anvil paperweight back in May '55!*" And she screamed, "*My God, you are right!*" And we walked into my office and the paperweight was there on my desk. **It was really some memento!**

The reason I finally *did* recognize her was this: At Purdue she was tall and always jogging (to stay in condition for the swimming team?) and that is how I remembered her. That and her Athenian beauty and wisdom, for she also planned to become a Nobel Laureate, I believe. *But even as a sophomore at Purdue she was a Goddess from Southern Illinois.*

And this is how I got invited to **all** of the Fall and Spring Dinner Dances at the Institute during their tenure there. Once I asked her, "Don't you get tired of us?" And she said, "*No*" and then she added this Kelshian fillip, "*You* at least show up."

So this is how I knew she had made the invitations. That and the fact that they stopped when Harry stepped down.

Think Question: Did I generously return the paperweight to Pat, or did I selfishly keep it as a memento?

---

[18]By an amazing coincidence, the Guy's daughter, Jennifer, called me up on June 10, 1998, to invite me to dinner. (Unknown to me, she had been living in Princeton for years!) Furthermore, she told me that although her father had had terminal cancer of the colon, the immediate cause of his death was peritonitis from a ruptured appendix.

Johnny von Neumann and "The Maniac"

This was the affectionate name that somebody gave the computer that John von Neumann devised and designed, and Julian Bigelow built in the ECP in the 40's. (To resonate with the Illiac built in Illinois or because it often went haywire?)[19]

The details can be found in the Smithsonian where it presently sits. But for the years I was ensconced in ECP, it was there in all its glory, 1000 vacuum tubes or more, with super heavy airconditioners to cool them off. I wonder how many tubes or disks were pirated away for souvenirs? Probably none, given the idealism of the place, and the sacred regard people had for von Neumann.

And what about the air conditioners? They were a mixed blessing—the offices closest to the Maniac froze and those farthest away and around the far corner sweltered.

Just for the record, Johnny von Neumann was President of the American Mathematical Society (1942–1943). Julian Bigelow is a permanent member of the Institute, and for the years I was there, had an office in ECP.

Who Got Einstein's Office?

A good book to read on the lore of the Institute is "Who Got Einstein's Office?" by Ed Regis. The jacket blurb says he is an Associate Professor of Philosophy at Howard University, and credits are given for his excerpts from Abraham Flexner's *"An Autobiography,"* Douglas Hofstadter's Gödel's diagonal argument adapted from his book "*Gödel, Escher and Bach*", P. R. Halmos' "*The Legend of John von Neumann*", Herman Goldstine's, "*The Computer from Pascal to von Neumann,*" Thomas Kuhn's *Oral History Interview of J. Robert Oppenheimer*, Richard P. Feynman's "*Surely you're joking Mr. Feynman: Adventures of a Curious Character,*" among other credits.

The Walkers, Frank Anderson, and Eben Matlis

Late in the summer of 1963 at the Institute I had the good fortune of meeting Carol and Elbert ("Tiger") Walker (New Mexico State U., Las Cruces). Professionally I knew Elbert from his paper [56] solving several of Kaplansky's Test Problems for Abelian groups. Sometime in the Spring of 1964, we began to seriously work on the paper containing the Faith-Walker Theorem [67], and my sequel [66a] became a "prequel" by the vagaries of journal backlogs.

Professor Frank W. Anderson also spent the year at the Institute posing difficult problems for us. Elbert started our collaboration with the question of when is the endomorphism ring H of a free R-module right self-injective. (See 3.33). This led to the question "when do injectives = projectives?" We were heavily indebted to him for lubricating the machinery at our frequent evening meetings at the Walker's house on Einstein Drive. We picked Eben Matlis' brains at tea times. (He had plenty left when we finished!)

[19] At first I didn't catch the pun "ill-iac.".

Carol and Elbert

Elbert and Carol were newly married, and their love for each other was a pleasant aspect of our collaboration. Carol had a mischievous streak in her which enlivened our lives, as with this repartee: "Mother always told me to be good. Was I good, Elbert?" And Elbert would *have* to agree "*Real* good!" And everybody would laugh like idiots at the innuendo. *Everybody loved Carol and Elbert so much.*

"Waiting for Gottfried"

Not long after Samuel Beckett's enigmatic play "Waiting for Godot" was staged (in 1955) I acted out a real-life version at the Penn Central Railroad Station in New Brunswick, waiting for Gottfried Köthe, our Colloquium speaker. As Colloquium Chairman of the department, I had invited Gottfried, who was a visiting professor at the University of Maryland where I had lectured at the Colloquium in the Fall of 1963. At that time, Gottfried had accepted my invitation for him to speak at Rutgers in the Spring of 1964.

When he failed to show at the railroad station, I got on the "blower" (navese for intercom) and found him in his office in College Park, Maryland: I had made a mistake–the talk was for the next week!

Not to worry, *I took everybody to my house where we devoured the feast we had prepared for Gottfried.* The next week Gottfried arrived at the scheduled time—no more waiting—and after his beautiful lecture on functional analysis, we followed the script of the week before: *I took everybody over to my house for the feast we had prepared for Gottfried.* Only *this time* Gottfried had the chance to enjoy it with us! (It was a *quid pro quo* for the one he served us in Heidelberg four years earlier!)

Isn't this a bit surreal, like the Beckett play?

Harish-Chandra

His full name has this hyphenated form, as listed in "A Community of Scholars," but people used Harish as a first name, i.e., given. I wish I had had the chance to have known Harish better. He had an ascetic quality and often stern visage that inhibited familiarity, although we often chatted with him at Institute socials, and his wife, Lily, was exceedingly friendly. (Harish may or may not have been a Brahmin, i.e., of the Indian ruling class: a carry-over of the Indian Caste system.) I have a photograph of Harish and Ed Nelson of the Princeton Mathematics Department, at an Institute dance. (They kindly posed for me.)

Because of my gregarious nature, I think I may have irritated Harish one day in the library, when he sharply asked me to please be quiet! (I had not seen him or anybody else there.) I could never have spoken that way to anyone: because of the *third child syndrome*, I yield (or maybe rebel?) to everybody on command. (Ha!)

But this is very unimportant—because of Harish's mathematical accomplishments, he is deservedly revered: he won the American Mathematical Society's prestigious Cole Prize in 1954 at age thirty-one, and some twenty years later he won India's Mathematical Society's Ramanujan prize.

Sadly, Harish suffered a severe heart attack late in his 40's, although he looked as thin as Mahatma Gandhi, and ate obviously low-fat food. (Nobody thought of those things, then!)

After that he was much more accessible. Because of his daily regimen of walking around the 0.7 mile oval drive in front of Fuld Hall, *and* my own regimen of running the oval 3 times daily, we bumped into each other a lot, and Ima Dyson who also jogged, or Freeman Dyson walking to his nearby house off Battle road.

Despite this, Harish died not many years after, in October 1983, just sixty years old. His death cast a deep pall over the Institute.[20] After the deaths of many great people, there is a sense of unreality, *a feeling that Harish still lives in a dimension beyond us.* Maybe as Proust has written: *People do not die immediately for us, but remain bathed in an aura of life...it is as if they are travelling abroad.*

I want to record here this deep gloom I felt on a personal level. Even today when I walk around the oval I half-expect to see Harish coming around the bend carrying his omnipresent briefcase.

Veblen, Tea, and the Arboretum

Tea was the only beverage served at the afternoon teas at the Institute because Oswald Veblen objected to stronger stimulants. As one of the first professors to come to the Institute, he selected the setting for the Institute in the 600 acres of woods and fields he loved to walk in. He also designated 100 acres of his estate to form Mercer County (formerly called Veblen) Arboretum off Snowden Lane. Veblen was President of the American Mathematical Society 1923–1925.

As a curiosity another professor of the Institute, Millard Meiss, in Historical Studies, left 100 acres of his land to the National Audubon Society, who ceded it to Princeton University for flora and fauna studies as well as preservation.

Once it was opened to the public, I tramped across Meiss' land with Professor Paul M. Cohn of the University of London who visited Rutgers in 69–70. Paul remarked it seemed such a waste of land! There is little open land in England since every available acre was used for farmland. I recalled the punning title of a hilarious movie "Tight Little Island", about smuggled whiskey.

"On the Banks of the Old Raritan" (School Song)

At Rutgers in Fall 1962, I continued the course based on the **Lectures**, and once again I was lucky to have an excellent group of auditors Dick Bumby (a Ph.D. student of D. C. Spenser at Princeton), Bill Caldwell, Dick Cohn (a former student of E. Kolchin), Dick Courter, Harry Gonshor, Bob Heaton, Barbara Osofsky, Joe Oppenheim, and Earl Taft (a former student of Nathan Jacobson), among others.

I ought to mention here how I came to Rutgers. Like so many I fell in love with the rolling fields, Stony Brook and woods of the Institute (now preserved for posterity by the Institute Woods Preservation Fund), the colonial village of

[20] I have Minoto Gangiolli, the mathematics Librarian and a friend of the Chandras, to thank for supplying these dates.

Princeton, and the proximity to the New Jersey shore on the East, the Poconos and Bucks County in Pennsylvania on the West, and New York and Philadelphia at the North and South Poles (so to speak).

It's a beautiful fact that the Institute's Stony Brook flows into Princeton's Lake Carnegie, formed by the confluence with the Millstone River which meanders north some 25 miles and empties into the Raritan on whose banks Rutgers University sits. Alongside these rivers run the Delaware and Raritan canals and the ancient towpath on which the commerce of New Jersey depended since Colonial Days.

A former UK classmate of mine, Wilson Zaring, suggested that I apply to Rutgers when he invited me to his house to watch the Kentucky Derby and sip mint juleps. I had heard that Rutgers University in New Brunswick was not in the same league as Princeton, that is, the Ivy League, and not even in the same league as Purdue, namely the Big Ten (which Penn State subsequently joined). The upshot was, would I be interested in a job: Rutgers was looking and so was I!

The Bumby-Osofsky Theorem

Two modules A and B over the same ring are **subisomorphic** if each is embeddable in the other. In 1940 Reinhold Baer generalized the concept of divisible Abelian groups to injective modules over any ring (i.e., a divisible Abelian group is an injective $\mathbb{Z}$-module). Subisomorphic divisible Abelian groups were known to be isomorphic, and in a parallel seminar on Abelian groups that Gonshor and I were conducting, the question arose as to the validity for injective modules. Bumby and Osofsky independently proved the answer was "yes," and the proof ran vaguely along the lines of the proof of the Cantor-Bernstein Theorem for sets.

Osofsky's Ph.D. Thesis

Shortly afterwards Osofsky also solved her main Ph.D. thesis problem: if R is a ring and every cyclic right R-module is injective, then R is semisimple Artinian. (For another one of her Ph.D. problems, see "Pere Menal" below.)

This was perhaps the record short time for a thesis problem! Yet the proof was highly ingenious, has had enormous ramifications in algebraic mathematics, and to-date no one has found a simpler proof. (As a matter of fact, the same is true of the Bumby-Osofsky Theorem.)

Just as Barbara was preparing her thesis defense, I discovered a paper by L. A. Skornyakov which purported to prove her main theorem. Not to worry, Barbara found a mistake in his proof, and subsequently published "A counterexample to a lemma of Skornyakov."

Was ever a thesis better defended?

In 1967–1968, Barbara was elected to membership in the Institute.

Yuzo

In the *Lectures*, I applied the Johnson-Utumi concepts of the maximal quotient ring of a ring to obtain a new proof of Goldie's and the Lesieur-Croisot Theorems

(3.13ff.). The idea of the Faith-Utumi theorem was first published in rudimentary form in my Abstract of the American Mathematical Society for a meeting held in Cincinnati in the Spring of 1961. By a stroke of luck Yuzo Utumi heard my lecture, and in the ensuing year and a half the Faith-Utumi Theorem (7.6A) was brought to light. It was the first in a series of 5 papers on which we collaborated and my second paper written at the Institute. Incredibly, this work was done totally via correspondence (via airmail not e-mail!). More than twenty-five years later we finally got together at Rutgers in Fall 1988 when he visited. At that time, Poobhalan Pillay, another collaborator of mine and I would be talking about Utumi's theorems in the Rutgers Hill Center lounge in Utumi's very presence but he evinced absolutely no interest. I believe that he was working in number theory, a subject relatively remote from our interest at the time, the little book listed in references published in Murcia in 1990. Shortly afterwards, Yuzo became ill and had to return to Japan.

At the Stockholm ICM (1962)

The value of the Faith-Utumi theorem lies in its application to Goldie's theorem, eliciting a graphic picture of how a Goldie semiprime ring R is positioned in its quotient ring $Q(R)$. I made the journey to the ICM in Stockholm to lecture on these theorems and by another amazing coincidence, A. W. Goldie was an auditor. And that's not all: after my lecture, he said to me "The Faith-Utumi theorem is false". Did that ever shock the living daylights out of me. Not to worry, later he not only changed his mind, but in a paper in the London Mathematical Society, he states, without proof, that the Faith-Utumi theorem was a Corollary of his theorem, despite the fact that while his theorem is much deeper, its use in the proof of our theorem only could be a "red herring."

There is a number of humorous jokes on mathematical aberrations: One is about a mathematician who reads another's paper and says "These results are false; besides I myself proved them first!" Greg Cherlin told me this one: he told a colleague about a theorem he proved on $\aleph_0$-categorical rings and his colleague replied, "I can prove a better theorem! But, *first*, tell me *What are* $\aleph_0$*-categorical rings?*"

Nathan Jacobson

Jake (Yale U.), as everybody calls him, revolutionized ring theory by his papers and books. The essence of his approach was to generalize important concepts to rings "without finiteness assumptions," e.g., in his paper [45b]. I attended many of his splendid lectures at the University of Chicago where he lectured frequently at AMS meetings. They were nonpareil expositions, as are his many books.

Jake was born in Warsaw in 1910, raised in Birmingham, Alabama, was a 1930 graduate of the State U. (home of The Crimson Tide), received his Princeton Ph.D. in 1934 under the direction of Wedderburn, was an assistant to Herman Weyl at the Institute in 1934–35, and lectured at Bryn Mawr in 1935–1936, replacing Emmy Noether, who died during an optional routine operation in April 1935.

Jake served as President of the American Mathematical Society (1971–73). A list of positions and honors bestowed and earned by him including National

Academy of Science membership in 1960, are given in each volume of Jacobson's Collected Papers [89]. (Somehow his tenure as Chairman was omitted.)

Jake received the Steele Prize for Distinguished Lifetime Achievement at the Annual Meeting of the Society in Baltimore in January 1998. Was there ever an award more deserved?

When he became Emeritus Professor in 1981, his wife Florie told a group of 100 well wishers at a celebratory dinner at Yale "The best thing Jake ever did was marry me!" The audience and Jake greeted this with applause and laughter.

Florie taught mathematics and "obtained her master's degree with Otto Schilling, and was working for a doctorate under Adrian Albert" when they married in 1942 (to quote Jake, *op.cit.*, vol. I, p.5). She taught mathematics at Albertus Magnus, a college near Yale.

How Jake Helped Me and Rutgers

Utumi and I were joyous when Jacobson's revised Colloquium Lectures [64] appeared with our theorem in Appendix B along with Goldie's theorems. This put the question of its falsity to rest, and brought us world-wide fame. Is it a coincidence that it was in the very same year (1964) that I received an offer from a Big 10 university as a full professor for a salary that almost doubled the maximum salary paid Rutgers professors? No, and I have Yitz to thank for recommending me.[21] The upshot was that Rutgers President Mason Gross was able to persuade the New Jersey legislature to revise Rutgers salaries to accommodate matching the Big 10 salaries. This in turn helped us win the NSF Center of Excellence Grant in 1966 of $4.5 million by convincing the NSF that Rutgers and New Jersey were committed to excellence.

Provost Richard Schlatter congratulated me on the offer, said he hoped I would stay, and that the Governor raised the question of whether New Jersey could afford a mathematics department! Dean Arnold Grobman wrote me a letter of commendation, and gave me a "Dean's increment" in salary when I returned from my year (1965–1966) as visiting scholar at the University of California at Berkeley where I wrote *Algebra I*, and helped recruit our present chair Professor Antoni Kosinski (maybe the increment was for the latter!)

Vic, John, Midge, and Ann

When I returned from Berkeley in Fall '66, I was assigned to teach Introduction to Abstract Algebra, and I quickly realized that Vic Camillo (U. of Iowa), Ann Koski Boyle and John and "Midge" Cozzens (the latter three presently are program Directors at the NSF: Algebra, Circuits and Signals, and Mathematics Education) were the perfect students to try out chapters in my Algebra I. They profoundly influenced my writing, and are properly acknowledged in the Introduction to the book. As Einstein has said, "a teacher should be an example—of what to avoid—if not of the other kind."

[21] This story is told about Mel Henriksen: when he heard of the offer, he rushed up to Kap (= Kaplansky) and said, "Faith isn't worth that much is he?" And Kap's Solomonlike reply: He is **now**!

A Problem of Bass and Cozzens' Ph.D. Thesis

In his Ph.D. thesis "Homological Properties of the Ring of Differential Polynomials," published as a research announcement in 1970 in the *Bulletin of the A.M.S.*, John solved an important problem posed by Bass in 1960 about rings over which R-module $M \neq 0$ has a maximal submodule, called *max rings* in Part I. John showed that the rings in his title over a universal differential field k with respect to a derivation D were not only max rings but were principal ideal domains with a much stronger property: *the intersection of all maximal submodules of every module is zero.* (These are called **V-domains**; see Part I, 3.19Aff.)

John also constructed simpler examples, e.g. the "localization" at (x) of the skew ring of polynomials $R = k[x, \alpha]$ with respect to an automorphism α of finite order over an algebraically closed field k. (Barbara subsequently constructed other examples.)

Consequently, John went to Columbia University as a Ritt Instructor, 1970–73, where Hy Bass was a professor, and, Ellis Kolchin, whose construction of universal differential fields enabled John to construct V-domains other than fields, and thus answer Bass' question.

Boyle's Ph.D. Thesis and Conjecture

Cozzens' thesis has had enormous influence in the development of ring theory and homological algebra, e.g. it led directly to Boyle's Ph.D. thesis in 1973 and Boyle's conjecture (see Part I, 3.9C) which is still open some twenty-five years later (Cf. 17.40ff).

Cozzens' thesis also led to our book in 1975, *Simple Noetherian Rings*, published by Cambridge, and several questions which stimulated a lot of productive research, e.g. the paper by Resco [87], which enabled Menal and the author to solve John's conjecture in the negative (see Example 13.18).

A Problem of Thrall and Camillo's Ph.D. Thesis

The original thesis problem was to solve Thrall's Problem, namely, determine all rings, called right QF-1, with the property that R was the bicentralizer (= biendomorphism ring) of every finitely generated faithful right R-module. This proved very difficult indeed, having lain unsolved even for finite dimensional algebras over a field ever since Thrall posed it in 1948.

Not to worry, when you can't solve a problem, you recast it in another form: suppose R induces in a natural way the bicentralizer of every M (and not just the finitely generated ones). Such rings are called **balanced**, and there Vic proved a decisive theorem, published in 1970 in the Transactions of the A.M.S. (see Theorem 13.29 in Part I).

Furthermore, Vic and Kent Fuller, and independently Dlab and Ringel proved a number of characterizations (see Theorems 13.30–13.30A–E).

I was very impressed with Vic's adaptation of results of Bass [60] that involved *interalia*, a *double induction*!

Avraham and Ahuva

Avraham Ornstein of Technion U. came to Rutgers in 64–65 to work with me, and while I went to Berkeley in Fall '65, he continued his research independently at Rutgers until we put on the **finale** at Berkeley in the summer of '65. This made his wife, Ahuva, very happy.

Abraham Zaks

Another happy consequence of my work with Avraham Ornstein was the invitation to visit Technion which is located in Haifa, and in January '77, I gave several lectures there on FPF rings. Was I ever surprised and pleased when Abraham Zaks (Technion) expressed interest in sponsoring Lecture Notes of the London Mathematical Society under the editorship of I. M. James and in due course it was published in 1984, with S. Page as co-author. Whatever influence "FPF Ring Theory"–the Lectures–has had is due in part to Zaks, and we duly acknowledged him in the book.

Professor Netanyahu

The late Professor Netanyahu, the father (or Uncle?) of the present Prime Minister of Israel of the same name, kindly drove us out to Caesarea for a delicious lunch and a valuable lesson in history: **It's everywhere in Israel**.

Jonathan and Hembda Golan

Avraham Ornstein and Abraham Zaks arranged the splendid opportunity to talk in Jerusalem. Jonathan Golan (Haifa U.), who had attended my lectures, gave me a harrowing drive over some mountains mostly on the wrong side of the road. When I complained, he just laughed and said "In Israel everybody expects to die tomorrow and drives crazy. *You have to drive crazy! They expect you to.*"

Jonathan also explained Israeli archeology. Compared to the Jewish, crusader ruins are not very old, ergo *they dig right past Christianity!*

Hembda Golan was an official in the Israeli Department of State who helped to write the Israel-Egypt Peace Treaty initiated by Anwar Sadat's overture to Israel's Prime Minister, Menachem Begin.

Shimshon Amitsur

After my lectures at Jerusalem, where I renewed my friendship with Shimshon Amitsur (whom I met when he lectured at Penn State in 1958), Jonathan invited me to drive to Beth Oren (Haifa), Ein Gedden (West Bank), the Dead Sea, and Beersheba (= seven wells, literally).

Soon after the mathematics department moved into Hill Center in 1973, Shimshon spent a summer at Rutgers where he lectured on PI-algebras. After his lecture I would drive him to his apartment in Highland Park (the Donaldson apartments) where we would walk along the banks of the Old Raritan.

About this time, Amitsur had proved the existence of uncrossed division algebras (§2), and when I told Atle Selberg about it, he arranged for Shimshon to lecture on this subject in Fuld Hall. Atle and Hetty gave a party for Shimson that evening (Atle told me that the Amitsurs had feted him on his visit to Israel.)

Here is an amusing and puzzling anecdote. On our walks, Shimson professed not to know the names of the flowers. "Flowers are just flowers to me," he protested. I found this attitude incredible (despite Shakespeare's famous line about "a rose"!). When I told Atle, who knew the name of every wildflower, he said, "Knowing Amitsur, this does not surprise me." (I am still puzzling over this twenty-five years later.)

Amitsur's "Absence of Leave"

Shimson told me that the Dean at Hebrew University refused to extend his leave at Yale to a second year (or semester, I forget which). "So I took an *Absence of Leave*!" he joked, but even though I chuckled, he said very seriously, "in Hebrew I am very funny!" I protested, "*You are very funny in English, too*," but he was unconvinced.

Miriam Cohen

Finally I arrived in Tel Aviv to be greeted by Miriam Cohen who gave me a tour of Jaffe, the harbor and other sights. There were many patrols that carried submachine guns even in those days, and there were many checkpoints where you had to identify yourself.

Miriam answered my question as to why Israeli breakfasts are so huge, including fish, cheeses, eggs (but never bacon!), salad, veggies, and any number of breads and condiments. "You need all the fuel you can get to get through the day" is what she said in essence. You could see what she meant: Like New York, Israel is very much a fast-forward video!

Joy Kinsburg

My (Covington, Ky) childhood sweetheart, Joy Kinsburg, became my (platonic) High School Sweetheart. She was my first (second grade!) Jewish friend and Zionist acquaintance. Moreover, my first knowledge of the Holocaust I experienced through her tears. Many of her relatives and friends were torn to pieces by the Nazis. I was apprehensive about our chances of defeating this maniac Hitler! In high school, 1941–1945, he seemed invincible until the very end. Joy's father tutored me in electronics to enable me to become a Navy R.T. (= radio technician). When I took the Eddy Test (named after a captain of that name), only 2 out of 41 passed, and I owed my passing to him, but the war ended in Germany just as I entered the service. (Hitler knew he was licked?)

The connection with this reminiscence is this: Joy lived in Beersheba. Sadly, I never got the chance to see her on this, the only trip I ever made to Israel. However, to this day, we sporadically call each other, occasionally on our birthdays. She has a Ph.D. from Ohio State U. in Agronomy, a subject of great benefit to Israel. Once, when I asked Joy if she knew Richard Schlatter who had become a Vice President of the University of Cincinnati, she said, "Why he's my Uncle Dick!" (Small world?)

Paul Erdős

Everybody knew the peripatetic Erdős: *Who travels lightest, travels farthest.* He came and went carrying just a briefcase, entrusting his hosts to supply the toilet articles and pajamas (somebody told me his PJ's were in the briefcase). Henriksen told me this story at Purdue about Erdős. Once having outstayed his welcome at Tarski's house in Berkeley, he returned one night to find himself locked out, and he banged at the doors and windows shouting: "Fascist!,"[22] "Capitalist!"[23]

Erdős was a member of the Institute 1938–1940.

I first met Erdős as a 2nd or 3rd year graduate student back at Purdue in 1953–54, where he gave a colloquium lecture on number theory. He and Selberg earlier had independently discovered an elementary proof of the Prime Number Theorem (By coincidence, a story about this appears in the *Intelligencer*, June 1997.) I can't recall if he was lecturing on it, but he used a term that sounded like oop-'sigh-lon, *which everything was less than.* At the lecture I had supposed it was some universal constant! Later, to my dismay, I found that it was *epsilon* being used familiarly as an *infinitesimal*! This meshed with his epithet for children—he called them epsilons!

Such was my earliest impression of one of the world's greatest geniuses, who wrote more than 1000 (some say 1200) mathematical papers: No doubt many were co-authored by the local mathematicians, when he visited a day or two, leaving them to do the typing!

What Is Your Erdős Number?

Because of Erdős' penchant for collaboration, it is conjectured that every mathematician has an **Erdős number**. Your Erdős number is 1 if you were a co-author with Erdős, and 2 if you weren't but were co-author with someone with Erdős' number 1. The definition proceeds by mathematical induction! The conjecture is that every mathematician has an Erdős number $< \infty$. When mathematicians tell me their Erdős number, I like to joke: well, what is your Gödel (girdle) number? (groan)

For many years Erdős offered cash prizes as inducements for solutions to unsolved problems. Ron Graham, a Rutgers Professor and Past President of the A.M.S., acts as Treasurer for Erdős' prize funds and expenses.

---

[22]In "The Meaning of Life" (David Friend and the Editors of Life, Little, Brown & Company, Boston, Toronto, London 1991) Erdős jokingly refers to "The Supreme Fascist (= s.f.) or in other words, God." "The aim of life is to prove a conjecture...and to keep the s.f. score low." Another mathematician, Raymond Smullyan, espouses a much happier "meaning of life" on the preceding page.

[23]L. Babai and J. Spencer expand on s.f. in their obituary of Erdős in the January 1998 *Notices of the A.M.S.*; especially see Erdős' Maxims on p.70 (The "Permanent Visiting Professor" title at Technion is also listed on that page!) Babai, C. Pomerance and P. Vértesi expound on "The Mathematics of Paul Erdős" on pp.19–31 in the same issue. See also Bollobás [98] and Graham [96] on Erdős.

Erdős also founded the Budapest Semesters for gifted undergraduates, which Molly's and my son, Japheth Wood, attended when he was a freshman at Washington U. in St. Louis. In May 1997 Japheth received his Ph.D. from the University of California at Berkeley in Universal Algebra under the guidance of Professor Ralph McKenzie. (His thesis was the proof that Type 2 and Type 4 minimal algebras are not computable.) I like to say that Erdős' Budapest Semesters helped mature him at an early stage in his development, and that Ralph polished him up (or off?)

Piatetski-Shapiro Is Coming!

One day during my visit in Israel, I bumped into Paul Erdős, at Technion where he was "Permanent Visiting Professor." He said to me enigmatically, "Piatetski-Shapiro is coming!" What did I know from this? Piatetski-Shapiro's coming was an event of great significance! In Israel this meant **coming from Russia, from behind the Iron Curtain**; and then only after years of waiting for permission to leave. There was a hilarious satirical movie "The Russians Are Coming! The Russians Are Coming!"

The next day, I saw Paul again and he said, "Piatetski-Shapiro is here!" I looked around and didn't see anybody, but I now knew that Piatetski-Shapiro was safe in Israel.

The very next day, the same thing happened, I saw Paul, and he said, "*Piatetski-Shapiro is leaving!*"

By this time the suspense was killing me, and I blurted out, "Where to?" And he said, "Yale", or maybe the "Institute," then "Yale," or vice versa. [Confession: throughout all these years, I didn't know how to spell Piatetski, and looked him up yesterday when I remembered this story. I looked him up in the Combined Membership List of the A.M.S., M.A.A., and S.I.A.M. under "New Haven; Yale" (and there he was! *Erdős wasn't lying!*)]

Gerhard Hochschild on Erdős

Gerhard Hochschild told me this story in Fall '65 at Berkeley about the prospect of collaborating with Paul Erdős. Erdős asked, *"Tell me Gerhard, what is this Galois theory you keep working on?"*

Gerhard gave a reply right out of Clint Eastwood's "Dirty Harry": *"Paul, don't even think about it!"*

Thus, Gerhard passed up Erdős Number 1!

Joachim Lambek

"Jim" Lambek (McGill U.) appears in the official photograph of the Canadian Mathematical Congress for 1945, number 28, along with G. D. Birkhoff (67), R. Brauer (69), H.M.S. Coxeter (32), S. H. Gould (120), I. Halperin (116), I. Kaplansky (97), and J. von Neumann (124). Also A. Pedoe, T. H. Hildebrandt, and M. Robertson are listed among the 124 participants, but I could not identify them.

Jim came to Canada via New England from Leipzig as a refugee from the Nazis and lived in a Displaced Person Camp, where Fritz Rothberger (Acadia U.,

Wolfville, N.S.) taught him mathematics to while away the hours.[24] (For the record Fritz was also at the 1945 CMS Congress, as was a teacher of mine at the University of Kentucky at Lexington, A. W. Goodman.)

Jim's book, "Lectures on Rings and Modules," with an Appendix by a student of his, Ian Connell, published by Blaisdell in 1966 (Chelsea published a revised edition in 1976), one of the most readable, hence influential, algebra books of the sixties and seventies. There were many exchange visits with Jim, his wife Hanna, and his three sons in Montreal, when I lectured at McGill; New York, where he visited his mother; New Brunswick where he frequently addressed our colloquia and seminars. I have learned a tremendous amount of mathematics from him, his papers, books, and his students at McGill, e.g. David Fieldhouse, and Mary Upham, whose Ph.D. theses I had the pleasure of reading as an outside examiner...also for John Lawrence, a student of Connell.

Jim devised a mathematical term "Dogma" as a pun on "Category."

I have a wonderful assortment of art post-cards from Jim from all over the world. His most recent one entitled "The Three Bares" is a sculpture of nudes at McGill by Gertrude Whitney, founder of the Whitney Museum on Madison Avenue in New York. (Jim is still punning at the age of "seventy-something".)

S. K. Jain and India

As soon as my **Lectures** were published, I had the good luck to have them read by S. K. Jain (now at Ohio U., Athens) 13,000 miles away in New Delhi. At this time, the government of India under Indira Gandhi (Nehru's daughter who married a Gandhi unrelated to the Mahatma) encouraged science development schools sponsored by the NSF and AID (= Agency for International development). "S.K.", as everybody calls him, organized a six weeks course (May-June) at the University of New Delhi, and as a consequence I was given the inestimable honor of lecturing to his students: Ram N. Gupta (now at Chandigahr), Saad Mohamed and Surjeet Singh, (both presently at Kuwait U.), Kamlesh Wasan, and *inter alios*, two students who came to Rutgers and in the early 70's earned Ph.D.'s under my direction: Saroj Jain (Butler State U.) and Ranga Rao (U. of Aurangabad). Bhama Srinivasan also attended the lectures, and showed me a bit of Old Delhi.

Professor Jain organized a train trip for the conferees to see the Taj Mahal in Agra and other Moslem buildings in nearby *Fatehpur Sikri.* Yes, the Taj is everything it is proclaimed to be—a veritable temple to love—by a prince to commemorate his wife who died. The original was bedecked with precious stones but these were stolen—like love often?

Kashmiri Gate at 5:00 P.M.

Bulls, kites
Bullocks, carts
Buses, cars
Bicycles, cripples

[24]Excerpt from a letter from Jim of July 23, 1997: "I went to England as a refugee. There I was interned and deported to Canada."

Bulls gaunt and children starving
British soldiers and fat cows
Black skin, and calves bawling
Barelegs and an orange sari
Beggar children and bawling calf
Blue saris and pierced ears
Blind man and white dhotis
Buddhists in red robes
Movie house disgorging occupants
Broken sewers stenching
Pedestrians, cows
Bandaged feet and lousy children
Gallant on horseback (racing madly between it all)
Bugles blaring at Red Fort
Businessmen (very rich)
Boy with one shoe and a cat
Bullock carts
Poetry in Hindi read by a Hindu (oblivious of it all)
Beefy women and skinny bulls
Blackskin leathered by the sun
Sandaled girls in white saris
Eyes everywhere staring at everything
Rickshaws running by, pulled by skeletons
Bandaged legs and bare feet
Girl in silver anklets
Pierced noses and pipes stenching
Movie house recovering digestion
All orchestrated by a policeman in khaki shorts and long woolen socks
Cripples on crutches
Bearded Sikh driving tattered taxis
One elephant (moving fast)
Two Buddhist monks
Red stockings on sandaled feet
A very small shapely ageless woman in soiled silk pajamas anxiously
staring out of deep sockets for someone who does not come
A mendicant mortifying himself
A Hindu with a King Cobra
Cowshit everywhere
Village idiots with long sticks gouging the oxen
(gets them moving, eh Sahib?)
A child at breast
Exhausted Haryjans sleeping on the sidewalk
Bengali Club and Atma Ram Bookstore
A dirty child, flies on his mouth, begging for rupees
(but he is weakening)
No whores, no pimps, and
A temple in a bare tree.

TOOT-TOOT FOR A DAY! TOOT-TOOT FOR AN AGE!

East of Calcutta's Black Hole
The bridge Howrah's crowded majesty
Spans spewing steamers
Swimming swirling tides
Wending West down India's sea.
The Houghly monster strides
Roiled rivers black with boats—
Morning air black with smoke—
Big black stacks belching boasts:

Toot-toot for a day!
Toot-toot for an age!

The Rupee Mountain

By another stroke of luck, there was a "rupee mountain" that needed to be melted. Because of India's precarious economic health then, money that India paid for food imports and other necessities was not permitted out of the country, and could be spent by the U.S. only in India. Thus the term: rupee mountain.

K. G. Ramanathan and Bhama Srinivasan (Bombay and Madras)

Accordingly, I was encouraged by Bill Orton, program director of the NSF in Delhi, to see a bit of India by lecturing in Jaipur, Bombay, Madras and Calcutta under the auspices of the NSF/USAID rupee mountain. The director of the Tata Institute, in Colaba, Bombay, was K. G. Ramanathan, whom I had met at the Institute (1961–62). Unknown to me, India was building an A-bomb, and the free atmosphere that I associated with universities was missing at Tata. As always, I took pictures. **That** provoked the guards, but somehow K. G. was able to reassure them that, like Inspector Clouseau of Peter Sellers' films, I was harmless, i.e., too stupid to be a spy! And he was right!!

After Bombay came Madras, at the invitation of Bhama Srinivasan (now at U. of Illinois at Chicago Circle). Bhama was a gracious host and drove me and my family to see the Hindu Temple at *Mahabili Purim*, south of Madras. I was to meet Bhama again at the Institute for Advanced Study in the summer of 1977.

After Madras, I visited Calcutta, and Banares, where they burn the bodies of the dead, and float them down the Ganges (Mother of India) River in funeral voyages.

The Indian Idea of Karma

Once I was travelling through New Delhi in a taxi which ran out of gas. As I was in a hurry, I started to flag down another, but the driver gave me a horrified look and said, "Don't do that! Don't you realize that it was fate that brought us together and you will upset your karma if you take another taxi—something bad will happen to you." So what could I do? Bowing to fate seemed the wisest thing to do. As we talked, the driver pushed his ultra lite cab around the corner, and as if by a miracle, there was a gasoline station that I hadn't seen before. Within minutes, the tank was topped off and we were on our way as fate commanded.

I had a similar experience in an Indian Coffee Shop where everybody spends hours in interminable debates over everything under the scorching Indian sun. (These are called Indian *muddles.*) I was suddenly approached by a young Indian who sat down and told me that he had been watching me at the coffee shop for several weeks. I said, "And?" "Well, you're all right!" was the reply. And often I've asked myself, *suppose* I hadn't been "all right"?

The concept of Karma is broader than just fate: **You can change your fate through your good karma**, that is, **your goodness**. This theory explains, I think, why some people "have all the luck."

Because of these experiences in India and others, even having the chance to visit India, and to see the Taj Mahal (one of the Seven Wonders of the World), I have come to believe in Karma and the ability of all people to have good luck through it. (It reminds me a bit of the Golden Rule.)[25]

Joan and Charles Neider

This couple, whom I made friends with at the Whitney party mentioned earlier, are scholars. Joan holds a Ph.D. from Columbia University—her thesis is a study of Thomas Mann whom they knew when he lived in Princeton.

Charley

Charles is a foremost Twain Scholar, and editor of Twain's, *The Complete Short Stories*, tales, novellas, novels (e.g. *The Adventures of Colonel Sellers, Huckleberry Finn), The Autobiography of Mark Twain, The Letters of Mark Twain* and other writings ("A Tramp Abroad").

He is the author of a distinguished biography of Mark Twain, and editor of "Papa", a biography of Mark Twain by Twain's thirteen-year-old daughter, Susy.

The Neider's named their own daughter Susy, and Charles wrote a memoir, "Susy, A Childhood," about her. Neider has written extensively on Robert Louis Stevenson ("Our Samoan Adventure"), Franz Kafka ("The Frozen Sea"), and Washington Irving.

In addition, Charles is an antarctic explorer and chronicler of the lives and exploits of Shackelton and Scott in his books on Antarctica; "Edge of the Earth: Ross Island Antarctica"; "Beyond the Horn: Travels in Antarctica", and "Antarctica."

He has written a number of novels, *The Authentic Death of Hendry Jones* which was the basis of a movie "One-Eyed Jacks" starring Marlon Brando, two comic novellas, *The Trip to Yazoo, Mozart and the Archbooby* published by Penquin-Viking, *Overflight* about the crash on Mt. Erebus that he survived in 1956, and recently, *The Left Eye Cries First.*

Neider has enjoyed considerable support in his writings, especially the National Science Foundation supported his humanist interest in Antarctica and was amply rewarded by his exquisite prose descriptions. He also has been a visitor at McDowell, Yaddo, and the University of California at Santa Cruz.

Molly and I, and his many friends and neighbors, were happy to wish him Happy Birthday on his eightieth birthday party thrown for him on January 15, 1995 by Joan in their home on Southern Way in Princeton. He was jubilantly showing his guests some beautiful amber, that had been found in New Jersey with embedded prehistoric mosquitoes. He later photographed the amber in a "dodge and burn" print which I purchased at a show of his photographs at Encore Books in 1996.

My Neider shelf consists of more than twenty of his books, and the ones I most treasure, inscribed by him, are gifts of the author.

[25]The American Heritage Dictionary derives *Karma: Hinduism and Buddhism.* "The sum of a persons actions...determining his destiny. [Hence] fate, destiny." It is related to the Turkish *Kismet*, from the Arabic.

Louis Fischer and Gandhi

I met Louis at tea at the Institute, and frequently bumped into him on Nassau street in Princeton walking with his beautiful secretary, when we would pause and chat. I admired his Old World Charm, his vast erudition and yes, he reminded me of my father! (He, too, was a kind and gentle man.) His books are listed in over thirty citations in the Rutgers Alexander and Douglass Libraries, including *Gandhi and Stalin*, 1947, *The Life of Gandhi*, 1950, *The Life and Death of Stalin*, 1952 and *The Life of Lenin*, 1964.

I took his *The Essential Gandhi, An Anthology*, 1962 and his *Gandhi*, 1954, with me to India in 1968. "*The Story of My Experiments with Truth*" (Gandhi's autobiography) I found to be very helpful. They were the most practical books imaginable, since Gandhi is revered there, and you need to know everything about him to understand India.[26] I am sure that Richard Attenborough did his homework for his film "Gandhi" (1982) by reading Fisher's books.[27]

Sputnik!

On October 4, 1957, the USSR launched an earth satellite **Sputnik**, which in Russian means travelling companion (or fellow traveller?). This amazing accomplishment jolted the US awake to the fact that not only was the USSR far ahead of us in rocketry but also in science and technology education across the spectrum. This lead to massive expenditures by the US and the National Science Foundation to shore up academia by paying for libraries, laboratories, science equipment for buildings, and a new cadre of science teachers perhaps lured into science by NSF summer research grants equal to–first 1/3d, then–2/9th of our academic salaries. At a conference at Oberwolfach in July 1993, A. B. Mikhalev told me that the US mathematical and science community ought to thank the USSR for the good fortune that Sputnik brought us!

At sunset with Heidi, now two years old, on my shoulder, we would stand in the back yard of our house in State College, Pennsylvania and marvel at the tiny bright orange Sputnik illuminated against the evening blue: *the world would never again be the same.*[28]

Govaru Po Russki? My Algebra Speaks Russian

In 1973, the MIR publishers of Moscow asked for and got permission from Springer-Verlag and me to translate my *Algebra*. Furthermore, MIR agreed to

[26] According to a story in the *New York Times* in March 1998 about a Hindu Nationalist who planned the assassination, Gandhi was killed because he was too lenient with Moslems, and allowed the partition of India that created the Moslem states of Pakistan and what is now Bangladesh. Gandhi's last words were "Oh, God!".

[27] According to the *Video Hound*, "Gandhi" won a record 8 Academy Awards including Best Film, Best Director, and Best Actor (Ben Kingsley).

[28] According to Heraclitus (circa 500 BC), my favorite Greek philosopher, the world is never the same twice. His famous anology is of a man who dips his toe into a river: the man, the toe, and the river are thereby irrevocably changed. (*This aspect of life I love.*) Zeno of Elea (495?–430? B.C.?) whose paradox that motion is impossible refutes Heraclitus, is another favorite of mine, as is the stoic Zeno of Citium (335?–263? B.C.) who taught that objects of desire are morally ambiguous, which led scholars to believe that some early Christians were stoics!

honor the International Copyright Law, and, for the very first time, agreed to pay royalties. My head swam at this stroke of luck! (Surely Karma was again working on my behalf.)

Because of the size of my *Algebra*, it was agreed that there would be not one but five translators who were outstanding mathematicians!

I am indebted to L. A. Skornyakov, L. A. Koifman, A. B. Mikhalev, T. C. Tolskaya, and G. M. Zuckerman for many improvements for the Russian translation, many of which I incorporated in the 2nd edition without specific acknowledgement. Moreover, the Russian edition contains numerous footnote references to pertinent Russian articles. I happily acknowledged them in a Foreword to the Russian Edition.

Walter Kaufmann and Nietzsche

I audited a course in Nietsche under this dynamic Professor of Philosophy at Princeton University. Although I already had read his Viking *Portable Nietzsche*, I wanted to study the Princeton University tutorial system. For Kauffman watching TV was a counter-example to Nietsche's *Will to Power* (*Wille zur Macht*). (More of a will to powerlessness?)

Years later Kaufmann published several volumes of brilliant color photographs of India that captured for me the India I had seen. This was one Philosophy Professor with a deeply sensual eye for visual beauty.

Hessy and Earl Taft

Earl is one of Jacobson's first five Ph.D. students (Yale 1956) and, like Jake, works in a broad spectrum of mathematics: associative and non-associative algebra, Lie algebras, Hopf algebras, cohomology, and quantum groups.

Starting in 1959 when Earl came to Rutgers from Columbia U., Hessy worked at the prestigious Waksman[29] Institute of Microbiology at Rutgers University, and later in the newly founded Rutgers Medical School (now the College of Medicine and Dentistry of New Jersey). Thereafter, she worked at the Educational Testing Service (ETS) in Princeton, where her expertise in chemistry and biology, and her fluency in five languages were highly prized. She worked unceasingly to maintain high standards for ETS (which meant bucking the trend that resulted in lower scores being adjusted upward!).

Earl is the founding editor of the "Lecture Notes and Monographs in Pure and Applied Mathematics"[30] and the journal "Communications in Algebra." He too is proficient in five languages, and he likes to surprise his communicating editors by writing to them in their own language (maybe not Oriental languages though!).

Hessy was born to Russian-speaking parents who fled via Germany to Paris, then to Cuba where she grew up. So counting English, there's five languages right there (not counting Yiddish and Hebrew!)

[29]Russian-born codiscoverer, with his graduate student, of streptomycin, for which he won a Nobel prize in 1952.

[30]These are two different series that he edits.

Earl is as New Yorker as you can be, having been born and raised there. Furthermore, they bought a house in Princeton with the smallest possible yard to minimize grass cutting! (I used to kid them about their love of nature!)

The Tafts are opera buffs who have subscribed to the Metropolitan Opera for forty years. They recently bought an apartment on Central Park West within walking distance of Lincoln Center.

Once at a birthday party Earl threw for Hessy, I heard Hessy complain about their two-story apartment stairs. So I slyly remarked that, when that happens, it means the person is over thirty! This made her furious: then "thirty-something" was a despised classification. Do you remember the slogan "Never trust anybody over thirty"? But now? As Oliver Wendell Holmes remarked on his seventy-fifth birthday (when someone asked him how it felt)... *Oh, to be seventy again!*

Earl was a member of the Institute in fall semesters 1973, 1978 and spring 1983. He also served as Treasurer for AMIAS for many years. As footnoted earlier, Earl was a "second cousin once removed" to the late Henny Youngman, master of the one-liner. For years, Earl kept me supplied for not only one-liners but his repertory of jokes. But my favorite story is one he told me about Hessy, when I complimented him on how vivacious she is: *she really wakes you up.* "Yes, Earl replied, "but she's vivacious about **everything**. She doesn't just say, "*Earl, take out the garbage,*" she says, "*Earl! Earl! Take out the garbage! Take out the garbage!*"

Earl tells a romantic saga about a Scottish forebearer named Taft who, with his brother were shipbuilders of Glasgow. On a business trip to Kiev, he met and fell in love with a Russian girl (*come to think of it, so did Earl!*).

The other Taft brother came to the American colonies to seek his fortune, and there is some evidence that his descendants include the Ohio Taft clan, including our twenty-seventh president, William Howard Taft, and Senator Robert Alphonso Taft, who also ran for the GOP presidential nomination, but lost out to New York's Thomas Dewey.[31]

As a remarkable coincidence, the Taft auditorium in Cincinnati, across the Ohio river from my native Covington, is a prominent cultural center, equal in its day, to the JFK Center for the Performing Arts in Washington, D.C.

Kenneth Wolfson, Antoni Kosinski, and Glen Bredon

Everybody called him "Ken." He was the Chair at Rutgers 1961–1976, and masterminded the development of the department. He asked each and every one in the department to recommend the best mathematicians to help build the Center of Excellence that the NSF contract with Rutgers funded. He would let you, as I did in the case of Antoni Kosinski, (who, by the way, is our present Chair) make the initial contacts, before he took over the heavy negotiations. The word genius is bandied about—how many can there be?—but Ken had a genius for talking to people, finding out what *they* wanted, letting them know what Rutgers **would** or **could** do for them. Obviously we couldn't hire everybody we wanted. In Antoni's case

[31] According to the *American Heritage Dictionary*, William served as President, 1909–13, and as Chief Justice of the Supreme Court, 1921–1930, and Robert was U. S. Senator, 1939–1953.

we were very lucky—how many people would leave Berkeley for New Brunswick? But actually, most of the people hired for the *Center of Excellence*, lived as Antoni and his wife Renate (Renate has a Ph.D. from Princeton) in Princeton, with its proximity to both Fine and Fuld Halls. And just one year after Antoni came, we raided Berkeley again and stole Glen Bredon!

Ken was a Ph.D. student of Rheinhold Baer at the University of Illinois, Urbana. *Two years after he obtained his Ph.D., Ken spent a year, 1954–5, at the Institute, exactly 20 years after Baer's membership.*

Paul Moritz Cohn

I recruited P. M. Cohn, who came for one year before taking a professorship at Bedford College of the University of London, and he also lived in Princeton. About the same time I contacted A. W. Goldie who turned us down, but he told me years later that he didn't think we were serious until P. M. Cohn's coming convinced him otherwise. By then we were looking elsewhere.

We did succeed in raising some Ivy League professor's salaries! Arun Jategaonkar (Cornell-Fordham) came for a year, but Steve Chase (Cornell) turned us down flat. Also, fruitless offers went out to the Brandeis "twins": Maurice Auslander and David Buchsbaum. For the record, we tried to persuade Jim Lambek to move to warmer climes but he declined without waiting for our offer. I think the fact that he and his three sons abhorred the Vietnam War, and U.S. militarism, was a factor in this decision.

Joanne Elliott, Vince Cowling, and Jane Scanlon

But we did lure Joanne Elliott from Barnard College. She had been a member of the Institute in Fall 1961, when I met her, and when I suggested her to Wolfson, he asked me to see if she would be interested. She *was*.

Vince Cowling, who had been a professor at Kentucky when I was an undergrad, was a member of the Institute in 1965–66, after becoming a Rutgers professor in 1962, the year I did. In those days I spent many afternoons at the Institute for lunch or tea, and one day I bumped into Vince in the Fuld Mathematics Library. (This must have been in Fall '65 of his year at the Institute.) I had been admiring a new book, a Surveys of the A.M.S. by Jane Scanlon at Brooklyn Poly, and I said "Vince, do you think she might be interested in an offer?" And Vince said, "I don't know but I'd sure like to find out!" And that is how we came to make Jane an offer, which she accepted! (Vince later left us to accept the Chair at SUNY, Albany.)

Rutgers Moves Up!

Under the Wolfson-Gorenstein Chairs, spanning twenty years, Rutgers was admitted to the Association of American Universities, which admits only 50 universities. It was the first time ever that Rutgers had been admitted, in its 220 year history (1766–1986). (On the other hand, Rutgers was founded way before AAU—by about 170 years.) About the same time, the National Academy of Sciences top-ranked Rutgers in the quality of its Mathematics Department, moving it up for the first time ever into the top 20.

Roz Wolfson

Anyone reading these "snapshots" of mathematical friends would think all mathematicians do is party. Well, at Rutgers for many years there were parties every Friday night following the Colloquium and dinner for the speaker, and most often Roz gave it.

The Wolfsons were a gregarious couple with a huge rec room in their home in Highland Park right across the Raritan from the Rutgers Boat House. Roz was a lovely dancer who could make an elephant seem like Fred Astaire–*she danced everybody off their shoes.*

Roz taught English, including Shakespeare, at Highland Park High School and she taught the mathematics department many things about life: how to dance, to forget mathematics for an evening, to laugh at yourself, in short, to enjoy life. How did she manage to keep our growing department of individualists together? Easy: *everybody loved Roz so much!*

The George William Hill Center

This building, for the Mathematical Sciences, was partially built by the **Centers of Excellence** grant, and there used to be a plaque outside the 7th floor lounge and Colloquium Room commemorating this fact. (The trouble is that students like to take down insignia as memorabilia of their happy years at Rutgers. Recently someone took a crowbar to the door trying to get in to study on a cushy sofa.)

When Peter Lax, of Courant Institute, was half-way through giving one of the four inaugural addresses, he stopped in his speech at a point where *Hill's Equation* was mentioned, and idly asked if the Center was named after *that* Hill! **Yes!** was the instantaneous answer, for George William Hill (1838–1914) was not only a Rutgers professor at the turn of the 20th century, but his Collected Mathematical Works (in 4 quarto size volumes) has a 12-page introduction by Henri Poincaré and was published by the Carnegie Institution of Washington, 1904.

Most people think that the Center gets its name from the knoll it stands on.

Three other speakers inaugurating Hill Center in 1973: Nathan Jacobson, Jacob Bronowski, and Arthur Grad of the NSF, who praised our Center of Excellence as the best built with NSF moola.

Daniel Gorenstein and the Classification of Simple Groups

Danny came to Rutgers from Clark U. in the late sixties, and became a driving force for excellence, not only in the mathematics department, but university wide, including all three campuses.

First, he organized the *Classification of Finite Simple Groups* (= CSG) which at one time counted several hundred theorists communicating their newest results through Danny and his Rutgers colleagues Richard Lyons, Mike O'Nan, and Charles Sims. (Gorenstein outlined the history of the Classification in the Introduction to his book "Finite Simple Groups," Plenum, 1982.) A Group Theory Year brought them all to Rutgers including Walter Feit and John Thompson, whose odd-order theorem [63] gave CSG its impetus. Every non-commutative simple finite group had even order. Gorenstein was immensely successful as a researcher as well, developing his induction technique on minimal counter-examples. (See Gorenstein, Lyons and Solomon [94,95] (= G.L.S.) and Aschbacher [94].) The final classification ran several thousand pages, whereas the odd-order theorem was $\approx$ 225 pages.

I remember a thick A.M.S. Memoir, vol. 147, (1974), by Gorenstein and K. Harada running 400 (?) pages *after* they found an error in their preprint whose correction, Danny told me, ran 90 pages!

The Monster Group

I attended Danny's CSG lectures at Rutgers, and became fascinated by the vast scale of the work, and its many deep theorems, including the monster group of order $2^{26} \cdot 3^{20} \cdot 5^9 \cdot 7^6 \cdot 11^2 \cdot 13^3 \cdot 17 \cdot 19 \cdot 23$ discovered by B. Fisher and R. L. Griess. (See Aschbacher [94], p.69, for the orders of the "sporadic groups.") As in "Hagar, the Horrible" cartoon, there is a "Baby Monster."

Was it Dyson who said that if the monster group could be found in the physical universe, e.g. as the group of symmetries of a crystal, or a pulsating star, it could be *another* proof of the existence of God? (I realize I am not expressing this nearly as well as the author of the comment.)

Danny and Yitz

I first met Danny at IDA in Princeton in summer 1964, where he talked about the influence that Yitz had in getting him to work in Group Theory: He got me to attend a conference in groups at Bowdoin, and "*I found something that I could do!*" Danny went on to attend a group theory year at Chicago in 1960–1961, the year he and Yitz wrote a paper on automorphisms of order 4. (I have Richard Lyons to thank for the last bit, including the date of the Bowdoin in 1958.)

I wish Yitz were still around so I could ask him: "Did you have any idea that Danny would become one of the world's greatest group theorists?" (Pst! See Gerhard Hochschild on Erdős above.) But, seriously, Yitz was exceedingly generous to his fellow workers in the mathematical vineyard, for, as he told me more than once, "A mathematician ought to be judged only by his best work."

Gorenstein Rings

Gorenstein Rings were named after Danny by Gröthendieck for a property of commutative rings that popped up in his thesis in algebraic geometry, written under Zariski, I believe.

A colloquium speaker at Rutgers spoke on the "Ubiquity of Gorenstein!" Gorenstein rings appear in Andrew Wiles solution of "Fermat's Last Theorem."

The irony is: Danny had often said, "I don't know what a Gorenstein Ring is."

"All the News That Is Fit To Print" - New York Times

Danny wrote the *New York Times* obituary for Richard Brauer, and *then* he was interviewed by the *New York Times* (by a Martha's Vineyard acquaintance of his); he stated that for twenty years he got up at 5 a.m. *each and every day*, and worked until noon on CSG. "I have a vested interest in group theory" he was quoted as saying in an enormous understatement!

The Gorenstein Report and "Dream Time"

I started out writing about Danny's contribution to excellence at Rutgers. Danny working closely with Rutgers University President Edward Bloustein, authored a master plan "Report on Higher Education in New Jersey," informally called "The Gorenstein Report." This was instrumental in establishing priorities for excellence in education. As Chair of the Department for many years, he did much to implement his own plan, e.g. I. M. Gelfand and his school came to Rutgers under Danny's tenure.[32]

Out of curiosity I asked how his being Chair of the Department affected his research. He thought for a moment, and said, *I miss the dream time.* (Richard Lyons' comment on this: "I bet he even dreamed at supersonic speed!")

Helen and Danny

Helen and Danny Gorenstein were a delightful couple who liked to stroll up and down Nassau Street in Princeton and do a bit of window shopping. *How many times I chatted with them on this or that corner.* Other times we'd duck into PJ's Pancake House for a "cuppa", and range over the art world (Danny was a self-proclaimed "culture-vulture"), literature (Helen wrote an insightful biography of Leo Tolstoy), and politics. We made it a point not to discuss mathematics when Helen was present.

As a couple, Helen and Danny were a harmonious duet, Danny *allegro, molto vivace*, and at times *mezzo forte*, while Helen was *cantabile, piano and dolce.* Their daughter Julia and my daughter Cindy were great pals in their teens, and we were guests at birthday parties for both Danny and Helen, in their home on Philip Drive. Later when they moved to Rocky Hill to free up capital to buy their summer house on the Vineyard, we *still* saw them strolling in Princeton, and kidded them about "life in the boonies". They both had a great sense of humor, but in that too the same musical terms were descriptive.

The Gorensteins' avid interest in art matched mine. It was not unusual for me to attend an art opening in New York—Modern Art, Metropolitan, Guggenheim, Whitney, or Princeton Art Museum, and find the two of them there. They also were familiars at McCarter Theater productions and Institute socials.

Danny was a member of the Institute in 1969–70—the year before his coming to Rutgers, and also the year of the first Moon Walk by Neil Armstrong on July 20, 1969.

Perhaps a defining moment in Danny and Helen's relationship came while we were having coffee in PJ's. I had been reading about the Tolstoy's stormy relations,

[32]Richard Lyons again corrects me: "Danny became such a legend that [like King David] he was given credit for things that happened after he left the chair, but he did recruit Robert Lee Wilson and James Lepowsky (both then new Yale Ph.D.'s)."

and something popped up that Helen didn't know about. This upset Danny a lot, and he verbally chastised Helen for not having been thorough enough in her research. Here, I would guess *he was applying to life and literature the same standard of rigor that he applied in mathematics*; while I didn't think that he should do that, a chastened Helen acquiesced. Not to worry—soon afterwards, at the Princeton Book Mart (now out-of-business), I saw Helen clutching a new book on Tolstoy by a well-known Russian literary critic (Ronald Hingley?), and with shining eyes Helen confided, "Danny said I could buy it!"

On Sundays out for a stroll, Helen and Danny often stopped by our house on Longview Drive on Lake Carnegie and would take a glass of wine with us. I treasure these memories of an affectionate friendship.

In spring 1972, I had the greatest pleasure of telling Danny of my having just served on the Fulbright committee in Washington, D. C. that awarded him his Fulbright to Oxford (1972–73). He said incredulously, "*You?...* were on the selection committee?"

Ken Goodearl, Joe Johnson, and John Cozzens

Ken Goodearl (University of California, Santa Barbara) visited us at Rutgers in Fall 1976 (the year of the U. S. Bicentennial), while he wrote his classic *Von Neumann Regular Rings.* A happy aspect of his visit was that he lived in Princeton at the nearby Vandeventer Apartments,[33] on the street with the same name, so occasionally we carpooled. Another happy aspect of his visit was his interaction with Joseph (Joe) Johnson and John Cozzens in differential algebra that they mutually exploited. They passed the differentials around and Joe combined a result of Goodearl that enabled him to solve a 50-year-old problem of M. Janet in differential equations (DE's). Goodearl, however, insisted that he owed the basic idea to a paper of Cozzens and Johnson on DE's. (A little bit like the double play combination. Tinkers to Evers to Chance.)[34]

Ken gave a memorable colloquium on von Neumann regular rings. After the talk, I asked him if all that derived from the question about simple self-injective rings that I raised in my (Springer) *Lectures.* Ken thought for a moment, and then smiled and said "Yes". I think he had completely forgotten the background. There is so much to remember and to relate that to give the history in detail would take up more time than we have in an hour's lecture. (If you want to know the question of mine that Ken answered, see the title of his 1974 paper.)

[33]I dubbed them the "Vandermeer Apartments" after Johnny Vandermeer of double no-hit fame of the Cincinnati Reds. The second of Vandermeers' two consecutive no-hitters was thrown at Ebbets Field against the Brooklyn Dodgers on June 15, 1938 during the first night game played in Brooklyn.

[34]Paraphrased excerpt from an e-mail letter of Ken Goodearl of February 12, 1998: "The theorem Joe Johnson used to prove Janet's conjecture (in addition to various results of his own in differential algebra) was a result of mine [= Goodearl] on projective dimension. Rosenberg and Rinehart proved the same result independently in greater generality [see 14.15 (11)]. Actually, all Joe needed was the case where R is a field, which ironically can be derived from his earlier work with Cozzens; but perhaps he didn't think of it in the right way until he saw my [Goodearl's] theorem." (The references, in square brackets are mine. See the paper of Johnson [76], and his joint paper with Cozzens [72].)

Hopkins and Levitzki

The same thing happened to me when at the 1964 Summer Meeting of the AMS in Boulder...I referred to a "Theorem of C. Hopkins". It so happened that S. A. Amitsur was an auditor and in the question period following the lecture he stated that Levitzki had proved it independently.

The reason for this glip was that the Hopkins' paper appeared in 1939 in the prestigious *Annals of Mathematics*, published jointly by the Institute and Princeton University, whereas Levitzki's paper, submitted to *Compositio Mathematica* in 1939, did not appear until after the end of world War II. However, when published, the journal carried the original fateful date—1939, the year Hitler invaded Poland and the Allies declared war on Nazi Germany.

Jakob Levitzki

For this reason many Americans knew *only* the work of Hopkins, and I was one! Later I expiated this *pecado* when I named a class of rings in Part I of my Dekker Lecture Notes (1982) after Jakob Levitzki, namely rings with the ascending chain condition (= acc) on annihilators, one of the many subjects he pioneered. (Polynomial identities being another—see Chapter 15.)

Unfortunately nobody followed my lead in this nomenclature, and later I adopted the shorter appellation: acc⊥ rings. Handy, but not as evocative as *Levitzki* rings.

A student of Levitzki at Technion, Benyamin Schwarz, wrote to me a moving letter, thanking me for my glowing references to Levitzki, saying that he had no idea that Levitzki was so revered outside Israel. (Both Levitzki and Amitsur were Israelis, and Levitzki was a teacher of Amitsur.)

Amitsur edited a posthumous paper of Levitzki in 1963 on acc⊥ rings, showing *inter alia* that nil ideals are nilpotent, a result proved independently by Herstein and Small in 1964. (See Theorem 3.41.)

Jacobson's Colloquium Lectures [55,64] introduces a goodly bit of Levitzki's work on I-rings and algebraic algebras, while his 1975 book "*PI-Algebras*," does the same for his work in that field. (Check out the Index of the text for references to other Levitzki theorems, e.g. Theorem 2.33.)

Chuck Weibel and Tony Geramita at the Institute (1977–1978)

The Institute is a mathematical paradise where each year you can meet 50 or 60 new visiting mathematicians plus the permanent faculty.[35] While visiting the Institute in 1977–1978, I was writing two papers, which later appeared as Volume 72 of *Lecture Notes in Pure and Applied Mathematics* (Marcel Dekker, 1982). Luckily for me, I soon discovered two members, Charles (Chuck) Weibel and Anthony Vito Geramita, who were inexhaustible sources of knowledge of commutative algebra. Chuck then was a new Chicago Ph.D. student (of Richard Swan) and already exhibited the incredibly deep knowledge that not many years later landed him the prestigious Editorship of the *Journal of Pure and Applied Algebra* (no connection

[35]The faculty in 1997-8 numbered eight, down from eleven in 1960-61.

with the Dekker Lecture Notes), the successor to Hy Bass, one of the founding editors.

In my *Lectures* (*ibid.*, p.66) I duly acknowledged "that I fleeced trade secrets off of Tony Geramita and Chuck Weibel [who let me know that they had plenty left!].

How I Helped Recruit Chuck

Chuck had applied to Rutgers for a job, which was more or less *de rigueur*, since in those days jobs did not grow on trees as they had in the 50's and 60's. But did my urging the Personnel Committee to grab Weibel have an effect? Well, no, but this was not Weibel's fault since they had a priority list: *let there be Algebraic Topology* (for instance), **or** (maybe) *Applied Mathematics.*

Bad luck! But next year the same priorities prevailed, and again we did not make Chuck the offer. Worse luck!

Not to worry. The third time was the charm. Barry Mitchell showed up at the Personnel Committee meeting, and said, "Hire Weibel!" *And so they did*! So thank Karma (or Kismet) again, and also Barry!

I can vouch for the fact that Chuck is one of the most popular faculty members in mathematics by the fact that *three years later the tenure vote was unanimous.* Not bad for a Terra Haute native! (Being a Kentuckian, I can get by with saying that about a Hoosier... .)

Poobhalan Pillay, Lalita, and Karma

Poo, as he liked to be called, came to visit me at Rutgers in the academic year 1979–1980. He was then a recent Ph.D. from the U. of Johannesberg, who taught at the segregated University of Durban, Westville. He came to Rutgers supported by a postdoctoral fellowship from South Africa, showing up on Labor Day weekend obviously expecting to find housing in the by then very tight college market! I lodged him in Nassau Inn in Princeton and gave him a telephone number of a local realtor. Within days he found a wonderful, big duplex house, with a large yard for his two children Khandon and Kanyakumari, on Mt. Lucas Road in Princeton. In this instance Karma exerted great force in my behalf, because the house was next door to Molly Sullivan, whom I met several days later standing in Poo's front yard balancing a mattress on her head, native style, that she had acquired for the Pillays at a nearby rummage sale.

A golden September light illuminated the trio, Molly, Poo, and his wife Lalita, and Molly was swaying slightly in a soft breeze that ruffled the skirt of my future wife. I might never had met Molly if I had not hosted Poobhalan at Rutgers on his Equal Rights Fellowship. Nor would I have had the opportunity to know her, had I not driven Poo to Rutgers thrice weekly, stopping by for tea on the return, where frequently we found Molly there chatting with Lalita.

It was another instance when the Indian Goddess Karma smiled down on us: after a long courtship, Molly and I were married in September 1987, eight years almost to the day we first met.

Poo and his family returned again in 1988–1989, when we wrote our little book *Classification of Commutative FPF Rings*, published in 1990 by the University of Murcia, Spain. Part of the book was written at the Autónoma University of Barcelona during the Algebra Semester in Fall 1989. By another bit of serendipity, Poo and Dolors Herbera collaborated on their joint paper [93] during the same conference. It negatively answered the question: if R has self-injective classical quotient ring, does the polynomial ring? (Of course, the answer is true in special cases, e.g. when R is a semisimple ring.)

"Tommy" Tominaga and "Tokyo Rose"

Tominaga visited us at Rutgers in Fall '81, and it was left to me to find him suitable housing. A friend of ours, Rose "Rosie" Mintz, who rented out rooms in her house on Forrester Drive in Princeton, agreed to give him Bread and Breakfast. Before long they were fast friends—Rosie called him "Tommy" and he loved it.

Tommy and I got on famously. As we drove in my car from Princeton to Hill Center in Piscataway, we would sing American songs of the 30's and 40's. When I asked him how come he knew so many of my favorites, he said, *"Why, from Tokyo Rose!"*

For the younger generation: Tokyo Rose was an American Radio Announcer for Japan propaganda in World War II. She sang nostalgic songs to undermine the American morale, but it had the opposite effect: the troops loved her, as I loved Tommy singing her songs.

Ted Faticoni, the Walkers and Me at Las Cruces

In spring 1982 I had a yen to see the Walkers again. I was due a leave from Rutgers at 80% of my salary, and I became enamored with thought of living in the desert and mountains of New Mexico. These had been romanticized for me in the writings of D. H. Lawrence, and the paintings of Georgia O'Keefe, both of whom had lived in Taos for many years.

Carol Walker, the Chair of the Mathematics Department became the enabling Angel...it was possible to hire me to teach a course in FPF Ring Theory. I was elated by the size of the class—a dozen or more graduate students and faculty showed up for the initial lectures. Many of the faculty I already knew: The Walkers, of course, Don Johnson and Joe Kist at Penn State, and Fred Richmond who had briefly visited the Walkers at the Institute sometime in 1963–1964. They came to listen to me mainly out of friendship and predictably dropped out one by one after several weeks.

I lectured from my notes on FPF Ring Theory from my lectures at Technion and Rutgers, and which were incorporated into the Lecture Notes of the London Mathematical Society, appearing two years later with Stan Page as co-author.

So you might say that I was well prepared for the course, but alas, except for Ted Faticoni, nobody was prepared to hear them! Ted was a student of Visonhaler at U Conn (as it's called), and took to this abstruse subject like the proverbial duck took to water. We had long wide-ranging conversations on the subject that proved immensely stimulating to me, and inspired Ted to write a series of important papers (see references, and also Theorem 5.46).[36]

Another happy aspect of my life in Las Cruces was that, after Molly and the children returned to Princeton in March, I was "invited" to move into the guestroom of the Walkers. (Invited may not be the proper word! Think of the anecdote about Erdős and Tarski that I related earlier.) To put it another way, I so much wanted to live with them that they kindly assented. The name of the street they live on–Imperial Ridge–conveys something of the spacious grandeur of the house. The guest room was off the first level rec room and dining room areas. The latter location got me invited to several fine dinners and Elbert's 50th birthday celebration!

Here is another instance of good karma! As we Kentuckians like to say—I was living in "hog heaven"! The Walkers are the most genial and generous of people—Elbert is a Texan to his boots and Huntsville roots, and Carol hails from Martinez, California. I owe them both a great debt of gratitude for their hospitality and kindness: I would not have been able to stay on in Las Cruces without the friendships of the Walkers and the Faticoni's.

New Mexico

Before saying *Adios* to Las Cruces, let me recommend a scenic tour that we took of the State of New Mexico that includes Taos, Santa Fe (Holy Faith!), Madrid [sic!], Bandiera National Monument, Albuquerque (named after a Spanish prince), Truth or Consequences, Las Cruces (the Crosses), El Paso (Texas), and Juarez (Mexico), all on or near the Rio Grande, Gila Cliff Dwellings National Monument on the Gila river (named after the Gila "monster"), Alamogordo (fat cottonwood or poplar), Los Alamos, White Sands National Monument,[37] and Carlsbad Caverns on the Pecos River. The list of attractions is well nigh inexhaustible: Acoma (city in the sky), Cañon de Chelly, Navajo Indian Reservations and Pueblos... Las Cruces is in a desert and the Organ (also called Needle) mountains rising out of the morning sun casts a daily dark shadow. One windy day in March we watched in amazement when during a sandstorm huge ominous tumbleweeds wheeled out of the desert. New Mexico, in particular, and the arid southwest USA in general, is not for sissies! The wind blows constantly, often with tornado force. But beautiful—I will never, ever, forget the awesome beauty of those six-foot high tumbleweeds.

Rio Grande

It's a long river, some 2000 miles flowing from the Rockies in Colorado, dividing New Mexico from Sante Fe to El Paso, forming the border of Texas adjacent to Mexico, as it winds its way to the Gulf of Mexico, via Eagle Pass, Del Rio and Matamoros . By the vagaries of life, in 1945, I was stationed in Corpus Christi at the Naval Air Base there, smack dab on the Gulf, not far from Matamoros! Like Tijuana, Matamoros was a place where sailors went to unwind (or wind up?).

[36]Ted subsequently was hired by Fordham University and went straight up the ladder: assistant prof, associate prof, full prof, and then lost his footing and slipped into Chair, and later Dean.

[37]And the Trinity Site where the first A-bomb was exploded on July 16, 1945.

Unlike Matamoros, the Rio Grande is often "bone dry," which presents little barrier to "wetbacks" from Mexico, but insurmountable problems to immigration officials! But, when it's running, it is quite shallow and wide...thus, **Rio Grande**.

Dolors Herbera and Ahmad Shamsuddin at Rutgers (1993–1994)

These two visited us at Rutgers in the academic year of 1993–1994. I attended Dolors' Ph.D. Thesis Defense at Autońoma U. of Barcelona in January 1992, and subsequently she was awarded a Fulbright Postdoctoral Fellowship to study at Rutgers, starting exactly one year later in January 1993. Ahmad (American U. of Beirut) also visited but kept a low profile in Fall '93, but finally in January 1994, I wrote him a polite note saying I would like to start a seminar with him, Dolors, and Barbara Osofsky. (The latter two were already "interacting" as mathematicians like to say.)

Dolors was working on a number of problems including my conjecture (= CC): *the endomorphism ring E of a linearly compact module M over any ring R is a semilocal ring.* The evidence for its truth was (1) CC was true when R is commutative, and (2) (CC) was true when M is Artinian (Camps-Dicks Theorem [93]—see 8.D in the text).

I gave Dolors' room number in Hill Center to Ahmad and invited him to introduce himself, since I hadn't managed to get them together. The very next week I found out they had proved (CC) using (A) the Camps–Dicks ingenious classification of semilocal rings via a dimension function; and (B) results on Goldie co-dimension.

In the text, I call the truth of (CC) "a very general Schur's lemma" (see their paper [95]). They went on to collaborate in another paper [96] on self-injective perfect rings, concerning another conjecture of mine, as did Dolors and Poobhalan (see below). The following year, her fellowship being renewed, Dolors and I wrote a paper on the Endomorphism Ring Conjecture (= $\mathcal{ERC}$) for linearly compact modules over commutative rings. (See References.)

Pere Menal

It would be impossible to fully express in this brief space the debts of friendship, both personal and mathematical, that I owe to the late Pere Menal Brufal.

It all began in 1981, when I wrote to Pere about his work [81] on algebraic regular rings in connection with tensor products. He expressed amazement and delight that it was connected with the Hilbert *Nullstellensatz.* (See, e.g. my paper in his Memorial Volume (Perelló [92].))

After some years of correspondence, we began contemplating an algebra semester under the auspices of Manuel Castellet's CRM, and they went on to organize one in Spring 1986; another much larger conference followed in Fall 1989.

Much creative mathematics flowed out of these conferences, particularly the theorem of Pere Menal and Peter Vámos [89] that realized a three decades-old dream of ring theory: an embedding of any ring in an FP-injective ring. In her Rutgers University Ph.D. thesis (1964), Barbara Osofsky had shown that in general the injective hull of a ring R could not be made into a ring containing R as a subring. It is also known that R can **not** always be embedded in a self-injective ring.

In the 1989 semester, Pere and I collaborated on a problem that eluded us individually for many years, and we finally found what we were looking for: "A counter-example to a conjecture of Johns" which appeared in the Proceedings of the American Mathematical Society in 1991, shortly after his tragic death in an automobile accident in April. Cf. the Pere Menal Memorial Volume (C. Perelló, ed.): Publ. Math. **36** (1992).

These are just two of the myriad collaborations that Pere Menal had with others: Jaume Moncasi, Pere Ara, Claudi Busqué, Ferran Cedó, Dolors Herbera, Rosa Camps (his talented students and colleagues) and Warren Dicks, Brian Hartley, Kenneth Goodearl, Robert Raphael, the aforementioned Peter Vámos, Boris Vaserstein (to mention but a few of his intense interactions with others).

In particular, Pere's paper with Dolors Herbera [89], Ferran Cedó's paper [91], Dolors Herbera's paper with Poobhalan Pillay [93], and her doctoral thesis, and that of Rosa Camps, at U.A.B., each greatly advanced our knowledge on subjects taken up in seminars during these conferences.

I am grateful that I happened to write to Pere back in 1981; otherwise, I might never have met this noble and gentle genius of Catalunya, who became an inspiration to many.

Alberto Facchini and More Karma

I met Alberto at the Exeter conference in June 1984 on "Direct Sum Decompositions" organized by Peter Vámos. When I was asked by Pere Menal to help organize the Algebra Conference at Centre Recerca Matematica (CRM) at Autónoma University of Barcelona in Spring 1986, I nominated both Alberto and Peter for my short list, which included Ken Goodearl and his mentor Robert Warfield, Barbara Osofsky, Roger Wiegand, and Sylvia Wiegand. (The latter five could not attend either because of previous engagements or family illness.)

Again, in the following conference at CRM in Fall 1989, we invited the same seven, in addition to the others listed above in my reminiscence about Pere Menal, only to receive the terrible news of the death of Robert Warfield. Pere, newly editor of Communications in Algebra, dedicated a volume in memory of Robert (this came out in 1991).

At the conference, Alberto answered a question that had been nagging me for several years by showing that any valuation ring R has FP-injective quotient ring $Q(R)$. Pillay and I included this result in our monograph on commutative FPF rings published by Murcia University in 1990. Later, after I found a counterexample to a conjecture of mine on FP^2F rings (which appeared in [92b]), the question naturally arose to determine all commutative rings R such that the quotient ring $Q(R/I)$ is FP-injective for all ideals I. Rings with this property are called fractionally (self) FP-injective (= FSFPI) rings, following Vámos' terminology for fractionally self-injective (= FSI) rings. The classification of FSFPI rings were obtained in our joint paper [96] which we wrote in 1995–96 communicating our results to each other entirely by fax! I had not yet succumbed to the allure of e-mail, but have since, due to that very thick sheaf of faxes that I accumulated.

To go back to the classification, yes, every FSFPI ring R is an arithmetic ring (in the sense that every local ring of R is a valuation ring), and the converse is proved, e.g. for semilocal rings (see Theorem 6.4 of the text). Moreover, FSFPI rings coincide with fractional p-injective rings (*vide.*).

Consider the good karma that derived from Peter Vámos' Exeter Conference: Vámos' theorem with Pere Menal embedding any ring R into an FP-injective ring, the joint work of Alberto Facchini and myself reported on above, and no doubt much more than I am aware of.

Barcelona and Bellaterra

Autónoma University of Barcelona (UAB) is located some 20 miles north of central Barcelona in a valley appropriately named *Bellaterra* (beautiful land). It was put there by Generalissimo Franco because student protests in central Barcelona not only disrupted traffic but also gave out too much anti-Franco propaganda.

First and foremost, Catalunya (also called Catalonia) relished its historical independence, language, and the culture that they had had before Spain conquered them. They considered it their patriotic duty to rebel against Franco. Second, they were fervid anti-fascists, at least against Spanish fascism, so their protests served to weaken the Franco *regime.* Consequently Franco put UAB out in the country, what we would call the *Boonies*, and said, "Let the cattle watch them protest!"

To get to Bellaterra, you take the *Ferrocarril* (literally, iron rail) through a breathtaking series of hills, valleys and tunnels that would do justice to Morse Theory! The little villages en route have poetical Catalán names such as Floresta (flowers) and Baixador de Valdoreix ("Reix" is pronounced "resh;" but don't forget to trill the "R": I once challenged Jaume Moncasi to pronounce R without a trill, and he managed only with great difficulty!)

After the beautiful train ride to Autónoma, the University comes as a shock. The gray buildings were all poured concrete monstrosities (called the *new brutal* in the USA) that only a fascist architect could dream up to dispirit the students and faculty. It looked more like a prison than a university (the new buildings in the 90's are of colorful brick).

Autónoma University differs from the University of Barcelona in central Barcelona in that Catalán is the language spoken at Autónoma, Spanish is spoken in the University of Barcelona. (Actually, Autónoma *could be* translated as "provincial," "regional," or "state," although autonomy is a cognate.) The Catalán history is of a proud and independent people who once ruled much of the Mediterranean and parts of the "New World," e.g., Cuba. Catalunya (Catalonia) had its own constitution over 1,000 years ago, and Catalán is recognized as an "autonomous" language.

American visitors to Barcelona recognize a kinship to the bristling industrial vitality of the region that more resembles nordic New York than romantic Madrid; Catalunya is by far the most prosperous of all the Spanish autonomous states, and attracts legions of job seekers from the rest of Spain.

Gaudí's Genius

Beyond its economic power, Barcelona is also noted for its artistic and cultural heritage. The great architect, Gaudí, left his indelible imprint on the city with his world famous *Familia Sagrada Cathedral*[38] that dominates the skyline, numerous revolutionary buildings with curving facades, gargoyles and multicolored tiles, and the exquisite Parque Güel that overlooks the port city from its perch in Lesseps, a barrio (suburb) where we lived.

During our annual visits to Barcelona in the decade 1986–1996, we lived in an apartment adjacent to Parque Güell, and like all who go there, fell under the spell of Gaudí's genius. The best way to describe Parque Güell is an enchanting fairyland of balustrades, winding paths, colonades, castles, grottoes, and delightful outdoor cafes where you can snack while you soak it all up.

How many days did we walk over to the park and watch the cargo ships floating in the Mediterranean due east of the city: *Mare Nostrum* (our sea) is what Caesar called it. Although on the same latitude as New York, Barcelona's climate resembles that of Los Angeles—semitropical, semiarid, sunny, and delightful except for the frequent smog.[39]

The Ramblas

The Ramblas may well be the world's most famous promenade. It originates in the harbor area and cuts the city in two with its broad mall that stretches a mile due west, ending a bit past Plaza Catalunya, where it bifurcates into two ramblas!

Words can hardly describe the cornucopia of bookstores, outdoor cafes, and businesses that dot the promenade. Each area has its own flavor—this one (Rambla de Flors) selling flowers, another parakeets, yet another selling jewelry!

In addition to the bustle of the pedestrians, there are the countless mimes, one imitates a stoic Roman soldier, while another mimics the gaits and facial expressions of passersby. The strains of Peruvian music fill the air with its prehistoric music. Monkeys or horses, or bears perform brilliant feats for their masters. Gypsies dance their ancient rituals of passionate love, while pickpockets prey on the unwary watchers.

You have to pay a small fee to sit in a nearby chair, but, if you did, then you wouldn't be able to see anything but the backs of the delighted throngs.

You don't need any license to perform on the Ramblas, and so impecunious students and travellers pull out their musical instruments, or just break out in song, hoping for a few pesetas to fall into the hat placed in front of them on the ramblas.

---

[38]Still under construction 70 odd years after Gaudı's accidental death in 1926. A book by Conrad Kent and Dennis Prindle on Gaudı's architecture, entitled "Hacia La Arquitectura de un Paraiso" published by Hermann Blume, Madrid, was praised by Raymond Carr in TLS, April 30, 1993.

[39]No explanation of Barcelona's "genius" can be complete—it has showered so many: Miró, Picasso, Utrillo, ... Robert Hughes' *Barcelona*, Alfred Knopf, 1992, is a cornucopia of Catalunya greats.

I have my wife to thank for supplying the Latin.

Once at Madison Square Garden in New York, the ringmaster of Ringling Brothers sang a catchy ditty "May Every Day be Circus Day." When you walk on the Ramblas—*every day is Circus Day*!

Norman Steenrod

By coincidence, Norman Steenrod, who was a Princeton neighbor of mine, rode the train back to Princeton after the AMS Cincinnati meeting in spring 1961 where I met Utumi. We shared an aversion to flying, and settled into an overnighter train back to Princeton. I made a voluminous entry into my diary covering the wide-ranging conversations that I had with him. From memory, however, I recall that he explained why the Princeton Mathematics Department did not have course requirements for the Ph.D. degree. "We don't want a student taking courses and making A's to think that is what we want him or her to do. We don't want him or her to limit their studies to just making good grades in courses. So our generals (exams), explore the depth and range of his or her inquiries."

The fact that Princeton perennially (as presently) ranks number 1 in their mathematics graduate school attests to their success. The ranking is done by the National Academy of Sciences by a poll of their members. The National Science Foundation has a ranking based on the productivity and positions achieved by Princeton, vis-a-vis other graduates.

Another recollection that I have of these conversations on the train is the origin of the term "injective" which according to Steenrod was first used in a book on algebraic topology in 1952, co-authored with Samuel Eilenberg. "We left a space in the text until we finally decided what to call it!", Steenrod explained.

Kaplansky, Steenrod and Borel

After a Colloquium lecture at Rutgers by Irving ("Kap") Kaplansky, we gave a party for him in Princeton, which both Steenrod and Borel attended. I distinctly remember the pleasure that Kap exuded when he discovered them *both* at my house after the colloquium dinner. "Norman *and* Armand!", he exclaimed with obvious delight. My graduate students Vic Camillo and John Cozzens beamed with pleasure at this "summit meeting" of intellectuals. In particular, Vic recalls Kap telling him apropos of topology and algebra "*It pays to know more than one subject*" Kap admonished.

Kap

Kaplansky regularly visited Rutgers over the thirty-five years that I was on the faculty. Sometimes his visits were unofficial in that he was visiting a relative in the philosophy department. But when he gave a colloquium, you had to get there early if you expected to get a seat—he was one of our most popular speakers. (Jack Milnor and Armand Borel were two others who lectured to overflow audiences.)

Kap was a member of the Institute in 1946–1947 and a professor at the University of Chicago for over thirty years before assuming the founding directorship of the Mathematical Sciences Institute at Berkeley, which he relinguished to David Eisenbud in 1997.

Kap was President of the American Mathematical Society 1985–87.

Kap's "Rings and Things"

At the Annual Meeting of the American Mathematical Society in Orlando in January 1996, Kap gave his (postponed) retiring presidential address with the above title. Here's a quote:

> As I look out at the audience, I wonder how many spotted the title? ...in Act II, Scene I, line 322 of "The Taming of the Shrew," Petruchio has successfully wooed Katherine (if "wooed" is the right word), and as the wedding impends, he tells her "We shall have rings and things and a fine array."
>
> The phrase was brought to my attention by my children. When they were young they at times told their friends that their daddy worked on rings and things. I believe that they did not know they were quoting Shakespeare.
>
> The late Einar Hille once remarked that wherever he looked in mathematics he managed to see semigroups of operators. I feel the same way about rings.[40]

Kap goes on to quote Bourbaki's historical note on commutative algebra saying that "the general notion of ring is probably that of Fraenkel in 1914."

The entire address is replete with history, and ought to be made to be required reading, if only we could persuade Kap to publish it! I want to pause here to thank a mutual friend at Berkeley (who prefers anonymity!) for sending a copy to me.[41]

"The World's Greatest Algebra Seminar"

In September 1982, on one of Kap's "unofficial" visits, he addressed the Algebra Seminar, and I was able to get a "snapshot" during the question-period at the end. Here is the list of attendees in seating order: (back row) David Rohrlich, Jim Lepowsky, Robert Lee Wilson, Charles Sims, R. Willet, Richard Lyons, Barry Mitchell; (middle row) Myles Tierney, Barbara Osofsky, Charles Weibel, Daniel Gorenstein, Justine Skalba, N. Adams (student of Tierney), Jack Towber, Wolmer Vasconcelos, Richard Moses Cohn, Roe Goodman, Solomon Leader and William Hoyt. For some reason Richard Bumby, Harry Gonshor, Eart Taft and Nolan Wallach did not appear in the photograph. The subject of the lecture was "*Virasoro Algebras*," and the day was September 15, 1982. His paper on the subject appeared the same year, and is #111 in his *Selected Papers*.

Samuel Eilenberg

Sammy Eilenberg (Professor Emeritus, Columbia) gave a lecture at Rutgers in the mid-seventies on his then new book *Automata, Languages and Machines* (Academic Press, 1974). Although this was a highly complicated subject, Sammy engaged the audience with eye contact, asking for and getting affirmative responses

---

[40]Evidently, so did Shakespeare!

[41]It was two years after I first asked Kap for a copy! However, according to Dr. "Anonymous," Kap graciously agreed to my request to quote from it, saying "*It would be an honor.*" On the contrary, it is **my** honor. I once told Kap that he paid my *Algebra* the greatest compliment when he wrote me back in 1973 in thanks for the copy I sent him..."*It's beautiful!*" (I never dared to quote him before now!)

to his propositions, simultaneously raising himself up on the ball of his right foot with his right arm outstretched in a balletic, almost physical, delivery.

As a balletomane, I took delight in the grace and skill of his movements which accentuated the *act of physically delivering his ideas to the audience.* Afterwards at dinner (at McAteers on Easton Avenue) he confirmed my conjecture as to his ballet training as a youth.

Regarding his zestful sexagenarian enjoyment of food and wine at dinner, Sammy confessed, "I just had my annual physical and the doctor gave me a clean bill of health, so **full speed ahead**!"

Myles "Tiernovsky"

Sammy was the epitome of wit and charm in this as in all things: I was quite delighted when, at a party following dinner, he browbeat big and Irish Myles Tierney into confessing his (imagined) Jewishness: *"Admit it, Myles, Tierney is short for Tiernovsky!"*

He howled with laughter at Myles' obvious discomfort with his joke.

Sammy Collects Indian Sculpture

Sammy was an internationally recognized connoisseur and avid collector of Indian and Southeast Asian art, which he saw for the first time on his trip to Bombay in 1953 to teach mathematics for a semester. Over the next 35 years, Sammy collected bronzes and stone artifacts in India, Pakistan, Indonesia, Thailand, as well as in Europe. He donated over 400 of these to the Metropolitan Museum of Art, and 187 were chosen for an exhibit in 1991–1992 at the Museum entitled "The Lotus Transcendent." Other artifacts were donated to Columbia University, who sold them to the Museum to endow a named professorship of mathematics in his honor (from an article in the *New York Times*, October 21, 1991, by Rita Rief, entitled "How a Trip to India Turned a Professor into a Collector").

Sammy told me that he had acquired so much Indian art that the airport customs officials were put on "red alert" whenever he entered the country in a vain attempt to thwart his smuggling. He was quite delighted by his ability to evade them.

"The Only Thing They Would Let Us Do"

At McAteers Restaurant, Sammy explained to everybody the success of Jewish mathematicians, beginning with Einstein, this way: "*It was the only thing they would let us do*!" I think this is a most profound insight into the astonishing quality of Jewish culture (which Einstein said made him proud to be Jewish),[42] and one might add, in retrospect of the holocaust: *and then they wouldn't even let us do that.*

---

[42]See, for instance *The Quotable Einstein*, collected and edited by Alice Calaprice with a foreword by Freeman Dyson, who "knew Einstein only secondhand" and whose "favorite babysitter" was Helen Dukas!

The late Maurice ("Moe") Auslander once told me "*They (the Nazis) wouldn't let you be anything else* [but Jewish]." The Holocaust has to be the saddest chapter in world history. Moe also had this to say about his name: it had become an adjective, as in the Auslander-Buchsbaum Theorem, or the Auslander Global Dimension Theorem.

Emil Artin

Emil Artin was the only mathematician to solve two of Hilbert's problems: the Ninth and the Seventeenth.[43]

Among his many great theorems is the Wedderburn-Artin Theorem, which systematized the study of "Artinian" rings, that is, rings with the descending chain condition on ideals.

Artin's delightful little book *Galois Theory* introduced a generation of mathematicians to the mysteries of solutions of polynomial equations over commutative fields. Danny Gorenstein told me that Artin's book was the model for his own book "Finite Simple Groups." "I tried to make the subject as accessible as Artin made Galois Theory," Danny said to me when I professed my debt to the latter, beginning with my Ph.D. thesis, for starters.[44]

In 1958, I wrote to Artin at Princeton University to ask if I could work with him. The reply was, Yes, but I would have to come to Hamburg, where he had returned. Like Reinhold Baer, and at about the same time, he had become repatriated in Germany.

The proposal did not work out because the NSF postdoctoral fellowship that I applied for was not to be, although Yitz Herstein had told me, when he came to Penn State for a colloquium, that I had been top-rated for the post-doctoral. For some reason the NSF had reversed the committee rankings.

Anyway, it may have worked out for the best, since I was able to obtain, again with Herstein's support, the Fulbright-Nato postdoctoral to Heidelberg in 1959–1960. Sadly, however, Emil Artin died in 1962, and I, and the rest of the world, forever lost the chance to work with this great mathematician.

Often I run into his widow, Natalie Jasny, in a supermarket, or just walking on Nassau Street, and when I do, I always ask after her children, Karin and Michael, who live in Cambridge. (Another son, Thomas, I never met.) She still lives in the family home on Evelyn Place, one of the architectural treasures of Princeton.

---

[43]The Seventeenth problem was solved by Artin [27b]: *Every positive* (=**definite** in [27b]) rational function $f(x_1, \dots, x_n)$ in n variables with rational number coefficients can be written as a sum of squares of rational functions. Hilbert [22],pp.106–107, proved it himself for $n = 1$, and also for $n = 2$ in *Acta Math.* vol. 17, p.169 (see Hilbert's *Collected Works* [32,33,35]).

[44]Tarski's theorem that any two real closed fields are elementarily equivalent (in a logical sense) can be applied to prove Artin's theorem solving Hilbert's Seventeenth Problem. See for example, Jensen and Lenzing [89],pp.8–9, where the model theoretic proof involving ultraproducts is discussed.

Michael Artin

Mike Artin (M.I.T.) is a counter-example to the conventional widsom that says mathematical talent is inherited—by the son-in-law![45] He is an engaging lecturer of the most informal school. At a Rutgers colloquium lecture some years ago he went to great pains to explain that his wife bought his beautiful sport shirt. Everybody enjoyed the comedy. Mike explained to me the Armenian origin of the family name, Artinian, which had been shortened in Germany and the United States. By the vagaries of life, the name Artinian was restored in adjectival form by Bourbaki: Artinian rings![46]

The Artins or Artinians are true mathematical royalty despite the assertion by Euclid: there is no royal road to geometry. Mike was President of the American Mathematic Society 1991–93.

I was privy to a conversation between Mike Artin and Bill Browder, who was then Chair of the Princeton University Mathematics Department. It went something like this:
Bill: Mike, are you ready to return to Princeton?
Mike: No, not yet.
Bill: Well, let me know when you are.
Mike: I will, Bill, I will.

This reminds me of the negotiations of the Institute's Founding Director Abraham Flexner with Albert Einstein regarding salary. Flexner asked Einstein how much he wanted. Einstein thought for a moment and said $3,000. That settled it—Flexner instead gave him $10,000. (This story is elaborated in Ed Regis' book *Who Got Einstein's Office*, p.22.)

University Towns

In these snapshots I have tried to convey impressions of people whom I met in a variety of university towns: Lexington, West Lafayette, East Lansing (Michigan State), State College (Penn State), Heidelberg, Princeton, Berkeley, New Delhi, Bombay, Madras, Las Cruces, and much later, Barcelona.

When I arrived at the Institute in September of 1960, my office was a huge one, which I believe originally was reserved for the director of the computer project—it was not only huge, but had lots of windows, and was right next door to the "maniac".

Thus did serendipity bring Bob Bonic and me together in the same office–all the other offices at the Institute were private. But, not to worry, the day after I arrived, Bob had managed two semiprivate offices by dividing the office with several tall bookcases.[47]

---

[45] This requires some explaining since John Tate (Harvard U.) is an ex-son-in-law!

[46] "I would like to call them Artinianian rings" (Jim Lambek in a letter of March 1998).

[47] The very first person I met in ECP, however, was not Bonic but Barry C. Mazur (Harvard) who was clearing out from his residence there in 1959-60. After we introduced outselves I slyly asked him if he was related to the "famous" Mazur, and he gave this tongue-in-cheek reply, "I *am* the famous Mazur!"

Still, how could two gregarious people manage to shut themselves up long enough to work? Again, trust in Karma, Bob discovered several restaurants *cum* coffee houses in Princeton. One was called the "Balt", short for Baltimore (it used to be a stagecoach stop on the Boston to Baltimore route). Another was Renwick's, which F. Scott Fitzgerald, or J. D. Salinger (or both) immortalized in their novels and writings. Fitzgerald's "An Afternoon of an Author" is a sketch of Princeton bucolic environs, including the Princeton "Dinky," the two-car train that connects Princeton with the junction for the main New York to Philly railroad (formerly the New York Central, now Amtrak).

Some Cafés and Coffee Houses

Bob Bonic and I frequently met in the Balt or Renwicks for coffee, rolls, or lunch, and I got to know people of Princeton beyond the Institute.

In later years (1968), after the Balt, and then Renwick's closed, and Bob left the Institute, I gravitated to the newly opened PJ's Pancake restaurant, and at least one other mathematician, Solomon Bochner, gravitated with me. Other habitués were Walter Kaufmann, Louis Fischer and Erich Kahler, who wrote *The Disinherited Mind.*

"Coffee housing" was a tradition that I adopted at the Corner Room in State College, when I taught at Penn State, letting the good conversation and the aroma of coffee percolate through the brain. Luckily the owner and manager, Herb Tuchman, is a pleasant indulgent man, and most of PJ's notorious charm owes to him. When the "Bunnamatic" coffeemaker was new, PJ's had the freshest coffee in town. (Besides that, they sold too much of it for it to become stale!)

In April 1997, PJ's unfortunately was gutted by fire, and then we discovered a delightful new coffee shop "The Small World Café" with old world charm, and serving 4 or 5 varieties of coffee and espresso.

Our favorite coffee spots in Barcelona were razed in 1995 to make space for a new hotel and convention center: Café Zurich Bar and Bracafe were were adjacent independent outdoor cafes, where everybody used to meet their friends. You could sit there for hours undisturbed by the waiters. Just as in France, your drink was the price you paid for the table where you sat. I've sat at the*Café des Deux Magots* in Paris for hours reading, e.g. Sartre's *Nausseé*, or *L'Etre et le Neant*, and *Camus' La Plague*, or *L'Etranger*, which I purchased at Gallimard's *Librairie* next door.[48] And since Sartre and Camus wrote at the café, the books may well have been written there.

In *The Psychology of Invention in the Mathematical Field*, (Princeton U. Press paperback) Jacque Hadamard relates how Henri Poincaré discovered "Theta Fuchsian Functions." After a long period of fruitless conscientious endeavor, the idea for them hit like a bolt of lightning one day as he stepped onto a tram going to work. The unconscious did the work in a relaxed moment. **Voilà: Cafés**.

In Café Mediterannean, Berkeley's first coffee house, inspirational classical music is constantly piped into the cafe for the benefit of the patrons who are a mixture

[48] I was disappointed to find out that the French word Magot means Barbary Ape, among several other *non sequiturs*. Naturally I thought it meant "The Café of the Two Maggots."

of professors, students, poets, artists, writers, and panhandlers (The latter line up on Telegraph Avenue asking for "spare change.") It's where I wrote much of Volume I of my *Algebra*. Later, I wrote much of Volume II and most of my part of *FPF Ring Theory* in PJ's. (I duly acknowledged PJ's in the latter book, published in 1984.)

Now that Starbucks and Bucks County Coffee Houses dot the Eastern Coast, one can expect a cornucopia of mathematical works. However, I have yet to encounter Andrew Wiles, who proved Fermat's last theorem, in PJ's, but one day Brooke ("Pretty Baby") Shields and a friend of hers slipped onto stools next to mine and chatted oblivious of the crackling mathematical thunderbolts. (I have seen Professor Wiles in less restful places in Princeton, e.g. Urken's Hardware.)

"Crazy Eddie," Svetlana, "Captain" Bill, and Jay

Two other wonderful places in Princeton, the Colonial Diner, and the College Diner, also burned down, the latter allegedly by "Crazy Eddie," an entrepreneur whose radio and TV sales pitch was "insane." (He was a co-owner, who allegedly needed the insurance money and fled to Israel to escape being arrested.)

The Colonial Diner was on Witherspoon Street which is named after a signer of the Declaration of Independence and a president of Princeton University (which it is perpendicular to). It was convenient to the Princeton Public Library where you would stock up on books to read at the diner. Two local people, "Captain" Bill, and Jay Wilson, used to engage in brilliant conversation in low tones for hours on end, and I could tune in or out at will.

One conversation that I tuned in on was Captain Bill's horology and how it destined him and Svetlana Alliluyeva to marry. Unfortunately for Bill, Svetlana was aloof to the idea, although it appears that they met several times socially. (Luckily for Svetlana, his horoscope was myopic.)

Captain Bill might well have been called "Crazy" Bill, since he often stood at the intersection of Nassau and Witherspoon in front of the University shouting at the top of his lungs. I once asked him why, and he gave this wonderfully Zen response: *to get an echo.*

Jay and Stan

Years later, Jay gravitated to PJ's where his sonorous voice regaled the patrons with his stories. Stanley Tennenbaum, a famous mathematician, also frequented PJ's during his many visits to the Institute. He grew up in Cincinnati (Zenzennati?), across the Ohio River from Covington where I grew up, and his family owned what I call the "Castro Convertible" of Cincinnati, that is, the best-known furniture store.[49]

I met Stanley while I was a graduate student at Purdue on mathematical junkets to Chicago, the mecca of the mathematical world of the midwest. Stanley never bothered to finish the requirements of the Ph.D. degree, but worked on important hard problems in logic with some of the best brains in mathematics, e.g. Kurt Gödel, with whom he discussed Souslin's problem dating from 1920.

[49] I like to joke "We were very Zen!"

I heard him lecture on the solution in the Sheraton Hotel in New York at a meeting of the American Mathematical Society, and nobody gave a better, funnier lecture in my then youthful experience.[50]

At PJ's Stanley took up conversing with Jay where Captain Bill left off, and I continued tuning in and out, periodically tuning onto the omnipresent radio that the cook needed to keep himself in tempo with the (mostly short) orders.

Occasionally I was waved into the conversations in order to render a judgment on some point of contention between them, or, more often, to share some fun they thought up. Jay, a Princeton grad, knew practically everybody, and had marvelous anecdotes to recount. (*That made him a raconteur?*)

Stanley had the gift of total recall down to the last minutia and minisecond. Often he quizzed me on my opinion on a conversation he had with somebody ten or twenty years ago that still lived in his mind like yesterday. This I found not as interesting as when they were conversing and telling stories, and I was placed in a quandary of how to avoid being interrogated by Stanley at too great a length. I hate myself for this, but I had thoughts of my own that I liked to think about, even when I went to a coffee house!

Not to worry, Stanley came in and out of PJ's so often, that allowing him five minutes here and five minutes there, he was able to get my reaction to his latest query. I never knew why he wanted it...maybe out of curiosity?

Stan suffered a serious heart attack about ten years ago, and I sent him a get-well telegram with an insider quip about his being wanted in Princeton for the theft of "Antoine's Necklace." To my dismay, I learned that this pun upset him a great deal, and a son of his (not Jonathan, the mathematician) chastised me for sending it. I never saw him after he convalesced and went to Germany to live with Jonathan.

Roy Hutson and Vic Camillo–Two Poet Mathematicians

Another denizen of PJ's was to become my co-Ph.D. student[51]–Holmes Leroy Hutson–but our friendship sprang from our love of literature, and poetry which Roy, a published poet, writes. Interestingly, he is just the second of my Ph.D. students to publish poetry—Vic Camillo not only publishes poetry but mounts exhibitions of his photographs. Vic is also a recipient of a "Iowa Young Artist" grant.

Marc Rieffel, Serge Lang, Steve Smale and Me

Marc Rieffel published an elegant theorem ("A General Wedderburn Theorem" in 1965 in the *Proceedings of the National Academy*) which gives a very simple proof of the Wedderburn-Artin theorem for simple rings with the descending chain condition, but also much more.

---

[50]Tennenbaum (1968) and Jech (1967) independently found a model of ZFC in which a "Souslin tree" exists. See Jech [97],p.22,Theorem 48, and p.584 for refinements in a joint work of Solovay and Tennenbaum (1971), and Jensen (1968,1972).

[51]Wolmer Vasconcelos did most of the important mentoring after I failed to make a ring theorist out of Roy! (His thesis appeared in 1988.)

However, a theorem of Morita [58] gave a much more general result to the effect that any generator M of the category of right R-modules satisfied the bicentralizer property. Marc's theorem was for the special case M is a nonzero right ideal and R is a simple ring.

I was just then writing up a simple proof that I had discovered for Morita's theorem that required just linear and matrix algebra. This was the centerpiece for Chapter 7 of Volume 1 of my *Algebra*.

Serge Lang who was also visiting Berkeley in the Spring of '65 met me in the Coffee Room at Coffee Hour and challenged me to prove it. I must confess my knees were shaking. I am not an extemporaneous *ad hoc* quick thinker like Serge! However, this time I was prepared (probably the first and last time ever). As I wrote the simple, elementary proof on the board, I was wondering if I hadn't made a mistake somewhere since it appeared uncharacteristically lucid. But no, Serge caught on right away and took the chalk out of my hand and finished the proof himself.

Parlez-Vous Français? My Proof Speaks French

But not only that: Serge got on the phone, called Sammy Eilenberg at Columbia University, and speaking in rapid French, told *him* the proof.[52] *Sammy blessed it*, and then Serge asked him to communicate it to the National Academy where Rieffel's paper appeared. More rapid French, and then Serge told me that he would type it up for me himself, and get Stephen Smale to submit it as a Research Announcement in the *Bulletin of the American Mathematical Society*!

And that is how it appeared (with the same title as Marc's) two years later in 1967. Now how's that for good Karma?

Mario Savio and The Berkeley Free Speech Movement (1964)

The Free Speech Movement (= FSM) began in Berkeley in 1964 in front of Sproul Hall during an anti-Vietnam War rally broken up by the police. The leader of FSM was Mario Savio. When he died in 1996, Wendy Lesser wrote in the *New York Times Book Review* (in Bookend, p.43, December 1996):

> "When Mario Savio clambered up on a police car to protest the arrest of a fellow student, he took off his shoes first so he wouldn't damage the car. Such gentleness did not prevail in American politics...
>
> He spoke without notes of any kind, and he spoke at length, directly addressing his audience with passion and imagination. The sentences he spoke were complicated and detailed, with clauses and metaphors and little byways of digression that together added up to a coherent grammatical whole..."
>
> "A complexity of language meant a complexity of thought, and [that] meant you couldn't win an election. But Mario Savio's politics... were about changing the way people thought about one another and the world."

The result of the Free Speech Movement was that students and townspeople won the right to congregate in front of Sproul Hall, Sather Gate and in other university and Berkeley sites, and exercise their First Amendment Rights.

---

[52] I recall his exact words "*Un module est balancé, si...*" I had coined the term balanced module and was delighted at how it sounded in French: It seemed strange and new.

Jerry Rubin

One day Steve Smale took me to lunch with Mario Savio and Jerry Rubin. During lunch I asked Jerry if he were related to Jerry Rubin who covered the Cincinnati Reds baseball team for the *Cincinnati Inquirer.* He was delighted to find a reader who remembered him. Several years later Jerry became famous for his exploits during the Republican Presidential convention, and the trial of The Chicago Seven, including Abbie Hoffman, in which they were exonerated. They formed a group called yippies identified by colorful headbands with turkey feathers sticking up, and dressed in Indian garb.

Twenty years later, in the 80's, Jerry became a Wall Street Broker,[53] saying that the real power in America flowed from the checkbook. He met a sad death. As he jaywalked across Wilshire Boulevard in Hollywood, he was hit and killed by an automobile. Invisibly he joined the ranks of fatalities of pedestrians by vehicles that included Pierre Curie, Antoni Gaudí, and Randall Jarrell.

Steve Smale

Steve was in the forefront of Vietnam War protesters who *interalia* stopped trains travelling through Oakland loaded with the ingredients from California forests used in napalm and other horrors. I deeply admired him for this, even if I did not follow his examples. In 1980 at the International Congress of Mathematicians he walked up the steps of Moscow University to a prearranged press conference in which he denounced both the USA and the USSR for their conduct in the Vietnam War. The *New York Times* gave it front page coverage. When my son, Japheth, refused to register for the draft, and while he was at the Budapest semester, he travelled to Moscow. I told him that Steve Smale had already denounced the USA, so he didn't have to! (This is logic?)

Some Undergraduate Gems at Rutgers and Penn State

An engineering student of elementary differential equations named Mary[54] was able to answer every question ever asked of her on the exams and every homework problem too. So following a tradition of the University of Kentucky, that straight A-students may be exempted from the final, I exempted her but collected her notebook instead, since we are required to keep final exams for a year. I noticed that she had worked out every problem in the book—not only the even numbered but also the odd. When I asked her why, she blurted out in class, "Oh, Professor Faith, you know it is so much fun!"

Another student, an Asian Indian, in a course in combinatorics showed little enthusiasm for the subject, but when I asked her which course she did like, she brightened and said "abstract algebra." Since this is perhaps the most hated and difficult undergraduate course, I evinced surprise and pleasure, and asked her why.

---

[53] On the other hand, Abbie Hoffman, who was arrested 52 times (according to Webster's New World encyclopedia), became a career protester much in demand on campuses across the United States as an anti-government icon.

[54] Unfortunately when I retired, I had to vacate my office that I occupied for 35 years, and I lost most of my class records in the move.

She said, "Oh, Professor Faith, the theorems are so dramatic!" I told her that was the most beautiful reason ever given by a student for the study of abstract algebra, and that it is one that I gave in the introduction to my *Algebra* (see **Envoi** below).

Another student of Abstract Algebra, Cindy Garrison, developed music theory using the theory of Abelian groups of order 12. The octave is equivalent to the unison in terms of the number of half steps from the starting point mod 12. (This means that an octave from B will give you a B.) Pythagoras, who originated the mathematical theory of chords and harmonics, would have been as proud of her as I was.

In 1996, Jennifer McKinney won an outstanding teacher award in a New Jersey public school, her first year of teaching. She had been one of the best students that I ever had in Foundations of Mathematics, which is another course with a high level of abstraction that stymies many students (too many).

In my very last class in Foundations, in Spring 1996, Lauren Altucher asked me if we were going to study infinite cardinal numbers. Well, we always aimed to reach this far in the text, but usually, as in the class with Jennifer, there was at most one student who could have coped at that level. But in Lauren's class, there was also Gabe Adamek, who went on to grad school at U. of Mass. the following fall, but I said "yes," *simply because she wanted it.* As a reward, I let her give the lecture-demonstration of the well-ordering theorem. Largely based on her wonderful grasp of the subject, she was chosen for the Budapest Semesters for 1997–1998 which had been organized for gifted students. (My son, Japheth, had attended circa Spring 1987.)

These gems, and hundreds of others, make teaching a joyous experience. On the other hand, not all students perform at this level, as the following poem by Billy Reeves, a student of mine at Penn State in the summer of 1959, attests.

"Carl, You Will Always Have Dumb Students!"

To whom do we owe all our bountiful knowledge?
To whom shall we fervently bow?

Towards what have we striven, and fiendishly laboured,
But math, with the sweat of our brow?

Here, in this place, let us drink to the man, and the math, and the cold perspiration,

In knowledge, not grades, let us revel this night, 'though hell be our sole consolation.

Tilt towards the heavens your goblets, young boys, (our host and the prof are included)

If in faith we're denied what we always desired,
In heaven we'll not be deluded.

Drink up, have a ball, it's not hard, said my friend,
You'll get it, said he, through your labours,

I got it, my dear, and right in the ear,
It's time you came through with the favors.

Oh–favors, good lord, are you really that bored?

Don't you see? It's really so simple –

The square root of this, over two times fifteen, times rho, and times "X", differentiated then by the Law of the mean and integrated twice over the integral between...oh - no - what happened?

What's the book say?

Then, Gaudeamus Igitur, and Juvenes dum Sumus,
Dear Carl, you'll not escape the fact...

...YOU'LL ALWAYS HAVE DUMB STUDENTS!

As an example of Billy Reeves' poem, let me cite a humorous experience in 1962 in my first semester of teaching at Rutgers. A student in freshman calculus (then taught in the basement classrooms in the cliff dormitories, e.g. Frelinghuysen, overlooking the Raritan) stood up in class and volubly complained about the "give 'em the keys" philosophy of teaching that I mentioned earlier. "Professor Faith, I've always learned things by memorization and I'm too old to change now!" I innocently asked, "How old are you?" "Eighteen", he replied, and the class roared with laughter, breaking the tension. He went on to do well enough to get a "C" grade. (This was eons before anyone heard of "grade inflation"!)

Envoi to My Century

"*Euclid Alone Has Looked on Beauty Bare*"
(Title of a poem by Edna St. Vincent Millay)

I have been working on this book for two years now. I do not pretend that this is a comprehensive study, far less an encyclopedic one. Some of the specialized articles in the Kluvver *Handbook of Algebra* run to half this book's size; and the *Cambridge Encyclopedia of Mathematics* is published in booksize bites! There are plenty of mathematicians including some of the "greats" who won't show up here. As I intimated by the Russian proverb on the first page of the Preface, my tastes, background and research interests have guided me throughout. Yet even beyond this, I have been seduced by the drama and beauty of the theorems. I am not alone in this:

> "Mathematics... possesses not only truth, but supreme beauty—a beauty cold and austere, like that of sculpture, without any appeal to any part of our weaker nature, without the gorgeous trappings of painting or music, yet sublimely pure, and capable of a stern perfection such as only the greatest art can show"—Bertram Russell in *"The Study of Mathematics"* (1902) (quoted in Bartletts).

Robert Graves wrote in a similar vein about poetry in his book *The White Goddess.* Rather than Graves, we shall quote a mathematician:

> "What motivates mathematicians is not the desire to solve practical problems or even problems of science. The important drives are aesthetic coming from the internal appeal of mathematics in itself"—Sir Roger Penrose in his review of *Fermat's Enigma* by Simon Singh in the *New York Times Book Review*, November 30, 1997.

Sir Roger had other important things to say about mathematics:

> "What is unusual about mathematics is that it is possible to settle differences of opinion by definitive argument—majority opinion counts for naught; so does authority. The sound mathematical reasoning of a single individual suffices."

I beg the readers to remember this good advice before citing me (or this book) as an authority. (Ha!)

According to legend, there were misprints on the first page of the first edition of the *King James Bible* despite the proofreading by thousands of scholars. King James wanted the perfect Bible as an *auto da fé*. Alas, King James ought to have read Juvenal's *The Vanity of Human Wishes.*

Therefore, I invite the reader to let me know of the inevitable egregious errors: I see a corrected reprint coming down the pike!

"The past is never dead. It is not even past." (William Faulkner)

"It's not over until it's over." (Yogi Berra)

"A work of art is never finished, only abandoned." (Picasso)

Index for Snapshots, Part II

Bibliography

Note: The notation [72] denotes 1972, the year the paper was published. Additional listings are in the author's *Algebra I* and *II*, and in the Russian translations published by MIR Publishers.

[72] S. S. Abhyankar, W. Heinzer and P. Eakin, *On the uniqueness of the coefficient ring in a polynomial ring*, J. Algebra **23** (1972), 310–342.

[73] J. Ahsan, *Rings all of whose cyclic modules are quasi-injective*, Proc. L.M.S. (3) **27** (1973), 425–439.

[69] T. Akiba, *Remarks on generalized rings of quotients*, III, J. Math., Kyoto U. **9002** (1969), 205–212.

[73,76] S. Alamelu, *Commutativity of endomorphism rings of ideals*, Proc.Amer.Math.Soc. **37** (1973), 29–31; II (1976) 271–274.

[39] A. A. Albert, *The Structure of Algebras*, Colloq. Pub, vol. XXIV, Amer. Math. soc., New York, 1939.

[40] ———, *On ordered algebras*, Bull. A.M.S. **46** (1940), 521–522.

[61] F. Albrecht, *On projective modules over semi-hereditary rings*, Proc. Amer. Math. Soc. **12** (1961), 638-639.

[74] T. Albu, *Sur la dimension de Gabriel des modules*, Algebra Berichte, Nr. 21, Seminar Kasch-Pareigeis, Munich, 1974.

[80] ———, *On commutative Grothendieck categories having Noetherian cogenerator*, Arch. Math. **34** (1980), 210–219.

[97] T. Albu and P. F. Smith, *Localization of modular lattices, Krull dimension, and the Hopkins-Levitzki Theorem* (II), Comm. Algebra **25** (1997), 1111–28.

[98] ———, *Dual Krull dimension and duality*, Rocky Mt. J. Math. (1998).

[98] T. Albu and P. Vámos, *Global Krull Dimension and Global Dual Krull Dimension of Valuation Rings*, in Dikranjan and Salce [98].

[96] R.A.M. Aldosray, *Note on rings with ascending chain condition on annihilator ideals*, Far East J. Math. Sci **4** (1996), 177–183.

[92] A. H. Al-Huzali, S. K. Jain, and S. R. López-Permouth, *Rings whose cyclics have finite Goldie dimension*, J. Algebra **153** (1992), 37–40.

[68] I. K. Amdal and F. Ringdal, *Catégories unisérielles*, C. R. Acad. Sci, Paris, Série A **267** (1968), 247–249.

[56] S. A. Amitsur, *Invariant submodules of simple rings*, Proc.Amer.Math.Soc. **7** (1956), 987–989.

[57a] ———, *A generalization of Hilbert's Nullstellensatz*, Proc.Amer.Math.Soc. **8** (1957), 649–656.

[57b] ———, *The radical of field extensions*, Bull. Res.Council of Israel (Israel J. Math.) **7F** (1957), 1–10.

[57c] ———, *Derivations in simple rings*, Proc. London Math.Soc. (3) **7** (1957), 87–112.

[59] ______, *On the semisimplicity of group algebras*, Mich. Math. J. **6** (1959), 251–253.

[61] ______, *Groups with representations of bounded degree*, II, Illinois J. Math. **5** (1961), 198–205.

[68] ______, *Rings with involution*, Israel J. Math **6** (1968), 99–106.

[71] ______, *Rings of quotients and Morita Contexts*, J. Algebra **17** (1971), 273–298.

[72a] ______, *On central division algebras*, Israel J. Math **12** (1972), 408–420.

[72b] ______, *Nil radicals: Historical notes and some new results*, in Kertesz [71].

[66] S. A. Amitsur and C. Procesi, *Jacobson rings and Hilbert algebras with polynomial identities*, Ann. Mat. Pura e Appl. (4) **71** (1966), 61–72.

[78] S. A. Amitsur and L. W. Small, *Polynomial rings over division rings*, Israel Math. J. **31** (1978), 353–358.

[96] ______, *Algebras over infinite fields revisited*, Israel Math. J. **96** (1996), part A,, 23–25.

[74] A. Z. Ananín and E. M. Zjabko, *On a question due to Faith*, Algebra i Logika **13** (1974), 125–131,234.

[77] D. D. Anderson, *Some Remarks on the ring $R(x)$*, Comment. Math. Univ., St. Paul (2) **26** (1977), 141–5.

[97] ______, *(ed.), Factorization in Integral Domains*, Proc. Iowa Conf. 1996, Lecture Notes in Pure and Appl. Math., vol. 189, Marcel Dekker, Basel, etc., 1997.

[85] D. D. Anderson, D. F. Anderson and R. Markanda, *The rings $R(x)$ and $R\langle x\rangle$*, J. Alg. **95** (1985), 96–115.

[87] D. D. Anderson and J. Huckaba, *The integral closure of a Noetherian ring, II*, Comm. Alg. **15** (1987), 287–295.

[88] D. D. Anderson and D. Dobbs, *Flatness, LCM-stability and related module theoretical properties*, J. Algebra **112** (1988), 139–150.

[72] F. W. Anderson and K. R. Fuller, *Modules with decompositions that complement direct summands*, J. Algebra **22** (1972), 241–253.

[73] ______, *Rings and Categories of Modules*, Springer-Verlag, New York, 1973.

[90] P.N. Ánh, *Morita duality for commutative rings*, Comm.Alg. **18** (1990), 1781–1788.

[97] ______, *Immediate extensions of rings and approximation of roots*, preprint, Math. Inst. Hungarian Acad. Sci., Budapest, 1997.

[97a] P. N. Ánh, D. Herbera, and C. Menini, *$AB-5^\star$ and linear compactness*, preprint, 1997.

[97b] ______, *Baer and Morita duality*, preprint, 1997.

[86] P. Ara, *On $\aleph_0$-injective regular rings*, J.Pure and Appl.Alg. **42** (1986), 89–115.

[88] ______, *Centers of maximal quotient rings*, Arch. Math. **50** (1988), 342–347.

[95] ______, *Strongly π-regular rings have stable range one*, Proc.Amer.Math.Soc. **124** (1997), 3293–3298.

[93] ______, *Extensions of exchange rings*, J. Algebra (1998 ??).

[97] P. Ara, K. R. Goodearl, K. C. O'Meara and E. Pardo, *Diagonalization of matrices over regular rings*, Linear Algebra and Appl. **265** (1997), 147–163.

[97] ______, *Separative cancellation for projective modules over exchange rings*, Israel J. Math. (1998).

[84] P. Ara and P. Menal, *On regular rings with involution*, Arch. Math. **42** (1984), 126–130.

[48] R.F. Arens and I.Kaplansky, *Topological representations of algebras*, Trans. Amer. Math.Soc. **63** (1948), 457–481.

[97] A. V. Arhangel'skii, K. R. Goodearl, and B. Huisgen-Zimmerman, *Kiiti Morita*, 1915–1995, Notices A.M.S. **44** (1997), 680–684.

[82] E. P. Armendariz, *On semiprime rings of bounded index*, Proc. Amer. Math. Soc. **85** (1982), 145–148.

[72] E. P. Armendariz and G. R. MacDonald, *Maximal quotient rings and S-rings*, Canad. J. Math. **24** (1972), 835–850.

[74] E. P. Armendariz and S. A. Steinberg, *Regular self-injective rings with polynomial identity*, Trans. Amer. Math. Soc. **190** (1974), 417–425.

[27] E. Artin, *Zur Theorie der hyperkomplexen Zahlen*, Abh. Math. Sem., Univ.Hamburg, Vol. 5 (1927), 251–260.

[27a] ———, *Über die Zerlegung definiter Funktionen in Quadrate*, Abh. Math. Sem. Univ., Hamburg **5** (1927), 100–115.

[27b] ———, *Beweis des Allgemeinen Reziprozitätsgesetzes*, *Ibid.*, 353–363.

[55] ———, *Galois Theory*, U. of Notre Dame, South Bend, 1955.

[57] ———, *Geometric Algebra*, John Wiley & Sons, New York, 1957.

[65] ———, *Collected Papers* (S. Lang and J. T. Tate, eds.), Addison-Wesley, Reading (Mass.), Palo Alto, London, Dallas, Atlanta, 1965.

[26] E. Artin and O. Schreier, *Algebraische Konstruktion reeller Körper*, Abh. Math. Sem., Hambrug **5** (1926), 83–115.

[50] E. Artin and J. Tate, *A note on a finite ring extensions*, J. Math. Soc. Japan **3** (1951), 74–77.

[69] M. Artin, *On Azumaya algebras and finite dimensional representations of rings*, J. Algebra **11** (1969), 532–563.

[91] ———, *Algebra*, Prentice-Hall, Englewood Cliffs, 1991.

[38] K. Asano, *Nichtkommutative Hauptidealringe*, Act. Sci. Ind. 696 Paris, 1938.

[49] ———, *Über Hauptidealringe mit Kettensatz*, Osaka Math. J. **1** (1949), 52–61.

[61] ———, *On the radical of quasi-Frobenius algebras; Remarks concerning two quasi-Frobenius algebras with isomorphic radicals; Note on some generalizations of quasi-Frobenius rings*, Kodai Math. Sem. Rpts. **13** (1961), 131–151; 224–226; 227–334.

[94] M. Aschbacher, *Sporadic Groups* (1994), Cambridge Tracts in Math. 104, Cambridge, Melbourne and New York.

[97] ———, *Review of Gorenstein, Lyons, and Solomon [94].*

[80] D. L. Atkins and J. W. Brewer, *D-automorphisms of power series and the continuous radical of D*, J. Algebra **67** (1980), 185–203.

[57] M. Auslander, *On regular group rings*, Proc. Amer. Math. Soc. **8** (1957), 658–664.

[74] ———, *Representation theory of Artin algebras, I,II*, Comm. Algebra **1** (1974), 177-268, 177-310.

[55] ———, *On the dimension of modules and algebras,* III, Nagoya Math.J. **9** (1955), 67–77.

[57] M. Auslander and D. Buchsbaum, *Homological dimension in local rings*, Trans. Amer. Math. Soc. **85** (1957), 390–405.

[59] ———, *Unique factorization in regular local rings*, Proc.Nat.Acad.Sci. U.S.A. **45** (1959), 733–734.

[60] M. Auslander and O. Goldman, *The Brauer group of a commutative ring*, Trans. A.M.S. **97** (1960), 367–409.

[77] C. Ayoub, *The additive structure of a ring, and the splitting of the torsion ideal*, in Ring Theory, Proc. of the Athens Conf. 1976 (S. Jain,ed.), Lecture Notes in Pure and Appl Math., vol. 25, Dekker, Basel and New York, 1977.

[48] G. Azumaya, *On generalized semiprimary rings and Krull-Remak-Schmidt's theorem*, Japan J.Math. **19** (1948), 525–547.

[49] ______, *Galois theory for uniserial rings*, J. Math.Soc. Japan, **1** (1949), 130–146.

[50] ______, *Corrections and supplementaries to my paper concerning Krull-Remak-Schmidt's theorem*, Nagoya Math. J. **I** (1950), 117–124.

[51] ______, *On maximally central algebras*, Nagoya Math. J. **2** (1951), 119–150.

[54] ______, *Strongly π-regular rings*, J. Fac. Sci. Kokkaido U. **13** (1954), 34–39.

[66] ______, *Completely faithful modules and self-injective rings*, Nagoya Math. J. **27** (1966), 697–708.

[74] ______, *Characterizations of semi-perfect and perfect modules*, Math. Z. **140** (1974), 95–103.

[80] ______, *Separable rings*, J. Algebra **63** (1980), 1–14.

[93] ______, *A characterization of semiperfect rings and modules*, in Ring Theory: Proc. Ohio State-Denison Conf. (1992) (S. K. Jain and S. T. Rizvi, eds.), World Scientific Publishers, River Edge, N.J., London and Singapore, 1993.

[89] G. Azumaya and A. Facchini, *Rings of pure global dimension zero and Mittag-Leffler modules*, J. Pure Appl. Algebra **62** (1989), 109–122.

[95] G. Baccella, *Semiartinian V-rings and semiartinian von Neumann regular rings*, J. Algebra **173** (1995), 587–612.

[37] R. Baer, *Abelian groups without elements of finite order*, Duke Math. J. **3** (1937), 88–122.

[40] ______, *Abelian groups that are direct summands of every containing Abelian group*, Bull. Amer. Math. Soc. **46** (1940), 800–806.

[43] ______, *Rings with duals*, Amer.J.Math. **65** (1943), 569–584.

[43b] ______, *Radical ideals*, Amer. J. Math. **65** (1943), 537–568.

[52] ______, *Linear Algebra and Projective Geometry*, Academic Press, New York, 1952.

[62] S. Balcerzyk, *On groups of functions defined on Boolean algebras*, Fund. Math. **50** (1962), 347–367.

[61] S. Balcerzyk, A. Bialynicki-Birula and J. Łoś, *On direct decompositions of complete direct sums of groups of rank 1*, Bull. L'Acad. Polonaise de Sci., série des sciences math. astr. et phys. **9** (1961), 451–454.

[76] B. Ballet, *Topologies lineares et modules Artiniens*, J. Algebra **41** (1976), 365–397.

[98] M. Barr, *Lambekfest at McGill*, Canad. Math. Soc. Notes **30** (1998), 12–13.

[92] F. D. Barrow, *Pi in the Sky*, Oxford U. Press, Oxford, New York, Tokyo, etal, 1992.

[77] J. Barwise, *Handbook of Mathematical Logic*, North-Holland, Amsterdam, New York, Oxford, 1977.

[60] ______, *Finitistic dimension and a homological generalization of semiprimary rings*, Trans. Amer. Math. Soc. **95** (1960), 466–488.

[62b] ______, *Injective dimension in Noetherian rings*, Trans.Amer.Math.Soc. **102** (1962), 18–29.

[63] ———, *Big projective modules are free*, Ill. J. Math. **7** (1963), 24–31.

[63b] ———, *The ubiquity of Gorenstein rings*, Math.Z. **83** (1963), 8–28.

[64a] ———, *Projective modules over free groups are free*, J. Algebra **1** (1964), 367–373.

[64b] ———, *K-theory and stable algebra*, Publications I.H.E.S., no. 22, 1964, 5–60.

[68] ———, *Algebraic K-Theory*, Benjamin, N. Y. and Amsterdam, 1968.

[71] H. Bass, *Descending chains and the Krull ordinal of commutative Noetherian rings*, J. Pure and Appl. Algebra **1** (1971), 347–360.

[71] J. A. Beachy, *On quasi-Artinian rings*, Pac. J. Math. (2) **3** (1971), 449–452.

[76] ———, *Some aspects of non-commutative localization,* Inter.Conf. Kent State U. 1975, Lecture Notes in Math., vol. 545, 1976, pp. 2–31.

[75] J. A. Beachy and W. D. Blair, *Rings when faithful left ideals are co-faithful*, Pac. J. Math. **58** (1975), 1–13.

[84] J. A. Beachy and W. D. Weakley, *Piecewise Noetherian rings*, Comm. Algebra **12** (1984), 2679–2706.

[72a] I. Beck, *Σ-injective modules*, J. Algebra **21** (1972), 232–249.

[72b] I Beck, *Projective and free modules, Math. Z.* **129** (1972), 231–234.

[78] I. Beck and P. J. Trosborg, *A note on free direct summands*, Math. Scand. **42** (1978), 34–38.

[69] J. L. Bell and A. B. Slomson, *Models and Ultraproducts: An Introduction*, North Holland, Amsterdam, New York, Oxford, 1977.

[91] L. P. Belluce, *Spectral spaces and noncommutative rings*, Comm. Algebra **19** (1991), 1855–1865.

[66] L. P. Belluce, S. K. Jain, and I. N. Herstein, *Generalized commutative rings*, Nagoya Math.J. **27** (1966), 1–5.

[64] G. M. Bergman, *A ring primitive on the right but not on the left*, Proc. Amer. Math. Soc. **15** (1964), 473–475.

[71] ———, *Hereditary commutative rings and centres of hereditary rings*, Proc. L.M.S. (3) **23** (1971), 214–236.

[74a] ———, *Modules over coproducts of rings*, Trans. Amer. Math. Soc. **200** (1974), 1–32.

[74b] ———, *Coproducts and some universal ring constructions*, Trans.Amer.Math.Soc. **200** (1974), 33–88.

[71] G. M. Bergman and P. M. Cohn, *The centres of 2-firs and hereditary rings*, Proc. L.M.S. (3) **23** (1971), 83–98.

[96] G. M. Bergman and A. O. Hausknecht, *Cogroups and Co-rings in Categories of Associative Rings*, Math. Surveys and Monographs, vol. 45, Amer. Math. Soc., Providence, 1996.

[61] A. Bialynicki-Birula, *See Balcerzyk.*

[87] G. F. Birkenmeier, *Quotient rings of rings generated by faithful cyclic modules*, Proc. Amer. Math. Soc. **100** (1987), 8–10.

[89] ———, *A generalization of FPF rings*, Comm. Algebra **17** (1989), 855–885.

[97] G. F. Birkenmeier, J.Y. Kim, and J.K. Park, *Splitting theorems and a problem of Mueller*, in Advances in Ring Theory (Jain and Rizvi, eds.), Birkhäuser, Boston, Basel and Berlin, 1997.

[48-67] G. Birkhoff, *Lattice Theory*, Amer.Math.Soc. Colloq., Publ vol. 25, Providence, 1948, new rev. ed. 1967.

[69] J. E. Björk, *Rings satisfying a minimum condition on principal ideals*, J. reine u. angew. Math. **236** (1969), 466–488.

[70] ______, *On rings satisfying certain chain conditions*, J. reine und angew. Math. **237** (1970), 63–73.

[72] ______, *The global homological dimension of some algebras of differential operators*, Invent.Math. **17** (1972), 67–78.

[72b] ______, *Radical properties of perfect modules*, J. reine u. angew. Math. **245** (1972), 78–86.

[73] ______, *Noetherian and Artinian chain conditions of associative rings*, Arch. Math. **24** (1973), 366-378.

[75] W. D. Blair, *See Beachy.*

[87] W. D. Blair and L. W. Small, *Embeddings in Artinian rings and Sylvester rank functions*, Israel J. Math. **58** (1987), 10–18.

[90] ______, *Embedding differential and skew polynomial rings into Artinian rings*, Proc. Amer. Math. Soc. **109** (1990), 881–886.

[86] D. Blessenohl and U. K. Johnsen, *Eine Verschärfung des Satzes von der Normalbasis*, J. Alg. **103** (1986), 141–159, (Editor's note: This paper contains most of the results of Faith [57] without mention.).

[91] ______, *Stabile Teilkörper galoisscher Erweiterungen und ein Problem von C. Faith*, Arch. Math. **56** (1991), 245–253.

[80] L. M. Blumenthal, *A Modern View of Geometry*, Dover, New York, 1980.

[67] L. A. Bokut, *The embedding of rings in fields*, Soviet Math.Dokl. **8** (1967), 901–904.

[98] B. Bollobás, *To prove and conjecture: Paul Erdős and his mathematics* **105** (1998), Amer. Math. Monthly, 209–237.

[69] A. Borel, *Linear Algebraic Groups*, Benjamin, New York, 1969.

[73] W. Borho and P. Gabriel, *Primideale in einhüllende auslösbaren Lie-Algebren*, Lecture Notes in Math., vol. 357, Springer-Verlag, Berlin, Heidelberg & New York, 1973.

[61a] N. Bourbaki, *Eléments de Mathématique (A.S.I.N° 1290)*, in Algèbre Commutative, Chapitre I (Modules Plats) et 2 (Localisation), Hermann, Paris 1961.

[61b] ______, *Eléments de Mathématique (A.S.I.N° 1293)*, in Algèbre Commutative, Chapitre 3 (Graduations, Filtrations, et Topologies) et 4 (Idéaux Premiers Associés et Décomposition Primaire), Hermann, Paris 1965.

[64] ______, *Eléments de mathématique (A.S.I.N° 1308),*, Algèbre Commutative, Chapitre 7 (Divseurs), Hermann, Paris 1965.

[65] ______, *Eléments de mathématique (A.S.I.N° 1314)*, Algèbre Commutative, Chapitre 7 (Diviseurs), Hermann, Paris 1965.

[67] A. J. Bowtell, *On a question of Mal'cev*, J.Algebra **7** (1967), 126–139.

[91] C. B. Boyer and U. C. Merzbach, *A History of Mathematics*, 2nd ed., Wiley, New York, Chichester, Brisbane, Toronto, Singapore, 1991.

[73] A. K. Boyle, *When projective covers and injective hulls are isomorphic*, Bull. Austral. Math. Soc. **8** (1973), 471–476.

[74] ______, *Hereditary QI-rings*, Trans. Amer. Math. Soc. **192** (1974), 115-120.

[73] W. Brandal, *Almost maximal domains and finitely generated ideals*, Trans. Amer. Math. Soc. **183** (1973), 203–222.

[79] ———, *Commutative rings whose finitely generated modules decompose*, in Lectures in Math., Springer, New York-Heidelberg-Berlin, 1979.

[29] R. Brauer, *Über Systeme hyperkomplexer Zahlen*, Math Z. **39** (1929), 79–107.

[49] R. Brauer, *On a theorem of H. Cartan*, Bull.Amer.Math.Soc. **55** (1949), 619–620.

[32] R. Brauer, H. Hasse and Emmy Noether, *Beweis eines Hauptssatzes in der Theorie der Algebren*, J. reine Angew. Math. **167** (1932), 399–404.

[80] J. W. Brewer, *See Atkins.*

[81] ———, *Power series over commutative rings*, in Lecture Notes in Pure and Applied Math, vol. 64, Marcel Dekker, New York and Basel, 1981.

[80] ———, *See Atkins.*

[72] J. W. Brewer and E. A. Rutter, *Isomorphic Polynomial Rings*, Arch. Math. XXIII (1972), 484–488.

[77] J. W. Brewer, E. A. Rutter and J. J. Watkins, *Coherence and weak global dimension of $R[[x]]$ when R is von Neumann regular*, J. Algebra **46** (1977), 278–289.

[80] J. W. Brewer and W. J. Heinzer, *R Noetherian implies $R\langle x\rangle$ is a Hilbert ring*, J. Algebra **67** (1980), 204–209.

[81] J. W. Brewer and M. K. Smith (eds.), *Emmy Noether, A Tribute to Her Life and Work*, M. Dekker, Inc., New York and Basel, 1981.

[72] G. M. Brodskii, *Steinitz modules*, (Russian), Mat. Issled. **7** (1972), 14–28, 284.

[50] B. Brown and N. McCoy, *The maximal regular ideal of a ring*, Proc.Amer.Math.Soc. **1** (1950), 165–171.

[82] K. A. Brown, C. R. Hajarnavis, and A. B. MacEacharn, *Noetherian rings of finite global dimension*, Proc. London Math. Soc. **44** (1982), 349–371.

[69] H. H. Brungs, *Generalized discrete valuation rings*, Canad. J. Math. **21** (1969), 1404–1408.

[70] ———, *Idealtheorie für eine Klasse noetherscher Ringe.*, Math. Z. **118** (1970), 86–92.

[73a] ———, *Non commutative Krull domains*, J. reine u. angew. Math. **264** (1973), 161–171.

[73b] ———, *Left euclidian rings*, Pac. J. Math. **45** (1973), 27–33.

[93] W. Bruns and J. Herzog, *Cohen-Macaulay Rings*, Cambridge Studies in Advanced Math, vol. 39, Cambridge, New York, Melbourne, Sydney, 1993.

[88] J. L. Bueso, B. Jara, B. Torrecillas (eds.), *Ring Theory Proceedings, (*Granada, 1986), Lecture Notes in Mathematics, vol. 1328, Springer-Verlag, Berlin, 1988.

[65] R. T. Bumby, *Modules which are isomorphic to submodules of each other*, Arch. Math. **16** (1965), 184–185.

[72] L. Burch, *Codimension and Analytic Spread*, Proc. Cambridge Phil. Soc. **72** (1972), 369–373.

[69] W. E. Burgess, *Rings of quotients of group rings*, Canadian Math. J. **21** (1969), 865–875.

[11] W. Burnside, *Theory of Groups of Finite Order*, Cambridge University Press, Cambridge, 1911, (2d. ed.).

[88] C. Busqué, *Directly finite $\aleph_0$-complete regular rings are unit regular*, in Ring Theory, pp. 38-49, see P. Bueso [88].

[93] ———, *Centers and ideals of right self-injective rings*, J. Algebra **175** (1993), 394–404.

[79] K. A. Byrd, *Right self-injective rings whose essential right ideals are two-sided*, Pac. J. Math. **82** (1979), 23–41.

[69] A. Cailleau, *Une caractérisation des modules Σ-injectifs*, C.R.Acad.Sci Paris **269** (1969), 997–999.

[70] A. Cailleau and G. Renault, *Etude des modules Σ-injective*, C.R.Acad.Sci. Paris **270** (1970), 1391–1394.

[68] W. Caldwell, *Hypercyclic rings*, Pac. J. Math. **24** (1967), 29–44.

[70] V. P. Camillo, *Balanced rings and a problem of Thrall*, Trans. A.M.S. **149** (1970), 143–153.

[74] ———, *Semihereditary polynomail rings*, Proc. Amer. Math. Soc. **45** (1974), 173–174.

[75a] ———, *Commutative rings whose quotients are Goldie*, Glasgow Math. J. **16** (1975), 32–33.

[75b] ———, *Distributive modules*, J. Algebra **36** (1975), 16–25.

[76] ———, *Rings whose faithful modules are flat*, Arch. Math. **27** (1976), 522–525.

[77] ———, *Modules whose quotients have finite Goldie dimension*, Pac. J. Math. **69** (1977), 337–338.

[78] ———, *On the homological independence of injective hulls of simple modules over commutative rings*, Comm. in Alg. **6** (1978), 1459–1469.

[84] ———, *Morita equivalence and infinite matrices*, Proc. A.M.S. **90** (1984), 186–188.

[85] ———, *On Zimmermann-Huisgen's splitting theorem*, Proc. Amer. Math. Soc. **94** (1985), 206–208.

[90] ———, *Coherence for polynomial rings*, J. Algebra **132** (1990), 72–76.

[98] ———, *Letter to the author, May 16, 1998; E-letter of June 8, 1998.*

[73] V. P. Camillo and J. H. Cozzens, *A nnote on hereditary rings*, Pac. J. Math. **4** (1973), 35–41.

[72] V. P. Camillo and K. R. Fuller, *Balanced and QF-1 algebras*, Proc. A.M.S. **34** (1972), 373–378.

[74] ———, *On Loewy lengths of rings*, Pac. J. Math. **53** (1974), 347–354.

[79] ———, *A note on Loewy rings and chain conditions on primitive ideals*, pp.75–86 in Faith and Wiegand (eds.), 1979.

[79] V. P. Camillo, K. R. Fuller, and E. R. Voss, *Morita equivalence and the fundamental theorem of projective geometry*, preprint, Univ. of Iowa, Iowa City, 1979.

[86] V. P. Camillo and R. Guralnick, *Polynomial rings are often Goldie*, Proc. Amer. Math. Soc. **98** (1986), 567–568.

[93] R. Camps and W. Dicks, *On semilocal rings*, Israel J. Math. **81** (1993), 203–211.

[91] R. Camps and P. Menal, *Power cancellation for Artinian modules*, Comm. Algebra **19** (1991), 2081–2095.

[96] X. H. Cao, etal (eds), *Rings, Groups, and Algebras*, Lecture Notes, Pure Appl Math. vol. 181, Marcel Dekker, Basel and New York, 1996.

[47] H. Cartan, *Theorie de Galois pour les corps non-commutatifs*, Ann.Sci.Ecole Norm.Sup. **64** (1947), 59–77.

[56] H. Cartan and S. Eilenberg, *Homological Algebra*, Princeton Univ. Press, Princeton, N.J. 1956.

[68] V. C. Cateforis and F. L. Sandomierski, *The singular submodule splits off*, J. Algebra **10** (1968), 149–165.

[76] G. Cauchon, *Les T-anneaux, la condition (H) de Gabriel, et ses consequences*, Comm. Algebra **4** (1976), 11–50.

[91] F. Cedó, *Zip rings and Mal'cev domains*, Comm.Algebra **19** (1991), 1983–1991.

[95] ———, *Power series over Kerr commutative rings*, U. Autónoma de Barcelona, preprint.

[96] ———, *The maximal quotient ring of regular group rings*, III, Publ. Mat. **40** (1996), 15–19.

[97] ———, *Strongly prime power series*, Comm. Alg. **25** (1997), 2237–2242.

[95] F. Cedó and D. Herbera, *The Ore condition for polynomial and power series rings*, Comm. Algebra **23** (1995), 5131–5159.

[95] ———, *The maximum condition on annihilators for polynomial rings*, Proc. Amer. Math. Soc. (1995).

[77] C. C. Chang and H. J. Kreisel, *Model Theory*, North-Holland, Ammsterdam, New York, Oxford, 1977.

[68] B. Charles (ed.), *Studies on Abelian groups*, Proc. Symp., Montpellier U. 1967, Springer-Verlag and Dunod, New York and Paris, 1969.

[60] S. U. Chase, *Direct products of modules*, Trans.Amer.Math.Soc. **97** (1960), 457–473.

[62a] ———, *On direct products and sums of modules*, Pac. J. Math. **12** (1962), 847–854.

[62b] ———, *A remark on direct products of modules*, Proc. Amer. Math. Soc. **13** (1962), 214–216.

[65] S.U. Chase and C. Faith, *Quotient rings and direct products of full linear rings*, Math. Z. **88** (1965), 250–264.

[65] S. U. Chase, D. K. Harrison, and A. Rosenberg, *Galois Theory and Galois Cohomology of Commutative Rings, Memoirs of A.M.S., vol. 52*, American Math. Soc., Providence, R.I., 1965.

[71] A. W. Chatters, *The restricted minimum condition in Noetherian hereditary rings*, J. London Math. Soc. **4** (1971), 83–87.

[72] ———, *A decomposition theorem for Noetherian hereditary rings*, Bull. Lond. Math. Soc. **4** (1972), 125–126.

[80] ———, *Rings with Chain Conditions*, Pitman, London, 1980.

[93] ———, *Non-isomorphic rings with isomorphic matrix rings*, Proc. Edinburgh Soc. **36** (1993), 339–348.

[96] ———, *Non-isomorphic maximal orders with isomorphic matrix rings*, preprint, U. of Bristol, 1998.

[77] A. W. Chatters and C. R. Hajarnavis, *Rings in which every complement right ideal is a direct summand*, Quart.J.Math. Oxford, (2) **28** (1977), 61–80.

[81] T. J. Cheatham and D. R. Stone, *Flat and projective character modules*, Proc. Amer. Math. Soc. **81** (1981), 175–177.

[76] G. L. Cherlin, *Model Theoretic Algebra-Selected Topics*, Lecture Notes in Math., vol. 521, Springer-Verlag, Berlin, Heidelberg, New York, 1976.

[80] ———, *On $\aleph_0$-categorical nilrings* II, J. Symbolic Logic **45** (1980), 291–301.

[98] ———, *Review of Dales and Woodin [96]*, Bull. A.M.S. **35** (1998), 91–98.

[36] C. Chevalley, *Demonstration d'une hypothèse de M. Artin*, Abb. Math. Sem., vol. 11, Hamburg U., Hamburg, 1936.

[55] ______, *The Construction and Study of Certain Important Algebras*, The Mathematical Society of Japan, Tokyo, 1955.

[56] ______, *Fundamental Concepts of Algebra*, Academic Press, New York, 1956.

[89] J. Clark, *A note on the fixed subring of an FPF ring*, Bull. Austral. Math. Soc. **40** (1989), 109–111.

[94] J. Clark and D. V. Huynh, *A note on perfect self-injective rings*, Quart. J. Math. Oxford **45** (1994), 13–17.

[94b] ______, *When is a self-injective semiperfect ring quasi-Frobenius?*, J. Algebra **165** (1994), 531–532.

[61-67] A. H. Clifford and G. B. Preston, *Algebraic Theory of Semigroups,* Vol. I., Surveys of the Amer. Math. Soc., Vol. 7, Providence 1961; vol. II, *ibid.*, 1967.

[46] I. S. Cohen, *On the structure and ideal theory of complete local rings*, Trans. Amer. Math. Soc. **59** (1946), 54–106.

[50] ______, *Commutative rings with restricted minimum condition*, Duke Math. J. **17** (1950), 2 7–42.

[51] I. S. Cohen and I. Kaplansky, *Rings for which every module is a direct sum of cyclic modules*, Math. Z. **54** (1951), 97–101.

[75] M. Cohen, *Addendum to semiprime Goldie centralizers*, Israel J. Math. **20** (1975), 89–93.

[75] M. Cohen and S. Montgomery, *Semisimple Artinian rings of fixed points*, Canad. Math. Bull. **18** (1975), 189–190.

[66] P. J. Cohen, *Set Theory and the Continuum Hypothesis*, Benjamin, New York, Amsterdam, 1966.

[61] P. M. Cohn, *Quadratic extensions of fields*, Proc.London Math.Soc. (3) **11** (1961), 531–556.

[66] ______, *Some remarks on the invariant basis property*, Topology **5** (1966), 215–228.

[71a] ______, *The embedding of firs in skew fields*, Proc.London Math.Soc. (3) **23** (1971), 193–213.

[71b] ______, *Free Ideal Rings*, Academic Press, New York 1971.

[74,77,91] ______, *Algebra,* 3 vols., John Wiley & Sons, Chichester, New York, Brisbane, Toronto, Singapore, 1974, 1977, 1991.

[75] ______, *Presentations of Skew Fields,* I, Existentially closed skew fields and the Nullstellensatz, vol. 77, Math.Proc.Cambridge Phil.Soc., 1975, pp. 7–19.

[81] ______, *Universal Algebra (2d Ed.)*, Reidel, Dordrecht 1981.

[91] ______, *See Cohn [74,77,91].*

[95] ______, *Skew Fields/General Theory of Division Algebras*, in Encyclopedia of Mathematics, Cambridge U. Press, Cambridge, New York and Melbourne, 1995.

[67] P. M. Cohn and E. Sasiada, *An example of a simple radical ring*, J. Algebra **5** (1967), 373–377.

[75] R. R. Colby, *Rings which have flat injectives*, J. Algebra **35** (1975), 239–252.

[71] R. R. Colby and E. A. Rutter, Jr., *II-Flat and II-Projective modules*, Arch. Math. **22** (1971), 246–251.

[63] I. Connell, *On the group ring*, Canad. J. Math. **15** (1963), 650–685.

[68] E. C. Copson, *Metric Spaces*, Cambridge Tracts in Math and Physics, **57**, Cambridge U. Press, Cambridge and New York.

[63] A.L.S. Corner, *Every countable reduced torsion-free ring is an endomorphism ring*, Proc. Lond. Math. Soc. (3) **13** (1963), 687–710.

[68,69,74] A. A. Costa, *Cours D'Algèbre Generale,* vols I,II,III; vol. I *(Ensembles, Treillis, Demi-Groupes, Quasi-Groupes);* vol. II *(Anneaux/Modules, Quasi-anneaux/Corps/Matrices /Algébres);* vol. III *(Demi-anneaux/Anneaux/Algébre Homologique/Representations /Algébres),* Fundcão Calouste Gulbenkian, Lisbon, 1968,1969,1974.

[77] F. Couchot, *Anneaux auto-fp-injectifs*, Acad. C.R. Sci. Paris Sér. A-B **284** (1977), A579–A582.

[82] ———, *Exemples d'anneaux auto-fp-injectifs*, Comm. Alg. **10** (1982), 339–360.

[65] R. C. Courter, *The dimension of a maximal commutative subalgebra of K_n*, Duke J. Math. **32** (1965), 225–332.

[69] ———, *Finite direct sums of complete matrix rings over perfect completely primary rings*, Canad. J. Math. **21** (1969), 430–446.

[73] S. H. Cox, Jr., *Commutative endomorphism rings*, Pac.J.Math. **45** (1973), 87–91.

[70] J. H. Cozzens, *Homological properties of the ring of differential polynomials*, Bull. Amer. Math. Soc. **76** (1970), 75–79.

[73] ———, *Twisted group rings and a problem of Faith*, Bull. Austral. Math. Soc. **9** (1973), 11–19.

[73] ———, *See Camillo.*

[72] J.H. Cozzens and J. Johnson, *Some applications of differential algebra to ring theory,* Proc. Amer. Math. Soc. **31** (1972), 354–356.

[75] J.H. Cozzens and C. Faith, *Simple Noetherian Rings*, in Cambridge Tracts in Mathematics and Physics, Cambridge Univ. Press, Cambridge, 1975.

[63] P. Crawley and B. Jonsson, *Direct decomposition of algebraic systems*, Bull. Amer. Math. Soc. **69** (1963), 541–547.

[64] ———, *Refinements for infinite direct decompositions of algebraic systems*, Pac. J. Math. **14** (1964), 797–855.

[59] R. Croisot, *See Lesieur.*

[62] C.W. Curtis and I. Reiner, *Representation Theory of Finite Groups and Associative Algebras*, Interscience, New York 1962.

[81] ———, *Methods of Representation Theory,* I, Wiley-Interscience, New York, Chichester, Brisbane, Toronto, 1981.

[65] C.W. Curtis and J.P. Jans, *On algebras with a finite number of indecomposable modules*, Trans. Amer. Math. Soc. **114** (1965), 122–132.

[71] E. C. Dade, *Deux groupes finis distincts ayant la mêne algèbre de groups sur tout corps*, Math. Z. **119** (1971), 345–348.

[81] ———, *Localization of injective modules*, J. Algebra **69** (1981), 416–425.

[96] H. G. Dales and W. H. Woodin, *Totally Ordered Fields with Additional Structure*, Clarendon Press, Oxford, 1996.

[79a] R. F. Damiano, *A right PCI ring is right Noetherian*, Proc. Amer. Math. Soc. **77** (1979), 11–14.

[79b] ———, *Coflat rings and modules*, Pac. J. Math. **81** (1979), 349–369.

[31] D. van Dantzig, *Studien über Topologische Algebra*, H. J. Paris, Amsterdam, 1931.

[94] J. Dauns, *Modules and Rings*, Cambridge U. Press, 1994.

[64] E. Davis, *Overrings of commutative rings* II, Trans.Amer.Math.Soc. **110** (1964), 196–212.

[1887] R. Dedekind, *Was Sind nd Was Sollen die Zahlen?*, Vieweg, Braunschweig, 1969 (reprint), Braunschweig, 1887.

[32] ———, *Gesammelte Mathematische Werke*, 3 vols., Braunschweig (Vieweg), 1932.

[64] ———, *Über die Theorie der ganzen algebraischen Zahlen*, reprint, Vieweg, Braunschweig, 1964.

[22] M. Dehn, Math. Annalen **85** (1922), 184–194.

[73] P. Deligne, *Variétés unirationnelles non-rationnelles*, Sem. Bourbaki 1971/72, Exp. 402, Lecture Notes in Math. vol. 317, Springer-Verlag, Berlin, Heidelberg, and New York, 1973.

[68] P. Dembowski, *Finite Geometries*, Springer-Verlag, Berlin, Heidelberg, and New York, 1968.

[81] F. DeMeyer, *Letter to the author,*, February, 1981.

[71] F. DeMeyer and E. Ingraham, *Separable Algebras over Commutative Rings*, Lecture Notes in Math., vol. 81, Springer-Verlag, New York-Heidelberg-Berlin, 1971.

[78] W. Dicks and E. D. Sontag, *Sylvester Domains*, J.Pure Appl.Algebra **13** (1978), 143–175.

[79] W. Dicks and P. Menal, *The group rings that are semifirs*, J. London Math. Soc. (2) **19** (1979), 288–290.

[93] W. Dicks, *See Camps.*

[05] L. E. Dickson, *On finite algebras*, Göttingen Nachrichten (1905), 358–393.

[23] ———, *Algebras and their Arithmetics*, U. of Chicago, Chicago 1923.

[75] ———, *The Collected Mathematical Papers* (A. A. Albert, ed.), Chelsea, New York, 1975.

[66] S. E. Dickson, *A torsion theory for Abelian categories*, Trans. A.M.S. **121** (1966), 223–235.

[69] S. E. Dickson and K. R. Fuller, *Algebras for which every indecomposable right module is invariant in its injective envelope*, Pac. J. Math. **31** (1969), 655-658.

[42] J. A. Dieudonné, *Sur le socle d'un anneaux et les anneux simples infinis*, Bull. Soc. Math. France **70** (1972), 46–75.

[48] ———, *La theorie de Galois des anneux simples et semi-simples*, Comment.Math.Helv. **21** (1948), 154–184.

[58] ———, *Remarks on quasi-Frobenius rings*, Illinois J.Math. **2** (1958), 346–354.

[84] D. Dikranjan and A. Orsatti, *On the structure of linearly compact rings and their dualities*, Rend.Accad.Naz.Sci., Memorie di Mat., vol. XIII (1984), 143–184.

[98] D. Dikranjan and L. Salce (eds.), *Abelian Groups, Module Theory and Topology*, Lecture Notes in Pure and applied Math., Marcel Dekker, Basel and New York, 1998.

[86] F. Dischinger and W. Müller, *Left PF is not PF*, Comm. Alg. **14** (1986), 1223–1227.

[68] J. Dixmier, *Sur les algèbres de Weyl*, Bull. Soc. Math. France **71** (1968), 209–242.

[72] V. Dlab and C. M. Ringel, *Balanced rings*, Lectures on Rings and Modules (Tulane Univ. Ring and Operator Theory Year, 1970–71, vol. I), pp.73–143, Lecture Notes in Math., vol. 246, Springer, Berlin, 1972.

[77] O. I. Domanov, *A prime but not primitive regular ring (*in Russian), Uspekhi. Mat. Nauk **32** (1977), 219–220.

[78] ———, *Primitive group algebras of polycyclic groups*, Sibirsk. Mat. Zhurn. **19** (1978), 37–43.

[57] M. P. Drazin, *Rings with nil commutator ideal*, Rend.Circ.Math. Palermo, II, Ser. 6 (1957), 51–64.

[98] V. Drensky, A. Gambruno, S. Sehgal (eds.), *Methods in Ring Theory*, in Lecture Notes in Pure and Appl. Math., vol. 198, Marcel Dekker, Basel, 1998.

[66] D. Dubois, *Modules of sequences of elements of a ring*, J. Lond. Math. Soc. **41** (1966), 177–180.

[82] M. Dugas and R. Göbel, *Every cotorsion-free algebra is an endomorphism ring*, Math. Z. **181** (1982), 451–470.

[89,90] N. V. Dung, *See Huynh.*

[92] N. V. Dung and P. F. Smith, *On Semiartinian V-modules*, J.Pure Appl.Algebra **82** (1992), 27–37.

[94] N. V. Dung, D. V. Huynh, and P. F. Smith, *Extending Modules*, Research Notices in Mathematical Series, 313, Pitman, London, 1994.

[95] N. V. Dung and P. F. Smith, *Rings over which certain modules are CS*, J. Pure and Appl. Algebra **102** (1995), 283–287.

[97] N. V. Dung and A. Facchini, *Weak Krull-Schmidt for infinite direct sums of uniserial modules*, J. Algebra **193** (1997), 102–121.

[88-89] P. Duren et al (eds), *A Century of Mathematics in America,* Parts I (1988), Part II (1988), Part III (1989), Amer. Math. Soc., Providence, R. I..

[72] P. Eakin and J. Silver, *Rings which are locally polynomial rings*, Trans. Amer. Math. Soc. **174** (1972), 425–449.

[73] P. Eakin and W. Heinzer, *A cancellation problem for rings*, Proc. Conf. in Comm. Alg. (Lawrence, Kansas, 1973); Lecture Notes in Math **311** (1973), 61–77.

[80] P. Eakin and A. Sathaye, *R-automorphisms of finite order in $R[x]$*, J. Algebra **67** (1980), 110–128.

[53] B. Eckmann and A. Schöpf, *Über injective Moduln*, Arch. Math. **4** (1953), 75–78.

[69] F. Eckstein, *On the Mal'cev theorem*, J. Algebra **12** (1969), 372–385.

[76] N. Eggert, *Rings whose overrings are integrally closed in their complete quotient ring*, J. Reine angew. Math. **262** (1976), 88–95.

[55] G. Ehrlich, *A note on invariant subrings*, Proc.Amer.Math.Soc. **6** (1955), 470–471.

[68] ———, *Unit regular rings*, Portugalia Math. **27** (1968), 209–212.

[76] ———, *Units and one-sided units in regular rings*, Trans. A.M.S. **216** (1976), 81–90.

[74] S. Eilenberg, *Automata, Languages and Machines*, Academic Press, New York, 1974.

[45] S. Eilenberg and S. MacLane, *General theory of natural equivalences*, Trans. Amer. Math. Soc. **58** (1945), 231–294.

[52] S. Eilenberg and N. Steenrod, *Foundations of Algebraic Topology*, Princeton University Press, Princeton, 1952.

[56] S. Eilenberg, H. Nagao, T. Nakayama, *On the dimension of modules and algebras,* IV, Nagoya Math. J. **10** (1956), 87–95.

[56] S. Eilenberg, *See Cartan.*

[57] S. Eilenberg, A. Rosenberg and D. Zelinsky, *On the dimension of modules and algebras,* VII, Dimension of tensor products, vol. 12, Nagoya Math.J., 1957, pp. 71–93.

[66] S. Eilenberg, D. K. Harrison, S. MacLane, and H. Röhrl (eds.), *Categorical Algebra*, Proc. La Jolla Conf. 1965, Springer-Verlag, New York, 1966.

[96] D. Eisenbud, *Commutative Algebra, Graduate Texts in Math.*, vol. 150, Springer-Verlag, Berlin, Heidelberg, New York, 1995, second corrected printing, 1996.

[70a] D. Eisenbud and J.C. Robson, *Modules over Dedekind prime rings*, J. Algebra **16** (1970), 67–85.

[70b] ______, *Hereditary Noetherian prime rings*, J. Algebra **16** (1970), 86–104.

[71a] D. Eisenbud and P. Griffith, *Serial rings*, J. Algebra **17** (1971), 389–400.

[71b] ______, *The structure of serial rings*, Pac. J. Math. **36** (1971), 109–121.

[71] P. C. Eklof and G. Sabbagh, *Model-completions and modules*, Ann.Math. Logic **7** (1971), 251–295.

[90] P. C. Eklof and A. H. Mekler, *Almost free modules*, North-Holland, New York, 1990.

[90] G. A. Elliott and P. Ribenboim, *Fields of generalized power series*, Arch. Math. **54** (1990), 365–371.

[62] S. Endo, *On flat modules over commutative rings*, J. Math. Soc., Japan **14** (1962), 284–291.

[67] ______, *Completely faithful modules and quasi-Frobenius algebras*, J. Math. Soc. Japan **19** (1967), 437–456.

[80] ______, *Letter to the author*, July, 1980.

[98] ______, *Letter to the author*, September 1998.

[71] E. G. Evans, Jr., *On epimorphisms to finitely generated modules*, Pac. J. Math. **37** (1971), 47–56.

[73] ______, *Krull-Schmidt and cancellation over local rings*, Pac. J. Math. **46** (1973), 115–121.

[81] A. Facchini, *Loewy and Artinian modules over commutative rings*, Ann. Mat. Pura Appl. **28** (1981), 359–374.

[83] ______, *On the structure of torch rings*, Rocky Mt. J. Math. **13** (1983), 423–428.

[85] ______, *Torsion-free covers and pure-injective envelopes over valuation domains*, Israel J. Math. **52** (1985), 129–139.

[87] ______, *Relative injectivity and pure-injective modules over Prüfer rings*, J. Algebra **110** (1987), 380–406.

[89] ______, *See Azumaya.*

[94] ______, *Generalized Dedekind domains and their injective modules*, J. Pure and appl. Algebra **94** (1994), 159–173.

[96a] ______, *Absolute pure modules and locally injective modules*, in Commutative Ring Theory, Lecture Notes in Pure and Appl. Math., vol. 153, Marcel Dekker, Basel and New York, 1996, pp. 105–109.

[96b] ______, *Krull-Schmidt fails for serial modules*, Trans. Amer. Math. Soc. **348** (1996), 4561–4575.

[97] ______, *See Dung.*

[90] A. Facchini and L. Salce, *Uniserial modules: sums and isomorphisms of subquotients*, Comm. Algebra **18** (1990), 499–517.

[95] A. Facchini, D. Herbera, L. Levy and P. Vámos, *Krull-Schmidt fails for Artinian modules*, Proc. Amer. Math. Soc. **123** (1995), 3587–3592.

[95] A. Facchini and G. Puninski, $\sum$*-pure-injective modules over serial rings*, in Abelian Groups and Modules (A. Facchini and C. Menini, eds.), Kluwer Academic Publishers, Dordrecht, 1995, pp. 145–162.

[95] Alberto Facchini and Claudia Menini (eds.), *Abelian Groups and Modules*, Proc. Padova Conf., 1994, Kluwer Acad. Pub., Dordrecht, 1995.

[97] A. Facchini and C. Faith, *FP-injective quotient rings and elementary divisor rings*, in Commutative Ring Theory (P-J. Cahen etal, eds.), Proc. of the International II Conf.; Lectures in Pure and Applied Math., Marcel Dekker, vol. 125, pp.293–302, Basel and New York, 1997.

[57] C. Faith, *Extensions of normal bases and completely basic fields*, Trans. A.M.S. **85** (1957), 406–427.

[58] ———, *On conjugates in division rings*, Canad. J. Math. **10** (1958), 374–380.

[58b] ———, *Galois extensions in which every element with regular trace is a normal basis element*, Proc. A.M.S. **9** (1958), 222–229.

[59a] ———, *Submodules of rings*, Proc.Amer.Math.Soc. **10** (1959), 596–606.

[59b] ———, *Rings with minimum conditions on principal ideals*, Archiv der Mathematik **10** (1959), 327–330.

[60] ———, *Algebraic division ring extensions*, Proc.Amer.math.Soc. **11** (1960), 43–53.

[61] ———, *On a theorem of Tsen*, Arch. der Math. **12** (1961), 333–335.

[61b] ———, *Radical extensions of rings*, Proc.Amer.Math.Soc. **12** (1961), 274–283.

[61c] ———, *On Herstein's theorem concerning three fields*, Nagoya Math.J. **19** (1961), 49–53.

[61d] ———, *Derivations and generations of finite extensions*, Bull. A.M.S. **67** (1961), 350–353.

[61e] ———, *Zwei-elemente-erzeugung und Endlichkeit der dimension von divisionalgebren*, Archiv der Mathematik **11** (1961), 405–406.

[61f] ———, *Rings with minimum condition on principal ideals*, II, Archiv der Mathematik **12** (1961), 179–182.

[62] ———, *Semialgebraic division ring extensions*, J.reine u. angew.Math. **209** (1962), 144–162.

[62b] ———, *Strongly regular extensions of rings*, Nagoya Math. J. **20** (1962), 169–183.

[64] ———, *Noetherian simple rings*, Bull. Amer. Math. Soc. **70** (1964), 73–731.

[64b] ———, *Baer modules*, Archiv der Math. **15** (1964), 266-270.

[64c] ———, *Intrinsic extensions of rings*, Pac. Jour. of Math. **14** (1964), 505–512.

[64d] ———, *On a new proof of Litoff's theorem*, Acta Hung. Math. **14** (1964), 369–371.

[65] ———, *See Chase.*

[66a] ———, *Rings with ascending condition on annihilators*, Nagoya Math. J. **27** (1966), 179–191.

[66b] ———, *On Köthe rings*, Math. Ann. **164** (1966), 207–212.

[67] ———, *Lectures on Injective modules and Quotient rings*, Lecture Notices in Mathematics, vol. 49, Springer, New York, Heidelberg, Berlin, 1967.

[67b] ———, *A general Wedderburn theorem*, Bull.Amer.Math.Soc. **73** (1967), 65–67.

[71a] ———, *Big decompositions of modules*, Notices of the Amer. Math. Soc. **18** (1971), 400.

[71b] ———, *A correspondence theorem for projective modules and the structure of simple Noetherian rings*, Symposia Matematica vol. XIII (Convegno sulle Algebre Associative, Indam, Roma, Nov. 1970) pp.309–345, Academic Press, London 1972.

[72a,81] ———, *Algebra I: Rings, Modules and Categories,* Grundl. der Math. Wiss, Vol. 190, Springer Verlag, Berlin, Heidelberg, New York, 1972, 1981.

[72b] ———, *Modules finite over endomorphism ring,* in Lecture Notes in Mathematics, Vol. 246, Springer, Berlin, Heidelberg, New York, 1972.

[72c] ———, *Galois subrings of Ore domains are Ore domains,* Proc.Amer.Math.Soc. **78** (1972), 1077–1080.

[73] ———, *When are proper cyclics injective?,* Pac. J. Math. **45** (1973), 97–112.

[75] ———, *See Cozzens.*

[76] ———, *Algebra II: Ring theory,* in Grundl. der math. Wiss., Vol. 191, Springer-Verlag, Berlin, Heidelberg, New York, 1976.

[76a] ———, *On hereditary rings and Boyle's conjecture,* Archiv Math. XXVII (1976), 113–119.

[76b] ———, *Galois subrings of commutative rings,* Math. J. Okayama U. **18** (1976), 113–116.

[76-77] ———, *Injective cogenerator rings and a theorem of Tachikawa,* Proc. Amer. Math. Soc. **60** (1976), 25–30, II; **62** (1977), 15–18.

[77b] ———, *Semiperfect Prüfer and FPF rings,* Israel Math. J. **26** (1977), 15–18.

[79a] ———, *Injective quotient rings of commutative rings,* Module Theory (see C. Faith and S. Wiegand (eds.) [79]) in Lecture Notes in Mathematics, vol. 700, 1979, pp.191–203.

[79b] ———, *Self-injective rings,* Proc. A. M. S. **77** (1979), 158–164.

[79c] ———, *The genus of a module and generic families of rings,* in Ring Theory, Proc. Antwerp Conf. 1978, pp.613–629; Lecture Notes in Pure and Appl Math., vol. 51, Marcel Dekker, Basel and New York, 1979.

[82a] ———, *Injective Modules and Injective Quotient Rings,* Lecture Notes in Pure and Applied Math., vol. 72, Marcel Dekker, Basel and New York, 1982.

[82b] ———, *Injective quotient rings of commutative rings,* II, in Lecture Notes in Pure and Applied Math., vol. 72, 1982, pp. 71–105.

[82c] ———, *On the Galois theory of commutative rings: Dedekind's Theorem revisited,* (in Algebraist's Hommage), Contemp. Math. **13** (1982), 183–192.

[82d] ———, *Subrings of self-injective and FPF rings,* in Advances in Non-Commutative Ring Theory, pp. 12–20. (See below).

[82e] ———, *Embedding modules in projectives: a report on a problem,* in "Advances in Non-Commutative Ring Theory" (P.Fleury,ed.), Proc. Hudson Symp., Plattsburgh, New York, 1981, Lecture Notes in Math., vol. 951, Springer-Verlag, Berlin, Heidelberg and New York, 1982.

[84a] ———, *Galois subrings of independent groups of commutative rings are quorite,* Math.J. Okayama U. **26** (1984), 23–25.

[84b] ———, *The structure of valuation rings,* J. Pure and Appl. Alg. **31** (1984), 7–27.

[84c] ———, *Commutative FPF rings arising as split-null extensions,* Proc.A.M.S. **90** (1984), 181–185.

[85] ———, *The maximal regular ideal of self-injective and continuous rings splits off,* Arch.Math. **44** (1985), 511–521.

[86a] ———, *Cozzens domains are hereditary,* Math. J. Okayama Univ. **28** (1986), 37–40.

[86b] ———, *Linearly compact injective modules and a theorem of Vámos,* Publ. Math. **30** (1986), 127–148.

[86c] ———, *The structure of valuation rings*, II, J. Pure and Appl.Math. **42** (1986), 37–43.

[87] ———, *On the Galois theory of commutative rings*, II: automorphisms induced in residue rings, Canad.J.Math. XXXIX (1987), 1025–1037.

[89a] ———, *Polynomial rings over Jacobson-Hilbert rings*, Publ.Mat. **33** (1989), 85–97, Addendum **34** (1990),223.

[89b] ———, *Rings with zero intersection property on annihilators: zip rings*, Publ.Mat. **33** (1989), 329–338.

[90a] ———, *Embedding torsionless modules in projectives*, Publ.Mat. **34** (1990), 379–387.

[90b] ———, *Review of Huckaba [89]*, Bull. Amer. Math. Soc. **22** (1990), 331–335.

[91a] ———, *Finitely embedded commutative rings*, Proc. Amer. Math. Soc. **112** (1991), 657–659, Addendum **118** (1993),331.

[91b] ———, *Annihilators, associated primes, and Kasch-McCoy quotient rings of commutative rings*, Comm.Algebra **19** (1991), 1867-1892.

[92a] ———, *Self-injective von Neumann regular subrings and a theorem of Pere Menal*, Publ.Mat. **36** (1992), 541–557.

[92b] ———, *Defeat of the FP^2F Conjecture: Huckaba's example*, Proc. Amer. Math. Soc. **116** (1992), 5–6.

[94] ———, *Polynomial rings over Goldie-Kerr commutative rings* II, Proc. Amer.Math.Soc. **120** (1994), 989–993.

[95a] ———, *Locally perfect commutative rings are those whose modules have maximal submodules*, Comm. Algebra **33** (1995), 4885–4886.

[95b] ———, *Rings whose modules have maximal submodules*, Publ.Mat. **39** (1995), 201–214, addendum **48** (1998), 265–6.

[96a] ———, *New characterizations of von Neumann regular rings and a conjecture of Shamsuddin*, Publ.Mat. **40** (1996), 383–385.

[96b] ———, *Polynomial rings over Goldie-Kerr commutative rings* II, Proc. Amer. Math. Soc. **124** (1996), 341–344.

[96c] ———, *Rings with few zero divisors are those with semilocal Kasch quotient rings*, Houston J.Math. **22** (1996), 687–690; Note: The result implied by the title is incorrect unless the word "Kasch" is deleted. The corrected statement is a result of E. Davis [64]. See Theorem 9.9 in the text.

[97a] ———, *See Facchini.*

[97b] ———, *Minimal generators over Osofsky and Camillo rings*, in Advances in Ring theory (S.K.Jain and T.Rizvi,eds.), Birkhäuser-Verlag, Boston, 1997.

[98a] ———, *Commutative rings with ascending chain conditions on irreducible ideals*, in Abelian Groups, Modules, and Topological Algebra; Padova Conf. 1997 in honor of Adalberto Orsatti's Sixtieth Birthday; Lecture Notes in Pure and Applied Math., Marcel Dekker, Basel and New York, 1998.

[98b] ———, *Quotient finite dimensional modules with acc on subdirectly irreducible submodules are Noetherian*, Comm. Algebra (1998).

[64] C. Faith and Y. Utumi, *Quasi-injective modules and their endomorphism rings*, Arch. Math. XV (1964), 166–174.

[64b] ———, *Baer modules*, Archiv der Math. **15** (1964), 266-270.

[64c] ———, *Intrinsic extensions of rings*, Pac. Jour. of Math. **14** (1964), 505–512.

[64d] ———, *On a new proof of Litoff's theorem*, Acta Hung. Math. **14** (1964), 369–371.

[65] ———, *Noetherian prime rings*, Trans. Amer. Math. Soc. **114** (1960), 53–60.

[67] C. Faith and E.A. Walker, *Direct sum representations of injective modules*, J. Algebra **5** (1967), 203–221.

[79] C. Faith and S. Wiegand (eds.), *Module Theory*, (Proc. Special Session, Amer.Math.Soc., U. of Washington, Seattle, 1977), Lecture Notes in Math. vol. 700, Springer-Verlag, Berlin, Heidelberg and New York, 1979.

[84] C. Faith and S. Page, *FPF Ring Theory: Faithful Modules and Generators of Mod-R*, Lecture Notes of the London Math. Soc., vol. 88, Cambridge U. Press, Cambridge, New York and Melbourne, 1984.

[90] C. Faith and P. Pillay, *Classification of Commutative FPF rings*, Notas de Matématica 4, Universidad de Murcia, Murcia, Spain, 1990.

[92] C. Faith and P. Menal, *A counter example to a conjecture of Johns*, Proc. Amer. Math. Soc. **116** (1992), 21–26.

[94] ———, *The structure of Johns rings*, Proc. Amer. Math. Soc. **120** (1994), 1071–1081; Erratum **125** (1997), p.127.

[95] ———, *A new duality theorem for semisimple modules and characterization of Villamayor rings*, Proc. Amer. Math. Soc. **123** (1995), 1635–1638.

[97] C. Faith and D. Herbera, *Endomorphism rings and tensor products of linearly compact modules*, Comm. Alg. **25** (1997), 1215–1256.

[73] D. Farkas, *Self-injective group rings*, J. Algebra **25** (1973), 313–315.

[76] D. R. Farkas and R. L. Snider, *K_0 and Noetherian group rings*, J. Algebra **42** (1976), 192–198.

[77] ———, *Noetherian fixed rings*, Pac.J.Math. **69** (1977), 347–353.

[84] T. Faticoni, *FPF rings, I: The Noetherian Case*, Comm. Alg. **13** (1985), 2119–2136.

[87] ———, *Semiperfect FPF rings and applications*, J. Algebra.

[88] ———, *Localization in finite dimensional FPF rings*, Pac. J. Math. **134** (1988), 79–98.

[63] W. Feit and J. G. Thompson, *The solvability of groups of odd order*, Pac. J. Math. **13** (1963), 775–1029.

[78] D. J. Fieldhouse, *Semihereditary polynomial rings*, Publ.Math.Debrecen **25** (1978), 211.

[58] G. D. Findlay and J. Lambek, *A generalized ring of quotients, I,II*, Canad. Math. Bull. **1** (1958), 77–85; 155–167.

[61] N. J. Fine, L. Gilman and J. Lambek, *Rings of quotients of rings of functions*, McGill University press, Montreal, 1965.

[82] M. F. Finkel Jones, *Flatness and f-projectivity of torsion free and injective modules*, pp.94–116, in "Advances in Non-Commutative Ring Theory," (P. Fleury, ed.) Lecture Notes in Math. vol. 951, Springer-Verlag, Berlin and New York, 1982.

[73] J. W. Fisher, *Structure of semiprime PI rings*, Proc. Amer. Math. Soc. **39** (1973), 465–467.

[74] J. W. Fisher and R. L. Snider, *Prime von Neumann regular rings and primitive group algebras*, Proc. Amer. Math. Soc. **44** (1974), 244–250.

[78] J. W. Fisher and J. Osterburg, *Semiprime ideals in rings with finite group actions*, J. Algebra **50** (1978), 488–502.

[33] H. Fitting, *Die Theorie der automorphismenringe Abelsher Gruppen und ihr Analogon bei nichtkommutativen Gruppen*, Math. Ann. **107** (1933), 514–542.

[35a] ———, *Über die direkten Productzerlegung eine Gruppe in direkt unzerlegbare Faktoren*, Math. Z. **39** (1935), 16–30.

[35b] ———, *Primärkomponenten zerlegung in Nichtkommutativen Ringen*, Math. Ann. **111** (1935), 19–41.

[74] P. Fleury, *A note on dualizing Goldie dimension*, Canad. Math. Bull. **17** (1974), 511–517.

[74] ———, *Hollow modules and local endomorphism rings*, Pacific J. Math. **53** (1975), 379–385.

[77] ———, *On local QF rings*, Aequationes Math. **16** (1977), 173–179.

[82] ———, *(ed.), Advances in non-commutative ring theory (*Proc. Hudson Symp., Plattsburgh, N.Y., 1981*)*, in Lecture Notes in Math. vol. 951, Springer-Verlag, Berlin, Heidelberg, New York, 1982.

[95] M. Fontana and N. Popescu, *Sur une classe d'anneaux qui generalisent les anneaux de Dedekind*, J. Algebra **173** (1995), 44–66.

[96] M. Fontana, J. Huckaba and I. Papick, *Prüfer Domains*, Monographs and Textbooks in Pure and Appl. Math., vol. 203, M. Dekker, Basel and New York, 1996.

[72] E. Formanek, *Central polynomials for matrix rings*, J. Algebra **238** (1972), 129–132.

[72] E. Formanek and R. L. Snider, *Primitive group rings*, Proc. A.M.S. **36** (1972), 357–360.

[74] E. Formanek and A. V. Jategaonkar, *Subrings of Noetherian rings*, Proc.Amer.Math.Soc. **46** (1974), 181–186.

[46] A. Forsythe and N. H. McCoy, *On the commutativity of certain rings*, Bull. A.M.S. **52** (1946), 523–526.

[73] R. M. Fossum, *The Divisor Class Group of a Krull Domain*, Springer-Verlag, New York, Heidelberg, Berlin, 1973.

[75] R. M. Fossum, P. A. Griffith and I. Reiten, *Trivial extensions of Abelian categories*, Lecture Notes in Math., vol. 456, Springer-Verlag, Berlin and New York, 1975.

[14] A. Fraenkel, *Über die Teiler der Null*, und die Zerlegung von Ringen, J. reine u. angew. Math. 145 (1914), 139–176.

[1885] G. Frattini, *Intorno alla generazione dei gruppi di operazione*, Rend.Att.Acad.Lincei (4) **1** (1885), 281–285; 455–457.

[95] R. Freese, J. Tezek and J. B. Nation, *Free Lattices*, Surveys of the Amer. Math., number 42, Providence 1995.

[64] P. Freyd, *Abelian Categories*, Harper and Row, New York 1964.

[1878] G. Frobenius and L. Stickelberger, J. reine angew.Math. **86** (1878), 217–262.

[79] P. Froeschl, *Chained rings*, Pac. J. Math. **65** (1979), 47–53.

[69a] L. Fuchs, *On quasi-injective modules*, Ann.Scuola Norm.Sup.Pisa (3) **23** (1969), 541–546.

[69b] ———, *Torsion preradicals and ascending Loewy series of modules*, J. reine U. angew. Math. **239/240** (1969), 169–179.

[73] ———, *Infinite Abelian Groups*, vol. I, Academic Press, New York (second edition), 1970.

[56] ———, *See Szele.*

[85] L. Fuchs and L. Salce, *Modules over Valuation Domains*, Marcel Dekker, Basel and New York, 1985.

[98] L. Fuchs, L. Salce, and P. Zanardo, *Note on the transitivity of pure extensions*, preprint, 1998.

[72a] J. D. Fuelberth and M. L. Teply, *The singular submodule of a finitely generated module splits off*, Pac. J. Math. **40** (1972), 78–82.

[72b] ———, *A splitting rings of global dimension two*, Proc.Amer.Math.Soc. **35** (1972), 317–324.

[68] K. R. Fuller, *Structure of QF-3 rings*, Trans. Amer. Math. Soc. **134** (1968), 343–354.

[69a] ———, *On indecomposable injectives over Artinian rings*, Pac. J. Math. **29** (1969), 113–135.

[69b] ———, *On direct sums of quasi-injectives and quasi-projectives*, Arch. Math. **20** (1969), 495–502, corrections, **Ibid 21** (1970), 478.

[76] ———, *Rings whose modules are direct sums of finitely generated modules*, Proc. Amer. math. Soc. **54** (1976), 39–44.

[72,75] K. R. Fuller, *See Anderson.*

[72,76] ———, *See Camillo.*

[75] K.R. Fuller and I. Reiten, *Note on rings of finite representation type*, Proc. Amer. Math. Soc. **50** (1975), 92–94.

[62] P. Gabriel, *Des catégories abeliennes*, Bull. Soc. Math., France **90** (1962), 323–448.

[64] ———, *See Popescu.*

[67] ———, *See Rentschler.*

[67] P. Gabriel and M. Zisman, *Calculus of Fractions and Homotopy Theory*, Springer, New York, Heidelberg, Berlin, 1967.

[98] A. Gambruno (ed.), *See Drensky.*

[60] E. Gentile, *On rings with one-sided fields of quotients*, Proc. Amer. Soc. **11** (1960), 380–384.

[67] ———, *A uniqueness theorem on rings of matrices*, J. Algebra **6** (1967), 131–134.

[71] D. T. Gill, *Almost maximal valuation rings*, J. Lond. Math. Soc. (no. 2) **4,** (1971), 140–146.

[60] L. Gilman and M. Jerison, *Rings of Continuous Functions*, Van Nostrand, New York and Princeton, 1960.

[65] L. Gilman, *See Fine.*

[72] R. Gilmer, *Multiplicative Ideal Theory*, M. Dekker, Basel and New York, 1972.

[84] ———, *Commutative Semigroup Rings*, U. of Chicago Press, 1984.

[92] ———, *Multiplicative Ideal Theory*, Queens University, Kingston, Ontario, 1992.

[97] ———, *An intersection condition for prime ideals*, pp. 327–331 in Anderson (ed.), [97].

[97] R. Gilmer and W. Heinzer, *Every local ring is dominated by a one-dimensional local ring*, Proc. Amer. Math. Soc. **125** (1997), 2513–2520.

[75] S. M. Ginn and P. B. Moss, *Finitely embedded modules over Noetherian rings*, Bull. Amer. Math. Soc. **81** (1975), 709–710.

[82] R. Göbel, *See Dugas.*

[64] K. Gödel, *The Consistency of the Axiom of Choice and the Generalized Continuum Hypothesis with the Axioms of Set Theory*, Ann. of Math. Studies, No. 3 (Sixth Printing), Princeton University Press, Princeton 1964.

[95] H. P. Goeters, *Warfield duality and module extensions over a Noetherian domain*, pp. 239–249, in Facchini and Menini [95].

[86] J. Golan, *Torsion Theories*, Pitman, Longman Scientific and Technical, John Wiley, New York, 1986.

[58] A. W. Goldie, *The structure of prime rings under ascending chain conditions*, Proc. Lond. Math. Soc. VIII (1958), 589–608.

[60] ———, *Semi-prime rings with maximum condition*, Proc. Lond. Math. Soc. **X** (1960), 201–220.

[62] ———, *Non-commutative principal ideal rings*, Arch. Math. **13** (1962), 213–221.

[64] ———, *Torsion-free modules and rings*, J. Algebra **1** (1964), 268–287.

[73] A. W. Goldie and L. W. Small, *A study in Krull dimension*, J. Algebra **25** (1973), 152–157.

[51] O. Goldman, *Hilbert rings and the Hilbert Nullstellensatz*, Math. Z. **54** (1951), 136–140.

[64] E. S. Golod, *On nil algebras and finitely approximable p-groups*, Izv.Akad.Nauk. SSSR Ser.Mat. **28** (1964), 273–276.

[64] E. S. Golod and I.R. Shafevitch, *On the class field tower*, Izv.Akad.Nauk. SSSR Ser.Math. **28** (1964), 261–272.

[85] J. L. Gómez Pardo, *Embedding cyclic and torsion free modules in free modules*, Arch. Math. **44** (1985), 503–510.

[89] ———, *Counterinjective modules and duality*, J. Pure Appl. Alg. **61** (1989), 165–179.

[91] ———, *Endomorphism rings with duality*, Comm. Alg. **19** (1991), 2097–2112.

[83] J.L. Gómez Pardo and N.R. González, *On some properties of IF rings*, Questiones Arithmeticae **5** (1983), 395–405.

[87] J. L. Gómez Pardo and J. M. Hernandez, *Coherence of endomorphism rings*, Arch. Math. **48** (1987), 40–52.

[97] J. L. Gómez Pardo and P.A.Guil Asensio, *Essential embedding of cyclic modules in projectives*, Trans. Amer. Math. Soc. (1997)).

[97b] ———, *When are all the finitely generated modules embeddable in free modules*, Algebraic and Geometric Methods in Ring Theory, SAGA IV, Antwerp, 1996, Lecture Notes in Pure and Appl. Math., Dekker, Basel and New York, 1997.

[97c] ———, *Indecomposable decompositions of pure-injectives*, J. Algebra **192** (1997), 200–8.

[97d] ———, *Rings with finite essential socle*, PAMS **125** (1997), 971–7.

[98] ———, *Embeddings in free modules and Artinian rings*, J. Algebra (1998).

[98b] ———, *Chain conditions on direct summands and pure quotient modules*, preprint.

[96] J. Z. Gonçalves and A. Mandel, *A commutativity theorem for division rings and an extension of a result of Faith*, Results in Math. **30** (1996), 302–309.

[83] N. R. González, *see Gómez Pardo.*

[72] K. R. Goodearl, *Singular Torsion and the Splitting Properties*, Memoirs of the Amer. Math., no. 124, Providence 1972.

[73a] ———, *Prime ideals in regular selfinjective rings*, Canad. J. Math. **25** (1973), 829–839.

[73b] ———, *Prime ideals in regular self-injective rings II*, Pure and Applied Algebra **3** (1973), 357–373.

[74] ______, *Global dimension of differential operator rings*, Proc.Amer.Math.Soc. **45** (1974), 315–322.

[74b] ______, *Simple self-injective rings need not be artinian*, Comm. Alg. **2** (1974), 83–89.

[75] ______, *Global dimension of operator rings,* II, Trans, A.M.S. **209** (1975), 65–85.

[78] ______, *Simple Noetherian rings—the Zalesski-Neroslavskii example*, Proc.Conf.Univ., Waterloo, 1978,pp. 118–130,, Lecture Notes in Math. 734, Springer, Berlin 1979.

[79,91] ______, *von Neumann Regular Rings*, Pitman, London-San Francisco-Melbourne, 1979, Second Ed. with a 40-page appendix, Krieger Publ. Co., 1991.

[80] ______, *Artinian and Noetherian modules over regular rings*, Comm. algebra **8** (1980), 477–504.

[97] ______, *See Ara, and see Arhangel'skii.*

[75] K.R. Goodearl and D. Handelman, *Simple self-injective rings*, Comm. Algebra **3** (1975), 797–834.

[76] K. R. Goodearl and R. B. Warfield, Jr., *Algebras over zero-dimensional rings*, Math. Ann. **223** (1976), 157–168.

[89] ______, *An Introduction to Non-commutative Rings*, London.Math.Soc. Student Texts, No. 16, Cambridge U. Press, Cambridge, New York, Melbourne and Sydney, 1989.

[89] K. R. Goodearl and J. Moncasi, *Cancellation of finitely generated modules over regular rings*, Osaka J. Math. **26** (1989), 679–685.

[86] K. R. Goodearl and B. Zimmermann-Huisgen, *Lengths of submodule chains versus Krull dimension in non-Noetherian modules*, Math. Z. **191** (1986), 519–527.

[66] N. S. Gopalakrishman and R. Sridharan, *Homological dimension of Ore-extensions*, Pacific J. Math. **19** (1966), 67–75.

[74] R. Gordon, *Gabriel and Krull dimension,*, in Ring theory, Proc. of the Oklahoma Conf., pp.2241–295, M. Dekker, Basel and New York, 1974.

[74b] ______, *Primary decomposition in right Noetherian rings*, Comm. Algebra **2** (1974), 491–524.

[73] R. Gordon and J. C. Robson, *Krull Dimension,* Memoir No. 133, Amer.Math.Soc., Providence, 1973.

[73] R. Gordon, T. H. Lenagan and J. C. Robson, *Krull dimension, nilpotency and Gabriel dimension*, Bull. Amer. Math. Soc. **79** (1973), 716–719.

[82] D. Gorenstein, *Finite simple groups: An Introduction to Their Classification*, Plenum Press, New York, 1982.

[94,95,97] D. Gorenstein, R. Lyons and R. Solomon, *The Classification of the finite simple groups*, Math. Surveys and monographs, vol. 40 (Parts I,II,III), Amer.Math.Soc., Providence, 1994,1995,1997.

[70] J. M. Goursaud, *Une caractérisation des anneaux unisériels*, C. R. Acad. Sci. Paris, Sér A-B **270** (1970), A364–7.

[75] J. M. Goursaud and J. Valette, *Sur l'envveloppe injective des anneaux de groupes réguliers*, Bull. Soc. Math. France **103** (1975), 91–102.

[81] J. M. Goursaud, J. Osterburg, J. L. Pascaud, and J. Valette, *Points fixes des anneaux reguliers, auto-injectifs*, C.R.Acad.Sci. Paris Sér. A-B **290** (1980), A985–987.

[65] V. E. Govorov, *On flat Modules*, (Russian) Sibirsk. Mat. Z **6** (1965), 300–304.

[96] R. L. Graham, *et al* (eds.), *The Mathematics of Paul Erdös, vol. I*, Springer-Verlag, New York-Heidelberg-Berlin, 1996.

[76] G. Grätzer, *Universal Algebra*, Springer-Verlag, Berlin,Heidelberg and New York, (rev.ed.), 1979.

[80] E. L. Green, *Remarks on projective resolutions*, in Lecture Notes in Math., vol. 832, pp.259–279, Springer-Verlag, Berlin, Heidelberg, and New York, 1980.

[70] M. Griffin, *Prüfer rings with zero divisors*, J. reine Angew. Math. **239/240** (1970), 55–67.

[74] ———, *Valuation rings and Prüfer rings*, Canad. J. Math. **26** (1974).

[70] P. Griffith, *On decomposition of modules and generalized left uniserial rings*, Math. Ann. **184** (1970), 300–308.

[76] ———, *A representation theorem for complete local rings*, J. Pure and Appl. Algebra **7** (1976), 303–315.

[71] ———, *See Eisenbud.*

[57] A. Gröthendieck, *Sur quelques points d'algèbre homologique*, Tohoku Math. J. **9** (1957), 119–221.

[65] ———, *Le Groupe de Brauer, Seminaire Bourbaki, exposé 20*, Hermann, Paris, 1965.

[71] L. Gruson, *See Raynaud.*

[97,98] P. A. Guil Asensio, *See Gómez Pardo.*

[73] T. H. Gulliksen, *A theory of length for Noetherian modules*, J. Pure and Appl. Alg. **3** (1973), 159–170.

[68a] R. N. Gupta, *Characterization of rings whose classical quotient rings are perfect rings*, Publ. Math. Debrecen **17** (1970), 215–222.

[68b] ———, *Self-injective quotient rings and injective quotient modules*, Osaka J. Math. **5** (1968), 69–87.

[69] ———, *On f-injective modules and semihereditary rings*, Proc. Nat. Inst. Sci. **35** (1969), 323–328.

[67] J. S. Haines, *A note on direct product of free modules*, Amer. Math. Monthly **74** (1967), 1079–1080.

[77,80] C. R. Hajarnavis, *See Chatters.*

[82] ———, *See Brown.*

[85] C. R. Hajarnavis and N. C. Norton, *On dual rings and their modules*, J. Algebra **93** (1985), 253–266.

[39] M. Hall, Jr., *A type of algebraic closure*, Ann. of Math. **40** (1939), 360–369.

[40] ———, *The position of the radical of an algebra*, Trans. Amer. Math. Soc. **48** (1940), 391–404.

[43] ———, *Projective planes*, Trans. A.M.S. **54** (1943), 229–277.

[67] R. Hamsher, *Commutative rings over which every module has a maximal submodule*, Proc. Amer. Math. Soc. **18** (1967), 1133–1137.

[75] D. Handelman, *See Goodearl.*

[75] D. Handelman and J. Lawrence, *Strongly prime rings*, Trans. Amer. Math. Soc. **211** (1975), 209–223.

[84] A. Hanna and A. Shamsuddin, *Duality in the Category of Modules. Applications*, Algebra Berichte 49, Verlag Reinhart-Fischer, 1984.

[80] J. Hannah, *Countability in regular self-injective rings*, Quart. J. Math. Oxford (2) **31** (1980), 315–327.

[74] G. Hansen, *Ph.D. Thesis*, Indiana Univ., Bloomington, 1974.

[56] M. Harada, *A note on the dimension of modules and algebras*, J. Inst. Polytech., Osaka City U. **7** (1956), 17–28.

[63] ______, *Hereditary orders*, Trans. A.M.S. **107** (1963), 273–290.

[64] ______, *Hereditary semiprimary rings and triangular matrix rings*, Nagoya Math. J. **27** (1964), 463–484.

[71] ______, *On categories of indecomposable modules,* II, Osaka J. Math. **8** (1971), 309–321.

[72a] ______, *A note on categories of indecomposable modules*, Publ. Dép. Math. (Lyon) **9** (1972), 11–25.

[72b] ______, *On the endomorphism ring of a Noetherian quasi-injective module*, Osaka J.Math. **9** (1972), 217–223.

[70] M. Harada and Y. Sai, *On categories of indecomposable modules*, I, Osaka J. Math **7** (1970), 323–344.

[72] M. Harada and Y. Ishii, *On the endomorphism ring of Noetherian quasi-injective modules*, Osaka J. Math. **9** (1972), 217–223.

[73] V. K. Harčenko, *Galois extensions and rings of quotients* (Russian), Algebra i Logika **13** (1974), 460–484,488.

[74] ______, *Generalized identities with automorphisms*, (Russian), Algebra i Logika **14** (1975), 459–465.

[77] ______, *Galois theory of semiprime rings*, Algebra i Logika **16** (1977), 313–363.

[96] ______, *See Kharchenko.*

[65] D. K. Harrison, *See Chase.*

[66] ______, *See Eilenberg.*

[67] R. Hart, *Simple rings with uniform right ideals*, J.London Math.Soc. **42** (1967), 614–617.

[71] ______, *Krull dimension and global dimension of simple Ore-extensions*, Math. Z. **121** (1971), 341–345.

[77] B. Hartley, *Injective modules over group rings*, Quart. J. Math. Oxford Ser. (2) **28** (1977), 1–29.

[52] A. Hattori, *On the multiplicative group of simple algebras and orthogonal groups*, J. Math. Soc., Japan **4** (1952), 205–217.

[77] G. Hauger, *Einfache Derivationspolynomringe*, Arch. Math. **29** (1977), 491–496.

[96] A. O. Hausknecht, *See Bergman.*

[96] M. Hazewinkel (ed.), *Handbook of Algebra*, North Holland-Elsevier, Amsterdam, Lausanne, New York, Oxford, Shannon, Tokyo, 1996.

[74] T. Head, *Modules, A Primer of Structure Theorems*, Brooks/Cole (A division of Wadsworth), Monterey, 1974.

[71] W. Heinzer, *See Abhyankar.*

[73] ______, *See Eakin.*

[80] ______, *See Brewer.*

[75] R. C. Heitman and L. S. Levy, *1-1/2 and 2 generator ideals in Prüfer domains*, Rocky Mt. J. of Math. **5** (1975), 361–373.

[76] A. Heller and M. Tierney (eds.), *Algebra, Topology and Category Theory*, A Collection of Papers in Honor of Samuel Eilenberg, Academic Press, New York, 1976.

[73] M. Henriksen, *On a class of rings that are elementary divisor rings*, Arch. Math. **24** (1973), 133–141.

[74] ———, *Two classes of rings generated by their units*, J. Algebra **31** (1974), 182–193.

[65] M. Henriksen and M. Jerison, *The space of minimal primes of a commutative ring*, Trans. A.M.S. **115** (1965), 110–130.

[08] K. Hensel, *Theorie der algebraischen Zahlen*, B. G. Teubner, Leipzig, 1908.

[91] D. Herbera, *Ph.D. Thesis* (in Catalan), Departáment de Matemàtiques, Univ. Autónoma de Barcelona, 08193 Bellaterra, Spain, 1991.

[95] ———, *See Cedó.*

[97] ———, *See Ánh.*

[97] ———, *See Faith.*

[89] D. Herbera and P. Menal, *On rings whose finitely generated faithful modules are generators*, J. Pure and Appl. Alg. **122** (1989), 425–438.

[93] D. Herbera and P. Pillay, *Injective classical quotient rings of polynomial rings are quasi-Frobenius*, J. Pure and Appl. Alg. **86** (1993), 51–63.

[95] D. Herbera and A. Shamsuddin, *Modules with semilocal endomorphism ring*, Proc. Amer. Math. Soc. **128** (1995), 3593–3600.

[87] J. M. Hernandez, *see Gómez Pardo.*

[53a] I. N. Herstein, *A theorem on rings*, Canad.J.Math. **5** (1953), 238–241.

[53b] ———, *The structure of a certain class of rings*, Amer.J.Math. **75** (1953), 864–871.

[54] ———, *On the Lie ring of a division ring*, Ann. Math. (2) **60** (1954), 571–575.

[55a] ———, *Two remarks on the commutativity of rings*, Canad.J.Math. **7** (1955), 411–412.

[55b] ———, *On the Lie and Jordan simplicity of a simple associative ring*, Amer.J.Math. **77** (1955), 279–284.

[55c] ———, *The Lie ring of a simple associative ring*, Duke Math.J. **22** (1955), 471–476.

[56] ———, *Conjugates in division rings*, Bull. Amer. Math. Soc. **7** (1956), 1021–1022.

[61] ———, *On a result of Faith*, Canad. Math. Bull. **18** (1975), 609.

[65] ———, *A counterexample in Noetherian rings*, Proc. Nat. Acad. Sci. USA **54** (1965), 1036–1037.

[66] ———, *See Belluce.*

[67] ———, *Special simple rings with involution*, J. Algebra **6** (1967), 369–375.

[68] ———, *Noncommutative Rings*, Math.Assoc.Amer. Carus Monograph, Number 15, 1968.

[69] ———, *Topics in Ring Theory*, U. of Chicago Press, Chicago, 1969.

[75] ———, *Topics in Algebra*, Wiley, New York, 1975.

[76] ———, *A commutativity theorem*, J. Algebra **38** (1976), 112-118.

[64,66] I. N. Herstein and L. W. Small, *Nil rings satisfying certain chain conditions*, Canad. J. Math. **16** (1964), 771–776, Addendum, **ibid. 18** (1966) 300–302.

[71] I. N. Herstein and S. Montgomery, *A note on division rings with involution*, Michigan Math. J. **18** (1971), 75–79.

[94a] I. Herzog, *Finitely presented right modules over a left pure-semisimple ring*, Bull. L.M.S. **26** (1994), 333–338.

[94b] ———, *A test for finite representation type*, J. Pure and Appl. Algebra **95** (1994), 151–182.

[97] ———, *The Ziegler spectrum of a locally coherent Grothendieck category*, Proc. L.M.S. (3) **74** (1997), 503–508.

[93] J. Herzog, *See Bruns.*

[54] D.G. Higman, *Indecomposable representations at characteristic p*, Duke Math. J. **21** (1954), 377–381.

[56] ———, *On a conjecture of Nagata*, Proc. Cambridge Phil.Soc. **52** (1956), 1–4.

[1892] D. Hilbert, *Über die irreduzibilität ganzer rationaler Funktionen mit ganzzahliger Koeffizienten*, J. reine angew. Math. **110** (1992), 104–1029.

[1897] ———, *Bericht über die Theorie der algebraischen Zahlkörper, Jahrbericht der Deutschen Matematischen Verein, iv,* reprinted in Hilbert [32].

[1898] ———, *Über die Theorie der algebraischen Formen*, Math. Ann. **36** (1898), 473–534.

[03] ———, *Grundlagen der Geometrie, 2. Aufl. (2nd ed.)*, Teubner, 1903.

[22] ———, *Grundlagen der Geometrie*, Leipzig, 1922.

[32,33,35] ———, *Gesamelte Abhandlungen*, (Collected Papers), Chelsea, New York, 1932, 1933, 1935.

[73] D. Hill, *Semiperfect q-rings*, Math. Ann. **200** (1973), 113–121.

[62] Y. Hinohara, *Projective modules over semilocal rings*, Tohoku Math. J. (2) **14** (1962), 205–211.

[93] Y. Hirano and J. K. Park, *On self-injective strongly π-regular rings* **21** (1993), 85–91.

[50] G. Hochschild, *Automorphisms of simple algebras*, Trans. amer. Math. Soc. **69** (1950), 292–301.

[56] ———, *Relative homological algebra*, Trans. A.M.S. **82** (1956), 246–269.

[58] ———, *Note on relative homological algebra*, Nagoya Math. J. **13** (1958), 89–94.

[69] M. Hochster, *Prime ideal structure in commutative rings*, Trans. Amer. Math. Soc. **142** (1969), 43–60.

[72] ———, *Nonuniqueness of coefficient rings in a polynomial ring*, Proc. Amer. math. Soc. **34** (1972), 81–82.

[39] C. Hopkins, *Rings with minimal condition for left ideals*, Ann. of Math. **40**, 712–730 (18-18).

[84] K. Hrbacek and T. Jech, *Introduction to Set Theory* (1984), Dekker, New York and Basel.

[49] L. K. Hua, *Some properties of a sfield*, Proc.Nat.Acad.Sci. **35** (1949), 533–537.

[73] J. Huckaba, *On valuation rings that contain zero divisors*, Proc. Amer. Math. Soc.

[76] ———, *On the integral closure of a Noetherian ring*, Trans. Amer. Math. Soc. **220** (1976), 375–379.

[88] ———, *Commutative Rings with Zero Divisors*, Dekker, Basel and New York, 1988.

[90] ———, *See Faith [90b].*

[79] J. Huckaba and J. Keller, *Annihilation in commutative rings*, Pac. J. Math. **83** (1979), 375–379.

[80] J. Huckaba and I. Papick, *Quotient rings of polymial rings*, Manuscr. Math. **31** (1980), 167–190.

[96] ———, *See Fontana.*

[75] A. Hudry, *Sur un problème de C. Faith*, J. Algebra **34** (1975), 365–374.

[76,78] B. Huisgen-Zimmermann, *See Zimmermann-Huisgen.*

[79,80] ———, *See Zimmermann-Huisgen.*

[97] ———, *See Arhangel'skii.*

[68] T. Hungerford, *On the structure of principal ideal rings*, Pac. J. Math. **25** (1968), 543–547.

[67] B. Huppert, *Endliche Gruppen, I*, Springer, Berlin, New York, Heidelberg, 1967.

[88] H. L. Hutson, *On zero dimensional rings of quotients and the geometry of minimal primes*, J. Algebra **112** (1988), 1–14.

[93] ———, *Higher dimensional rings of quotients*, Publ. Mat. **37** (1993), 239–243.

[76] D. V. Huynh, *Über einen Satz von A. Kertész*, Acta Math.Acad.Sci.Hungar. **28** (1976), 73–75.

[77] ———, *Die Sparkenheit von MHR-Ringen*, Bull. Acad. Polon. Sci. Sér. Sci. Math. Ast. Phys. **25** (1977), 939–941.

[95] ———, *A right sigma-CS ring with ACC or DCC on projective principal right ideals is left Artinian and QF-3*, Trans. Amer. Math. Soc. **347** (1995), 3131–3139.

[97] ———, *Letter to the author*, December 1997.

[89] D. V. Huynh, N.V. Dung, and P. F. Smith, *Rings characterized by their cyclic submodules*, Proc. Edinburgh Math. Soc. **32** (1989), 355–362.

[90] ———, *A characterization of rings with Krull dimension*, Journal of Algebra **132** (1990), 104–112.

[91] D. V. Huynh, N. V. Dung, P. F. Smith, and R. Wisbauer, *Extending Modules*, Pitman, London, 1994.

[96] D. V. Huynh, S. K. Jain and S. R. López-Permouth, *When is a simple ring Noetherian?*, J. Algebra **184** (1996), 786–794.

[98] ———, *On the symmetry of the Goldie and CS-conditions*, preprint, 1998.

[96] D. V. Huynh, S. T. Rizvi and M. F. Yousif, *Rings whose finitely generated modules are extending*, J. Pure and Appl. Alg. **111** (1996), 325–328.

[97] D. V. Huynh and S. T. Rizvi, *An approach to Boyle's Conjecture*, Proc.Edinburgh Math.Soc. **40** (1997), 267–273.

[94] ———, *See Clark; also Dung.*

[51] M. Ikeda, *Some generalizations of quasi-Frobenius rings*, Osaka J. Math. **3** (1951), 227–239.

[52] ———, *A characterization of quasi-Frobenius rings*, Osaka J. Math. **4** (1952), 203–210.

[54] M. Ikeda and T. Nakayama, *On some characteristic properties of quasi-Frobenius and regular rings*, Proc. Amer. Math. Soc. **5** (1954), 15–19.

[71] E. Ingraham, *See DeMeyer.*

[72] Y. Ishii, *See Harada.*

[70] G. Ivanov, *Rings with zero singular ideal*, J. algebra **16** (1970), 340–346.

[72] ______, *Ph.D. Thesis*, Australian National University, Canberra, 1972.

[74] ______, *Left generalized uniserial rings*, J. Algebra **31** (1974), 166–181.

[75] ______, *Decomposition of modules over uniserial rings*, Comm. Algebra **3** (1975), 1031–1036.

[71] H. Jacobinski, *Two remarks about hereditary orders*, Proc. A.M.S. **28** (1971), 1–8.

[36] N. Jacobson, *Totally disconnected locally compact rings*, Amer. J. Math. **58** (1936), 433–445.

[37] ______, *Abstract derivations and Lie Algebras*, Trans.A.M.S. **42** (1937), 206–224.

[42] ______, *The Theory of Rings, Surveys of the A.M.S., vol. 2*, American Math. Soc., Providence, 1942.

[45a] ______, *The radical and semisimplicity for arbitrary rings*, Amer. J. Math. **67** (1945), 300-342.

[45b] ______, *The structure of simple rings without finiteness assumptions*, Trans. Amer. Math. Soc. **57** (1945), 228–245.

[45 c] ______, *Structure theory for algebraic algebras of bounded degree*, Ann. of Math. **46** (1945), 695–707.

[47] ______, *A note on division rings*, Amer.J.Math. **69** (1947), 27–36.

[50] ______, *Some remarks on one-sided inverses*, Proc. Amer. Math. Soc. **1** (1950), 352–355.

[51,53,64] ______, *Lectures in Abstract Algebra, vols. I–III*, Van Nostrand Company, New York and Princeton, 1951,1953,1964.

[55,64] ______, *Structure of Rings*, Colloquium Publication, Vol. 37, Amer. Math. Soc., Providence, 1955, rev. 1964.

[74,80] ______, *Basic Algebra,* I,II, W. H. Freeman, San Francisco, 1974, 1980.

[75] ______, *PI-Algebras*, Lecture Notes in Math., vol. 441, Springer-Verlag, Berlin, Heidelberg and New York, 1975.

[89] ______, *Collected Mathematical Papers* (Gian-Carlo Rota, ed.), 3 vols., Birkäuser, Boston, Basel and Berlin, 1989.

[96] ______, *Finite Dimensional Division Algebras*, Springer, Berlin, Heidelberg and New York, 1996.

[35] N. Jacobson and O. Taussky, *Locally compact rings*, Proc. Nat. Acad. Sci., U.S.A. **21** (1935), 106–108.

[73] Saroj Jain, *Flat and FP-injectivity*, Proc. Amer. Math. Soc. **41** (1973), 437–442.

[77] S. K. Jain (ed.), *Ring Theory* (Proc. Conf. Ohio Univ. Athens, Ohio, 1976), Lecture Notes in Pure and Applied Math., vol. 25, Marcel Dekker, Basel and New York, 1977.

[66] ______, *See Belluce.*

[92] ______, *See Al-Huzali.*

[96,98] ______, *See Huynh.*

[66] ______, *See Belluce.*

[69] S.K. Jain, S. H. Mohamed and S. Singh, *Rings in whose every right ideal is quasi-injective*, Pac.J.Math. **31** (1969), 73–79.

[75] S. K. Jain and S. Singh, *Quasi-injective and pseudo-injective modules*, Canad. Math. Bull. **18** (1975), 359–366.

[89] S. K. Jain and S. R. López-Permouth, *A general Wedderburn theorem*, PAMS **106** (1989), 19–23.

[90] ———, *Rings whose cyclics are essentially embeddable in projectives*, J. Algebra **128** (1990), 257–269.

[90b] ———, *Non-commutative ring theory*, Proc. of the Athens Conf. 1989, Lecture Notes in Math. vol. 1448, Springer-Verlag, Berline, Heidelberg, New York, 1990.

[90] S. K. Jain, S. R. López-Permouth and S. T. Rizvi, *Continuous rings with acc on essentials*, Proc. Amer. Math. Soc. **108** (1990), 583–586.

[92] S. K. Jain, S. R. López-Permouth and S. Singh, *On a class of QI-rings*, Glasgow Math. J. **34** (1992), 75–81.

[93] S. K. Jain, S. R. López-Permouth and M. A. Saleh, *Weakly projective modules*, in Ring Theory: Proceedings of the OSU-Denison Conference, World Scientific Press, New Jersey, 1993, pp. 200–208.

[93] S. K. Jain and S. T. Rizvi (eds.), *Ring Theory*, in Proc. of the Ohio State-Denison Conf. (1992), World Science Publishers, River Edge, N.J., London and Singapore, 1993.

[97] ———, *Advances in Ring Theory*, Birkhäuser, Boston, Basel, Berlin, 1997.

[98] I. M. James (ed.), *A History of Topology*, North Holland, Amsterdam, 1998.

[56] J. P. Jans, *On the indecomposable representations of an algebra*, Ann. of Math. **66** (1967), 418–429.

[69] ———, *On co-Noetherian rings*, J. London Math. Soc. **1** (1969), 588–590.

[56] J. P. Jans and T. Nakayama, *On the dimension of modules and algebras*, Nagoya Math. J. **11** (1956), 67–76.

[65] ———, *See Curtis.*

[88] W. Jansen, *See Zelmanowitz.*

[69] G. J. Janusz, *Indecomposable modules for finite groups*, Ann. of Math. **89** (1969), 209–241.

[70] ———, *Faithful representations of p-groups at characteristic p*, I, J. Algebra **15** (1970), 335–351.

[72a] ———, *Faithful representations of p-groups at characteristic p*, II, J. Algebra **22** (1972), 137–160.

[72b] ———, *Some left serial algebras of finite type*, J. Algebra **23** (1972), 404–411.

[88] P. Jara Martinez, *See Bueso.*

[68] A. V. Jategaonkar, *Left principal ideal domains*, J. Algebra **8** (1968), 148–155.

[69] ———, *A counter-example in ring theory and homological algebra*, J. Algebra **12** (1969), 418–440.

[70a] ———, *Left principal ideal rings*, Lecture Notes in Mathematics, vol.123, Springer, Berlin-Heidelberg-New York, 1970.

[70b] ———, *Orders in Artinian rings*, Bull. Amer. Math. Soc. **75** (1970), 1258–1259.

[72a] ———, *Structure and classification of hereditary Noetherian prime rings*, pp.171–229 (Proceedings) Ring Theory, Academic Press, New York, 1972.

[72b] ———, *Skew polynomials over orders in Artinian rings*, J. Algebra **21** (1972), 51–59.

[74] ———, *See Formanek.*

[74a] ———, *Injective modules and classical localization in non-commutative Noetherian rings*, Trans. Amer. Math. Soc. **79** (1973), 152–157.

[74b] ———, *Jacobson's conjecture and modules over fully bounded Noetherian rings*, J. Algebra **30** (1974), 103–121.

[74c] ———, *Relative Krull dimension and prime ideals in right Noetherian rings*, Comm. Algebra **2** (1974), 429–468.

[74d] ———, *Integral group rings of polycyclic-by-finite groups*, J. Pure and Appl Algebra **4** (1974), 337–343.

[75] ———, *Principal ideal theorem for Noetherian P.I. rings*, J. algebra **35** (1975), 17–22.

[86] ———, *Localization in Noetherian Rings*, London Math. Soc. Lecture Notes, No. 98, Cambridge U. Press, Cambridge and New York, 1986.

[84] T. Jech, *See Hrbacek.*

[97] ———, *Set Theory, second corrected edition*, Springer-Verlag (2nd corrected ed.), New York-Heidelberg-Berlin, 1997.

[41] S. A. Jennings, *The structure of the group ring of a p-group over a modular field*, Trans.Amer.Math.Soc. **50** (1941), 175–185.

[63] C. U. Jensen, *On characterization of Prüfer rings*, Math.Scand. **13** (1963), 90–98.

[66a] ———, *A remark on flat and projective modules*, Canad. J. Math. **18** (1966), 945–949.

[66b] ———, *A remark on semi-hereditary local rings*, J. Lond. Math. Soc. **41** (1966), 479–482.

[66c] ———, *Arithmetical rings*, Acta Math. Acad. Sci. Hungar **17** (1966), 115–123.

[89] C.U. Jensen and H. Lenzing, *Module Theoretic Algebra*, Gordon and Breach, New York-London-Paris-Tokyo-Melbourne, 1989.

[60] M. Jerison, *See Gilman.*

[65] ———, *See Henriksen.*

[98] E. Jespers, *Units in integral group rings: a survey*, pp.141–169 in Drensky et al [98].

[95] J. Ježek, *See Freese.*

[77] B. Johns, *Annihilator conditions in Noetherian rings*, J. Algebra **49** (1977), 222–224.

[86,90] U. K. Johnsen, *See Blessenohl.*

[72] J. Johnson, *See Cozzens.*

[78] ———, *Systems of n partial differential equations in n unknowns: the conjecture of M. Janet*, Trans. A.M.S. **242** (1978), 329–334.

[51a] R. E. Johnson, *The extended centralizer of a ring over a module*, Proc. A.M.S. **2** (1951), 891–895.

[51b] ———, *Prime rings*, Duke Math. J. **18** (1951), 799–809.

[59] R. E. Johnson and E. T. Wong, *Self-injective rings*, Canad. Math. Bull. **2** (1959), 167–173.

[61] ———, *Quasi-injective modules and irreducible rings*, J. Lond. Math. Soc. **36** (1961), 260–268, (see also Wong and Johnson).

[70] S. Jøndrup, *On finitely generated flat modules*, Math. Scand. **26** (1970), 233–240.

[71] ———, *P.p. and finitely generated ideals*, Proc. Amer. Math. Soc. **28** (1971), 431–435.

[76] ———, *Projective modules*, Proc. A.M.S. **59** (1976), 217–2212.

[63] B. Jónsson, *See Crawley.*

[64] ———, *See Crawley.*

[72] ———, *Topics in Universal Algebra*, Lecture Notes in Math., vol. 250, Springer-Verlag, Berlin, Heidelberg and New York, 1972.

[1878] C. Jordan, *Memoire sur les équations differentiellees linéares à integrale algébrique*, J. für Math. **84** (1878), 89–215.

[71] V. S. Kahlon, *Problem of Krull-Schmidt-Azumaya-Matlis*, J. Indian Math.Soc. (N.S.) **35** (1971), 255–261.

[88a] M. A. Kamal and B. J. Müller, *Extending modules over commutative domains*, Osaka J. Math. **25** (1988), 531–538.

[88b] ———, *The structure of extending modules over Noetherian rings*, Ibid, 539–551.

[88c] ———, *Torsionfree extending modules*, Ibid, 825–832.

[76] M. Kamil, *On quasi-Artinian rings*, Rend. Mat. **9** (1976), 617–619.

[42] I. Kaplansky, *Maximal fields with valuations*, Duke Math. J. **9** (1942), 303–321.

[45] ———, *Maximal fields with valuations*, II, Duke Math. J. **12** (1945), 243–248.

[46] ———, *On a problem of Kurosch and Jacobson*, Bull. Amer. Math. Soc. **52** (1946), 496–500.

[48] ———, *Rings with polynomial identity*, Bull. A.M.S. **54** (1948), 575–580.

[48b] ———, *Locally compact rings*, Amer. J. Math. **70** (1948), 447–459, 9–562.

[49] ———, *Elementary divisors and modules*, Trans. Amer. Math. Soc. **66** (1949), 464–491.

[50] ———, *Topological representations of algebras II*, Trans. Amer. Math. Soc. **68** (1950), 62–75.

[52] ———, *Modules over Dedekind rings and valuation rings*, Trans. Amer. Math. Soc. **72** (1952), 327–340.

[55] ———, *Any orthocomplemented complete modular lattice is a continuous geometry*, Ann. Math. (2) **61** (1955), 524–41.

[57] ———, *An Introduction to Differential Algebra*, Actualités Sci.Indust., no. 1251, Hermann, Paris, 1957.

[58a] ———, *Projective modules*, Ann. of Math. **68** (1958), 372–377.

[58b] ———, *On the dimension of modules and algebras, X. A right hereditary ring which is not left hereditary*, Nagoya Math. J. **13** (1958), 85–88.

[60] ———, *A characterization of Prüfer rings*, J. Indian Math. Soc. **24** (1960), 279–281.

[62] ———, *The splitting of modules over integral domains*, Arch. Math. **13** (1962,), 341–343.

[68] ———, *Rings of Operators*, Benjamin, New York, 1968.

[69] ———, *Infinite Abelian Groups (*Second Edition*)*, Univ. of Michigan Press, Ann Arbor, 1969.

[69b] ———, *Fields and Rings*, Univ. of Chicago Press, Chicago, Ill., 1969.

[70-74] ———, *Commutative Rings*, Allyn and Bacon, Inc., Boston, 1970, rev. ed., Univ. of Chicago, 1974.

[79] ———, *(ed.), Selected Papers of Saunders Maclane, 1979.*

[94] ———, *Letter to the author of October 12, 1994.*

[94b] ———, *Review of Warner* [93], Bull. A.M.S. **31** (1994), 146–147.

[95a] ———, *Selected Papers and Other Writings*, Springer, New York-Berlin-Heidelberg, 1995.

[95b] ———, *Commutativity Theorems revisited*, in Collected Papers [95a].

[96] ———, *Rings and Things. Retiring Presidential Address*, Annual Meeting of the Amer. Math. Soc. at Orlando, Jan. 1996, Providence, R.I.

[48] ———, *See Arens.*

[51] ———, *See Cohen.*

[79] M. I. Kargapolov and Ju. I. Merzlyjakov, *Fundamentals of the Theory of Groups*, Springer-Verlag, New York, 1979.

[89] G. Karpilovsky, *Units of Group Rings*, Pitman Monographs and Surveys in Pure and Appl. Math., vol. 47, Longman, Harlow, 1989.

[54] F. Kasch, *Grundlagen einer Theorie der Frobenius–Erweiterungen*, Math. Ann. (1954), 453–474.

[61] ———, *Dualitätseigenschaften von Frobenius Erweiterungen*, Math. Z. **77** (1961), 229–337.

[57] F. Kasch, M. Kneser and H.Kupisch, *Unzerlegbare modulare Darstellungen endlicher Gruppen mit zyklischer p-Sylow*, Gruppe. Arch. Math. **8** (1957g), 320–321.

[66] F. Kasch and E. Mares, *Eine Kennzeichnung semiperfekter Moduln*, Nagoya Math. J. **27** (1966), 525–529.

[67] T. Kato, *Self-injective rings*, Tohoku Math. J. **19** (1967), 485–494.

[68] ———, *Torsionless modules*, Tohoku Math. J. **20** (1968), 234–243.

[63] O. H. Kegel, *Zur Nilpotenz gewisser assoziativer Ringe*, Math. Ann. **149** (1963), 258–260.

[64] ———, *On rings that are sums of two subrings*, J. Algebra **1** (1964), 103–109.

[93] A. V. Kelarev, *A sum of two locally nilpotent rings may be not nil*, Arch. Math. **60** (1993), 431–435.

[97] ———, *A primitive ring which is a sum of two Wedderburn radical subrings*, Proc. A.M.S. **125** (1997), 2191–2193.

[79] J. Keller, *See J. Huckaba.*

[80] R. Kennedy, *Krull rings*, Pac. J. Math. **89** (1980), 131–136.

[79] J. W. Kerr, *An example of a Goldie ring whose matrix ring is not Goldie*, J. Algebra **61** (1979), 590–592.

[90] ———, *The polynomial ring over a Goldie ring need not be Goldie*, J. Algebra **134** (1990).

[92] ———, *The power series ring over an Ore domain need not be Ore*, J. Algebra **75** (1982), 175–177.

[64] A. Kertész, *Zur Frage der Spartenkeit von Ringen,*, Bull. Acad. Polonaise Sci. Sér. Math. Ast. Phys. **12** (1964), 91–93.

[72] A. Kertész (ed.), *Rings, Modules and Radicals*, Proc. Int. Colloq. on Associative Rings, Modules and Radicals, Keszthely, 1971, Colloq. Math. Soc. Janos Bolyai, vol. 6, North Holland, Amsterdam-London, and Janos Bolyai Math.Soc., Budapest, 1972.

[96] K. Kharchenko, *Simple, prime, and semiprime rings*, in Handbook of Algebra, vol. I, pp. 761–814, North-Holland, Amsterdam, New York, Tokyo, 1996.

[73,77] ———, *See Harčenko.*

[89] S. M. Khuri, *Correspondence theorems for modules and their endomorphism rings*, J. Algebra **122** (1989), 380–396.

[90] ———, *Modules with regular, perfect, Noetherian, or Artinian endomorphism rings,* pp.7–18, in Jain-López-Permouth [90].

[67] R. Kiełpiński, *On Γ-pure injective modules*, Bull. Acad. Pol. Sci. **15** (1967), 127–131.

[97] J. Y. Kim, *See Birkenmeier.*

[76/77] Y. Kitamura, *Note on the maximal quotient ring of a Galois subring*, Math.J. Okayama U. **19** (1976/77), 55–60.

[69] G.B. Klatt and L.S. Levy, *Pre-self-injective rings*, Trans. Amer. Math. Soc. **122** (1969), 407–419.

[69] A. A. Klein, *Necessary conditions for embedding rings in fields*, Trans.Amer.Math.Soc. **137** (1969), 141–151.

[94] ———, *The sum of nil one-sided ideals of bounded index of a ring*, Israel J. Math. **88** (1995), 25–30.

[96] I. Kleiner, *The genesis of the abstract ring concept*, Amer. Math. Monthly **103** (1996), 417–424.

[57] M. Kneser, *See Kasch.*

[82] S. Kobayashi, *A note on f-injective modules*, Math. Sem. Notes, Kobe U. **10** (1992), 127–134.

[85] ———, *On non-singular FPF rings,* I,II, Osakaya J. Math. **22** (1985), 787–795; 797–803.

[70] A. Koehler, *Quasi-projective covers and direct sums*, Proc. Amer. Math. Soc. **24** (1970), 655–658.

[66] K. Koh, *On simple rings with maximal annihilator right ideal*, Canad. Math. Bull. **9** (1966), 667–668.

[58] C. W. Kohls, *Prime ideals in rings of continuous functions*, Illinois J. Math. **2** (1958), 505–536.

[70] L.A. Koifman, *Rings over which each module has a maximal submodule (*Russian*)*, Mat. Zametki **7** (1970), 359–367.

[71a,b] ———, *Rings over which singular modules are injective*, I,II (Russian) **6** (1971), 85–104,161; 62–84,199–200.

[82] K. Kosler, *On hereditary rings and Noetherian V-rings*, Pac. J. Math. **103** (1982), 467–473.

[30a] G. Köthe, *Die Struktur der Ringe deren Restklassenring nach dem radical vollständig reduzibel ist*, Math. Z. **32** (1930), 161–186.

[30b] ———, *Über maximale nilpotente Unterringe and Nilringe*, Math. Ann. **103** (1930), 359–363.

[35] ———, *Verallgemeinerte Abelsche Gruppen mit Hyperkomplexem Operatorenring*, Math. Z. **39** (1935), 31–44.

[70] G. Krause, *On the Krull dimension of left Noetherian left Matlis rings*, Math. Z. **118** (1970), 207–214.

[72] ———, *On fully bounded left Noetherian rings*, J. Algebra **23** (1972), 88-99.

[77] H. J. Kreisel, *See Chang.*

[72] J. Krempa, *Logical connections between some open problems concerning nil rings*, Fund. Math. **76** (1972), 121–130.

[1895] Kronecker, *Werke*, Teubner, Leipzig , 1985.

[24] W. Krull, *Die verschiedenen Arten der Hauptidealringe*, Sitzungsber. Heidelberger Akad. **67** (1924).

[25] ———, *Über verallgemeinerte endliche Abelsche Gruppen*, Math. Z **23** (1925), 161–196.

[26] ———, *Theorie and Anwendung der verallgemeinerten Abelschen Gruppen*, Sitzungsber. Heidelberger Akad. **7** (1926), 1–32.

[28a] ———, *Primketten in allgemeinen Ringbereichen*, Sitz.-Bereich Heidelberg. Akad. Wiss. 7 Abhandl, Heidelberg, 1928.

[28b] ———, *Zur theorie der allgemeinen Zahlringe*, Math. Ann. **99** (1928), 51–70.

[32] ———, *Allgemeine Bewertungstheorie*, J. reine angew. Math. **167** (1932), 160–196.

[35–48] ———, *Idealtheorie*, Ergebnisse der Math. u. Ihrer Grenzegebiete, Springer 1933, Chelsea (Reprint), New York, 1948.

[32] ———, *Matrizen, Moduln, und verallgemeine Abelsche Gruppen*, in Bereich der Ganzen algebraischen Zahlen, Heidelberger Akad. der Wissenshaften (1932), pp. 13–38.

[38] ———, *Dimensiontheorie in Stellenringen*, J. reine und angew. Math. **149** (1938), 204–226.

[50] ———, *Jacobsonscher Ringe und Hilbertscher Nullstellensatz*, Dimensionstheorie, Proceedings of the International Congress of Mathematicians, vol. 2, (Cambridge, Mass. 1950), Amer. Math. Soc., Providence, 1952.

[51] ———, *Jacobsonscher Ringe, Hilbertscher Nullstellensatz*, Math. Z. **54** (1951), 354–387.

[58] ———, *Über Laskersche Rings*, Rend.Cir.Mat. Palermo (2) **7** (1958), 155–166.

[52] L. Kulikov, *On direct decompositions of groups*, Ukrain. Math. Z. **4** (1952), 230–275; 347–372 (Russian), AMS translation, 1956.

[56] H. Kupisch, *Über geordnete Gruppen von reelen Functionen*, Math. Z. **64** (1956), 10–40.

[75] ———, *Quasi-frobenius algebras of finite representation type*, in Lecture Notes in Math., Vol. 488, 1975, pp. 184–199.

[57] ———, *See F. Kasch.*

[91] A. Kurata, *Note on dual bimodules,* First China-Japan Symp. on Ring Theory, Okayama U., Okayama, 1991.

[63] A. G. Kurosh, *General Algebra*, Chelsea, New York, 1963.

[70] R. P. Kurshan, *Rings whose cyclic modules have finitely generated socle*, J. Algebra **15** (1970), 376–386.

[69] J.P. Lafon, *Sur un problème d'Irving Kaplansky*, C.R. Acad. Sci. Paris, Série A **268** (1969), 1309-1311.

[73] ———, *Modules de presentation finie et de tyle fini sur un anneau arithmétique*, Symposia Math. **11** (1973), 121–141.

[95] T. Y. Lam, *A lifting theorem, and rings with isomorphic matrix rings*, in "Five Decades as a Mathematician and Educator: On the 80th Birthday of Y. C. Wong, (K. Y. Chan and M. C. Liu, eds.) World Sci. Pub. Co., Singapore-London-Hong Kong, 1995.

[98a] ———, *Lectures on Modules and Rings*, Graduate Texts in Math., Springer, Berlin, Heidelberg and New York, 1998.

[98b] ———, *Faxes to the author,* May 18, June 20 and September 15, 1998.

[98c] ———, *Modules with isomorphic multiples, and rings with isomorphic matrix rings,* Center for Pure and Appl. Math. PAM-736, U. Calif. at Berkeley, 1998, to appear in L'Enseg. Math.

[51] J. Lambek, *The immersibility of a semigroup in a group*, Canad.J.Math. **3** (1951), 34–43.

[63] ———, *On Utumi's ring of quotients*, Canad. J. Math. **15** (1963), 77–85.

[64] ———, *A module is flat iff its character module is injective*, Canad. Math. Bull. **7** (1964), 237–243.

[66] ———, *Rings and Modules*, Blaisdell, New York 1966, reprint Chelsea, New York.

[71] ———, *Torsion Theories, Additive Semantics and Rings of Quotients*, Lecture Notes in Math., vol. 177, Springer, Berlin, Heidelberg and New York, 1971.

[58] ———, See Findlay.

[65] ———, *See Fine.*

[76] J. Lambek and G.O. Michler, *On products of full linear rings*, Pub. Math., Debrecen 1976.

[98] ———, *Letters to the author, January and March, 1998.*

[65] S. Lang, *See Artin.*

[69] C. Lanski, *Nil subrings of Goldie rings are nilpotent*, Canad. J. Math. **21** (1969), 904–907.

[74] M.D. Larsen, W.J. Lewis and T.S. Shores, *Elementary divisor rings and finitely presented modules*, Trans. Amer. Math. Soc. **187** (1974), 231–248.

[74] J. Lawrence, *A singular primitive ring*, Proc. A.M.S. **45** (1974), 59–63.

[75] ———, *Semilocal group rings and tensor products*, Michigan Math. J. **22** (1975), 309–314.

[75] ———, *See Handelman.*

[77] ———, *A countable self-injective ring is quasi-Frobenius*, Proc. Amer. Mat. Soc. **65** (1977), 217–220.

[76] J. Lawrence and S. M. Woods, *Semilocal group rings in characteristic* 0, Proc. Amer. Math. Soc. **60** (1976), 8–10.

[64] D. Lazard, *Sur les modules plats*, C. R. Acad. Sci. Paris **258** (1964), 6313–6316.

[74] ———, *Liberté des gros modules projective covers*, J. Algebra **31** (1974), 437–451.

[92] F. C. Leary, *Dedekind finite objects in module categories*, J. Pure & Appl. Algebra **82** (1992), 71–80.

[70] B. Lemonnier, *Quelques applications de la dimension de Krull*, C.R. Acad. Sci. Paris. Sér., A-B **270** (1970), A1395–1397.

[72] ———, *Déviation des ensembles et groupes abeliens totalement ordonnés*, Bull. Sci. Math. **96** (1972), 289–303.

[79] ———, *$AB5^\star$ et la dualité de Morita*, C. R. Acad. Sci. Paris, Sér. A **289** (1979), 47–50.

[73] T. H. Lenagan, *Nil radical of rings with Krull dimension*, Bull. Lond. Math. Soc. **5** (1973), 307–311.

[73] ———, *See Gordon.*

[73] ———, *Nil radical of rings with Krull dimension*, Bull. London Math. Soc. **5** (1973), 307–311.

[75] ———, *Artinian ideals in Noetherian rings*, Proc. Amer. Math. Soc. **51** (1975), 499–500.

[74] H. W. Lenstra, *Rational functions invariant under a finite abelian group*, Invent. Math. **25** (1974), 299–325.

[69] H. Lenzing, *Endlich präesentierbare Moduln*, Arch. Math. **20** (1969), 262–266.

[71] ———, *A homological characterization of Steinitz rings*, Proc. A.M.S. **29** (1971), 269–271.

[76] ———, *Direct sums of projective modules as direct summands of the direct product*, comm. alg. **4** (1976), 681–691.

[89] ———, *See Jensen.*

[55] H. Leptin, *Linear Kompakte Moduln und Ringe,* I, Math. Z. **62** (1955), 241–267.

[57] ———, *Linear Kompakte Moduln und Ringe,* II, Math. Z. **66** (1957), 289–327.

[59] L. Lesieur and R. Croisot, *Sur les anneaux premiers noethériens à gauche*, Ann. Sci. École Norm. Sup. **76** (1959), 161–183.

[31] J. Levitzki, *Über nilpotente Unterringe*, Math.Ann. **105** (1931), 620–627.

[35] ———, *On automorphisms of certain rings*, Ann. of Math. **36** (1935), 984–992.

[39] ———, *On rings which satisfy the minimum condition for right-hand ideals*, Compositio Math. **7** (1939), 214–222.

[44] ———, *On a characteristic condition for semiprimary rings*, Duke Math. J. **11** (1944), 367–368.

[45a] ———, *Solution of a problem of G. Köthe*, Amer. J. Math. **67** (1945), 437–442.

[45b] ———, *On three problems concerning nil rings*, Bull. A.M.S. **51** (1945), 913–919.

[46] ———, *On a problem of Kurosch*, Bull. Amer. Math. Soc. **52** (1946), 1033–1035.

[51] ———, *Prime ideals and the lower radical*, Amer. J. Math. **73** (1951), 25–29.

[63] ———, *On nil subrings* (Posthumous paper ed. by S. A. Amitsur), Israel J. Math. **1** (1963), 215–216.

[63a] L.S. Levy, *Unique direct sums of prime rings*, Trans. Amer. Math. Soc. **106** (1963), 64–76.

[63b] ———, *Torsion-free and divisible modules over non-integral domains*, Canad. J. Math. **15** (1963), 132–151.

[66] ———, *Commutative rings whose homomorphic images are self-injective*, Pac. J. Math. **18** (1966), 149–153.

[75] ———, *Matrix equivalence and finite representation type*, Comm. algebra **3** (1975), 739–748.

[69] ———, *See Klatt.*

[95] ———, *See Facchini.*

[75] ———, *See Heitman.*

[82] L. S. Levy and P. F. Smith, *Semiprime rings whose homomorphic images are serial*, Canad. J. Math. XXXIV (1982), 691–695.

[74] W. J. Lewis, *See Larsen, also Shores.*

[84] R. Lidl and G. Pilz, *Applied Abstract Algebra*, Undergraduate Texts in Math., Springer-Verlag, Berlin, Heidelberg, New York, 1984.

[96] S. X. Liu, *See Cao.*

[92] S. R. López-Permouth, *Rings characterized by their weakly-injective modules*, Glasgow Math. J. **34** (1992), 349–353.

[98] ———, *Letter to the author of January 19, 1998.*

[92] ———, *See Al-Huzali.*

[90-93] ———, *See Jain.*

[96] ———, *See Huynh.*

[91] J. Łós, *See Balcerzyk.*

[76] K. Louden, *Maximal quotient rings of ring extensions*, Pac. J. Math. **62** (1976), 489–496.

[89] T. G. Lucas, *Characterizing when $R[X]$ is integrally closed, II*, J. Pure and Appl. Alg. **6** (1989), 49–52.

[92] ———, *The complete integral closure of $R[X]$*, Trans. A.M.S. **330** (1992), 757–768.

[93] ———, *Strong Prüfer rings and the ring of finite fractions*, J. Pure and Appl. Alg. **84** (1993), 59–71.

[97] ———, *The integral closure of $R(x)$ and $R\langle x\rangle$*, Comm. Algebra **25** (1997), 847–872.

[94] R. Lyons, *See Gorenstein.*

[72] G. R. MacDonald, *See Armendariz.*

[82] A. B. MacEacharn, *See Brown.*

[39] S. MacLane, *The universality of formal power series rings*, Bull. Amer. Math. Soc. **45** (1939), 888–890.

[63] ———, *Duality in groups*, Bull. Amer. Math. Soc. **56** (1950), 485–516.

[79] ———, *Selected Papers* (I. Kaplansky, ed.), Amer. Math. Soc., Providence, R. I., 1979.

[45] ———, *See Eilenberg.*

[66] ———, *See Eilenberg, etal., (eds.).*

[73] K. MacLean, *Commutative principal ideal rings*, Proc. London Math. Soc. **26** (1973), 249–273.

[94] A. Magid (ed.), *Rings, Extensions and Cohomology*, Proc. of the Zelinsky Retirement Conference 1993, Northwestern U., Marcel Dekker, Basel, etc., 1994.

[36] A. I. Mal'cev, *On the immersion of an algebraic ring into a field*, Math. Ann. **113** (1936), 686–691.

[42] ———, *On the representation of an algebra as a direct sum of its radical and semi-simple algebra*, Dokl. Akad. Nauk. SSSR **36** (1942), 42–45.

[67] M. E. Manis, *Extension of valuation theory*, Bull. Amer. Math. Soc. **73** (1967), 735–736.

[63] E. Mares, *Semiperfect modules*, Math. Z. **82** (1963), 347–360.

[66] ———, *See Kasch.*

[85] L. Marki and R. Wiegandt, *Radical Theory*, North Holland, Amsterdam, Oxford, New York, 1985.

[68] J. Marot, *Extension de la notion d'anneaux valuation*, Dept. Mat. Fac. des Sci. de Brest, Brest (1968), 1–46 et une complément de 33 p.

[58] W. S. Martindale, III, *The structure of a special class of rings*, Proc. Amer. Math. Soc. **9** (1958), 714–721.

[69] ———, *Rings with involution and polynomial identities*, J. Algebra **11** (1969), 186–194.

[1898] H. Maschke, *Über den arithmetischen Charakter der Coefficienten der Substitutionen endlicher linearer Substitutionsgruppen*, Math. Ann. **50** (1898), 483–498.

[58] E. Matlis, *Injective modules over noetherian rings*, Pac. J. Math. **8** (1958), 511–528.

[59] ______, *Injective modules over Prüfer rings*, Nagoya J. Math. **15** (1959), 57–69.

[66] ______, *Decomposable modules*, Trans. Amer. math. Soc. **125** (1966), 147–179.

[72] ______, *Torsion-Free Modules*, Univ. of Chicago Press, Chicago and London, 1972.

[85] ______, *Commutative semi-coherent and semiregular rings*, J. Algebra **95** (1985), 343–372.

[80] H. Matsumura, *Commutative Algebra*, (2nd ed.), Benjamin/Cummings, Reading, 1980.

[70] S. McAdam, *Primes and annihilators*, Bull. Amer. Math. Soc. **76** (1970), 92.

[73] P.J. McCarthy, *The ring of polynomials over a von Neumann regular ring*, Proc. Amer. Math. Soc. **39** (1973), 253–254.

[84] J. C. McConnell, *On the global and Krull dimensions of Weyl algebras over affine coefficient rings*, J. London Math. Soc. (2) **29** (1984), 249–253.

[87] J. C. McConnell and J. C. Robson, *Non-commutative Noetherian rings*, John Wiley Interscience, Chichester, New York, Brisbane, Toronto, Tokyo, 1987.

[48] N. H. McCoy, *Rings and Ideals*, Carus Mathematical Monographs, No. 8, Math. Assoc. Amer., Washington, D.C., 1948.

[49] ______, *Prime ideals in general rings*, Amer. J. Math. **71** (1949), 823–833.

[57a] ______, *Finite unions of ideals and groups*, Proc.Amer.Math.Soc. **8** (1957), 633–637.

[57b] ______, *Annihilators in polynomial rings*, Math.Assoc.Amer.Monthly **64** (1957), 28–29.

[64–73] ______, *The Theory of Rings, Reprint of a book originally published by MacMillan in 1964*, Chelsea, New York, 1973.

[50] ______, *See Brown.*

[70] C. Megibben, *Absolutely pure modules*, Proc. Amer.Math.Soc. **26** (1970), 561–566.

[82] ______, *Countable injectives are Σ-injective*, Proc. Amer. Math. Soc. **84** (1982), 8–10.

[87] R. McKenzie, G.F.McNulty, and W. F. Taylor, *Algebras, Lattices, Varieties,* vol. I, Wadsworth & Brooks, Cole Advanced Books, Belmont and Monterey, 1987.

[58] J. E. McLaughlin, *A note on regular group rings*, Mich. Math. J. **5** (1958), 127–128.

[87] G. F. McNulty, *See McKenzie.*

[90] A. H. Mekler, *See Eklof.*

[81] Pere Menal, *On tensor products of algebras being von Neumann regular or self-injective*, Comm. Alg. **9** (1981), 691–697.

[81b] ______, *On π-regular rings whose primitive factor rings are Artinian*, J.Pure and Appl.Alg. **20** (1981), 71–81.

[82] ______, *On the Jacobson radical of algebraic extensions of regular algebras*, Comm. Algebra **10** (1982), 1125–1137.

[82] ______, *On the endomorphism ring of a free module*, Publ. Mat. **27** (1982), 141–154.

[86] ______, *Cancellation Modules over regular rings*, in Ring Theory Proceedings, Granada, 1986; Lecture Notes in Math., vol. 1328, 1988, 187–209.

[84] ______, *See Ara.*

[86] ______, *Cancellation modules over regular rings*, pp.187–204 in Bueso *et al* [88].

[89] ______, *See Herbera.*

[94] ———, *Collected Works*, Soc. Catalana de Matemátiques (M. Castellet, W. Dicks and J. Moncasi, eds.), Barcelona 1994, also distributed by Birkäuser Publ. Ltd., Basel-Boston.

[79] ———, *See Dicks.*

[91] ———, *See Camps.*

[92,94,95] ———, *See Faith.*

[81] P. Menal and J. Moncasi, *Letter to the author, 1981. (See Faith [81e]).*

[84] P. Menal and R. Raphael, *On epimorphism final rings*, Comm. Alg. **12** (1984), 1871–1876.

[84] P. Menal and P. Vámos, *Pure ring extensions and self-FP-injective rings*, Math. Proc. Cambridge Phil. Soc. **105** (1989), 447–458.

[86] C. Menini, *Jacobson's conjecture, Morita duality and related questions*, J. Algebra **103** (1986), 638–655.

[95] ———, *See Facchini.*

[97] ———, *See Ánh.*

[98] ———, *Orsatti's contribution to module theory*, Proc. Padova Conf. 1997 in honor of A. Orsatti, Lecture Notes in Pure and Appl. Math., Marcel Dekker, Basel and New York, 1998.

[91] U. C. Merzbach, *See Boyer.*

[85] A. Mewborn, *Some conditions on commutative semiprime rings*, J. Algebra **13** (1969), 422–431.

[77] K. Meyberg and B. Zimmermann-Huisgen, *Rings with descending chain conditions on certain principal ideals*, Proc. Konink. Nederl. Akad. Weten. **80** (1977), 225–229.

[66] G. O. Michler, *On maximal nilpotent subrings of right Noetherian rings*, Glasgow Math. J. **8** (1966), 89–101.

[73] G. O. Michler and O.E. Villamayor, *On rings whose simple modules are injective*, J. Algebra **25** (1973), 185–201.

[94] S. V. Mihovski, *Linearly independent automorphisms of semiprime rings*, Bulgaria Scientific Works, vol. 31, Book 3, 1994, pp. 61–67.

[94] S. V. Miklovski, *See Mihovski.*

[79] R. W. Miller, *See Teply.*

[65] B. Mitchell, *Theory of Categories*, Academic Press, New York 1965.

[80] J. Mitchell (ed.), *A Community of Scholars, 1930–1980*, Institute for Advanced Study, Princeton, 1980.

[64,65] Y. Miyashita, *On quasi-injective modules, a generalization of completely reducible modules*, J. Fac. Hokkaido U. Ser. I **18** (1964/65), 158–187.

[69] S. H. Mohamed, *See Jain.*

[70a] ———, *q-rings with chain conditions*, J.London Math.Soc. (2) **2** (1970), 455–460.

[70b] ———, *Semilocal q-rings*, Indian J. Pure and Appl.Math. **1** (1970), 419–424.

[70c] ———, *Rings whose homomorphic images are q-rings*, Pac.J.Math. **35** (1970), 727–735.

[89] A. Mohammed and F. L. Sandomierski, *Complements in projective modules*, J. Algebra **127** (1989), 206–217.

[90] S.H. Mohamed and B.J. Müeller, *Continuous and Discrete Modules*, Lecture Notes of London Math.Soc., vol. 147, Cambridge U. Press, Cambridge, New York, Melbourne and Sidney, 1990.

[1893] T. Molien, *Über Systeme köherer complexer Zahlen*, Math. Ann. **41** (1893), 83–165; Berichtigung, **42** (1893), 308–312.

[84] J. Moncasi, *Stable range in regular rings,* Ph.D. Thesis, U. Autonoma Barcelona, 1984.

[89] ______, *See Goodearl.*

[94] ______, *See Menal [81c] and [94].*

[72] G. S. Monk, *A characterization of exchange rings*, Proc. Amer. Math. Soc. **35** (1972), 349–353.

[70] S. Montgomery, *Lie structure of simple ring of characteristic 2*, J. Algebra **15** (1970), 387–407.

[79] ______, *Automorphism groups of rings with no nilpotent elements*, J. Algebra **60** (1979), 238–248.

[80] ______, *Fixed rings of automorphism groups of associative rings*, Lecture Notes in Mathematics, vol. 818, Springer-Verlag, Berlin, Heidelberg, and New York, 1980.

[71] ______, *See Herstein.*

[75] ______, *See Cohen.*

[83] ______, *Von Neumann finiteness of tensor products*, Comm. Algebra **11** (1983), 595–611.

[84] S. Montgomery and D. Passman, *Galois theory of prime rings*, J. Pure and Appl. Alg. **31** (1984), 139–184.

[56] K. Morita, *On group rings over modular fields which possess radicals expressible as principal ideals*, Sci. Rpts., Tokyo Daigaku **4** (1956), 155–172.

[58] ______, *Duality for modules and its applications to the theory of rings with minimum condition*, Sci. Rpts. Tokyo Kyoiku Daigaku **6** (1958), 83–142.

[76] ______, *Localizations of categories,* IV, cited by Onodera [76].

[64,68] B. J. Müller (Also spelled Mueller), *Quasi-Frobenius Erweiterungen* I,II, Math. Z. **85** (1964), 345–368; **88** (1968), 380–409.

[70] ______, *Linear compactness and Morita duality*, J. Algebra **16** (1970), 60–66.

[70b] ______, *On semiperfect rings*, Illinois J. Math. **14** (1970), 464–467.

[73] ______, *All duality theories for linearly topologized modules come from Morita dualities*, Colloq. Math.Soc. Janos Bolyai, vol. 6 (1973), 357–360.

[88] ______, *See Kamal.*

[90] ______, *See Mohamed.*

[86] W. Müller, *See Dischinger.*

[63,64] I. Murase, *On the structure of generalized uniserial rings*, I., Sci Papers of the College of Gen. Edu., U. of Tokyo **13** (1962), 1–22; II, *ibid.* (1963) 131–158; III, *ibid.* **14** (1964), 11–25.

[64] J. Mycielski, *Some compactifications of general algebras*, Colloq. Math. **13** (1964), 1–9.

[70] T. Nagahara, *See Tominaga.*

[52] M. Nagata, *Nilpotency of nil algebras*, J.Math.Soc., Japan **4** (1952), 296–301.

[60] ———, *On the fourteenth problem of Hilbert*, in Proc. Int'l. Congress of Math. (ICM), 1958,pp.459–462, Cambridge U. Press, London, 1960.

[62] ———, *Local Rings, Interscience Tracts in Mathematics, Number 13*, Wiley, New York, 1962.

[39,41] T. Nakayama, *On Frobeniusean algebras* I, II, Ann. of Math. **40** (1939), 611–633, **42** (1941), 1–21.

[40] ———, *Note on uniserial and generalized uniserial rings*, Proc. Imp. Acad. Tokyo **16** (1940), 285–289.

[41b] ———, *Algebras with antiisomorphic left and right ideal lattices*, Proc. Imp. Acad. Tokyo **17** (1941), 53–56.

[49] ———, *Galois Theory for general rings with minimum condition*, J.Math.Soc. Japan **1** (1949), 203–216.

[50b] ———, *On two topics in the structural theory of rings*, in Galois theory and Frobenius algebras, Proc. ICM Vol. II, 1950, pp. 49–54.

[51] ———, *A remark on finitely generated modules*, Nagoya Math. J. **3** (1951), 139–140.

[53] ———, *On the commutativity of division rings*, Canad.J.Math. **5** (1953), 290–292.

[55] ———, *Über die Kommutativität gewisser Ringe*, Abh.Math.Sem., Univ. Hamburg **20** (1955), 20–27.

[56] ———, *See Jans.*

[59] ———, *A remark on the commutativity of algebraic rings*, Nagoya Math.J. **12** (1959), 39–44.

[80] C. Nastasescu, *Theóreme de Hopkins pour les catégories de Grothendieck*, pp. 88–93, in Lecture Notes in Math., vol. 825, Springer-Verlag, Berlin, Heidelberg, New York, 1980.

[68] C. Nastasescu and N. Popescu, *Anneaux semi-Artiniens*, Bull. Math. Soc. France **95** (1968), 357–368.

[95] B. Nation, *See Freese.*

[77] O. M. Neroslovskii, *See Zalesski.*

[46] C. J. Nesbitt and R. M. Thrall, *Some ring theorems with applications to modular representations*, Ann. of Math. (2) **47** (1946), 551–567.

[49] B. H. Neumann, *On ordered division rings*, Trans. Amer. Math. Soc. **66** (1949), 202–252.

[75] ———, *On semiperfect modules*, Canad. Math. Bull. **18** (1975), 77–80.

[77] ———, *Lifting idempotents and exchange rings*, Trans.Amer.Math.Soc. **229** (1977), 269–278.

[95a] W. K. Nicholson and M. F. Yousif, *On a theorem of Camillo*, Comm. Alg. **23** (1995), 5309–5314.

[95b] ———, *Principally injective rings*, J. Algebra **174** (1995), 77–93.

[95c] ———, *Annihilators and the CS-condition*, Glasgow Math. J. **10** (1998), 213–222.

[68] G. Nöbeling, *Verallgemeinerung eines Satzes von Herrn E. Specker*, Inventiones Math. **6** (1968), 41–55.

[21] Emmy Noether, *Idealtheorie in Ringbereichen*, Math. Ann. **83** (1921), 24–66.

[26] ———, *Abstracter Aufbau der Idealtheorie*, in Algebraischen Zahl und Funktionen Körpern, Math. Ann. **96** (1926).

[29] ______, *Hyperkomplexe Grössen und Darstellungstheorie*, Math. Z. **30** (1929), 641–692.

[32] ______, *See Brauer.*

[33] ______, *Nichtkommutative Algebren*, Math. Z. **37** (1933), 514–541.

[83] ______, *Collected Papers* (N. Jacobson, ed.), Springer-Verlag, Berlin, Heidelberg, New York, 1983.

[60] D. G. Northcott, *Homological Algebra*, Cambridge U. Press, Cambridge, 1960.

[85] N. C. Norton, *See Hajarnavis.*

[62] R. J. Nunke, *On direct products of infinite cyclic groups*, Proc. Amer. Math. Soc. **13** (1962), 66–71.

[71] U. Oberst and H. J. Schneider, *Dre Struktur von projectiven Moduln*, Invent. Math. **13** (1971), 295–304.

[69] J. Ohm, *Semi-valuations and groups of divisibility*, Canad. J. Math. **21** (1969), 576–591.

[68] J. Ohm and R. L. Pendleton, *Rings with Noetherian spectrum*, Duke J. Math. **35** (1968), 631–640.

[75] K. C. O'Meara, *Right orders in full linear rings*, Trans. Amer. Math. Soc. **203** (1975), 299–318.

[97] ______, *See Ara.*

[76] J. D. O'Neill, *Rings whose additive subgroups are subrings*, Pac. J. Math. **66** (1976), 509–522.

[77] ______, *Survey of rings whose additive subgroups are subrings or ideals*, pp.161–167 in Jain [77].

[84] ______, *Noetherian rings with free additive groups*, Proc. Amer. Math. Soc. **92** (1984), 323–324.

[87] ______, *A theorem on direct products of slender modules*, Rend. Sem. Mat. Univ. Padova **78** (1987), 261–266.

[90] ______, *Direct summands of vector groups*, Acta. Math. Hung. **55** (1990), 207–209.

[91] ______, *An unusual ring*, J. London Math. Soc. **44** (1991), 95–101.

[92] ______, *Examples of non-finitely generated projective modules*, in Methods in Module Theory, Lecture Notes in Pure and Appl. Math., vol. 140, Marcel Dekker, Basel and New York, 1992, pp. 271–278.

[93a] ______, *When an infinite direct product of modules is a free module of finite rank*, Communications in Algebra **21** (1993), 3829–3837.

[93b] ______, *When a ring is an F-ring*, Journal of Algebra **156** (1993), 250–258.

[94] ______, *Direct summands of* $\mathbb{Z}^\kappa$ *for large* κ, Contemp. Math. **171** (1994), 313–323.

[95] ______, *A result on direct products of copies of the integers*, Comm. Alg. **23** (1995), 4925–4930.

[96a] ______, *A new proof of a theorem of Balcerzyk, Bialynicki-Birula and Łoś*, Colloq. Math. LXX (1996), 191–194.

[96b] ______, *Linear algebra over various rings*, International J. of Math. Educ. Sci. Technol. **27** (1996), 561–564.

[97] ______, *On infinite direct products of copies of a Dedekind domain*, Journal of Algebra **192** (1997), 701-712.

[68] T. Onodera, *Über Kogeneratoren*, Arch. Math. **19** (1968), 402–410.

[68B] ———, *On modules flat over their endomorphism rings*, Hokkaido Math. J. **7** (1978), 179–182.

[72] ———, *Linearly compact modules and cogenerators*, J. Fac. Sci, Hokkaido U. Ser. I **22** (1972), 116–125.

[73] ———, *Koendlich erzeugte Moduln und Kogeneratoren*, Hokkaaido Math. J., II (1973), 69–83.

[76] ———, *On balanced projectives and injectives over linearly compact rings*, Hokkaido Math. J. **5** (1976), 249–256.

[81] ———, *On a theorem of W. Zimmermann*, Hokkaido Math. J. **10** (1981), 564–567.

[51] T. Onodera and H. Tominaga, *On strictly Galois extensions of primary rings*, J. Fac. Hokkaido U. Ser. I. **16** (1961), 193–194.

[31] O. Ore, *Linear equations in non-commutative fields*, Anns. of Math. **32** (1931), 463–477.

[33a] ———, *Theory of non-commutative polynomials*, Anns. of Math. **34** (1933), 480–508.

[33b] ———, *On a special class of polynomials*, Trans.Amer.Math.Soc. **35** (1933), 559–584.

[67] A.J. Ornstein, *Rings with restricted minimum conditions*, Rutgers Ph.D. Thesis 1967.

[68] ———, *Rings with restricted minimum conditions*, Proc. Amer. Math. Soc. **19** (1968), 1145–1150.

[81] A. Orsatti and V. Roselli, *A characterization of discrete linearly compact rings by means of a duality*, Rend.Sem.Math., Univ. Padova **64** (1981), 29–43.

[84] A. Orsatti, *See Dikranjan.*

[87] K. Oshiro, *Structure of Nakayama rings*, in Proc. of the 20th Symposium in Ring Theorey, pp. 109–133, Okayama U., 1987.

[96] K. Oshiro and S. T. Rizvi, *The exchange property of quasi-continuous modules with the finite exchange property*, Osaka J. Math. **33** (1996), 217–234.

[64a] B. L. Osofsky, *Rings all of whose finitely generated modules are injective*, Pac. J. Math. **14** (1964), 646–650.

[64b] ———, *On ring properties of injective hulls*, Canadian Math. Bull. **7** (1964), 405–413.

[65] ———, *A counter-example to a lemma of Skornyakov*, Pac. J. Math. **15** (1965), 985–987.

[66] ———, *A generalization of Quasi-Frobenius rings*, J. Algebra **4** (1966), 373–387.

[66b] ———, *Cyclic injective modules of full linear rings*, Proc. A.M.S. **17** (1966), 247–253.

[67] ———, *A non-trivial ring with non-rational injective hull*, Canad. Math. Bull. **10** (1967), 275–282.

[68a] ———, *Endomorphism rings of quasi-injective modules*, Canad. J. Math. **20** (1968), 895–903.

[68b] ———, *Noncommutative rings whose cyclic modules have cyclic injective hulls*, Pac. J. Math. **25** (1968), 331–340.

[68c] ———, *Homological dimension and the continuum hypothesis*, Trans.Amer.Math.Soc. **132** (1968), 217–230.

[69] ———, *Commutative local rings with finite global dimension and zero divisors*, Trans. Amer. Math. Soc. **141** (1969), 377–385.

[69b] ———, *Global dimension of commutative rings with linearly ordered ideals*, J. London Math. Soc. **44** (1969), 183–185.

[70] ______, *A remark on the Krull-Schmidt-Azumaya theorem*, Canad. Math. Bull. **13** (1970), 501.

[71] ______, *On twisted polynomial rings*, J. Algebra **18** (1971), 597–607.

[71b] ______, *Loewy length of perfect rings*, Proc. A.M.S. **28** (1971), 352–354.

[73] ______, *Homological Dimensions of Modules*, Amer. Math. Soc., Providence, 1973.

[74] ______, *The subscript of* $\aleph_n$, *projective dimension and the vanishing of* $\lim^{(n)}$, Bull. A.M.S. **80** (1974), 8–26.

[78] ______, *Projective dimension of "nice" directed unions*, J. of Pure and Appl. Algebra **13** (1978), 179–219.

[84] ______, *A semiperfect one-sided injective ring*, Comm. alg. **12** (1984), 2037–2041.

[91] ______, *Minimal cogenerators need not be unique*, Comm. Alg. (7) **19** (1991), 2072–2080.

[91b] ______, *A construction of nonstandard uniserial modules over valuation domains*, Bull. A.M.S. (1991).

[92] ______, *Chain conditions on essential submodules*, Proc. A.M.S. **114** (1992), 11–19.

[91] B. L. Osofsky and P. F. Smith, *Cyclic modules whose quotients have all complement submodules direct summands*, J. Algebra **139** (1991), 342–354.

[78] J. Osterburg, *See Fisher.*

[80] ______, *See Goursaud.*

[35] A. Ostrowski, *Untersuchungen zur arithmetischen Theorie der Körper*, Math. Zeit **39** (1935), 269–404.

[84] A. Page, *On the centre of hereditary P.I. rings*, J. London Math. Soc. (2) **30** (1984), 193–196.

[78] S. Page, *Regular FPF rings*, Pac. J. Math. **79** (1978), 169–176; corrections and Addendum **97** (1981), 488–490.

[82] ______, *Semiprime and non-singular FPF rings*, Comm. Alg. **11** (1982), 2253–2259.

[83] ______, *FPF and some conjectures of C. Faith*, Canad. Math. Bull. **26** (1983) 257–259.

[84] ______, *See Faith.*

[80] I. Papick, *See Huckaba.*

[96] ______, *See Fontana.*

[59] Z. Papp, *On algebraically closed modules*, Pub. Math., Debrecen (1958), 311-327.

[97] E. Pardo, *See Ara.*

[93] J. K. Park, *See Hirano.*

[97] ______, *See Birkenmeier.*

[83] K. H. Parshall, *In search of the finite division algebra theorem, and beyond: Joseph H. M. Wedderburn, Leonard Dickson and Oswald Veblen*, Archives Internationales d'histoire des sciences **35** (1983), 274–299.

[84] ______, *E. H. Moore and the founding of a mathematical community in America*, Annals of Science **41** (1984), 313–333, reprinted in Duren [88], Part II.

[85] ______, *Joseph H. M. Wedderburn and the structure theory of algebras*, Archiv for History of Exact Sciences **32** (1985), 223–249.

[79] J. L. Pascaud and J. Valette, *Group actions on QF-rings*, Proc. Amer. Math. Soc. **76** (1979), 43–44.

[80] J. L. Pascaud, *See Goursaud.*

[77] D. S. Passman, *The Algebraic Structure of Group Rings*, Wiley-Interscience, New York, London, Sydney, Toronto, 1977.

[97] ———, *Semiprimitivity of group algebras: past results and recent progress*, in Trends Ring Theory, Conf. Proc. Canadian Math. Soc., vol. 22, American Math. Soc., Providence, 1997.

[98] ———, *Semiprimitive of group algebras*, pp.199–212 in Drensky *et al* [98].

[84] ———, *See Montgomery.*

[68] R. L. Pendleton, *See Ohm.*

[92] C. Perelló (Ed.), *Pere Menal Memorial Volume*, (M. Castellet, W. Dicks, and J. Moncasi, eds.), Publ. Mat. **36** (1992).

[42] S. Perlis, *A characterization of the radical of an algebra*, Bull. A.M.S. **48** (1942), 128–132.

[50] S. Perlis and G. L. Walker, *Abelian group algebras of finite order*, Trans. A.M.S. **68** (1950), 420–426.

[67] R. S. Pierce, *Modules over Commutative Regular Rings*, Memoirs of the Amer. Math. Soc., vol. 70, Providence, 1967.

[71] ———, *The submodule lattice of a cyclic module*, Algebra Universalis **1** (1971), 192–199.

[80] P. Pillay, *On semihereditary commutative polynomial rings*, Proc. Amer. Math. Soc. **78** (1980), 473–474.

[84] ———, *Polynomial rings over non-commutative rings*, Pub. Mat. (UAB) **203** (1984), 24–49.

[90] ———, *See Faith.*

[93] ———, *See Herbera.*

[84] G. Pilz, *See Lidl.*

[39] L. Pontryagin, *Topological Groups*, Princeton University Press, Princeton, 1939.

[42] ———, *Über stetische algebraischen Körper*, Ann. of Math. **33** (1942), 163–174.

[84] N. Popescu, *On a class of Prüfer domains*, Rev. Roumaine Pures Appl. **29** (1984), 777–786.

[68] ———, *See Nastasecu.*

[95] ———, *See Fontana.*

[64] N. Popescu and P. Gabriel, *Caractérisations des catégories abeliennes avec générateurs et limites inductives exactes*, C. R. Acad. Sci. Paris **258** (1964), 4188–4190.

[60] E. Posner, *Prime rings satisfying a polynomial identity*, Proc.Amer.Math.Soc. **11** (1960), 180–184.

[88] M. Prest, *Model Theory and Modules*, London Math. Soc. Lecture Notes, vol. 130, Cambridge U., Cambridge, 1988.

[61] G. B. Preston, *See Clifford.*

[63] C. Procesi, *On a theorem of Goldie concerning the structure of prime rings with maximal condition (in Italina)*, Acad. Naz. Lincei Rend. **34** (1963), 372–377.

[65] ———, *On a theorem of Faith and Utumi (in Italian)*, Rend. Math. e. appl. **24** (1965), 346–347.

[65] C. Procesi and L. Small, *On a theorem of Goldie*, J. Algebra **2** (1965), 80–84.

[66] ______, *See Amitsur.*

[71] ______, *On a theorem of M. Artin*, J. Algebra **22** (1972), 309–315.

[73] ______, *Rings with Polynomial Identities*, Marcel Dekker, Basel and New York, 1973.

[23] H. Prüfer, *Untersuchungen über die Zerlegbarkeit der abzählbaren primären abelschen Gruppen*, Math. Z. **17** (1923), 35–61.

[95] G. Puninski, R. Wisbauer and M. Yousif, *On p-injective rings*, Glasgow Math. J. **37** (1995), 373–378.

[96] G. Puninski and Wisbauer, $\sum$*-injective modules over left duo and left distributive rings*, J. Pure and Appl. Algebra **113** (1996), 55–66.

[95,96] G. G. Puninski, *See Facchini.*

[96] D. Pusat-Yilmaz and P. F. Smith, *Chain conditions in modules with Krull dimension*, Comm. Algebra **24** (1996), 4123–4133.

[71] Y. Quentel, *Sur la compacité du spectre minimal d'un anneaux*, Bull. Soc. Math. France **99** (1971), 265–272; Erratum, *ibid* **100** (1972), 461.

[62] F. Quigley, *Maximal subfields of an algebraically closed field not containing a given element*, Proc. A.M.S. **13** (1962), 562–566.

[76] D. Quillen, *Projective modules over polynomial rings*, Invent.Math. **36** (1976), 167–171.

[73] V. S. Ramamurthi and K. M. Rangaswamy, *On finitely injective modules*, J. Austral. Math. Soc. **16** (1973), 239–248.

[74] M. Ramras, *Orders with finite global dimension*, Pac. J. Math. **50** (1974), 583–587.

[73] K.M. Rangaswamy, *Abelian groups with self-injective endomorphism ring*, Proc. Second Internat. Conference, Theory of Groups, Springer, Berlin-Heidelberg-New York 1973, pp. 595–604.

[71] R. Raphael, *Rings of quotients and* π*-regularity*, Pac.J.Math. **39** (1971), 229–233.

[92] ______, *On algebraic closures*, Publ. Math. **36** (1992), 913–923.

[84] ______, *See Menal.*

[72] L. Ratliff, *On prime divisors of the integral closure of a principal ideal*, J. reine angew. Math. **255** (1972), 210–220.

[71] M. Raynaud and L. Gruson, *Critères de platitude et de projectivité*, Invent.math. **13** (1971), 1–89.

[73] J. P. Razmyslov, *A certain problem of Kaplansky*, Izv. Akad. Nauk SSSR Ser.Mat. **37** (1973), 483–501; Russian.

[72] A. Regev, *Existence of identities in* $A \otimes B$, Israel J. Math. **11** (1972), 131–152.

[87] E. Regis, *Who Got Einstein's Office?*, Addison-Wesley, Reading, MA, New York, 1987.

[70] C. Reid, *Hilbert*, Springer-Verlag, New York, Heidelberg and Berlin, 1970.

[61] I. Reiner, *The Krull-Schmidt theorem for integral representations*, Bull. Amer. Math. Soc. **67** (1961), 365–367.

[75] ______, *Maximal Orders*, Academic Press, New York 1975.

[62] ______, *See Curtis.*

[70] C. M. Reis and T. M. Viswanathan, *A compactness property for prime ideals in Noetherian rings*, PAMS **25** (1970), 353–356.

[75] I. Reiten, *see Fossum; also Fuller.*

[67] G. Renault, *Sur les anneaux A, tels que tout A-module a gauche non nul continient un sous-module maximal*, C. R. Acad. Sci. Paris Sér. A **264** (1967), 622–624.

[73] ———, *Sur les anneaux des groupes*, in Rings, Modules and Radicals (Proc. Colloq. Keszthely, 1971), Colloq. Math. Soc. Janos Bolyai, vol. 6, North Holland, Amsterdam, 1973, pp. 391–396.

[70] ———, *See Cailleau.*

[67] R. Rentschler and P. Gabriel, *Sur la dimension des anneaux et ensembles ordonnés*, C. R. Acad. Sci. Paris, Sér. A **265** (1967), 712–715.

[79] R. Resco, *Transcendental division algebras and simple Noetherian rings*, Israel J. Math. **32** (1979), 236–256.

[80] ———, *A dimension theorem for division rings*, Istrael J. Math. **35** (1980), 215–221.

[87] ———, *Division rings and V-domains*, Proc.Amer.Math.soc. **99** (1987), 427–431.

[69] P. Ribenboim, *Rings and Modules*, Wiley-Interscience, New York, London, Sydney, Toronto, 1969.

[97] ———, *Semisimple rings and von Neumann regular rings of generalized power series*, J. Algebra (1997).

[90] ———, *See Elliott.*

[65] F. Richmond, *Generalized quotient rings*, Proc. Amer. Math. Soc. **16** (1965), 794–799.

[65] M. Rieffel, *A general Wedderburn theorem*, Proc.Nat.Acad.Sci. USA **54** (1965), 1513.

[62] G. S. Rinehart, *Note on the global dimension of a certain ring*, Proc. Amer. Math. Soc. **13** (1962), 341–346.

[76] G. S. Rinehart and A. Rosenberg, *The global dimension of Ore extensions and Weyl algebras*, in Algebra, Topology and Category Theory, pp. 169–180, Academic Press, New York, 1976.

[68] F. Ringdal, *See Amdal.*

[88] S. T. Rizvi, *Commutative rings for which every continuous module is quasi-injective*, Arch. Math. **50** (1988), 435–442.

[90] S. T. Rizvi and M. Yousif, *On continuous and singular modules*, in Non-Commutative Ring Theory, Proc. Athens, 1989, Lecture Notes in Math. vol. 1448,pp.116–124, Springer, Berlin, New York and Heidelberg, 1990.

[90,93,97] ———, *See S. K. Jain.*

[96] S. T. Rizvi, *See Huynh; also Oshiro.*

[97] ———, *See Huynh.*

[51] A. Robinson, *On the Metamathematics of Algebra*, North Holland, Amsterdam, 1951.

[63] ———, *Introduction to Model Theory and the Metamathematics of Algebra*, North Holland, Amsterdam, 1963.

[67a] J. C. Robson, *Artinian quotient rings*, Proc. Lond. Math. Soc. **17** (1967), 600–616.

[67b] ———, *Pri rings and Ipri rings*, Quarterly J. Math. **18** (1967), 125–145.

[74] ———, *Decompositions of Noetherian rings*, Comm. Algebra **4** (1974), 345–349.

[74] J. C. Robson and L. W.Small, *Hereditary P.I. rings are classical hereditary orders*, J. London Math. Soc. (2) **8** (1974), 499-503.

[70a] J. C. Robson, *See Eisenbud.*

[73] ______, *See Gordon.*

[87] ______, *See McConnell.*

[87] K. Roggenkamp and L. Scott, *Isomorphisms of p-adic group rings*, Ann. Math. **126** (1987), 593–647.

[66] H. Röhrl, *See Eilenberg, etal, (eds.).*

[89] M. Roitman, *On Mori domains and commutative rings with* $CC^{\perp}$, I, J. Pure and Appl. Algebra **56** (1989), 247–268; II, Ibid. **61** (1989) 53–77.

[72] J. E. Roos, *Detérmination de la dimension homologique globale des algèbres de Weyl*, C. R. Acad. Sci. Paris Sér. A-B **274** (1972), A23–A26.

[73] J. E. Roseblade, *Group rings of polycyclic groups*, J.Pure and Applied Algebra **3** (1973), 307–328.

[81] V. Roselli, *See Orsatti.*

[56] A. Rosenberg, *The Cartan-Brauer-Hua Theorem for matrix and locally matrix rings*, Proc. Amer. Math. Soc. **7** (1956), 891–898.

[57] ______, *See Eilenberg.*

[65] ______, *See Chase.*

[76] ______, *See Rinehart.*

[59] A. Rosenberg and D. Zelinsky, *On the finiteness of the injective hull*, Math. Z. **760** (1959), 372–380.

[61] ______, *Annihilators*, Portugalia Math. **20** (1961), 53–65.

[76] A. Rosenberg and J. T. Stafford, *Global dimension of Ore extensions*, in Algebra, Topology and Category Theory, (pp.181–188), Academic Press, New York, 1976.

[82] J. Rosenstein, *Linear Orderings*, Academic Press, New York, 1992.

[89] G. C. Rota (ed.), *See Jacobson.*

[88] L. H. Rowen, *Ring Theory, Volume I*, Academic Press, New York, 1988.

[89] ______, *On Koethe's conjecture*, Israel Math. Conf. Proc., vol. 1, U. of Jerusalem, 1989.

[73] R. A. Rubin, *Absolutely torsion-free rings*, Pac. J. Math. **46** (1973), 503–514.

[69] E. A. Rutter, *Two characterizations of quasi-Frobenius rings*, J. Algebra **4** (1966), 777–784.

[74] ______, *A Characterization of QF-3 rings*, Comm. Alg. **3** (1974), 533–536.

[75] ______, *Rings with the principal extension property*, Comm. Algebra **3** (1975), 203–212.

[71] ______, *See Colby.*

[70] G. Sabbagh, *Aspects logiques de la pureté dans les modules*, C. R. Acad. Sci. Paris Sér. A-B **271** (1970), A909–A912.

[54,56] Šafarevič, *See Shafarevitch.*

[69] I. I. Sahaev, *Rings over which every finitely generated flat module is projective*, Isv. Vyssh. Uchebn. Zaved. Mat. **9** (1969), 65–73.

[77] ______, *Finite generation of projective modules*, Isv. Vyssh. Uchebn. Zaved. Mat. **184** (1977), 69–79, [Russian].

[70] Y. Sai, *See Harada.*

[85] L. Salce, *See Fuchs.*

[90] ———, *See Facchini.*

[98] ———, *See Fuchs.*

[98] ———, *See Dikranjan.*

[81] L. Salce and P. Zanardo, *On a paper of I. Fleischer*, in Abelian Group Theory, Lecture Notes in Math. vol. 874, Springer-Verlag, Boston, Heidelberg and Berlin, 1981, pp. 76–86.

[93] M. A. Saleh, *See Jain.*

[82] J. Sally, *See Srinivasan.*

[57] P. Samuel, *La notion de la place dans un anneau*, Bull. Soc. Math. France **85** (1957), 123–133.

[58-60] ———, *See Zariski.*

[67] F. L. Sandomierski, *Semisimple maximal quotient rings*, Trans. Amer. Math. Soc. **128** (1967), 112–120.

[68] ———, *Nonsingular rings*, Proc. A.M.S. (1968), 225–230.

[72] ———, *Linearly compact modules and local Morita duality*, Ring Theory, Academic Press, New York, 1972.

[77] ———, *Classical localizations at prime ideals of fully bounded Noetherian rings*, in Ring Theory (Prof. Conf. Ohio U. Athens, 1976); Lecture Notes in Pure and Appl. Math. vol. 25, 1977, 169–181.

[68] ———, *See Cateforis.*

[89] ———, *See A. Mohammed.*

[96] C. Santa-Clara and P. F. Smith, *Extending modules which are direct sums of injective modules and semisimple modules*, Comm. Alg. **24** (1996), 3641–3651.

[74] B. Sarath and K. Varadarajan, *Injectivity in direct sums*, Comm. Alg. **1** (1974), 517–530.

[61] E. Sasiada, *Solution of a problem of the existence of a simple radical ring*, Bull. Acad. Polon. Sci **9** (1961), 25F.

[67] ———, *See Cohn [67].*

[80] A. Sathaye, *See Brewer.*

[50] O.F.G. Schilling, *The Theory of Valuations*, Math. Surveys No. 4, Amer.Math.Soc., Providence, R.I., 1950, R.I. 1957.

[28] O. Schmidt, *Über unendliche Gruppen mit endlicher Kette*, Math. Z. **29** (1928), 34–41.

[95] S. E. Schmidt, *Grundlagen Zu Einer Algemeinen Affinen Geometrie*, Birkhäuser, Basel, Boston, Berlin, 1975.

[71] H. Schneider, *See Oberst.*

[85a] A. H. Schofield, *Artin's problem for skew fields*, Math.Proc. Cambridge Phil.Soc. **97** (1985), 1–6.

[85b] ———, *Representatiions over Skew Fields*, Lecture Notes of London Math.Soc., vol. 92, Cambridge U. Press, Cambridge, New York, Melbourne and Sidney, 1985.

[67] ———, *See Cohn.*

[37] A. Scholz, *Konstruktion algebraischer Zahlkörper mit beliebiger Gruppe von Primzahlpotenzordnung*, I, Math. Z. **42** (1937), 161–188.

[28] O. Schreier, *Über den Jordan-Hölderschen Satz.*, Abh. Math. Sem., Univ. Hamburg **6** (1928), 300–302.

[26] ———, *See Artin.*

[04] I. Schur, *Über die Darstellung der endlichen Gruppen durch gebrochene lineare Substitutionem*, J. Reine u. Angew. Math. **127** (1904), 20–50.

[93] S. K. Sehgal, *Units in integral group rings, Pitman Surveys*, in Pure and Appl. Math., vol. 69, Longman, Harlow, 1993.

[98] ———, *(ed.) See Drensky.*

[47] H. Seifert and W. Threlfall, *Lehrbuch der Topologie*, Leipzig 1947; reprint: Chelsea, New York, 196–.

[55] J.P. Serre, *Sur la dimension homologique des anneaux et des modules Noethériens*, Proc. Internat. Sympos. Algebraic Number Theory, Tokyo 1955.

[58] C. Seshadri, *Triviality of vector bundles over the affine space* K^2, Proc. Nat. Acad. Sci. USA **44** (1958), 456–458.

[54,56] I. R. Shafarevitch, *Construction of fields of algebraic numbers with given solvable Galois group*, Izv. Akad. Nauk. SSSR Ser. Mat. **18** (1954), 389–418; Amer. Math. Soc. Transl. **4** (1956), 151–183 (Shafarevich is a variant transliteration in English).

[64] ———, *See Golod [64].*

[91] A. Shamsuddin, *Minimal pure epimorphisms*, Comm. Algebra **19** (1991), 325–331.

[98] ———, *Rings with Krull dimension 1*, Communications in Algebra (1998).

[84] ———, *See Hanna.*

[95] ———, *See Herbera.*

[77] R. Y. Sharp and P. Vámos, *The dimension of the tensor product of two field extensions*, Bull. L.M.S. **9** (1977), 42–48.

[85] ———, *Baire category theorem and prime avoidance in complete local rings*, Arch. Math. **44** (1985), 243–248.

[71] D.W. Sharpe and P. Vámos, *Injective Modules*, Cambridge Univ. Press, Cambridge 1971.

[51] J. C. Shepherdson, *Inverses and zero-divisors in matrix rings*, Proc. L.M.S. (3) **1** (1951), 71–85.

[91] P. Shizhong, *Commutative quasi-Frobenius rings*, Comm. Alg. 19 **2** (1991), 663–667.

[71a] R. C. Shock, *Nil subrings in finiteness conditions*, Amer. Math. Monthly **78** (1971), 741–748.

[71b] ———, *Essentially nilpotent rings*, Israel J. Math. **9** (1971), 180–185.

[72] ———, *Polynomial rings over finite dimensional rings*, Pac. J. Math. **42** (1972), 251–258.

[74] ———, *Dual generalizations of Artinian and Noetherian conditions*, Pac. J. Math. **54** (1974), 227–235.

[32-33] K. Shoda, *Uber die Galoissche Theorie der halbeinfachen hyperkomplexen systeme*, Math.Ann., vol. 107 (1932–1933), 252–258.

[71] T.S. Shores, *Decompositions of finitely generated modules*, Proc. Amer. math. Soc. **30** (1971), 445–450.

[74] T. S. Shores and W. Lewis, *Uniserial modules and endomorphism rings*, Duke Math. J. **41** (1974), 889–909.

[73] T.S. Shores and R. Wiegand, *Decompositions of modules and matrices*, Bull. Amer. Math. Soc. **79** (1973), 1277–1280.

[74] ———, *Rings whose finitely generated modules are direct sums of cyclics*, J. Algebra **32** (1974), 57–72.

[74b] ———, *On generalized valuations*, Michigan Math. J. **21** (1974), 405–409.

[96] K. P. Shum, *See X. H. Cao.*

[67] J. Silver, *Noncommutative localization and applications*, J.Alg. **7** (1967), 44–76.

[72] ———, *See Eakin.*

[72] D. Simson, *On the structure of flat modules*, Bull. Acad. Polonaise Sci. (2) **20** (1972), 115–120.

[97] Simon Singh, *Fermat's Enigma*, Walker and Company, New York, 1997.

[75–76] Surjeet Singh, *Modules over hereditary Noetherian prime rings*, Can. J. Math. **27** (1975), 867–883; II, *ibid.* **28** (1976), 73–82.

[76b] ———, *Some decomposition theorems in Abelian groups and their generalizations*, Lecture Notes in Math **25** (1976), 183–189.

[82] ———, *On a Warfield theorem on hereditary rings*, Arch. der Math. **39** (1982), 306–311.

[84] ———, *Serial right Noetherian rings*, Can. J. Math, XXXVI (1984), 22–37.

[69,75,92] ———, *See Jain.*

[60] L. A. Skornyakov, *Modules with self-dual lattice of submodules* (Russian), Sibirsk Mat. Z. **1** (1960), 238–241.

[65] ———, *Einfache bikompakte Ringe*, Math. Z. **87** (1965), 241–251.

[69] ———, *When are all modules semi-chained* (Russian), Mat. Zametki **5** (1969), 173–182.

[76] ———, *Decomposability of modules into a direct sum of ideals* (Russian), Mat. Zametki **20** (1976), 187–193.

[77] A. B. Slomson, *See Bell.*

[66a] L. W. Small, *Hereditary rings*, Proc. Nat. Acad. Sci. U.S.A. **55** (1966), 25–27.

[66b] ———, *Some questions in Noetherian rings*, Bull. A.M.S. **72** (1966), 853–857.

[66-68] ———, *Orders in Artinian rings*, J. Algebra **4** (1966), 13-41; Addendum 505-507; II *ibid.* **9** (1968) 206-273.

[69] ———, *The embedding problem for Noetherian rings*, Bull. Amer. Math. Soc. **75** (1969), 147–148.

[71] ———, *Localization in PI-rings*, J. Algebra **18** (1971), 269–270.

[73] ———, *Prime ideals in Noetherian PI-rings*, Bull. Amer. Math. Soc. **79** (1973), 421–422.

[64,66] ———, *See Herstein.*

[65] ———, *See Procesi.*

[73] ———, *See Goldie.*

[87,90] ———, *See Blair.*

[96] ———, *See Amitsur.*

[81,85] ——— (ed.), *Reviews in Ring Theory, 1940–1979, 1980–1984*, Amer. Math. Soc., Providence, 1981, 1985.

[81] L. W. Small and A. R. Wadsworth, *Some examples of rings*, Comm. Algebra **9** (1981), 1105–1118.

[1861] H.J.S. Smith, *Phil. Trans. of the Royal Soc. of London* **151** (1861), 293–326.

[79] P. F. Smith, *Rings characterized by their cyclic submodules*, Canad. Math. J. XXXI (1979), 93–111.

[81] ______, *The injective test in fully bounded rings*, Comm. Alg. **9** (1981), 1701–1708.

[90] ______, *CS-modules and weak CS-modules*, in Lecture Notes in Math., vol. 1448, 1990, pp. 99–115.

[97] ______, *Nonsingular extending modules*, in Advances in Ring Theory (Jain and Rizvi, eds), Birkhäuser, Boston, Basel, Berlin, 1997.

[82] ______, *See Levy.*

[89.90] ______, *See Huynh.*

[91] ______, *See Osofsky.*

[92] ______, *See Dung.*

[96] ______, *See Pusat-Yilmaz; also Santa Clara.*

[97,98] ______, *See Albu.*

[71] W. W. Smith, *A covering condition for prime ideals*, PAMS **30** (1971), 451–2.

[74] R. L. Snider, *Group algebras whose simple modules are finite dimensional over their commuting rings*, Comm. Algebra **2** (1974), 15–25.

[76] ______, *The singular ideal of a group algebra*, Comm. Algebra **4** (1976), 1087–1089.

[88] ______, *Noncommutative regular local rings of global dimension* 3, Proc. A.M.S. **104** (1988), 49–50.

[72] ______, *See Formanek.*

[73] ______, *See Farkas.*

[94] R. Solomon, *See Gorenstein.*

[95] ______, *On finite simple groups and their classification*, Notices AMS **42** (1995), 231–239.

[78] E. D. Sontag, *See Dicks.*

[50] E. Specker, *Additive Gruppen von Folgen ganzer Zahlen*, Portugaliae Math. **9** (1950), 131–140.

[60] B. Srinivasan, *On the indecomposable representations of a certain class of groups*, Proc.London Math.soc. **10** (1960), 4970–513.

[82] B. Srinivasan and J. Sally (eds.), *Emmy Noether in Bryn Mawr*, Springer-Verlag, Berlin, New York, 1982.

[77] J. T. Stafford, *Stable structure of non-commutative Noetherian rings*, J. Algebra **47** (1977), 244–267.

[78] ______, *A simple Noetherian ring not Morita equivalent to a domain*, Proc. Amer. Math. Soc. **68** (1978), 159–160.

[79] ______, *Morita equivalence of simple Noetherian rings*, Proc.Amer.Math.Soc. **74** (1979), 212–214.

[82] ______, *Noetherian full quotient rings*, Proc. London Math. Soc. **44** (1982), 385–404.

[76] ______, *See Rosenberg.*

[52] N. Steenrod, *See Eilenberg.*

[74] S. A. Steinberg, *See Armendariz.*

[10,50] E. Steinitz, *Algebraische Theorie der Körper*, J. Rein U. angew. Math., 1910; reprint, Chelsea, New York, 1950.

[11,12] ———, *Math.Am.* **71** (1911), 328–354; **72** (1912), 297–345.

[70] B. Stenström, *Coherent rings, and FP-injective modules*, J. Lond. Math. Soc. **2** (1970), 323–329.

[75] ———, *Rings of quotients* in Grundl. der Math. Wiss., Vol. 217, Springer, Berlin-Heidelberg-New York 1975.

[69] W. Stephenson, *Lattice isomorphisms between modules*, J. London Math. Soc. **2** (1969), 177–183.

[1878] L. Stickelberger, *See Frobenius.*

[86] J. Stock, *On rings whose projective modules have the exchange property*, J. Algebra **103** (1986), 437–453.

[81] D. R. Stone, *See Cheatham.*

[68] H. H. Storrer, *Epimorphisinen von kommutativen Ringen*, Comm.Math.Helv. **43** (1968), 378–401.

[73] ———, *Epimorphic extensions of non-commutative rings*, Comm. Math.Helv. **48** (1973), 72–86.

[87] D. J. Struik, *A Concise History of Mathematics,* 4th rev.ed., Dover Publs., New York, 1987.

[76] A. A. Suslin, *On the structure of the general linear group over polynomial rings*, Mat.Sbornik **135** (1974), 588–595.

[77] ———, *Projective modules over polynomial rings are free*, Dokl.Nauk. S.S.S.R **229** (1976), (= Soviet Math. Doklady **17** (1976), 1160–1164).

[60] R. G. Swan, *Induced representations and projective modules*, Ann. of Math. **71** (1962), 55–61.

[62] ———, *Projective modules over group rings and maximal orders*, Ann. of Math. **76** (1962), 55–61.

[69] ———, *Invariant rational functions and a problem of Steenrod*, Invent. Math. **7** (1969), 148–158.

[68] R. G. Swan and E. G. Evans, *Algebraic K-theory*, Lecture Notes in Math., vol. 76, Springer, Berlin-Heidelberg-New York, 1968.

[81] F. A. Szász, *Radical of Rings*, Wiley Interscience, John Wiley & Sons, New York, 1981.

[56] T. Szele and L. Fuchs, *On Artinian rings*, Acta. Sc. Math. Szeged **17** (1956), 30–40.

[69] H. Tachikawa, *A generalization of quasi-Frobenius rings*, Proc. Amer. Math. Soc. **20** (1969), 471–476.

[73] ———, *Quasi-Frobenius Rings and Generalizations of $QF-3$ and $QF-1$ Rings*, in Lecture Notes in Mathematics, vol. 351, Springer, Berlin,Heidelberg,New York, 1973.

[57] E. J. Taft, *Invariant Wedderburn factors*, Illinois J. Math. **1** (1957), 565–573.

[64] ———, *Orthogonal conjugacies in associative and Lie algebras*, Trans.Amer.Math.Soc. (1964), 18–29.

[68] ———, *Cohomology of groups of algebra automorphisms*, J. Algebra **10** (1968).

[49] A. Tarski, *Arithmetical classes and types of algebraically closed and real closed fields*, Bull. A.M.S. **54** (1949), 64.

[56] ______, *Logic Semantics, Metamathematics, Papers from 1923-1938*, Transl. by J. H. Woodger, Oxford U. Press, 1956.

[50] J. T. Tate, *See Artin.*

[65] ______, *See Artin.*

[87] W. F. Taylor, *See McKenzie.*

[37] O. Teichmüller, *Diskrete bewertete perfekte Körper mit unvollkommenem Restklassenkörper*, J. reine angew. Math. vol 176 (1937), 141–152.

[70] M. L. Teply, *Homological dimension and splitting torsion theories*, Pac. J. Math. **34** (1970), 193–205.

[72] ______, *See Fuelberth.*

[75] ______, *Pseudo-injective modules which are not quasi-injective*, Proc.Amer.Math.Soc. **49** (1975), 305–310.

[79] M.L. Teply and R.W. Miller, *The descending chain condition relative to a torsion theory*, Pac. J. Math. **83** (1979), 207–219.

[48] R.M. Thrall, *Some generalizations of quasi-Frobenius algebras*, Trans. Amer. Math. Soc. **64** (1948), 173–183.

[46] ______, *See Nesbitt.*

[76] M. Tierney, *See Heller.*

[70] T. S. Tol'skaya, *When are cyclics essentially embedded in free modules*, Mat. Issled **5** (1976), 187–192.

[61] H. Tominaga, *See Onodera.*

[73] ______, *Note in Galois subrings of prime Goldie rings*, Math. J. Okayama Univ. **16** (1973), 115–116.

[79] ______, *A generalization of a theorem of A. Kertész*, Acta Math.Acad.Sci.Hung. **33** (1979), 333.

[70] H. Tominaga and T. Nagahara, *Galois Theory of Simple Rings*, Okayama Mathematical Lectures, Okayama U., Okayama, Japan 1970.

[88] B. Torrecillas, *See Bueso.*

[93] J. Torrecillas and B. Torrecillas, *Flat torsionfree modules and QF-3 rings*, Osaka J. Math. **30** (1993), 529–542.

[96] J. Trilfaj, *Two problems of Ziegler and uniform modules over regular rings*, in "Abelian groups and Modules," pp.373–383, Lecture Notes in Pure & Appl Math., vol. 182, Marcel Dekker, Basel and New York, 1996.

[78] P. J. Trosborg, *See Beck.*

[34] C. C. Tsen, *Algebren Über Funktionenkörpern*, Ph.D. Dissertation, Göttengen, 1934.

[36] ______, *Zur Stufentheorie der quasi-algebraisch–Abegeschlossenheit kommutativer Körper*, J. Chinese Math. Soc. **1** (1936), 81–92.

[77] A. A. Tuganbaev, *Quasi-injective and poorly injective modules (*Russian*)*, Vestnik. Moskov. U. Sér. I Meh. (1977), 61–64.

[82] ______, *Weakly injective rings*, Uspekhi Mat.Nauk **37** (1982), 201–202.

[97] B. Ulrich, *Review of Vasconcelos [94]*, Bull. Amer. Math. Soc. **34** (1997), 177–181.

[56] Y. Utumi, *On quotient rings*, Osaka Math. J. **8** (1956), 1–18.

[57] ———, *On ξ-rings*, Proc.Japan Acad. **33** (1957), 63–66.

[60] ———, *On continuous regular rings and semi-simple self-injective rings*, Canad. J. Math. **12** (1960), 597–605.

[61] ———, *On continuous regular rings*, Canad. Math. Bull. **4** (1961), 63–69.

[63] ———, *Prime J-rings with uniform one-sided ideals*, Amer. J. Math. **85** (1963), 583–596.

[63a] ———, *A Theorem of Levitzki*, Math.Assoc. of Amer. Monthly **70** (1963), p. 286.

[63b] ———, *On rings of which any one-sided quotient rings are two-sided*, Proc. Amer. Math. Soc. **14** (1963), 141–147.

[63c] ———, *A note on rings of which any one-sided quotient rings are two-sided*, Proc. Japan Acad. **39** (1963), 287–288.

[65] ———, *On continuous rings and self-injective rings*, Trans.Amer.Math.Soc. **118** (1965), 158–173.

[66] ———, *On the continuity and self-injectivity of a complete regular ring*, Canad. J. Math **18** (1966), 404–412.

[67] ———, *Self-injective rings*, J. Algebra **6** (1967), 56–64.

[64,65] ———, *See Faith.*

[63] A. I. Uzkov, *On the decomposition of modules over a commutative ring into a direct sum of cyclic submodules*, Math. Sbornik **63** (1963), 469–475.

[87] K. G. Valente, *The p-primes of a commutative ring*, Pac. J. Math. **126** (1987), 385–400.

[75,81] J. Valette, *See Goursaud.*

[68] P. Vámos, *The dual of the notion of finitely generated*, J. Lond. Math. Soc. **43** (1968), 643–646.

[71] ———, *Direct decompositions of modules*, Algebra Seminar Notes, Dept. of Math., Univ. of Sheffield, 1971G.

[75] ———, *Classical rings*, J. Algebra **34** (1975), 114–129.

[75b] ———, *Multiply maximally complete fields*, J. Lond. Math. Soc. (2) **12** (1975), 103–111.

[77a] ———, *Rings with duality*, Proc. Lond. Math. Soc. (3) **35** (1977), 175–184.

[77b] ———, *The decomposition of finitely generated modules and fractionally self-injective rings*, J. London Math. Soc. **16** (1977), 209–220.

[77c] ———, *The Nullstellensatz and tensor products of fields*, Bull. L.M.S. **9** (1977), 273–278.

[79] ———, *Sheaf-theoretical methods in the solution of Kaplansky's problem*, in Applications of sheaves (Proc. Res. Sympos. Appl. Sheaf Theory to Logic, Algebra and Anal., Durham, 1977); Lecture Notes in Math. 753, Springer, Berlin-Heidelberg-New York, pp. 732–738.

[71] ———, *See Sharpe.*

[84] ———, *See Menal.*

[77,85] ———, *See Sharp.*

[95] ———, *See Facchini.*

[98] ———, *See Albu.*

[31] B. L. Van der Waerden, *Moderne Algebra,* vols. I,II; Grundl. Math. Wiss. Bd. 33,34, Springer-Verlag, 1931.

[48,50] ______, *Modern Algebra,* vols. I,II, Frederick Ungar (English Translation of [31]), New York, 1948,1950.

[85] ______, *A History of Algebra (*from Al-Khwārizmī to Emmy Noether), Springer-Verlag, Berlin, Heidelberg and New York, 1985.

[79] F. Van Oystaeyen (ed.), *Ring Theory*, Proceedings of the 1978 Antwerp Conference, Lecture Notes in Pure and Appl. Math., vol. 51, Marcel Dekker, Basel, 1979.

[74] K. Varadarajan, *See Sarath.*

[79] ______, *Dual Goldie dimension*, Comm. Algebra **7** (1979), 565–610.

[69] W. V. Vasconcelos, *On finitely generated flat modules*, Trans.Amer.Math.Soc. **138** (1969), 505–512.

[70a] ______, *Flat modules over commutative Noetherian rings*, Trans.Amer.Math.Soc. **152** (1970), 137–143.

[70b] ______, *Simple flat extension*, J. algebra **16** (1970), 106–107.

[70c] ______, *On commutative endomorphism rings*, Pac. J. Math. **35** (1970), 795–798.

[70d] ______, *Injective endomorphisms of finitely generated modules*, Proc. A.M.S. **25** (1970), 900–901.

[72] ______, *The local rings of global dimension two*, Proc.Amer.Math.Soc. **35** (1972), 381–386.

[73a] ______, *Coherence of one polynomial ring*, Proc.Amer.Math.Soc. **41** (1973), 449–456.

[73b] ______, *Finiteness in projective ideals*, J. Algebra **25** (1973), 269–278.

[73c] ______, *Rings of global dimension two*, in Proc.Conf.Comm. Algebra (Lawrence, 1972), Springer-Verlag, Berlin-Heidelberg-New York, 1973.

[76] ______, *Rings of Dimension Two*, Marcel Dekker, Basel and New York, 1976.

[94] ______, *Arithmetic of Blow-up Algebras*, Lond.Math.Soc.Lecture Notes, vol. 195, Cambridge U. Press, Cambridge, 1994.

[78] W. Vasconcelos and R. Wiegand, *Bounding the number of generators of a module*, Math. Z. **164** (1978), 1–7.

[71] L. N. Vasershtein, *Stable ranks of rings and dimensionality of topological spaces*, Functional Anal. Appl. **5** (1971), 17–27; translation 102–110.

[87] J. P. Vicnair, *On valuation rings*, Rocky Mountain J. Math. **17** (1987), 55–58.

[92] N. Vila, *On the inverse problem of Galois theory*, Publ. Mat. **36** (1992), 1053–1073.

[73] O. E. Villamayor, *See Michler.*

[73] C. Vinsonhaler, Supplement to the paper: *"Orders in QF-3 rings"*, J. Algebra **17** (1971), 149–151.

[75] J. E. Viola-Prioli, *On absolutely torsion-free rings*, Pac. J. Math. **56** (1975), 275–283.

[86] M. Vitulli, *Letter to the author of September 5, 1986*, Dept. Math., U. of Oregon, Eugene.

[36a] J. von Neumann, *On regular rings*, Proc. Nat. Acad. Sci (USA) **22** (1936), 707–713.

[36b] ______, *Examples of continuous geometries*, Proc. Nat. Acad. Sci (USA) **22** (1936), 101–108.

[60] ———, *Continuous Geometry*, Princeton mathematical Series No. 25, Princeton Univ., Princeton, N.J., 1960.

[81] A. R. Wadsworth, *See Small.*

[37] W. Wagner, *Über die Grundlagen der Projective Geometry and allgemeine Zahlensysteme*, Math. Ann. **113** (1937), 528–567.

[56] E. A. Walker, *Cancellation in direct sums of groups*, Proc.Amer.Math.Soc. **7** (1956), 898–902.

[67] ———, *See Faith.*

[66] C. L. Walker and E. A. Walker, *Quotient categories of modules*, pp. 404–420 in Eilenberg, etal. (eds.) [66].

[72] ———, *Quotient categories and rings of quotients*, Rocky Mt. J. Math. **2** (1972), 513–555.

[50] G. L. Walker, *See Perlis.*

[66] D. W. Wall, *Characterizations of generalized uniserial algebras* III, Proc. Edinburgh Math. Soc. (2) **15** (1966), 37–42.

[69a] R.B. Warfield, Jr., *Purity and algebraic compactness for modules*, Pac. J. Math. **28** (1969), 699–710.

[69b] ———, *Decompositions of injective modules*, Pac. J. Math. **31** (1969), 263–276.

[69c] ———, *A Krull-Schmidt theorem for infinite sums of modules*, Proc. Amer. Math. Soc. **22** (1969), 460–465.

[70] ———, *Decomposability of finitely presented modules*, Proc. Amer. Math. Soc. **25** (1970), 167–172.

[72a] ———, *Rings whose modules have nice decompositions*, Math. Z. **125** (1972), 187–192.

[72b] ———, *Exchange rings and decompositions of modules*, Math. Ann. **199** (1972), 31–36.

[73] ———, *Review of Kahlon* [71], Math. Reviews **46** (1973), #5388, Reprinted in Small [81].

[75] ———, *Serial rings and finitely presented modules*, J. Algebra **37** (1975), 187–222.

[78] ———, *Large modules over Artinian rings*, in Representation Theory of Algebras, Proceedings of the Philadelphia Conference, pp. 451–483; Lecture Notes in Pure and appl. Math., vol. 47, M. Dekker, Basel and New York, 1978.

[79a] ———, *Bezout rings and serial rings*, Comm. Algebra **7** (1979), 533–545.

[79b] ———, *Modules over fully bounded Noetherian rings*, in Ring Theory, Waterloo 1978, pp. 339–552, Lecture Notes in Math., vol. 734, Springer Verlag, Berlin, Heidelberg, New York, 1979.

[89] ———, *See Goodearl.*

[93] S. Warner, *Topological Rings*, North Holland, Amsterdam, 1993.

[86] J. Waschbüsch, *Self-duality of serial rings*, Comm. Algebra **14** (1986), 581–589.

[84] W. Weakley, *Modules whose proper submodules are finitely generated*, J. Algebra **84** (1984), 189–219.

[70] D. B. Webber, *Ideals and modules of simple Noetherian hereditary rings*, J. Algebra **16** (1970), 239–242.

[05] J.H.M. Wedderburn, *A theorem on finite algebras*, Trans. Amer. Math. Soc. **6** (1905), 349–352.

[08] ——, *On hypercomplex numbers*, Proc. Lond. Math. Soc (2) **6** (1908), 77–117.

[94] J. A. Wehlen, *Lifting properties of the extensions of the Wedderburn Principal Theorem*, pp. 223–240, in Magid (ed.) [94].

[94] C. A. Weibel, *An Introduction to Homological Algebra*, Cambridge Studies in Advanced Mathematics, vol. 38, Cambridge, New York,, and Melbourne, 1994.

[98] ——, *A History of Homological Algebra*, in "The History of Topology" (I.M. James, ed.), North-Holland, Amsterdam, 1998.

[1898] A. N. Whitehead, *Universal Algebra*, Cambridge University Press, Cambridge, 1898.

[92] H. Whitney, *Collected Papers (J. Eels and D. Toledo, eds.)*, Birkhäuser, Boston, Basel and Berlin, 1992.

[71–73] R. Wiegand, *See Shores.*

[78] ——, *See Vasconcelos.*

[98] ——, preprint.

[77] R. Wiegand and S. Wiegand, *Commutative rings whose finitely generated modules are direct sums of cyclics*, in Lecture Notes in Math. 616, pp. 406-423, Springer-Verlag, New York, 1977.

[75] S. Wiegand, *Locally maximal Bezout domains*, Proc. Amer. Math. Soc. **47** (1975), 10–14.

[79] ——, *See Faith and Wiegand (eds.) [79].*

[74] R. Wiegandt, *Radical and Semisimple Classes of Rings*, Queen's Papers in Pure and Appl. Math., vol. 37, Queen's U., Kingston, Ont., 1974.

[85] ——, *See Marki.*

[73] R. W. Wilkerson, *Finite dimensional group rings*, Proc. A.M.S. **41** (1973), 10–16.

[75] ——, *Twisted polynomial rings over finite dimensional rings (*Afrikaans summary), Tydskr. Natur Wetenskap **15** (1975), 103–106.

[78] ——, *Goldie dimension in power series rings*, Bull. Malaysian Math. J. (2) **1** (1978), 61–63.

[88] R. Wisbauer, *Grundlagen der Modul und Ringtheorie*, Verlag R. Fischer, Munich 1988.

[91] ——, *See Huynh.*

[95] ——, *See Puninski.*

[31] E. Witt, *Über die Kommutativikeit endlicher Schiefkörper*, Abh. Math. Sem., Hamburg, **8** (1931) 413.

[53] Kenneth G. Wolfson, *Ideal theoretic characterization of the ring of all linear transformations*, Amer. J. Math. **75** (1953), 358–385.

[55] ——, *A class of primitive rings*, Duke Math. J. **22** (1955), 157–164.

[56b] ——, *Annihilator rings*, J. London Math. Soc. **31** (1956), 94–104.

[56c] ——, *Anti-isomorphisms of the ring and lattice of a normed linear space*, Portugal. Math. **7** (1956), 852–855.

[61] ——, *Baer rings of endomorphisms*, Math. Ann. **143** (1961), 19–28.

[62] ——, *Isomorphisms of the endomorphism ring of a free module over a principal left ideal domain*, Michigan Math. J. **9** (1962), 69–75.

[61] E. T. Wong, *See Johnson.*

[59] E. T. Wong and R. E. Johnson, *Self-injective rings*, Canad. Math. Bull. **2** (1959), 167–173, E.T. Wong (see Johnson).

[97] J. Wood, *Minimal Algebras of Type 4 Are Not Computable*, Ph.D. Thesis, Berkeley, 1997.

[96] W. H. Woodin, *See Dales.*

[71] S. M. Woods, *On perfect group rings*, Proc. Amer. Math. Soc. **27** (1971), 49–52.

[76] ———, *See Lawrence.*

[80] H. Woolf (Director), *Foreword to a Community of Scholars (*J. Mitchell, ed.), The Institute for Advanced Study, 1930–1980, Princeton, New Jersey, 1980.

[73] T. Würfel, *Über absolut reine Ringe*, J. reine Angew. Math. 262/263 (1973), 381–391.

[92] W. Xue, *Rings with Morita Duality*, Lecture Notes in Math., vol. 1523, Springer, Berlin-Heidelberg-New York 1992.

[96] ———, *Polynomial rings over commutative linear compact rings*, Chinese Sci. Bull. **41** (1996), 459–461.

[96b] ———, *Quasi-duality, Linear Compactness and Morita duality for power series rings*, Canad. Bull. Math. **39** (1996), 250–256.

[96c] ———, *Recent developments in Morita theory,* pp. 277–900, "Rings, Groups and Algebras" (X. H. Cao *etal*,, eds.), Lecture Notes, Pure and appl Algebra, vol. 181, Marcel Dekker, New York, 1996.

[98] ———, *Two questions on rings whose modules have maximal submodules*, Fujian Normal U., Fuzhou, Fujian 350007, P. R. China, and U. Iowa, Iowa City, 1998.

[73] K. Yamagata, *A note on a problem of Matlis*, Proc. Japan Acal. **49** (1973), 145–147.

[74] ———, *The exchange property and direct sums of indecomposable injective modules*, Pac. J. Math. (1974), 301–317.

[75] ———, *On rings of finite representation type and modules with the finite exchange property*, Sci Reps. tokyo Koiku Daigaku, Sect. A **13** (1975), 347–365, 1-6.

[96] C. C. Yang, *See Cao.*

[56] T. Yoshii, *On algebras of bounded representation type*, Osaka Math. J. **8** (1956), 51–105.

[91] H. Yoshimura, *On finitely pseudo-Frobenius rings*, Osaka J. Math. **28** (1991), 285–294.

[94] ———, *On FPF rings and a result*, Proc. of the 27th Symposium in Ring Theory, Okayama U., 1994.

[95] ———, *On rings whose cyclic faithful modules are generators*, Osaka J. Math. **32** (1995), 591–611.

[97] ———, *FPF rings which are characterized by 2-generated faithful modules,* preprint.

[98] ———, *Rings whose finitely generated ideals correspond to finitely generated over modules*, Comm. Algebra **26** (1998), 997–1004.

[90] M. Yousif, *See Rizvi.*

[91] ———, *On semiperfect FPF rings*, Canad. Math. Bull. **37** (1994), 287–288.

[91] ———, *See Camillo.*

[95] ———, *See Nicholson.*

[95] ———, *See G. Puninski.*

[96] ———, *See Huynh.*

[98] ______, *See Nicholson.*

[94] H. P. Yu, *On modules for which the finite exchange property implies the countable exchange property*, Comm. Alg. vol 22 (1994), 3887–3901.

[97] ______, *On the structure of exchange rings*, Comm.Alg. **25** (1997), 661–670.

[81] R. Yue Chi Ming, *On injective and p-injective modules*, Riv. Math. Univ., Parma **7** (1981), 187–197.

[95] ______, *On injectivity and p-injectivity*, II, Soochow J. Math. **21** (1995), 401–412.

[88] D. Zacharia, *A characterization of Artinian rings whose endomorphism rings have finite global dimension*, Proc. A.M.S. **104** (1988), 37–38.

[68] A. Zaks, *Semiprimary rings of generalized triangular type*, J. Algebra **9** (1968), 54–78.

[69] ______, *Injective dimension of semiprimary rings*, J. Algebra **13** (1969), 63–86.

[71a] ______, *Dedekind subrings of $k[x_1 \dots, x_n]$ are rings of polynomials*, Israel J. Math. **9** (1971), 285–289.

[71b] ______, *Some rings are hereditary*, Israel J. Math. **10** (1971), 442–450.

[72] ______, *Restricted left principal ideal rings*, Israel J. Math. **11** (1972), 190–215.

[74] ______, *Hereditary Noetherian rings*, J. Algebra **30** (1974), 513–526.

[72] A. E. Zalesski, *On a problem of Kaplansky*, Soviet Math. **13** (1972), 449–552.

[75] A. E. Zalesski and O. M. Neroslavskii, *On simple Noetherian rings (*Russian*)*, Izv. Akad. Nauk., USSR.

[77] ______, *There exist simple rings with zero divisors but without idempotents*, (Russian, English Summary) Comm.Alg. **5** (1977), 231–244.

[81] P. Zanardo, *See Salce.*

[85] ______, *Valuation domains without pathological modules*, J. Algebra **96** (1985), 1–8.

[98] ______, *See Fuchs.*

[58] O. Zariski, *On Castelnuovo's criterion of rationality $p_1 = p_2 = 0$ of an algebraic surface*, Illinois J. Math. **2** (1958), 303–315.

[58-60] O. Zariski and P. Samuel, *Commutative Algebra, Vols. I and II*, Van Nostrand, Princeton and New York, 1958 and 1960.

[53] D. Zelinsky, *Linearly compact modules and rings*, Amer. J. Math. **75** (1953), 79–90.

[54] ______, *Every linear transformation is a sum of nonsingular ones*, Proc. Amer. Math. Soc. **5** (1954), 627–630.

[57] ______, *See Eilenberg.*

[59] ______, *See Rosenberg.*

[61] ______, *See Rosenberg.*

[67] J. Zelmanowitz, *Endomorphism rings of torsionless modules*, J. Algebra **5** (1967), 325–341.

[69] ______, *A shorter proof of Goldie's theorem*, Canad. Math. Bull. **12** (1969), 597–602.

[72] ______, *Regular modules*, Trans. Amer. Math. Soc. **163** (1972), 341–355.

[76a] ______, *An extension of the Jacobson Density Theorem*, Bull. Amer. Math. Soc. **82** (1976), 551–553.

[76b] ______, *Finite intersection property on annihilator right ideals*, Proc. Am. Math. Soc. **57** (1976), 213–216.

[77] ———, *Dense rings of linear transformations*, in Ring Theory II, Lecture Notes in Pure and Appl. Math., vol. 26, 1977, pp. 281–295.

[81] ———, *Weakly primitive rings*, Comm. Algebra **9** (1981), 23–45.

[82] ———, *On Jacobson's density theorem*, Contemp.Math. **13** (1982), 155–162.

[84] ———, *Representations of rings with faithful monoform modules*, J. London Math. Soc. (2) **29** (1984), 238–248.

[82] ———, *See Fahy.*

[88] J. Zelmanowitz and W. Jansen, *Duality for module categories*, Algebra Berichte, Nr.59, pp.1–33, Reinhard Fisher Verlag, Munich, 1988.

[04] E. Zermelo, *Beweis dass jede Menge wohlgeordnet werden kann*, Math. Ann. **59** (1904), 514–516.

[08a] ———, *Neuer Beweis für die Möglichkeit einer Wohlordnung*, Math. Ann. **65** (1908), 107–128.

[08b] ———, *Untersuchungen über die Grundlagen der Mengenlehre*, I., Math. Ann. **65** (1908), 261–281.

[84] H. Ziegler, *Model theory of modules*, Ann. Pure and Appl. Logic **26** (1984), 149–213.

[76] B. Zimmermann-Huisgen, *Pure submodules of direct products of free modules*, Math. Ann. **224** (1976), 233–245.

[79] ———, *Rings whose right modules are direct sums of indecomposable modules*, Proc. Amer. Math. Soc. **77** (1979), 191–197.

[80] ———, *Direct products of modules and algebraic compactness*, Habilitationschrift Tech. Univ. Munich, 1980.

[77] ———, *See Meyberg.*

[97] ———, *See Arhangel'skii.*

[78] B. Zimmermann-Huisgen and W. Zimmermann, *Algebraically compact rings and modules*, Tech. U. München (TUM), Munich, 1980; also Math. Z. **161** (1978), 81–93.

[72] W. Zimmermann, *Einige Charakterisierungen der Ringe, über denen reine Untermoduln direkte Summanden sind*, Bayer. Ada. Wiss. Math.-Natur.Kl. Sitzungsber. (1972), 77–79.

[76] ———, *Über die aufstiegende Kettenbedingung für Annulatoren*, Arch. Math. **27** (1976), 319–328.

[77] ———, *Rein injektive direkte summen von Moduln*, Comm. Algebra **5** (1977), 1083–1117.

[82] ———, *(Σ)-algebraic compactness of rings*, J. Pure and Appl. Algebra **23** (1982), 319–328.

[35] L. Zippin, *Countable torsion groups*, Ann. of Math. **36** (1935), 86–99.

[81] H. Zöschinger, *Projective Moduln mit endlich erzeugten Radicalfaktorenmoduln*, Math. Ann. **255** (1981), 199–206.

[83] ———, *Linear-kompakte moduln über Noetherschen Ringen*, Arch. Math. **41** (1983), 121–130.

Register of Names

Notes: (1) To avoid unnecessary duplication, authors of theorems appearing in the index, with few exceptions, are not given page citations here; (2) Part II is omitted from the Register (see the Index to Snapshots).

* P denotes the Preface

Index of Terms and Authors of Theorems

Note:
This index does not cover Part II.
1.2f denotes below 1.2
1.2ff denotes below 1.2 and more
1.2s denotes above 1.2
1.2ss denotes above 1.2 and more
P denotes the preface

Selected Titles in This Series

(Continued from the front of this publication)

35 **Shreeram S. Abhyankar,** Algebraic geometry for scientists and engineers, 1990
34 **Victor Isakov,** Inverse source problems, 1990
33 **Vladimir G. Berkovich,** Spectral theory and analytic geometry over non-Archimedean fields, 1990
32 **Howard Jacobowitz,** An introduction to CR structures, 1990
31 **Paul J. Sally, Jr. and David A. Vogan, Jr., Editors,** Representation theory and harmonic analysis on semisimple Lie groups, 1989
30 **Thomas W. Cusick and Mary E. Flahive,** The Markoff and Lagrange spectra, 1989
29 **Alan L. T. Paterson,** Amenability, 1988
28 **Richard Beals, Percy Deift, and Carlos Tomei,** Direct and inverse scattering on the line, 1988
27 **Nathan J. Fine,** Basic hypergeometric series and applications, 1988
26 **Hari Bercovici,** Operator theory and arithmetic in H^∞, 1988
25 **Jack K. Hale,** Asymptotic behavior of dissipative systems, 1988
24 **Lance W. Small, Editor,** Noetherian rings and their applications, 1987
23 **E. H. Rothe,** Introduction to various aspects of degree theory in Banach spaces, 1986
22 **Michael E. Taylor,** Noncommutative harmonic analysis, 1986
21 **Albert Baernstein, David Drasin, Peter Duren, and Albert Marden, Editors,** The Bieberbach conjecture: Proceedings of the symposium on the occasion of the proof, 1986
20 **Kenneth R. Goodearl,** Partially ordered abelian groups with interpolation, 1986
19 **Gregory V. Chudnovsky,** Contributions to the theory of transcendental numbers, 1984
18 **Frank B. Knight,** Essentials of Brownian motion and diffusion, 1981
17 **Le Baron O. Ferguson,** Approximation by polynomials with integral coefficients, 1980
16 **O. Timothy O'Meara,** Symplectic groups, 1978
15 **J. Diestel and J. J. Uhl, Jr.,** Vector measures, 1977
14 **V. Guillemin and S. Sternberg,** Geometric asymptotics, 1977
13 **C. Pearcy, Editor,** Topics in operator theory, 1974
12 **J. R. Isbell,** Uniform spaces, 1964
11 **J. Cronin,** Fixed points and topological degree in nonlinear analysis, 1964
10 **R. Ayoub,** An introduction to the analytic theory of numbers, 1963
9 **Arthur Sard,** Linear approximation, 1963
8 **J. Lehner,** Discontinuous groups and automorphic functions, 1964
7.2 **A. H. Clifford and G. B. Preston,** The algebraic theory of semigroups, Volume II, 1961
7.1 **A. H. Clifford and G. B. Preston,** The algebraic theory of semigroups, Volume I, 1961
6 **C. C. Chevalley,** Introduction to the theory of algebraic functions of one variable, 1951
5 **S. Bergman,** The kernel function and conformal mapping, 1950
4 **O. F. G. Schilling,** The theory of valuations, 1950
3 **M. Marden,** Geometry of polynomials, 1949
2 **N. Jacobson,** The theory of rings, 1943
1 **J. A. Shohat and J. D. Tamarkin,** The problem of moments, 1943

ISBN 0-8218-0993-8